The Arnold-Gelfand Mathematical Seminars

V. I. Arnold
I. M. Gelfand
V. S. Retakh
M. Smirnov
Editors

Birkhäuser
Boston • Basel • Berlin

V. I. Arnold
Ceremade
Université Paris-Dauphine
75775 Paris Cedex 16
France

I. M. Gelfand
Department of Mathematics
Rutgers University
New Brunswick, NJ 08903

V. S. Retakh
Department of Mathematics
Harvard University
Cambridge, MA 02139

M. Smirnov
Department of Mathematics
Columbia University
New York, NY 10027

Printed on acid-free paper

© 1997 Birkhäuser Boston

Birkhäuser ®

ISBN 0-8176-3883-0
ISBN 3-7643-3883-0

Typeset by TEXniques, Boston, MA
Printed and bound by Quinn-Woodbine, Woodbine, NJ.
Printed in the U.S.A.
9 8 7 6 5 4 3 2 1

Contents

Preface

It is very tempting but a little bit dangerous to compare the style of two great mathematicians or of their schools. I think that it would be better to compare papers from both schools dedicated to one area, geometry and to leave conclusions to a reader of this volume.

The collaboration of these two schools is not new. One of the best mathematics journals *Functional Analysis and its Applications* had I.M. Gelfand as its chief editor and V.I. Arnold as vice-chief editor. Appearances in one issue of the journal presenting remarkable papers from seminars of Arnold and Gelfand always left a strong impact on all of mathematics.

We hope that this volume will have a similar impact. Papers from Arnold's seminar are devoted to three important directions developed by his school: Symplectic Geometry (F. Lalonde and D. McDuff), Theory of Singularities and its applications (F. Aicardi, I. Bogaevski, M. Kazarian), Geometry of Curves and Manifolds (S. Anisov, V. Chekanov, L. Guieu, E. Mourre and V. Ovsienko, S. Gusein-Zade and S. Natanzon). A little bit outside of these areas is a very interesting paper by M. Karoubi *Produit cyclique d'espaces et opérations de Steenrod*.

Papers from Gelfand's seminar are complimentary to a recent Gelfand seminar volume and are more or less related by the notion of integral transforms. An application of Radon transforms to a solution of Hilbert's fourth problem is given by J.C. Alvarez, I. Gelfand and M. Smirnov, nonlinear integrable equations and nonlinear Fourier transform are considered by A. Fokas, I. Gelfand and M. Zyzkin. Two papers (I. Gelfand, M. Graev and A. Postnikov, A. Kazarnovski-Krol) are devoted to hypergeometric functions — a cornerstone of modern theory of such functions developed by Gelfand's school — is also a Radon transform of homogenous functions.

The Arnold-Gelfand
Mathematical Seminars

Discriminants and local invariants
of planar fronts

Francesca Aicardi

Introduction

The aim of this study is the description of all the local additive invariants of the plane wave fronts. A generic wave front is a curve whose only singularities are the transversal self-intersections on the semicubical cusps (see fig. 1a,b). The invariants that we shall find are "dual" to different strata of the discriminant formed by the nongeneric wave fronts.

For instance, our results imply the following minoration of the numbers of different events of nongenericity on any path connecting the two curves $\omega_{3,0}$ and $K_{3,0}$ shown in fig. 1b:

i) the number of self-tangencies with parallel coorientations is at least 2;

ii) the number of self-tangencies with antiparallel coorientations is at least one;

iii) the number of cusp crossings is at least 4, provided that the triple points are avoided;

iv) the number of triple points is at least 2, provided that the cusp crossings are avoided.

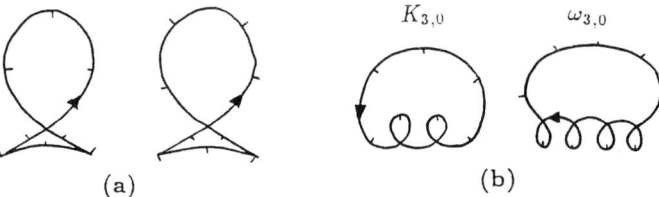

(a) (b)

Figure 1. Oriented and cooriented planar fronts

The proofs are based on the introduction of some topological invariants of the generic wave fronts, whose jumps at the crossing of the corresponding nongeneric events are fixed. The calculation of the values of these invariants on the curves $\omega_{3,0}$ and $K_{3,0}$ provides the above minoration.

There are five basic invariants (independent of the orientations of the front and plane as well as of the coorientation of the front: n (the number of double points), λ (the number of cusp points), J^+ (responsible for the parallel coorientations tangencies), J^- (responsible for the antiparallel coorientations tangencies), Sp (responsible of the triple points and cusp crossing).

The invariants J^+, J^- and Sp of wave fronts reduce to the invariants J^+, J^- and St for those wave fronts which have no cusps, i.e., to the Arnold invariants of smooth closed plane curves.

The number of the independent local additive invariants of generic oriented and cooriented wave fronts on the oriented plane (counting both those invariants, like $J^\pm$, which are independent of the orientations and those which depend on the orientations of the front and plane as well as of the coorientation of the front) is equal to 10. One of these invariants is the Maslov index μ. We introduce 6 new basic invariants $(f^+, f^-, \lambda^\uparrow, \lambda^\downarrow, p^\uparrow, p^\downarrow)$.

The invariant f^+ (f^-) is counting the double points where two positive (negative) branches of the front do intersect (the branch is positive if its orientation and coorientation define the positive orientation of the plane).

The invariant $\lambda^\uparrow$ ($\lambda^\downarrow$) is counting the number of cusps at which the orientation of the plane defined by the two branches of the oriented curve leaving the cusp point (first along the coming branch, second along the outgoing branch) is positive (negative).

The invariants $p^\uparrow$, $p^\downarrow$ are responsible for the cusp crossings taking some sign rule into account (see §3). These six new invariants verify the relations:

$$\lambda^\uparrow + \lambda^\downarrow = \lambda, \qquad f^+ + f^- - J^+ + J^- - \frac{1}{2}\lambda = n,$$

hence only 4 of them are really new.

A *planar front* is the projection to $\mathbf{R}^2$ (with coordinates x, y) of a legendrian curve. A *legendrian curve* is the image of a C^1-immersion of S^1 in the space M^3 (with coordinates $x, y, \phi (\mathrm{mod} 2\pi)$ for cooriented fronts, $(\mathrm{mod} \pi)$ for noncooriented fronts) of the contact elements of the plane, with its natural contact structure

$$(cos\phi)dx + (sin\phi)dy = 0.$$

We call such an immersion of S^1 into M^3 an *L-immersion*.

The front is *cooriented,* if the contact element is cooriented, i.e., if a choice of one of two half-planes into which it divides the tangent plane is made. We shall now consider the cooriented fronts, and M^3 will be the space of the cooriented contact elements.

The front is *oriented* if the preimage circle S^1 is oriented.

A *front class* is a class of L-immersions having generic fronts up to diffeomorphisms of the plane and preimage circle.

In the space of L-immersions the immersions having nongeneric fronts form a hypersurface (a subvariety of codimension one) called the *discriminant.*

A coorientation on a smooth hypersurface in a functional space is a choice of one of the two parts separated by this hypersurface in a neighborhood of any of its points. This part is called positive. The discriminant (or a connected component of it) is called *coorientable* if the coorientations of its parts separated by strata of codimension 1 on it (of codimension 2 in the ambient space) are consistent. This means that the intersection index of any generic small oriented closed curve with the cooriented hypersurface should vanish.

In the functional space the connected components of the complement to the discriminant are the front classes. A *local invariant* of a front class is a function constant over each of such components. *Local* means that its increment under a generic crossing of the discriminant depends only on the behavior of the family of immersions in the neighborhoods of the points of the preimage circle sent to the points involved in the singularity.

A local invariant is thus defined up to an additive constant in terms of its jump at the crossing of the discriminant, provided that such discriminant is coorientable and that it defines in the ambient space a trivial 1-cocycle.

In fact every cooriented stratum of codimension 1, provided with a jump at its crossing in the positive direction (i.e., from the negative to the positive side defined by the coorientation), gives a 1-cocycle. A 1-cocycle is a function defined on generic curves in the ambient space: the value of a 1-cocycle on a generic curve is defined by the sum of the jumps (with signs) on all the crossings of the curve with the corresponding stratum.

The study using this approach of the immersion of a circle in the plane [2] leads to the definition of three basic invariants. One proves that the space of immersions with a given index is connected. In this space the discriminant has three different components, all coorientable, corresponding to three different types of degenerations (see fig.2):

1) self-tangency point with parallel orientations,

2) self-tangency point with antiparallel orientations,
3) triple point.

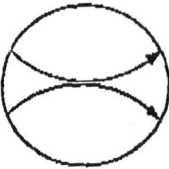 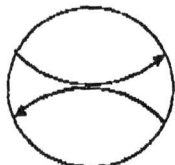 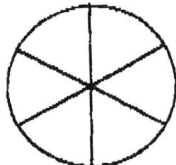

Figure 2. Parallel self-tangency, antiparallel self-tangency
and triple point

The choice of the jumps at these strata thus defines three 1-cocycles. These cocyles are trivial. The corresponding invariants are called J^+, J^- and St; the jumps are chosen so that $J^+ - J^-$ gives the number of double points of the immersions while the jump at the stratum of the triple points is 1 by a rule explained below.

The additive constants are chosen so that the three invariants result in being additive under the connected sum of immersions: it is sufficient to require that the values of the three basic invariants be all equal to zero on the class of the circle.

The values of the three invariants are thus well defined on the basic curves of every index. The values of the invariants of a generic immersion α of a given index are thus defined by whatever path in the space of immersions connecting the basic curve with α. Their increment along this path is in fact proportional to the intersection index of the path with the corresponding stratum. It follows that the difference of the values of a local invariant on two different immersions minorates the number of singular events of the corresponding type that occur in the course of a generic perestroika transforming one into the other.

Any linear combination of the basic invariants is again an invariant; some of them are interesting, for example the number of double points $(J^+ - J^-)$ and $\epsilon = J^+ + 2St$ which vanishes on the tree-like curves [2].

Coming back to the fronts, a smooth plane closed curve, provided with a normal vector at every point, i.e., with an unitary vector orthogonal to the tangent vector and forming with it a frame positively orienting the plane, is the front of a legendrian curve which is nowhere vertical (nowhere tangent to the fibre of the fibration $M^3 \longrightarrow \mathbf{R}^2$). Thus the classification of the fronts must contain the classification of the immersions of a circle in the plane.

The space of L-immersions is not so simple as the space of the immersions of a circle in the plane, and the study of its topology is preliminary to the definition of the local invariants.

To every oriented legendrian curve in the space of cooriented contact elements of the plane one associates two integers: the *index i* and the *Maslov index* μ (see [1]). Both these indices can be calculated in terms of the front of the legendrian curve.

The index i of an oriented legendrian curve in the space M^3 is equal to the total angle (divided by 2π) of the rotation of the coorienting normal vector of its front when a point on the front makes a full turn along the curve.

The Maslov index μ of a generic oriented legendrian curve is equal to the angle (divided by π) of the rotation of the tangent vector to the curve in the contact plane during a complete turn of the preimage point along the preimage circle. In terms of the front, the Maslov index is equal to the difference between the number μ^+ of positive cusps and the number μ^- of negative cusps: $\mu = \mu^+ - \mu^-$. (A cusp of an oriented and cooriented front is called *positive* if the coorienting normal vector at the cusp point belongs to the half-plane bounded by the tangent line at the cusp and containing the cusp branch with the orientation going away from the cusp point).

The sign of the index does not depend on the front coorientation and changes when the orientations of the plane or front are changed. The Maslov index changes its sign when the orientation or the coorientation of the front is changed and is independent of the orientation of the plane.

The space of all L-immersions with given index i and given Maslov index μ is denoted by $\Omega_{i,\mu}$.

One proves that $\Omega_{i,\mu}$ is connected but not simply connected [3].

An L-immersion is called *marked* if a given (once and for all) point of the preimage circle is sent to a given (once and for all) cooriented contact element of the oriented Euclidean plane.

The group G_3 of Euclidean motions of the plane acts without fixed points on the space of L-immersions, which is thus homeomorphic to the direct product of the space of the marked L-immersions $\Omega_{i,\mu}^0$ with $G_3 \approx S^1 \times \mathbf{R}^2$.

The discriminants and their coorientations are G_3-invariant. Hence in the construction of the invariants one can consider only the marked immersions. The space $\Omega_{i,\mu}^0$ is (weakly) homotopy equivalent to a circle, namely the circle of all possible directions of the marked legendrian curve in the contact plane at the chosen point.

It follows that a hypersurface of the discriminant stratum of codimension 1 in $\Omega_{i,\mu}^0$ defines a corresponding invariant if it satisfies the following two conditions:

I) it is cooriented,

II) the cooriented hypersurface is cohomologous to zero in $\Omega^0_{i,\mu}$ (its intersection indices with all the closed loops vanish).

The study of the discriminant of the self-tangencies, called stratum K (from the Russian word *kasanie* meaning tangency), has been done by V. Arnold [1]. The two components K^+ and K^- corresponding to the tangencies with parallel and antiparallel *normal* vectors are coorientable. Moreover, it is proved that the 1-cocycle defined by the jump at the cooriented stratum K^+ is cohomologous to zero. Then the values of the corresponding invariant J^+ (which is a natural generalization of the invariant J^+ defined for the immersions of a circle in the plane) are chosen on the basic curves belonging to the spaces $\Omega_{i,\mu}$, so that the resulting invariant is additive under the connected summation of two oriented and cooriented fronts.

The invariant J^+, unlike the index and the Maslov index, is independent of the orientation and of the coorientation of the front and of the orientation of the plane.

This work answers to the questions:

What are all the possible local additive invariants of fronts, independent of all the orientations and coorientation?

Are there local additive invariants of the fronts which depend on their orientation or on their coorientation or on the orientation of the plane?

To answer these questions we proceed in the following way:

i) We classify all the codimension 1 strata of the discriminant with and without the condition of independence of the orientations and the coorientation. We provide every such strata with a local coorientation and we define the value of the *elementary jump* of a corresponding local function, which is nonzero only at the crossing of that *elementary local stratum*.

ii) We impose the coorientability condition at every stratum of codimension 2: the increment of the local invariants along a small close path linked with this stratum must vanish. So we obtain a system of linear equations (*coherence system*) for the jumps defining the 1-cocycles.

iii) We choose a simple set of linearly independent generators of the space of solutions of the coherence system. The corresponding linear combinations of the elementary jumps define the *basic 1-cocycles*.

iv) We verify the vanishing of cohomology classes of the 1-cocycles. The jump of every basic 1-cocycles is thus the jump of a *basic invariant*.

v) We define the values of the basic invariants on the basic curve in $\Omega_{i,\mu}$ for any i and μ, so that the invariants are additive under the connected summation of two fronts. We prove that these invariants are unambiguously defined by the additivity condition.

The results are the following:

Theorem 1. *Every additive local invariant of the fronts that does not depend on their orientation, their coorientation and on the orientation of the plane is a linear combination of five linearly independent basic invariants.*

Theorem 2. *Every local additive invariant of the fronts (that may depend on orientation, coorientation or on the orientation of the plane) is a linear combination of 10 (including the Maslov index) linearly independent basic invariants.*

Remark. The investigation of the local basic invariants starting from the local strata of the discriminant dependent on the orientations leads to the definition of all the invariants, which are dependent on the orientations in different ways. However, I divide the problem into two parts, finding first the basic local invariants independent of the orientations. This seems to me more useful.

1. The codimension 1 local strata of the discriminant

1.1. Classification of the codimension 1 local strata of the discriminant

In the space of L-immersions, the discriminant, consisting of legendrian curves having nongeneric fronts, has four components corresponding to different singular events: self-tangencies, triple points, cusp crossing and cusps births. They are denoted [1] by the letters K, T, Π and Λ respectively (Π for the russian *prokhozhdenie* meaning passing and Λ from the Russian *lastochki khvost* meaning swallow tail).

One can, however, on every component locally distinguish different events, depending on the different possibilities of the orientation of the plane and of the orientation and coorientation of the segments of the curve containing the point involved in the singularity.

Consider the three following involutions acting on the space of L-immersions:

the involution a which reverses the orientation of the plane (x, y);

the involution b which reverses the orientation of the preimage circle, so reversing the orientation of the front;

the involution c which reverses the coorientation of the contact elements, so reversing the coorientation of the front.

These commuting involutions generate a group, called RO (for *ori-*

entation reversing)[4]. All other elements of the group different from the identity (having still order two) are ab, ac, bc, abc.

Given a particular nongeneric event occurring on a front, the action of RO transforms it into different events. The number of distinguishable events among them (up to rotations of the plane) depends on the symmetry type of the event. These different events are identified in the classification which is invariant under all involutions, and are distinguished in the classification which takes the orientations into account. In the following tables both classifications are shown. The subgroup $R_{(x)(y)}$ of RO indicates the symmetry type of the event, i.e., the subgroup generated by the involutions x and y with respect to which the event is invariant. There are no singular local events invariant for the entire group RO. We shall call *RO-independent* a class of local strata which is invariant under the actions of the group RO. Moreover, we shall say "orientations" for short instead of "orientation of the plane, orientation and coorientation of the front."

Lemma 1.1. *The discriminant containing L-immersions having nongeneric fronts with a self-tangency contains 10 different local strata* $(K_{C+}^+, \ldots, K_{D-,i}^-)$ *dependent on the orientations, corresponding to 4 RO-independent local strata* $(K_C^+, \ldots, K_D^-)$, *shown in table 1.*

The discriminant containing L-immersions having nongeneric fronts with a triple point contains 24 different local strata $(T_{A\longrightarrow}^{+++}, \ldots, T_{B\longleftarrow}^{---})$ *dependent on the orientations, corresponding to 5 RO-independent local strata* $(T_A^I, \ldots, T_B^{II})$, *shown in table 2.*

The discriminant containing L-immersions having nongeneric fronts with a cusp crossing contains 16 different local strata $(\Pi_{C\uparrow}^{-+}, \ldots, \Pi_{D\downarrow}^{+-})$ *dependent on the orientations, corresponding to 2 RO-independent local strata* (Π_C, Π_D), *shown in table 3.*

The discriminant containing L-immersions having nongeneric fronts with a cusps birth contains 4 different local strata $(\Lambda_+^\uparrow, \ldots, \Lambda_-^\downarrow)$ *dependent on the orientations, corresponding to 1 RO-independent local stratum* (Λ), *shown in table 4.*

Proof. We start with the explanation of the notations used to specify the local strata.

Definition 1.1. A smooth segment of an oriented and cooriented front is called *positive (negative)* if the pair (orienting tangent vector, coorienting normal vector) orients the plane positively (negatively).

Definition 1.2. A cusp point of an oriented front is called of type *up* or *down* if looking at the segment containing it so that the orientation

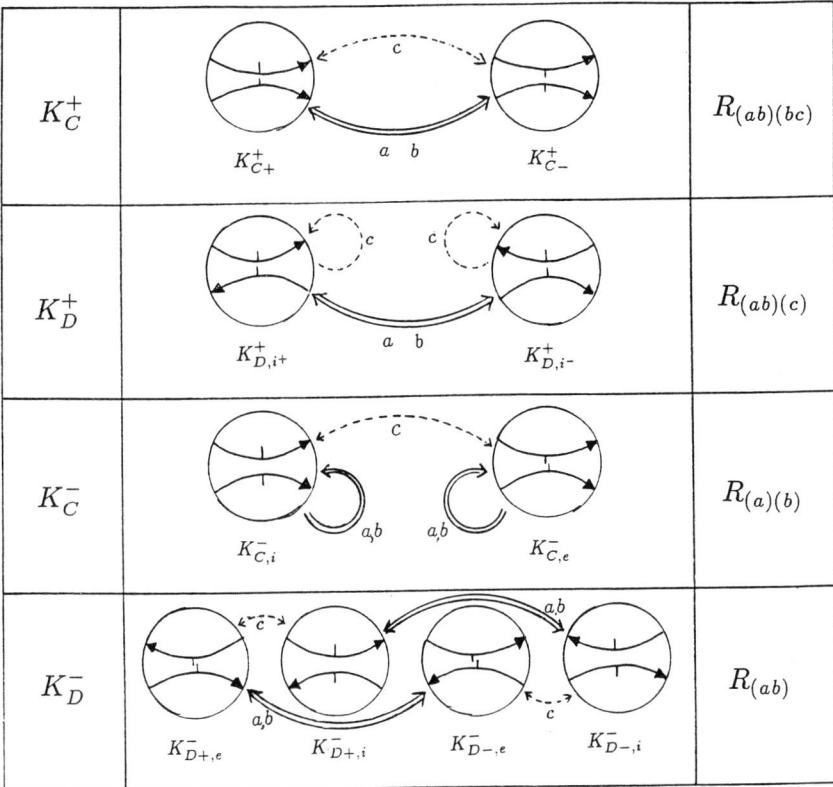

Table 1. Local strata of the discriminant K RO-independent, RO-dependent and their symmetries.

goes from left to right, the cusp is pointing upward or downward. (The plane is here supposed to be oriented by the frame (to the right, up)).

Remark 1.1. The character *up* or *down* is coorientation independent.

Remark 1.2. A cusp point is positive either if it is of type *up* and the branch going away from it is positive or if it is of type *down*, the outgoing branch being negative.

The stratum K of self-tangencies. The upper symbols $+$ and $-$, chosen in [1], mean respectively the parallelism and the antiparallelism at the tangency point of the coorienting normal vectors of the two tangent segments. The lower symbol C or D means that the orienting vectors at the tangency point are parallel (concordant) or antiparallel (discordant). The lower symbol $+$ or $-$ is the sign of both tangent segments, when they coincide.

For the other symbols consider the area bounded by the two segments crossing each other in two points after the tangency event. The lower symbol i (internal) or e (external) means that the coorienting vectors on both segments lie inside or outside this area. The lower symbol i^+ or i^- means that the only normal vector lying inside this area belongs to a positive or negative segment. (equivalent notations for i^+ and i^- should be e^- and e^+, respectively).

The involutions a and b (whose action on the local strata is shown in table 1 by a double arrow) reverses the lower indices $(+, -)$ of the strata K (including the signs of i^+ and i^-). The involution c (whose action on the strata is shown in table 1 by a dotted arrow) reverses the same signs and interchanges i with e, so that i^+ and i^- remain unchanged. Only the characters C and D and the upper signs meaning "parallel" and "antiparallel coorientations" are RO-independent.

The stratum T of triple points. Consider the three segments meeting at the triple point as the diagonals of a regular hexagon. The triple point is of type B or A if the union of the three oriented segments has or does not have the symmetry of order three.

The upper symbol I means that the signs of the three outgoing branches coincide, II and III means that the signs do not coincide, and in the case A of no rotation symmetry, the signs may alternate (III) or not (II).

The right or left arrow in the notations has the following meaning: consider the three outgoing directions as three points on the circle centered on the triple point. The cyclic order of the three visits of the corresponding branches defines an orientation of the circle. We use the right arrow if this orientation is positive and the left arrow if it is negative. (A circle is positively oriented in the oriented plane if the frame (tangent vector, external normal vector) orients the plane positively).

The three signs are the signs of the three segments, in the clockwise order of the outgoing branches starting at the left on the representatives of the triple points shown in table 2, the plane being oriented by the frame (to the right, up).

The involution a, reversing the orientation of the plane, reverses the arrows, the signs of the segments and their order; the involution b, reversing the orientation of the front, reverses the arrows and the signs of the segments; the involution c reverses only the signs. The characters A and B, as well as the coincidence of the three signs and (in the case A) their alternative, are RO-independent.

The stratum Π of cusp crossings. The symbol C or D means that the orientations of the segment containing the cusp point and that of

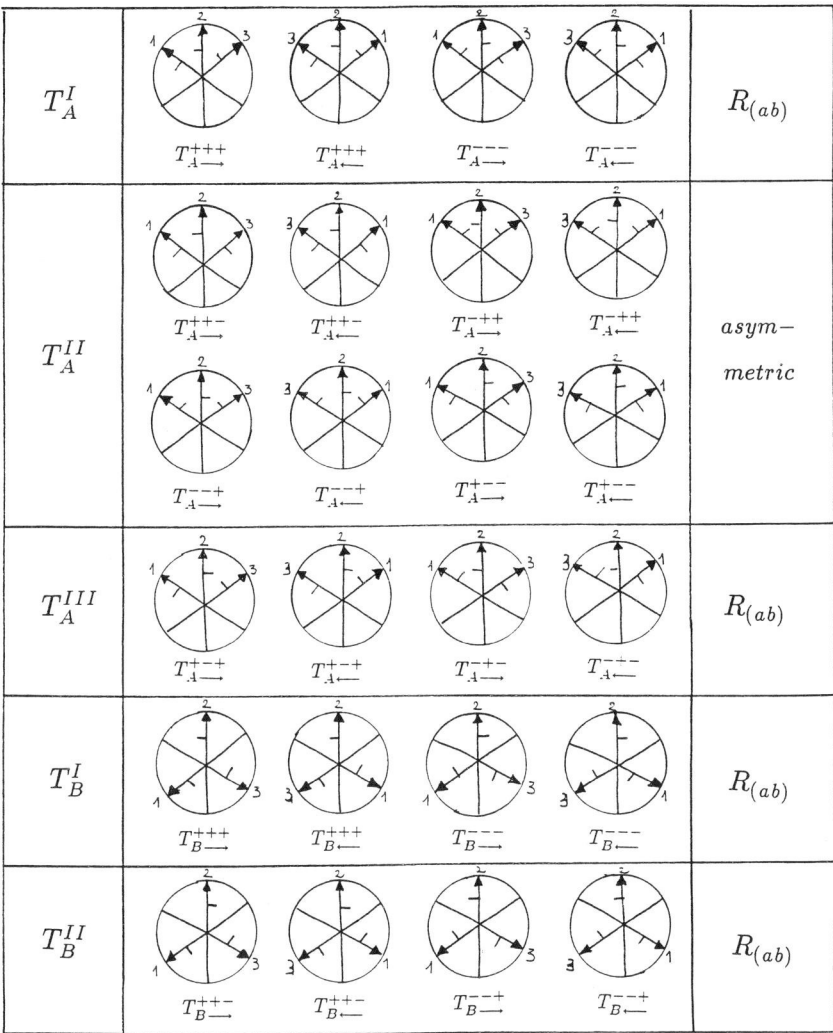

Table 2. Local strata of the discriminant T
RO-independent, RO-dependent and their symmetries

the smooth one are concordant or discordant.

The up or down arrow means that the cusp is of type up or down. (We orient the plane by the frame (to the right, up)).

The first of the two signs is the sign of the segment going away from the cusp point, and the second one is the sign of the smooth segment crossing the cusp.

The involution a, reversing the orientation of the plane, reverses the arrow and all the signs; the involution b, reversing the orientation

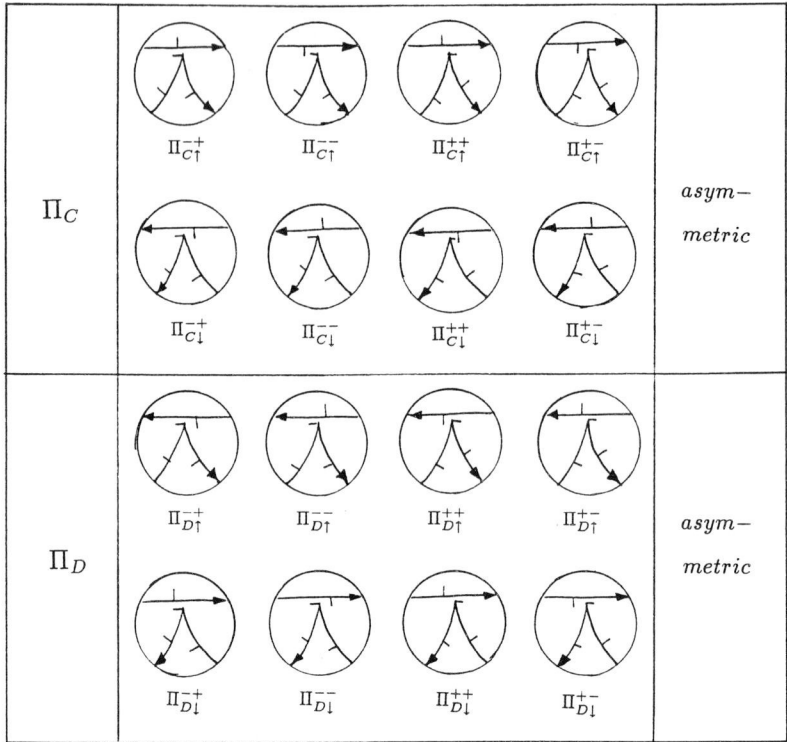

Table 3. Local strata of the discriminant Π
RO-independent, *RO*-dependent and their symmetries

of the front, reverses the arrow and the second sign; the involution c, reversing the coorientation of the front, reverses only the signs. Only the character C and D are *RO*-independent.

The stratum Λ of cusps births. The up or down arrow means that the two new-born cusps are both of type up or of type down. The sign is that of the smooth segment before the birth of the cusps. Involutions a and b reverse the arrow and the sign; the involution c reverses the sign only.

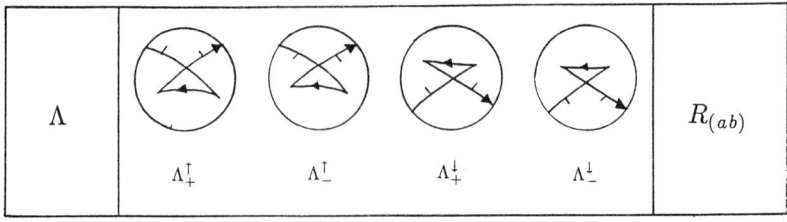

Table 4. Local strata of the discriminant Λ
RO-independent, *RO*-dependent and their symmetries

1.2. Coorientations of the local strata

For every local stratum we define a quantity so that its variation (called *elementary jump*) at the generic crossing of the discriminant is always zero except at the crossing of that specific stratum. For the notations I follow [1] in the case of self-tangencies and triple points. The elementary jumps not vanishing only at the crossings of the strata K and T are denoted by ΔJ and by ΔSt respectively. These jumps are locally constant functions defined along the smooth part of the discriminant. These functions are defined in table 5. For the elementary jumps at the strata Π and Λ, I use the corresponding small greek letters $\Delta\pi$ and $\Delta\lambda$. All the indices (letters, signs and arrows) are the same as for the strata.

The local coorientations of the strata are shown in table 5 (next page). They are independent of the orientations. The choice shown for a RO-independent stratum holds also for all the corresponding local strata dependent on the orientations. Moreover, for the strata K, Π and Λ the corresponding elementary jumps are always chosen in terms of the variation Δn of the number of double points. For all the local strata of type T the coorientation is the same, and was defined in [2]: one has to look at the two triangles existing before and after the crossing of the stratum T, i.e., the triangles vanishing in the triple point. The cyclic order of their sides, given by the order of visit of a point running along the curve following its orientation, defines the orientations of the vanishing triangles. The sign of a triangle is $s = (-1)^q$ where q is the number of sides whose directions coincide with that given by the cyclic orientation. The signs of the two vanishing triangles are always opposite. The crossing of T is called positive if the sign of the triangle changes from $-$ to $+$. The elementary jump in this case is defined by $\Delta St = 2$ (differently from the Arnold "strangeness" St of smooth curves, for which this jump is equal to 1).

In the figures the local coorientation of a stratum is indicated by the normal vectors to the stratum pointing in the positive direction. The positive direction is defined by the corresponding elementary jump which is positive at the crossing of the stratum from the negative to the positive side.

2. The 5 basic local RO-independent invariants

2.1 Results

Here we show how the basic local invariants independent of the orientations are defined. The proofs are in §§2.2, 2.3, 2.4.

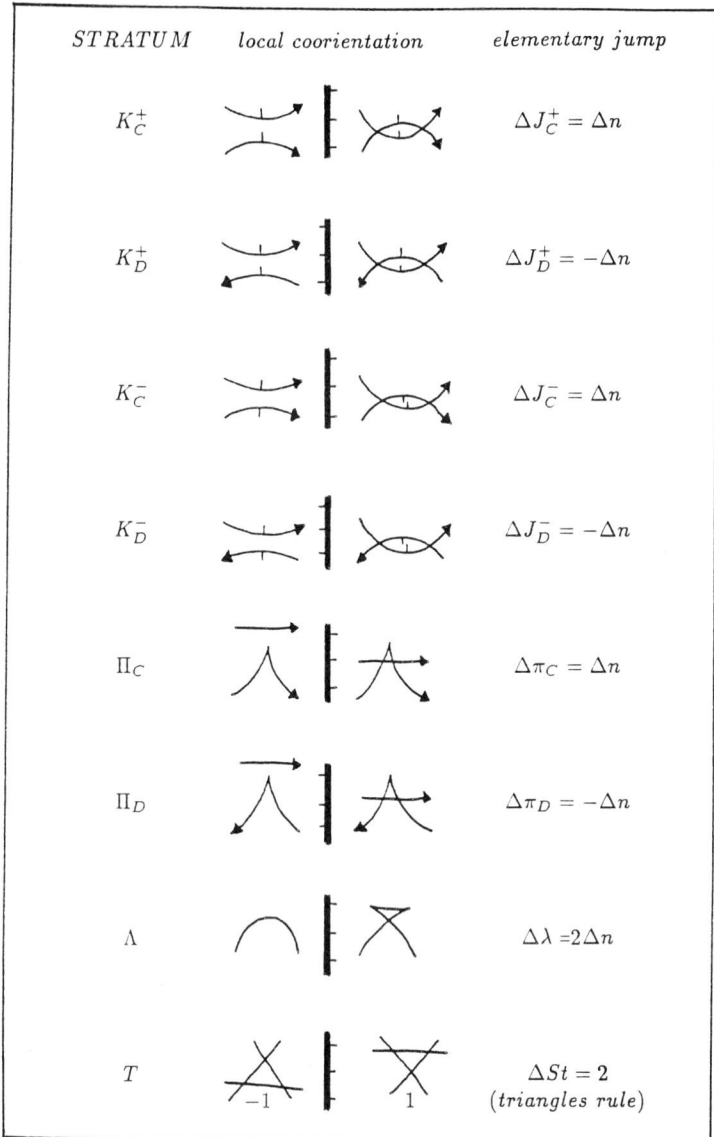

Table 5. The local coorientations of the strata of codimension one

We impose the coorientability condition at the codimension 2 strata of the discriminant.

The set ∇ of 12 numbers

$$(\nabla J_C^+, \nabla J_D^+, \nabla J_C^-, \nabla J_D^-, \nabla \pi_C, \nabla \pi_D,$$
$$\nabla St_A^I, \nabla St_A^{II}, \nabla St_A^{III}, \nabla St_B^I, \nabla St_B^{II}, \nabla \lambda)$$

is the vector of the coefficients of the RO-independent jump

$$\xi = \nabla J_C^+ \cdot \Delta J_C^+ + \ldots + \nabla \lambda \cdot \Delta \lambda.$$

This jump ξ defines a 1-cocycle if and only if the vector ∇ verifies a system of linear equation

$$M\nabla = 0 \qquad\qquad (I)$$

called the coherence system.

Lemma 2.1. *The linear operator M is defined by the following matrix*

$$M = \begin{pmatrix}
1 & -1 & 0 & 0 & -1 & 1 & 0 & 0 & 0 & 0 & 0 & 0 \\
0 & 0 & 1 & -1 & -1 & 1 & 0 & 0 & 0 & 0 & 0 & 0 \\
0 & 0 & 0 & 0 & 0 & 0 & 0 & 1 & -1 & 0 & 0 & 0 \\
0 & 0 & 0 & 0 & 0 & 0 & 1 & 0 & 0 & -1 & 0 & 0 \\
0 & 0 & 0 & 0 & 0 & 0 & 0 & 1 & 0 & 0 & -1 & 0 \\
0 & 0 & 0 & 0 & 0 & 0 & 0 & 0 & 1 & 0 & -1 & 0 \\
0 & 0 & 0 & 0 & 0 & 0 & 1 & 0 & 0 & 0 & -1 & 0 \\
0 & 0 & 0 & 0 & 0 & 0 & 0 & 0 & 1 & -1 & 0 & 0 \\
0 & 0 & 0 & 0 & 0 & 0 & 1 & -1 & 0 & 0 & 0 & 0 \\
0 & 0 & 0 & 0 & 1 & 1 & -1 & 0 & 0 & 0 & 0 & 0 \\
0 & 0 & 0 & 0 & 1 & 1 & 0 & -1 & 0 & 0 & 0 & 0
\end{pmatrix}$$

Lemma 2.2. *The space of solutions of the coherence system has dimension 5. As the 5 $\mathbf{R}$-basic solutions one may choose, for instance, the five jump vectors*

$$\Delta J^+ = \Delta J_C^+ + \Delta J_D^+$$
$$\Delta J^- = \Delta J_C^- + \Delta J_D^-$$
$$\Delta Sp = \Delta \pi_C + \Delta \pi_D + 2\Delta(St_A^I + St_A^{II} + St_A^{III} + St_B^I + St_B^{II})$$
$$\Delta h = \Delta(J_C^+ + J_C^- + \pi_C) - \Delta(J_D^+ + J_D^- + \pi_D)$$
$$\Delta \lambda$$

(For example, the coefficients of $\xi = \Delta J^+$ are $\nabla J_C^+ = 1$, $\nabla J_D^+ = 1$, all the other coefficients being zero).

Definition 2.1. The following cooriented hypersurfaces, defined by the generators of the solutions of eq.(I) as linear combinations of the local *RO*-independent strata, are called *basic RO-independent cooriented strata*:

$$K^+ = K_C^+ + K_D^+$$
$$K^- = K_C^- + K_D^-$$
$$Tp = \Pi_C + \Pi_D + 2(T_A^I + T_A^{II} + T_A^{III} + T_B^I + T_B^{II})$$
$$H = K_C^+ + K_C^- + \Pi_C - K_D^+ - K_D^- - \Pi_D$$
$$\Lambda$$

Lemma 2.3. *The basic RO-independent 1-cocycles, whose jumps are defined in Lemma 2.2, are cohomologous to zero in the space $\Omega_{i,\mu}$.*

Theorem 2.1. *Every local invariant of fronts independent of the orientations and additive under the connected summation is a real linear combination of the following 5 basic invariants:*

$$J^+, \quad J^-, \quad Sp, \quad h, \quad \lambda,$$

whose jumps are defined by Lemma 2.2, and whose values on the basic curves are shown in table 6.

Example. The number n of double points is a local invariant independent of the orientations. It is a linear combination of the basic invariants

$$n = h + \frac{1}{2}\lambda$$

Corollary 2.1. *Any generic path in $\Omega_{i,0}$ connecting two L-immersions whose fronts belong to the front classes of $\omega_{i,0}$ and $K_{i,0}$ (defined in fig. 3, see [1,2]), contains at least $i - 1$ events of type "self-tangency with parallel coorientations" and at least $i - 2$ events of type "self-tangency with antiparallel coorientations." If we want avoid the occurrence of triple points, then we will have at least $2i - 2$ cusp crossings. If we*

basic curves	J^+	J^-	Sp	h	λ
$n = i+1$ $\omega_{i,0}$	0	$-(i+1)$	0	$i+1$	0
$\mu = 2k$ $\omega_{0,\mu}$	$-k$	0	$2k$	$-k$	$2k$
$n = i+1$ $\omega_{i,\mu}$ $\mu = 2k$	$-k$	$-(i+1)$	$2k$	$i+1-k$	$2k$

Table 6. The value of the basic invariants on the basic fronts

want avoid the occurrence of cusp crossings, then we will have at least $i-1$ triple points.

Example. In fig. 3 two paths connecting the front $\omega_{3,0}$ with $K_{3,0}$ are shown. In the path avoiding cusp crossings, there are two triple points events; in the path avoiding triple points there are four cusp crossings.

Proof. One has $J^+(\omega_{i,0}) - J^+(K_{i,0}) = 2(i-1)$, and $J^-(\omega_{i,0}) - J^-(K_{i,0}) = 2i - 4$. Moreover, $Sp(\omega_{i,0}) - Sp(K_{i,0}) = 4i - 4$. But $Sp(\omega_{i,0}) - Sp(K_{i,0}) = \sum_j (\Delta_j \pi + 2\Delta_j St)$ (summation on the events) by Lemma 2.2. The triple points avoiding condition $\Delta_j St = 0$ implies $\sum_j \Delta_j \pi = 4i-4$. Each cusp crossing contributes ± 2 to $\sum_j \Delta_j \pi$. Hence the number of the cusp crossings is at least $2i - 2$. Similarly, the cusp crossings avoiding condition implies that the triple points events are at least $i-1$.

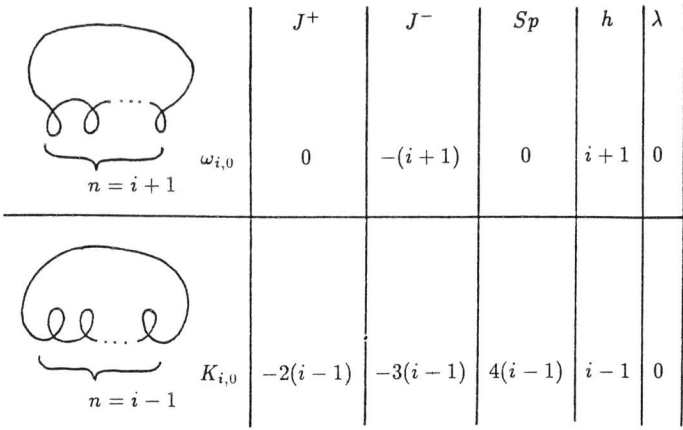

	J^+	J^-	Sp	h	λ
$\omega_{i,0}$	0	$-(i+1)$	0	$i+1$	0
$K_{i,0}$	$-2(i-1)$	$-3(i-1)$	$4(i-1)$	$i-1$	0

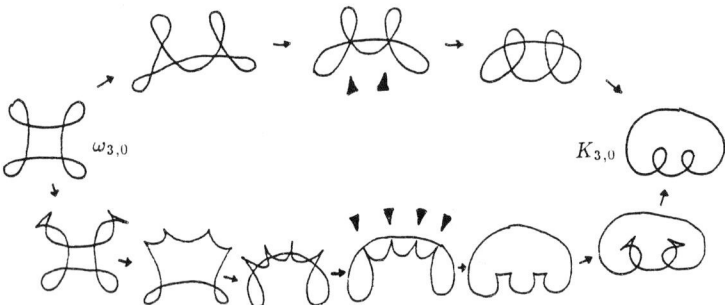

Figure 3. The fronts $\omega_{i,0}$ and $K_{i,0}$ with their basic invariants and two paths from $\omega_{3,0}$ to $K_{3,0}$.

2.2 The RO-independent 1-cocycles
We give now the

Proof of Lemma 2.1. Consider the strata of codimension 2 in the space of L-immersions . Some of these strata correspond to two non-generic events happening at different points of the front and others to the higher degeneration at one point. At a point of the first type, the germs of the discriminant hypersurface consist of two transversal hypersurfaces. Whenever the coorientations of both hypersurfaces are chosen, we obtain the coorientation of their union. Hence no coherence conditions arise at such strata.

There are 8 types of codimension 2 strata corresponding to the higher degenerations at one point [1]. They contain the L-immersions whose fronts have one of the following singular points:

KK: point of cubic self-tangency of the front;

TT: quadruple point;

$\Pi\Pi$: superposition of two cusp points (the tangents are supposed to be different);

$\Lambda\Lambda$: singular point of order $5/4$;

$K\Pi$: point of self-tangency coinciding with a cusp point;

KT: point of self-tangency coinciding with a triple point;

$T\Pi$: superposition of a double point and a cusp point (the tangents are supposed to be all different);

$\Pi\Lambda$: double point coinciding with a singular point of order $4/3$.

We require that the intersection index of a close path linked with every such strata having the codimension 1 discriminant, provided with the local coorientations defined above, vanishes. These conditions provide the equations of the system (I). Every equation defines a row of the matrix M. To obtain all the rows we shall consider case by case all the eight types of strata.

First we prove that the coherence conditions at the strata KK, TT, $\Pi\Pi$ and $\Lambda\Lambda$ are automatically satisfied and contribute nothing to the matrix M.

Stratum KK. There are 4 types of strata KK because the two meeting branches K are of the same type (both of type K_C^+, K_D^+, K_C^- or K_D^-). In fact (see fig. 4) the situations at two generic points a and b are the same, looking at the parallelism or antiparallelism of the normal and the tangent vectors; moreover also the coorientations of the strata coincide in KK. The coherence condition at the strata KK is thus automatically satisfied.

Stratum TT. It is not necessary to analyze all the possible cases: the following observation is sufficient to conclude that for this stratum as well the coorientability condition is automatically satisfied. Consider two generic points (a) and (b) lying on a stratum T symmetrically with respect to the stratum TT (see fig. 5). What happens at the triple point in the cases (a) and (b) is completely independent of the position of the fourth segment, so that the coorientations of T in (a) and in (b) coincide, whatever the stratum T is.

Stratum $\Pi\Pi$. To distinguish the strata of cusp crossing Π_C and Π_D one has to look only at the orientations (concordant or discordant). Then there are 2 distinguishable cases (see fig. 6) given by the two possibilities: (a) both cusps are of the same type (*up* or *down*) and (b) the cusps are of different types. Both cases give the same trivial equation

$$\nabla\pi_C - \nabla\pi_D + \nabla\pi_D - \nabla\pi_C = 0.$$

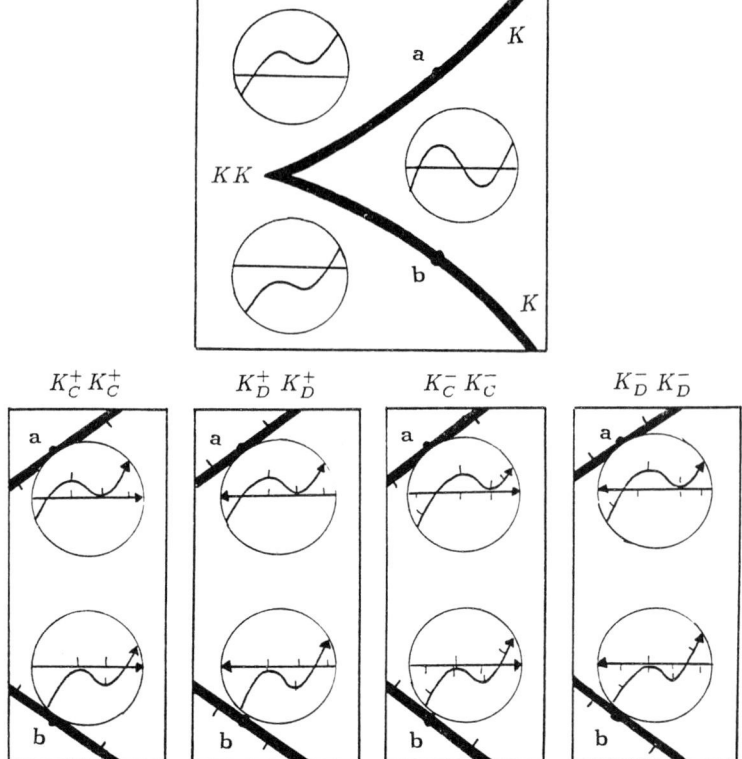

Figure 4. Local coherence at the strata KK

Hence the coorientability condition is automatically satisfied.

Stratum $\Lambda\Lambda$. Figure 7 shows that the two strata Λ meeting at the stratum $\Lambda\Lambda$ are coherently cooriented. The two strata Π are always of type Π_C, and are also coherently cooriented. Again the coorientability equation is trivially fulfilled.

Stratum $K\Pi$. Here two local strata of type K and the two strata of type Π meet. The K-strata are either both of type K^+ or both of type K^-. However one of them is of type C and the other of type D. In fact (see fig. 8) the segment crossing the cusp and successively tangent to one of its branches reverses its orientation moving from the left to the right in the figures (being vertical at the center, at the stratum $K\Pi$). Hence the sides of the narrower angles formed by the strata K and Π at the point $K\Pi$ – the K-side and the Π-side – are both of type C or both of type D. Since the number of double points is higher inside the two narrow angles, the coorienting vectors of the strata of type C

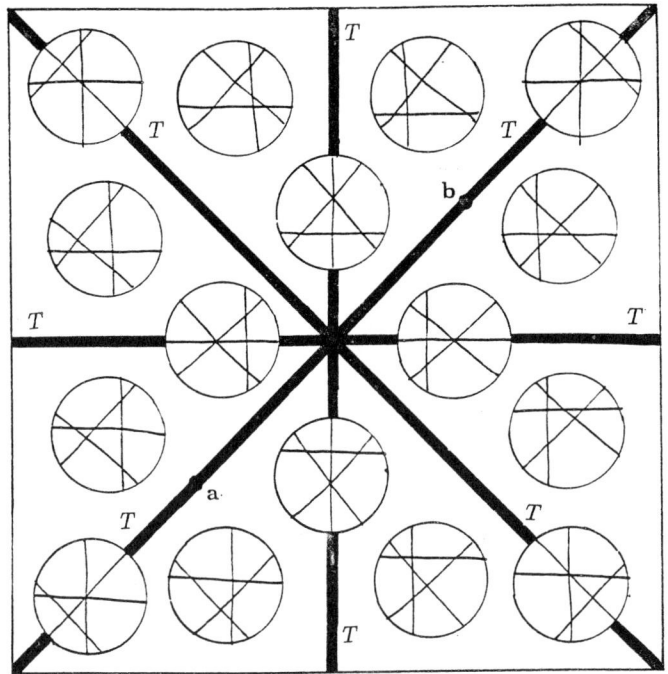

Figure 5. Local coherence at the stratum TT

point these angles inward while those of type D point outward. The two cases when the strata K are of type K^+ (a) and of type K^- (b) are responsible of the rows (1) and (2) of the matrix M.

Stratum KT. The two local strata of type K are of the same type and their coorientations are of course coherent (the type of self-tangency does not depend on the position of the third segment). Hence the resulting equations involve only the two strata T. The coorientations of the two strata of type T are compatible [2], and they can be arbitrary since the cyclic order of visit of the triple point is not fixed. The distinguishable cases are given only by the different types of strata T. They are shown in fig. 9. In (a) and (b) the types are the same, so the compatibility is automatically satisfied. In the other cases (c), (d), (e), and (f) we obtain the rows (3), (4), (5) and (6) of the matrix M.

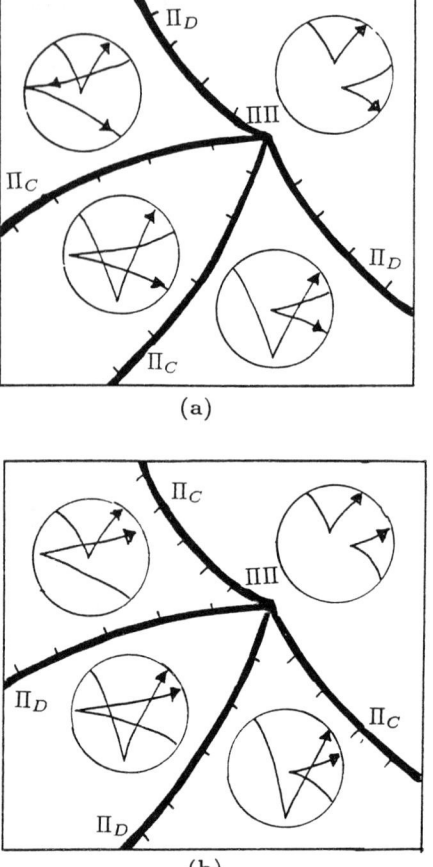

Figure 6. Local coherence at the strata ΠΠ

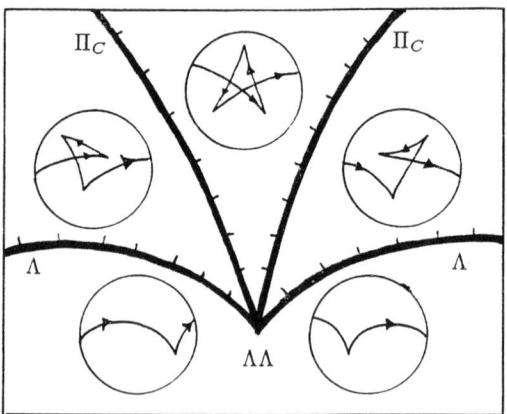

Figure 7. Local coherence at the stratum ΛΛ.

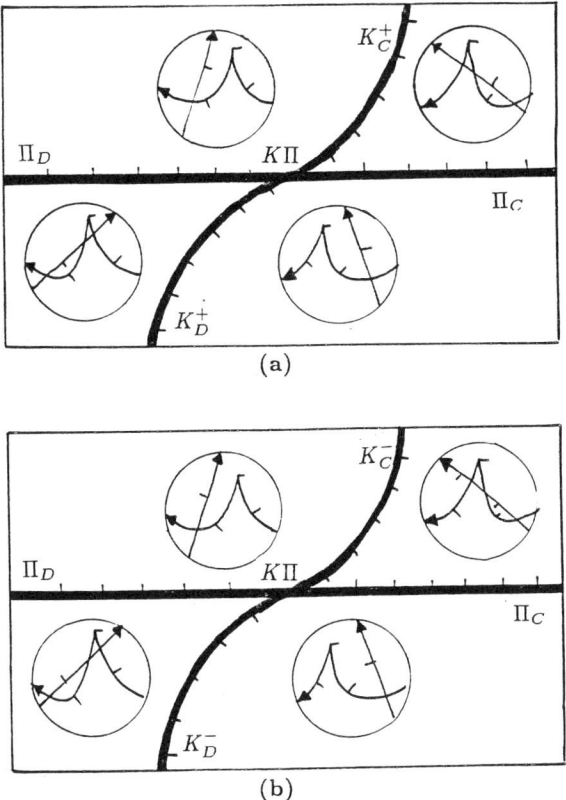

(a)

(b)

Figure 8. Local coherence at the strata $K\Pi$

Stratum $T\Pi$. The two lines of type Π in fig. 10 can be independent of any of the types Π_C and Π_D. Their types, as their coorientations, coincide in the two opposite branches. Indeed in two opposite points (for examples the points (a) and (b) in fig. 10) the types of the cusp crossing are the same. Hence this stratum gives again equations only for the jumps of the type ΔSt.

The two branches of the strata T are coherently cooriented, as fig. 11 shows. In the four small triangles vanishing at the two triple points in (c) and (d) of fig. 10, the cyclic orders of the visits of the sides are the same. The orientations of two sides are the same. The orientation of the third side, being a cusp branch, is opposite at (c) and at (d) (fig. 11). Hence at the left of the strata T near (c) and near (d) the vanishing triangles have opposite signs. Similarly the vanishing triangles at the right of the strata T at (c) and at (d) have opposite

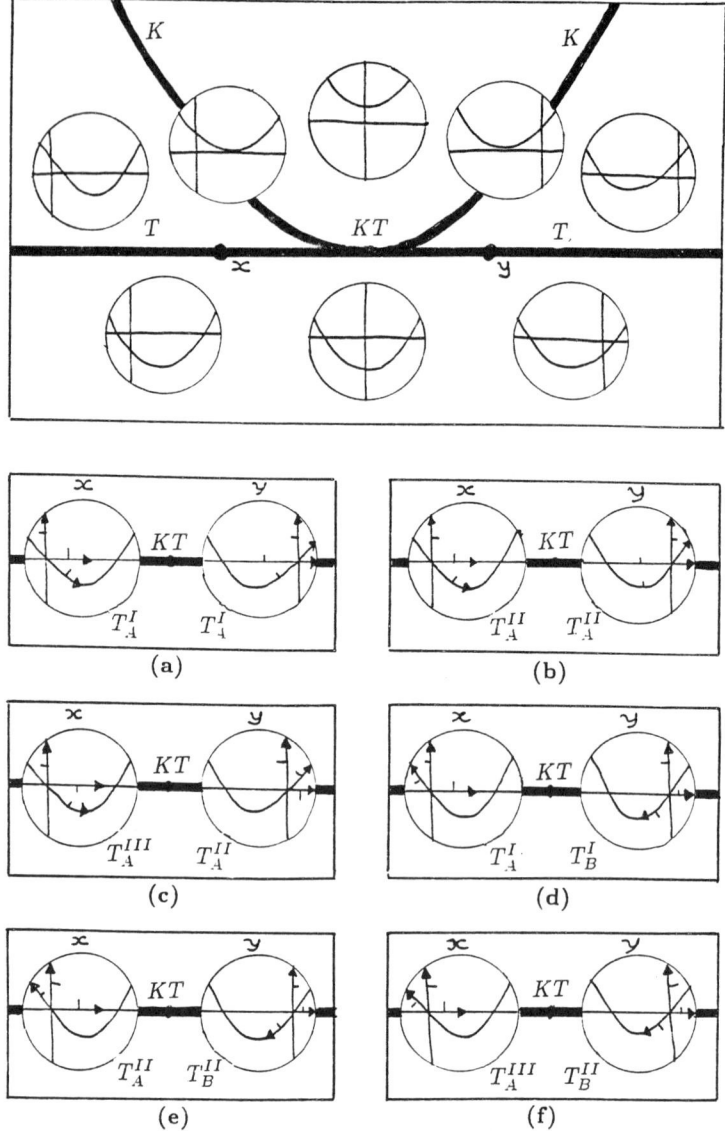

Figure 9. Local coherence at the strata KT

sign. Therefore the coorientations in (c) and (d) are mirror-symmetric, i.e., they are compatible at the stratum $T\Pi$.

The distinguishable cases for the strata $T\Pi$ are shown in fig. 12. In the cases (a) and (b) the RO-independent strata are the same. The

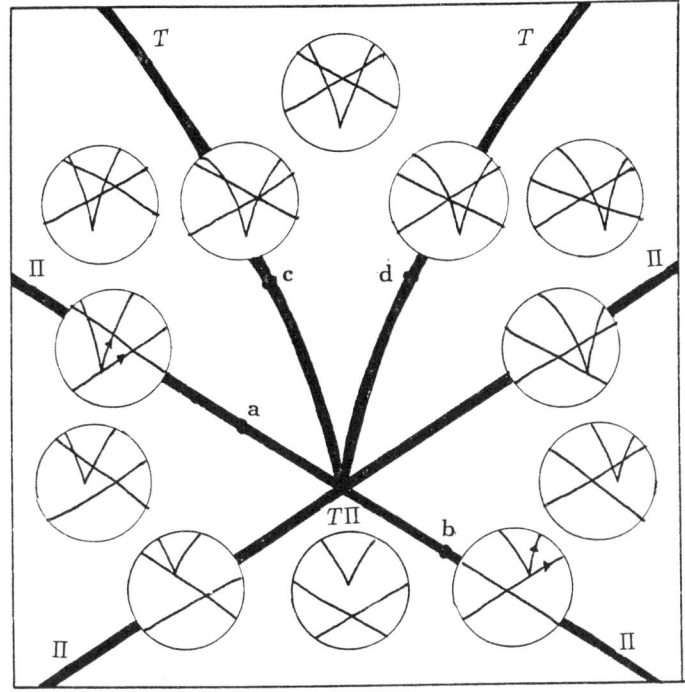

Figure 10. Local coherence of the strata Π at a stratum *T*Π

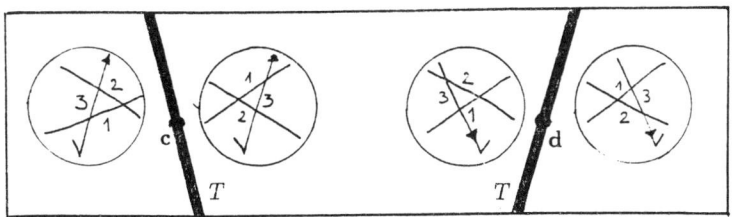

Figure 11. Coorientation of the two strata *T* near *T*Π

cases (c), (d), (e) and (f) give the rows (5), (7), (8) and (9) of *M*.

Stratum ΠΛ. The coorientation of the stratum Λ does not change at stratum ΠΛ. The two strata of type Π are always one of type Π_C and one of type Π_D. They are cooriented in the opposite way (since the number of double points above the Π strata in fig. 13 is higher than below it).

The stratum of triple points cannot be of any of the types T_B and T_A^{III}. The cyclic order of the visit of the triple point is fixed by the

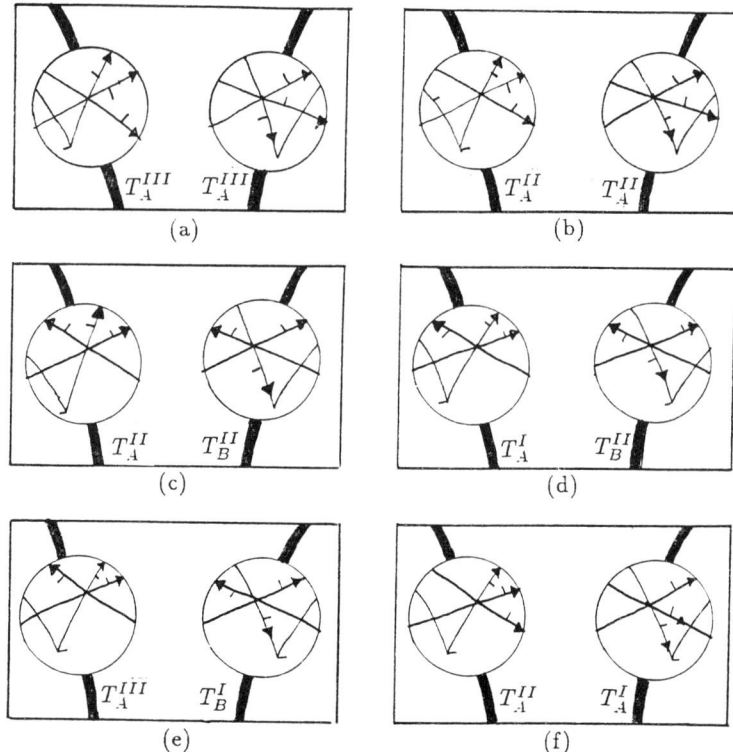

Figure 12. Local coherence at the strata $T\Pi$

orientation of the front. Hence the coorientation of the stratum T is always as shown in fig. 13, i.e., compatible with the orientations of Π_C and Π_D. The four distinguishable cases are shown in fig. 13, where (a) and (b) are responsible for the rows (10) and (11) of M.

Lemma 2.1 is proved.

2.3 The cohomological triviality of the basic RO-independent 1-cocycles

Every solution of system (I) defines the jump of a 1-cocycle. This jump can be written in terms of the jumps given by the generators defined in Lemma 2.2. Every generator represents a 1-cocycle (*basic cocycle*), whose value at a generic closed curve is the sum of the corresponding jumps at the events along this curve. We now have to verify that these 1-cocycles are trivial.

Proof of Lemma 2.3. We have to prove that the value of any of the

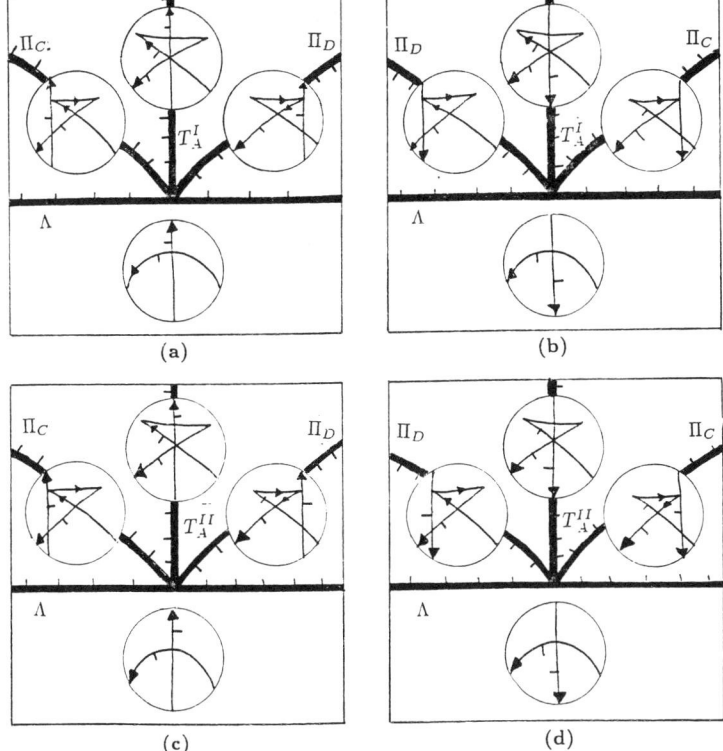

Figure 13. Local coherence at the strata $\Pi\Lambda$

5 basic cocycles on a generic closed path in the space of the marked L-immersions vanishes. In fact every connected component $\Omega_{i,\mu}$ is (weakly) homotopy equivalent to a circle, so we have to do such a verification only for one noncontractile path.

This is only a generalization to our 5 1-cocycles of the Arnold proof of the same lemma proved in [1] for the stratum K^+.

One has to distinguish two cases:

Case 1. The Maslov index is different from zero.

In this case, given a generic front in $\Omega_{i,\mu}^0$ ($\mu \neq 0$), consider the family of marked fronts obtained from it marking successively all its points in the order given by its orientation. This means that we have simply to move the front in the plane so that the marked point and the coorienting normal vector to it remain fixed (see fig. 14).

When the turn of the front is completed, i.e., the front is again in the initial position, the tangent vector to the L-immersions at the marked point has done a number $k = \mu/2$ of full rotations in the contact

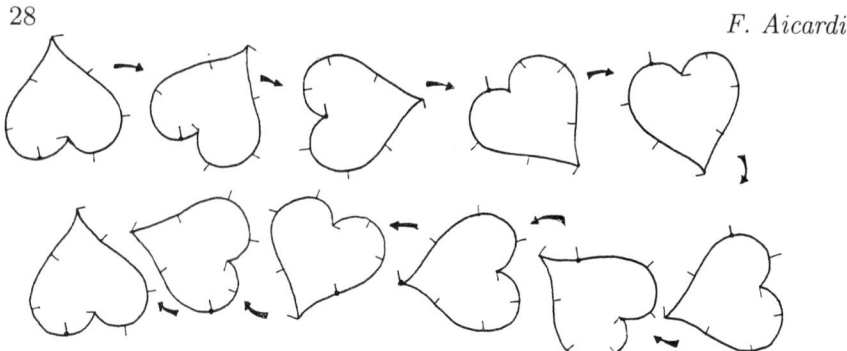

Figure 14. A noncontractile path in $\Omega_{1,2}$

plane, so that this family of marked L-immersion is a noncontractile path in $\Omega^0_{i,\mu}$. This path does not intersect any stratum of the discriminant. Hence the values on it of basic RO-independent 1-cocyles are trivially vanishing.

Case 2. The Maslov index is equal to zero.

In this case generic noncontractile paths in $\Omega^0_{i,0}$ are not so trivial. We verify that the sum of the jumps of the generators defined by Lemma 2.2 in terms of the elementary jumps along a generic noncontractile path in the space $\Omega_{i,0}$ for every value i of the index vanish.

A generic noncontractile path in $\Omega^0_{1,0}$ is shown in fig. 15.

All the generic crossings of the discriminant along this path are numbered (1-14), and the corresponding elementary jumps are indicated. The noncontractility is visible in the last part of the path, where there are no crossings of the discriminant but the marked points experience the same perestroikas as in the case $\mu \neq 0$. Denote by (x) the value of the elementary jump at the crossing numbered by x. The increments δ of the generators along the path are given by the following formulas:

$$\delta(J^+) = (6) + (10) = 0$$
$$\delta(J^-) = (1) + (4) = 0$$
$$\delta(Sp) = (3) + (5) + (9) + (11) = 0$$
$$\delta(h) = -(1) - (3) - (4) - (5) - (6) + (9) + (10) + (11) + (12) - (14) = 0$$
$$\delta(\lambda) = (2) + (7) + (8) + (13) = 0$$

In the case of $\Omega_{n-1,0}$, consider the path shown in fig. 16.

From (a) to (b) the perestroikas are similar to perestroikas 1-14 shown in fig. 15. The only difference consists of the n loops added near the marked point, which remain fixed and are not involved in

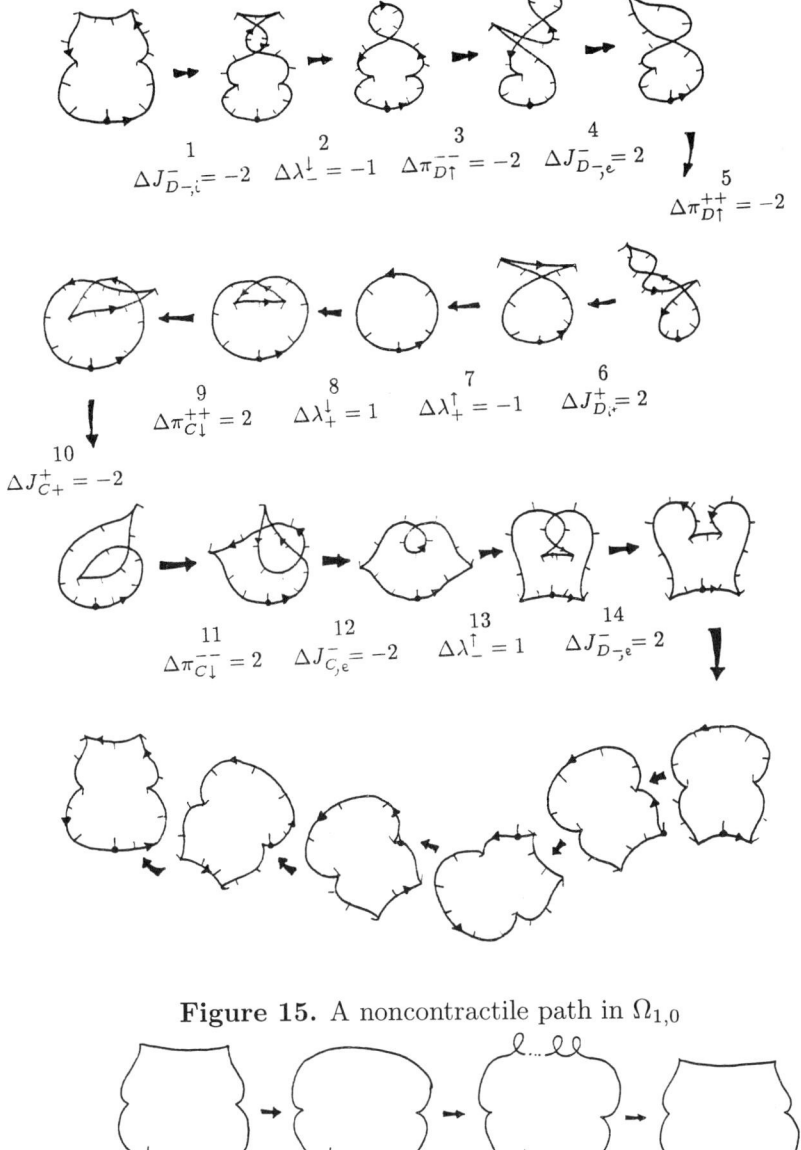

Figure 15. A noncontractile path in $\Omega_{1,0}$

Figure 16. A noncontractile path in $\Omega_{n-1,0}$

these perestroikas. We go from (b) to (c), transporting the n loops beyond two cusps points. In fig. 17 it is shown that in the operation of the bypassing by a pair of cusps of the same sign of a loop, all the

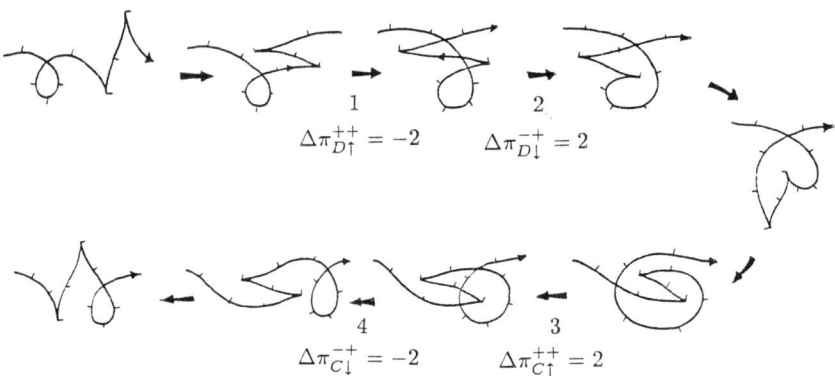

Figure 17. The bypassing of a loop by a pair of cusps

intersection indices with the discriminant vanish.

The crossings of the discriminants are numbered and the values of the elementary jumps indicated. Using the same notation as before, we calculate the only possible non null increments at the crossing of the strata Tp and H respectively:

$$\delta(Sp) = (1) + (2) + (3) + (4) = 0$$
$$\delta(h) = -(1) - (2) + (3) + (4) = 0$$

The other strata of the discriminant are not crossed in the path from (b) to (c).

From (c) to (d) (see again fig. 16) the front is moved on the plane to the original position, so there are no other crossings of the discriminant in the noncontractile closed path. The lemma is proved.

2.4 Definition of the basic RO-independent invariants

The previous lemmas ensure that the functions J^+, J^-, Sp, h and λ define, up to additive constants depending only on the absolute values of the index and of the Maslov index, the basic local invariants of the fronts independent of orientations. This means that their variation along a generic path connecting two L-immersions does not depend on the path, but only on the initial and final L-immersions.

Proof of Theorem 2.1 I) Existence of the additive invariants.

We show that the choice of values of the basic invariants on the basic curves makes these invariants additive under the connected summation of fronts.

A *RO*-independent local invariant X is additive under the connected sum of fronts if and only if it satisfies:

i) $X(\theta) = 0$, θ being the front class of the circle with every orientation and coorientation.

ii) $X(\alpha \# \beta) = X(\alpha) + X(\beta)$ for every choice of the generic points at the boundaries of the two fronts α and β where the gluing is made (when the orientations and coorientations allow this gluing).

From (i) we obtain that $\omega_{1,0}$ has all the basic invariants equal to zero but $J^- = -2$ and $h = 2$.

From (ii) we obtain that $\omega_{0,0}$ has all the basic invariants equal to zero but $J^- = -1$ and $h = 1$ (see fig. 18).

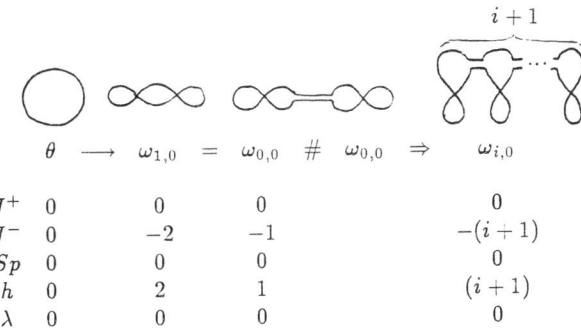

	$\theta \longrightarrow$	$\omega_{1,0}$	$=$	$\omega_{0,0}$	$\#$	$\omega_{0,0}$	$\Rightarrow$	$\omega_{i,0}$
J^+	0	0		0				0
J^-	0	-2		-1				$-(i+1)$
Sp	0	0		0				0
h	0	2		1				$(i+1)$
λ	0	0		0				0

Figure 18. The basic curves $\omega_{i,0}$

We obtain $\omega_{i,0}$ as the connected sum of $(i+1)$ copies of $\omega_{0,0}$.

To define the values of the basic invariants on the curves $\omega_{1,\mu}$ we proceed in this way. Start again from the curve θ, transform it into γ and calculate its invariants (see fig. 19). Then write γ as the connected sum of two fronts, both in the class of $\omega_{1,2}$ up to the c involution.

The invariants of the curve $\omega_{1,2}$ are thus defined, and also those of the curves $\omega_{i,\mu}$ obtained as the connected sum of $k = \mu/2$ copies of $\omega_{1,2}$. The values of the invariants for the basic curve $\omega_{i,\mu}$ are finally obtained as the sums of the invariants of $\omega_{i,0}$ and of $\omega_{1,\mu}$, since $\omega_{i,\mu} = \omega_{i,0} \# \omega_{1,\mu}$.

To conclude the proof we have to verify that the condition (ii) holds for the basic curves. This implies that condition (ii) is verified for every sum of every fronts α and β. Indeed, suppose that p and q are the points respectively at the boundary of α and β where the

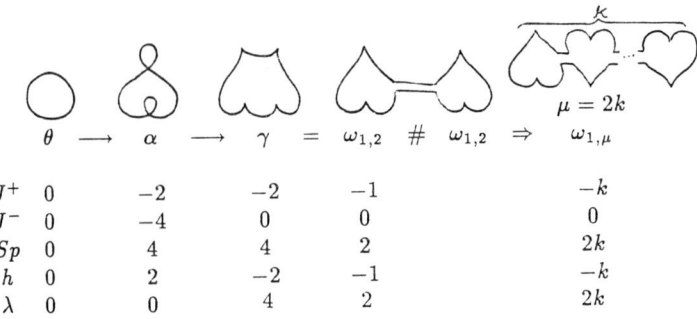

$$\mu = 2k$$

	θ	$\longrightarrow$	α	$\longrightarrow$	γ	$=$	$\omega_{1,2}$	$\#$	$\omega_{1,2}$	$\Rightarrow$	$\omega_{1,\mu}$
J^+	0		-2		-2				-1		$-k$
J^-	0		-4		0				0		0
Sp	0		4		4				2		$2k$
h	0		2		-2				-1		$-k$
λ	0		0		4				2		$2k$

Figure 19. The invariants of the curves $\omega_{1,\mu}$

gluing occurs. Choosing a point p at the boundary of α, it is always possible to transform the front α into another front having the same indices (in particular into the basic front) leaving the point p always generic and lying at the boundary and far from the points involved in the perestroikas (the boundary of a front is the boundary of the infinite domain of the plane belonging to the complement to the front into the plane). This is proved in [5]. Hence we can transform two fronts α and β into the basic curves leaving p and q at their boundary, and glue the two basic curves at these points so obtaining a front which can be reached from $\alpha\#\beta$ exactly by means of the same perestroikas (see fig. 20).

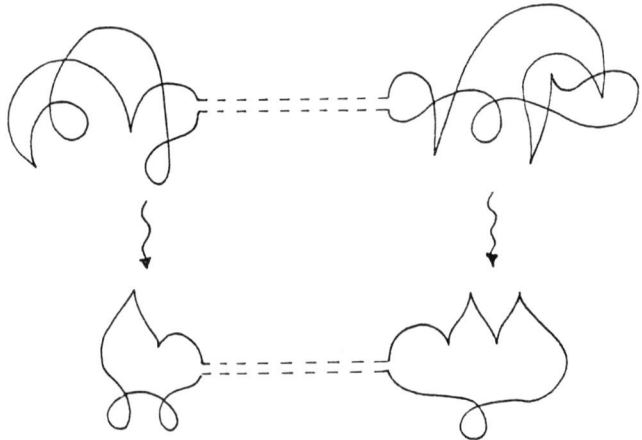

Figure 20. The connected sum during the perestroikas

The validity of (2) for generic α and β depends thus on the validity of (2) for the basic fronts.

To transform this into a basic curve of type $\omega_{i+j+1,\mu+\nu}$, a connected sum of two basic curves $\omega_{i,\mu}$ and $\omega_{j,\nu}$ we have in general to do a set of perestroikas. We verify that the increments of all the basic invariants at these perestroikas agree with the rule of summation. The complete set of perestroikas is formed by the following elementary perestroikas and by the perestroikas obtained from these by the action of the *RO*-involutions (see fig.21):

P_1: bypassing of a loop by a cusp point (two cases fig.21-1a,b). In both cases one obtains the inverse perestroika by an *RO*-involution. Hence the values before and after the perestroika of an *RO*-independent invariant cannot be different.

P_2: bypassing of a double point by a pair of cusps of the same sign (fig.21-2). One has two opposite crossings of Π_C or of Π_D, so all the values of the invariants remain unchanged.

P_3: bypassing of a double point by a loop (fig.21-3). In this perestroika, as one can verify from the figure, one has the following increments of the invariants

$$\delta(J^+) = -2; \quad \delta(J^-) = -2; \quad \delta(Sp) = 4; \quad \delta(h) = 0; \quad \delta(\lambda) = 0.$$

P_4: suppression of two cusp points of opposite signs by the creation of a loop (fig. 21-4). The nonzero increments are

$$\delta J^- = -2; \qquad \delta \lambda = -2.$$

P_5: annihilation of an internal-external pair of loops (fig.21-5). In this perestroika (see figure):

$$\delta(J^+) = 2; \quad \delta(J^-) = 4; \quad \delta(Sp) = -4; \quad \delta(h) = -2; \quad \delta(\lambda) = 0.$$

We are now ready to put together these perestroikas to transform every connected sum of basic curves into a basic curve. Consider all the 4 possibilities for the positions of the points p at the boundary of a basic curve: (1) between a loop and a cusp; (2) on a loop; (3) between two cusp points; (4) between two loops. In the cases (3) and (4) by means of perestroikas P_1 leaving p fixed, it is possible to transport all the cusps at one side and all the loops at the other side with respect to p, so going to the case (1). The invariants are unchanged, so we forget the third and the fourth possibilities for the gluing point.

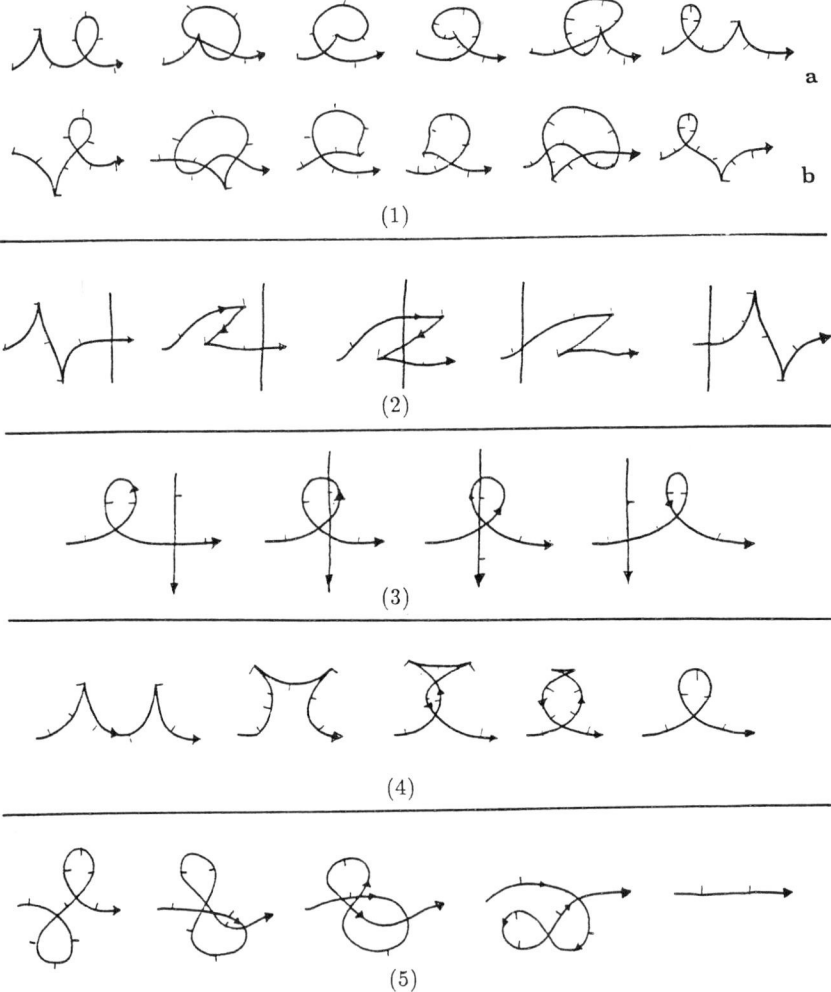

Figure 21. Perestroikas occurring in the transformation into a basic front of a connected sum of basic fronts

The choice of points p and q belonging to the basic curves $\omega_{i,\mu}$ and $\omega_{j,\nu}$ can be of different types: type (1)-(1), type (2)-(2) and type (1)-(2).

In the case (1)-(1), let the indices i and j of the two fronts be greater than zero (they must have the same sign to allow the gluing). After the gluing there are two possibilities (see fig. 22). If the Maslov indices of the two curves have also the same sign (a), we have obtained exactly the basic curve $\omega_{i+j+1,\mu+\nu}$ whose invariants are exactly the sum of the invariants of the glued fronts. If not (b), let $\mu = 2m > 0$

and $\nu = -2k < 0$ be the values of the Maslov indices and moreover let $k < m$. Then near p and q after the gluing there are two cusps of opposite signs. We do a perestroika P_4, obtaining a loop, internal or external. By means of perestroikas of type P_1, we transport this loop near the other loops (if any), i.e., far from the sets of cusps. We have again two cusps of opposite signs. With a perestroika P_4 we transform them into a loop, which result in their being of different types with respect to the previous one (i.e., external or internal). We do a number of these perestroikas equal to $2k$ obtaining k pairs of internal-external loops. To obtain the basic curve $\omega_{i+j+1,2m-2k}$ we have to do k times the perestroika P_5. At the end of this process, the variations of the invariants which have occurred in $2k$ times P_4 and in k times P_5, are:

$$\delta(J^+) = 2k; \quad \delta(J^-) = 4k - 2(2k) = 0; \quad \delta(Sp) = -4k;$$
$$\delta(h) = -2k; \quad \delta(\lambda) = -4k.$$

For every of such variations δX $(X = J^+, J^-, Sp, h, \lambda)$ we verify easily:

	$X(\omega_{i,2m})$	$+$	$X(\omega_{j,2k})$	$+$	$\delta(X)$	$=$	$X(\omega_{i+j+1,2m-2k})$
J^+	$-m$		$-k$		$2k$		$-(m-k)$
J^-	$-(i+1)$		$-(j+1)$		0		$-(i+j+2)$
Sp	$2m$		$2k$		$-4k$		$2(m-k)$
h	$i+1-m$		$j+1-k$		$2k$		$i+j+2-(m-k)$
λ	$2m$		$2k$		$-4k$		$2m-2k$

In the case (2)-(2), the signs of the indices still have to be equal to allow the connected sum (see fig. 23). This case is reduced to the previous one by means of type P_1 perestroikas and of a perestroika where one has

$$\delta J^- = +2; \quad \delta h = -2.$$

The basic front so obtained has index equal to $i+j-1$. This agrees with the differences δJ^- and δh with respect to the sum of the invariants:

$$J^-(\omega_{i,2m}) + J^-(\omega_{j,2k}) - J^-(\omega_{i+j-1,2m+2k}) = -2$$
$$h(\omega_{i,2m}) + h(\omega_{j,2k}) - h(\omega_{i+j-1,2m+2k}) = 2$$

In the case (1)-(2) the signs of the indices are opposite. Let their values be i and $-j$ and let $-j$ be the index of the curve where the

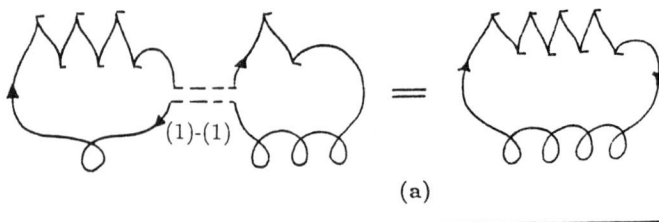

(a)

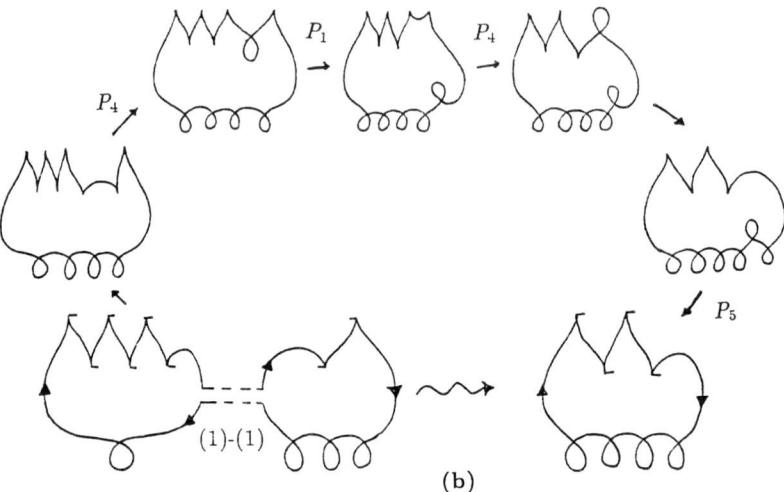

(b)

Figure 22. Connected sums of type (1)-(1) of two basic fronts

gluing point is of type (2). Moreover let $i \geq j$. Suppose that the Maslov indices are both positive (see fig. 24).

We transport the j loops and the $2k$ cusps as shown in fig. 24 making exactly j times the perestroika P_3 and k times the perestroika P_2. We obtain a curve with $\mu + \nu$ cusps of the same sign, $i+1$ external loops and j internal loops. By means of perestroikas of type P_1 we assemble together the cusps and the loops. Then we eliminate the j internal loops with j external loops by means of j perestroikas of type P_5. Since the Maslov indices have the same sign, we have obtained a basic curve with index $i - j + 1$, which has indeed $i - j + 2$ double

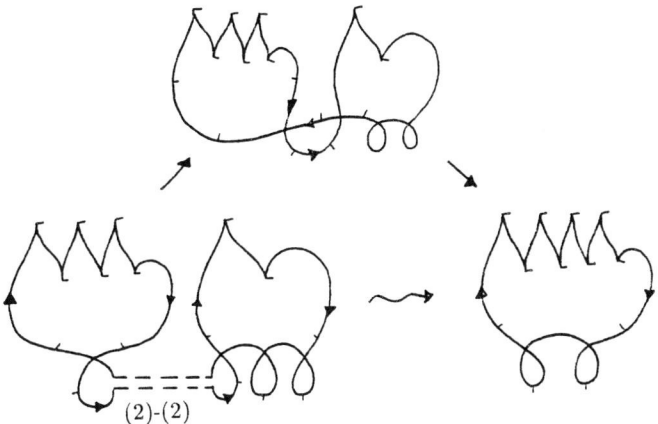

Figure 23. Connected sums of type (2)-(2) of two basic fronts

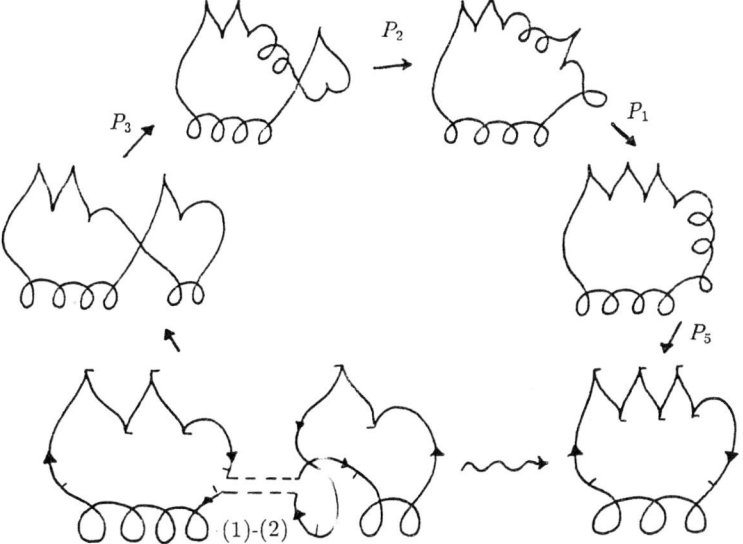

Figure 24. Connected sums of type (1)-(2) of two basic fronts

points. The total variations of the values of the invariants are

$$\delta(J^+) = -2j + 2j = 0; \quad \delta(J^-) = -2j + 4j = 2j;$$
$$\delta(Sp) = 4j - 4j = 0; \quad \delta(h) = -2j; \quad \delta(\lambda) = 0.$$

These increments coincide with the differences between the values of
the sum of the invariants of the added basic curves and the values of
the invariants of the basic curve $\omega_{i-j+1,\mu+\nu}$, as shown:

$$X(\omega_{i,2m}) \quad + \quad X(\omega_{-j,2k}) \quad + \quad \delta(X) \quad = \quad X(\omega_{i-j+1,2m+2k})$$

	$X(\omega_{i,2m})$	$X(\omega_{-j,2k})$	$\delta(X)$	$X(\omega_{i-j+1,2m+2k})$
J^+	$-m$	$-k$	0	$-(m+k)$
J^-	$-(i+1)$	$-(j+1)$	$2j$	$-(i-j+2)$
Sp	$2m$	$2k$	0	$2(m+k)$
h	$i+1-m$	$j+1-k$	$-2j$	$i-j+2-(m+k)$
λ	$2m$	$2k$	0	$2m+2k$

The case $i < j$ is similar. If the Maslov indices do not have the same
sign, then we come back to the situation of the case (1)-(1). The
existence of the additive invariants is proved.

II) Uniqueness of the additive local invariants.

An additive invariant is unambiguously defined by its jumps (see
Lemma 3.6 below). Hence we have listed *all* the additive local *RO*-
independent invariants.

2.5. Some remarks on the *RO*-independent invariants

Remark 2.1. The nonexistence of the invariant St in the case of the
fronts is due to codimension 2 stratum $\Pi\Lambda$, showing that the stratum
of triple points is not coorientable.

If we start from a smooth front in $\Omega_{i,0}$ and we forbid the crossing
of the stratum Λ, we are moving just in the connected space of the
smooth immersions of a circle in the plane: the invariant λ is zero, the
invariant Sp becomes the invariant St (up a constant factor) and the
invariant h becomes n.

If now we forbid the crossing of the stratum Λ starting from ·a
generic front with a given number of cusps (and Maslov index equal or
not to zero), we obtain for the elementary jumps of the local invariants
a new coherence system $M'\nabla = 0$, where M' is the linear operator
obtained from M erasing the last two rows, given by the coorientability
conditions at the codimension 2 stratum $\Pi\Lambda$. The new system has the
set of solutions of dimension 7, i.e., the two basic cooriented strata Tp
and H split in two different cooriented strata, respectively:

$$Tp\Big\langle \begin{array}{l} T = T_A^I + T_A^{II} + T_A^{III} + T_B^I + T_B^{II} \\ \Pi = \Pi_C + \Pi_D \end{array}$$

and

$$H \left\{ \begin{array}{l} H_C = K_C^+ + K_C^- + \Pi_C \\ H_D = K_D^+ + K_D^- + \Pi_D \end{array} \right.$$

The cohomological triviality of the corresponding 1-cocycles has however to be proved, because the homotopy type of the space of L-immersions whose fronts have a fixed number of cusp points is not investigated yet.

It is interesting to note that, in the hypothesis that the 1-cocycles defined by the jumps at H_C and H_D are trivial, the corresponding invariants h_C and h_D should be another generalization for the fronts of the invariants J^+ and J^- defined for the immersions of a circle in the plane.

Their definition in fact is totally independent of the coorientation of the front. Moreover the local coorientations of the strata Π_C and Π_D coincide with the coorientations of the corresponding strata $K_C^\pm$ and $K_D^\pm$ so that the invariants h_C and h_D, forgetting the coorientation of the front, should operate as a sort of "smoothening," i.e., their values should coincide with the values J^+ and J^- of a corresponding immersion where all cusp points are smoothened.

This is the meaning of the identity $\Delta h = \Delta n$ when λ is fixed, where $h = h_C - h_D$ (corresponding to $J^+ - J^-$ for the immersions). Moreover these conclusions should hold also for the noncooriented fronts with any number (even or odd) of cusps.

Remark 2.2. The strata K^+ and K^- are not irreducible [1]. Indeed one can associate to every self-tangency event a nonordered pair of numbers (called the *partial indices* [1][5]) which is invariant along the components of the strata. The local invariants J^+ and J^- are the sums of an infinity of invariants, defined in the same way (thus independent of the orientations), but indexed by the partial indices. This fact does not contradict Theorem 2.1 since these invariants are not additive under the connected summation of two fronts.

3. The basic RO-dependent local invariants

3.1. Results

The proofs of the following facts are in §§3.2, 3.3, 3.4 and 3.5.

In this part we repeat the procedure followed in the search for the RO-independent basic invariants starting from the elementary jumps dependent on the orientations, defined in Lemma 1. Their number is 54, so it is not convenient to write the entire system of the local coherence equations at the codimension 2 strata.

Definition 3.1. A *partial coherence system* is a set of equations belonging to the entire coherence system involving only the jumps at the strata of only one type $(K, T, \Pi$ or $\Lambda)$. Its solutions are called *reduced* elementary jumps.

We find first a set of reduced elementary jumps, then for these new variables we write the other equations of local coherence at the strata of codimension 2 involving different types of jumps.

Lemma 3.1. *The following reduced elementary jumps form the fundamental system of the solutions of the union of the partial coherence systems for the strata K, T and Π.*

$$1) \quad \Delta J_{C+}^{+}$$

$$2) \quad \Delta J_{C-}^{+}$$

$$3) \quad \Delta J_{D}^{+} = \Delta J_{D,i+}^{+} + \Delta J_{D,i-}^{+}$$

$$4) \quad \Delta J_{C}^{-} = \Delta J_{C,i}^{-} + \Delta J_{C,e}^{-}$$

$$5) \quad \Delta J_{D+}^{-} = \Delta J_{D+,e}^{-} + \Delta J_{D+,i}^{-}$$

$$6) \quad \Delta J_{D-}^{-} = \Delta J_{D-,e}^{-} + \Delta J_{D-,i}^{-}$$

$$7) \quad \Delta St = \sum_{\alpha,\beta,\gamma,\eta} \Delta St_{A\ \eta}^{\alpha\beta\gamma} + \sum_{\alpha,\beta,\eta} \Delta St_{B\ \eta}^{\alpha\alpha\beta}$$

$$8) \quad \Delta\pi_{\uparrow}^{-+} = \Delta\pi_{C\uparrow}^{-+} + \Delta\pi_{D\uparrow}^{--}$$

$$9) \quad \Delta\pi_{\uparrow}^{--} = \Delta\pi_{C\uparrow}^{--} + \Delta\pi_{D\uparrow}^{-+}$$

$$10) \quad \Delta\pi_{\uparrow}^{++} = \Delta\pi_{C\uparrow}^{++} + \Delta\pi_{D\uparrow}^{+-}$$

$$11) \quad \Delta\pi_{\uparrow}^{+-} = \Delta\pi_{C\uparrow}^{+-} + \Delta\pi_{D\uparrow}^{++}$$

$$12) \quad \Delta\pi_{\downarrow}^{-+} = \Delta\pi_{C\downarrow}^{-+} + \Delta\pi_{D\downarrow}^{--}$$

$$13) \quad \Delta\pi_{\downarrow}^{--} = \Delta\pi_{C\downarrow}^{--} + \Delta\pi_{D\downarrow}^{-+}$$

$$14) \quad \Delta\pi_{\downarrow}^{++} = \Delta\pi_{C\downarrow}^{++} + \Delta\pi_{D\downarrow}^{+-}$$

$$15) \quad \Delta\pi_{\downarrow}^{+-} = \Delta\pi_{C\downarrow}^{+-} + \Delta\pi_{D\downarrow}^{++}$$

$$16) \quad \Delta\pi_{C} = \sum_{\alpha,\beta,\eta} \Delta\pi_{C\ \eta}^{\alpha\beta}$$

where the indices α, β, γ run on the set of signs $(+,-)$, and η runs on the set of arrows.

Every reduced elementary jump defines a 1-cochain whose value at the 1-chain positively crossing transversally the corresponding stratum of the discriminant is equal to 2, being zero if the crossing takes place at any other stratum.

Definition 3.2. We define some linear combinations of reduced jumps of the above list, which will be useful in the sequel:

$$\Delta\pi_D = \sum_{\alpha,\beta,\eta} \Delta\pi_{D\ \eta}^{\alpha\beta} = \sum_{\alpha,\beta,\eta} \Delta\pi_{\eta}^{\alpha\beta} - \Delta\pi_C$$

$$\Delta\pi_{\uparrow}^{1+} = \sum_{\beta} \Delta\pi_{\uparrow}^{+\beta}$$

$$\Delta\pi_{\uparrow}^{1-} = \sum_{\beta} \Delta\pi_{\uparrow}^{-\beta}$$

$$\Delta\pi_{\downarrow}^{1+} = \sum_{\beta} \Delta\pi_{\downarrow}^{+\beta}$$

$$\Delta\pi_{\downarrow}^{1-} = \sum_{\beta} \Delta\pi_{\downarrow}^{-\beta}$$

$$\Delta\pi^{2+} = \Delta\pi_C - \sum_{\alpha,\eta} \Delta\pi_{\eta}^{\alpha-}$$

$$\Delta\pi^{2-} = \Delta\pi_C - \sum_{\alpha,\eta} \Delta\pi_{\eta}^{\alpha+}$$

where α and β run on the set of signs $(+,-)$, η on the set of arrows $(\uparrow,\downarrow)$.

Remark 3.1. $\Delta\pi_C$ and $\Delta\pi_D$ are the elementary RO-independent jumps defined at the strata of the cusp crossings.

$\Delta\pi_{\uparrow}^{1+}$ is the jump at the cusp crossing strata which is nonzero only if the cusp is of type *up* and the branch going away from the cusp point is positive. Analogously for $\Delta\pi_{\uparrow}^{1-}$ ($\Delta\pi_{\downarrow}^{1+}$, $\Delta\pi_{\downarrow}^{1-}$) the jump is non zero only in the cases (negative, *up*), (positive, *down*) and (negative, *down*) respectively.

$\Delta\pi^{2+}$ ($\Delta\pi^{2-}$) is the jump at the cusp crossings strata nonzero only if the smooth segment crossing the cusp is positive (negative). The jump $\Delta\pi^{2+}$ ($\Delta\pi^{2-}$) has the opposite value on the positive crossings of the strata Π_C and Π_D.

Remark 3.2. There are no partial coherence systems for jumps of type $\Delta\lambda$, so we have to add to the set of 16 independent reduced elementary jumps all the 4 elementary jumps of type $\Delta\lambda$. The 54 unknowns in the system of the coherence equations are thus reduced to the 20 new unknowns.

Definition 3.3. For every reduced jump (and for every linear combination described in definition 3.2) we define the *reduced local stratum* to be the linear combination of the local strata corresponding to the elementary jumps which compose this reduced jump or combination.

To write the matrix of the coherence system we denote the unknown jump ξ (defining the 1-cocycle) by

$$\xi = \nabla J_{C+}^+ \cdot \Delta J_{C+}^+ + \ldots + \nabla\lambda_-^{\downarrow} \cdot \Delta\lambda_-^{\downarrow}$$

and denote by $\nabla^{(r)}$ the vector of coefficients:

$$\begin{aligned}
\nabla^{(r)} =&(\nabla J_{C+}^+, \nabla J_{C-}^+, \nabla J_D^+, \nabla J_C^-, \nabla J_{D+}^-, \nabla J_{D-}^-, \nabla St,\\
&\nabla\pi_{\uparrow}^{-+}, \nabla\pi_{\uparrow}^{--}, \nabla\pi_{\uparrow}^{++}, \nabla\pi_{\uparrow}^{+-}, \nabla\pi_{\downarrow}^{-+}, \nabla\pi_{\downarrow}^{--}, \nabla\pi_{\downarrow}^{++}, \nabla\pi_{\downarrow}^{+-}, \nabla\pi_C,\\
&\nabla\lambda_+^{\uparrow}, \nabla\lambda_-^{\uparrow}, \nabla\lambda_+^{\downarrow}, \nabla\lambda_-^{\downarrow})
\end{aligned}$$

Lemma 3.2. *The vector $\Delta^{(r)}$ defines a 1-cocycle if and only if it verifies the coherence system*

$$M^{or}\nabla^{(r)} = 0 \tag{II}$$

the linear operator M^{or} being defined by the following matrix

$$
\begin{pmatrix}
-1 & 0 & 1 & 0 & 0 & 0 & 0 & 0 & 0 & 1 & -1 & 0 & 0 & 0 & 0 & 1 & 0 & 0 & 0 & 0 \\
0 & 0 & 0 & -1 & 1 & 0 & 0 & 0 & 0 & 1 & -1 & 0 & 0 & 0 & 0 & 1 & 0 & 0 & 0 & 0 \\
0 & -1 & 1 & 0 & 0 & 0 & 0 & -1 & 1 & 0 & 0 & 0 & 0 & 0 & 0 & 1 & 0 & 0 & 0 & 0 \\
0 & 0 & 0 & -1 & 0 & 1 & 0 & -1 & 1 & 0 & 0 & 0 & 0 & 0 & 0 & 1 & 0 & 0 & 0 & 0 \\
-1 & 0 & 1 & 0 & 0 & 0 & 0 & 1 & -1 & 0 & 0 & 0 & 0 & 0 & 0 & 1 & 0 & 0 & 0 & 0 \\
0 & 0 & 0 & -1 & 1 & 0 & 0 & 1 & -1 & 0 & 0 & 0 & 0 & 0 & 0 & 1 & 0 & 0 & 0 & 0 \\
0 & -1 & 1 & 0 & 0 & 0 & 0 & 0 & 0 & -1 & 1 & 0 & 0 & 0 & 0 & 1 & 0 & 0 & 0 & 0 \\
0 & 0 & 0 & -1 & 0 & 1 & 0 & 0 & 0 & -1 & 1 & 0 & 0 & 0 & 0 & 1 & 0 & 0 & 0 & 0 \\
-1 & 0 & 1 & 0 & 0 & 0 & 0 & 0 & 0 & 0 & 0 & 0 & 1 & -1 & 1 & 0 & 0 & 0 & 0 & 0 \\
0 & 0 & 0 & -1 & 1 & 0 & 0 & 0 & 0 & 0 & 0 & 0 & 1 & -1 & 1 & 0 & 0 & 0 & 0 & 0 \\
0 & -1 & 1 & 0 & 0 & 0 & 0 & 0 & 0 & 0 & -1 & 1 & 0 & 0 & 1 & 0 & 0 & 0 & 0 & 0 \\
0 & 0 & 0 & -1 & 0 & 1 & 0 & 0 & 0 & 0 & -1 & 1 & 0 & 0 & 1 & 0 & 0 & 0 & 0 & 0 \\
-1 & 0 & 1 & 0 & 0 & 0 & 0 & 0 & 0 & 0 & 1 & -1 & 0 & 0 & 1 & 0 & 0 & 0 & 0 & 0 \\
0 & 0 & 0 & -1 & 1 & 0 & 0 & 0 & 0 & 0 & 1 & -1 & 0 & 0 & 1 & 0 & 0 & 0 & 0 & 0 \\
0 & -1 & 1 & 0 & 0 & 0 & 0 & 0 & 0 & 0 & 0 & 0 & -1 & 1 & 1 & 0 & 0 & 0 & 0 & 0 \\
0 & 0 & 0 & -1 & 0 & 1 & 0 & 0 & 0 & 0 & 0 & 0 & -1 & 1 & 1 & 0 & 0 & 0 & 0 & 0 \\
0 & 0 & 0 & 0 & 0 & 0 & 0 & 0 & 1 & -1 & 0 & 0 & 0 & 0 & 0 & -1 & 1 & 0 & 0 & 0 \\
0 & 0 & 0 & 0 & 0 & 0 & 0 & -1 & 1 & 0 & 0 & 0 & 0 & 0 & 0 & 1 & -1 & 0 & 0 & 0 \\
0 & 0 & 0 & 0 & 0 & 0 & 0 & 0 & 0 & 0 & 0 & 1 & -1 & 0 & 0 & 0 & -1 & 1 & 0 & 0 \\
0 & 0 & 0 & 0 & 0 & 0 & 0 & 0 & 0 & 0 & -1 & 1 & 0 & 0 & 0 & 0 & 1 & -1 & 0 & 0 \\
0 & 0 & 0 & 0 & 0 & 0 & -1 & 1 & 0 & 0 & 0 & 0 & 0 & 1 & 0 & 0 & 0 & 0 & 0 & 0 \\
0 & 0 & 0 & 0 & 0 & 0 & -1 & 0 & 1 & 1 & 0 & 0 & 0 & 1 & 0 & 0 & 0 & 0 & 0 & 0 \\
0 & 0 & 0 & 0 & 0 & 0 & -1 & 0 & 0 & 0 & 0 & 1 & 0 & 1 & 1 & 0 & 0 & 0 & 0 & 0 \\
0 & 0 & 0 & 0 & 0 & 0 & -1 & 0 & 0 & 0 & 0 & 1 & 1 & 0 & 1 & 0 & 0 & 0 & 0 & 0
\end{pmatrix}
$$

The fundamental system of 1-cocycles derived from the solutions of eq. (II) is formed by the following basic jumps:

$$\Delta J^+ = \Delta J^+_{C+} + \Delta J^+_{C-} + \Delta J^+_D$$

$$\Delta J^- = \Delta J^-_C + \Delta J^-_{D+} + \Delta J^-_{D-}$$

$$\Delta Sp = 2\Delta St + \Delta \pi_C + \Delta \pi_D$$

$$\Delta f^+ = 2\Delta J^+_{C+} - 2\Delta J^-_{D+} + \Delta \pi^{2+} + \Delta \lambda^\uparrow_+ + \Delta \lambda^\downarrow_+$$

$$\Delta f^- = 2\Delta J^+_{C-} - 2\Delta J^-_{D-} + \Delta \pi^{2-} + \Delta \lambda^\uparrow_- + \Delta \lambda^\downarrow_-$$

$$\Delta p^\uparrow = \Delta \pi^{1+}_\uparrow - \Delta \pi^{1-}_\uparrow$$

$$\Delta p^\downarrow = \Delta \pi^{1+}_\downarrow - \Delta \pi^{1-}_\downarrow$$

$$\Delta \lambda^\uparrow = \Delta \lambda^\uparrow_+ + \Delta \lambda^\uparrow_-$$

$$\Delta \lambda^\downarrow = \Delta \lambda^\downarrow_+ + \Delta \lambda^\downarrow_-$$

Remark 3.3. ΔJ^+, ΔJ^- and ΔSp are the jumps of the *RO*-independent basic invariants (Theorem 2.1), so in the following we analyze only the other six new possible invariants.

Lemma 3.3. *The RO-dependent jumps defined by Lemma 3.2 satisfy the following RO symmetries:*

Δf^+ *and* Δf^- *are* $R_{(ab)(bc)}$*-invariant. Indeed the involutions a, b and c interchange the values of these jumps.*

$\Delta p^\uparrow$ *and* $\Delta p^\downarrow$ *are invariant with respect to* $R_{(abc)}$. *Indeed the involution b interchange the values of these jumps, while the involution c reverses the signs of the values of these jumps. The involution a acts as the involution bc.*

$\Delta \lambda^\uparrow$ *and* $\Delta \lambda^\downarrow$ *are* $R_{(ab)(c)}$*-invariant: the involutions a and b interchange the values of these jumps.*

Definition 3.4. The following cooriented hypersurfaces, defined by the solutions of (II) as linear combinations of reduced local strata, are called *basic RO-dependent cooriented strata:*

$$F^+ = 2K_{C+}^+ - 2K_{D+}^- + \Pi^{2+} + \Lambda_+^\uparrow + \Lambda_+^\downarrow$$
$$F^- = 2K_{C-}^+ - 2K_{D-}^- + \Pi^{2-} + \Lambda_-^\uparrow + \Lambda_-^\downarrow$$
$$P^\uparrow = \Pi_\uparrow^{1+} - \Pi_\uparrow^{1-}$$
$$P^\downarrow = \Pi_\downarrow^{1+} - \Pi_\downarrow^{1-}$$
$$\Lambda^\uparrow = \Lambda_+^\uparrow + \Lambda_-^\uparrow$$
$$\Lambda^\downarrow = \Lambda_+^\downarrow + \Lambda_-^\downarrow$$

Lemma 3.4. *The basic 1-cocycles defined by the jumps* Δf^+, Δf^-, $\Delta p^\uparrow$, $\Delta p^\downarrow$, $\Delta \lambda^\uparrow$ *and* $\Delta \lambda^\downarrow$ *are cohomologous to zero in the space of L-immersions .*

Theorem 3.1. *Every local invariant of the fronts additive under the connected summation is a linear combination of the RO-independent invariants, of the Maslov index and of the following 6 basic invariants:*

$$f^+, \quad f^-, \quad p^\uparrow, \quad p^\downarrow, \quad \lambda^\uparrow, \quad \lambda^\downarrow,$$

whose jumps are defined in the lemma 3.2, and whose values on the basic fronts are shown in table 7.

basic curves	f^+	f^-	$p^\uparrow$	$p^\downarrow$	$\lambda^\uparrow$	$\lambda^\downarrow$
$n = i+1$ $\omega_{i,\mu}$ $\mu = 2k$	$2(i+1)$	0	$-k$	$-k$	k	k
$n = i+1$ $\omega_{i,-\mu}$ $\mu = 2k$	0	$2(i+1)$	k	k	k	k
$n = i+1$ $\omega_{-i,\mu}$ $\mu = 2k$	$2(i+1)$	0	k	k	k	k
$n = i+1$ $\omega_{-i,-\mu}$ $\mu = 2k$	0	$2(i+1)$	$-k$	$-k$	k	k

Table 7. The basic fronts with the values of their
RO-dependent basic invariants

Remark 3.4. The basic fronts with $k \neq 0$ are asymmetric. Hence the action of RO transforms each of them into 8 fronts of 8 different classes. In table 7 only 4 classes are shown, distinct by the values of the indices. In fact to every of such classes $\omega_{i,\mu}$ there correspond another one with the same values as the indices, namely the class $abc\omega_{i,\mu}$ (for instance, the class $\omega_{i,\mu}$ shown in table 6). By Lemma 3.3, the class

$abc\omega_{i,\mu}$ has the same values as the invariants of the class $\omega_{i,\mu}$ but f^+ and f^-, which are interchanged.

On the other hand, the class $ac\omega_{i,\mu}$, having the opposite signs of the indices, has the same values as all the invariants of the class $\omega_{i,mu}$. In fact the involution ac preserves the values of the invariants f^+ and f^- and interchanges the values of $p^\uparrow$ and $p^\downarrow$, and the values of $\lambda^\uparrow$ and of $\lambda^\downarrow$. But these interchanged values on all the basic curves coincide.

Corollary 3.1. *There are exactly 10 (including the Maslov index) basic local additive invariants of the fronts.*

Proof. The 10 invariants are the Maslov index, the 6 RO-dependent invariants of Theorem 3.1 and the invariants J^+, J^- and Sp, by Lemma 3.2. The other RO-independent basic invariants of Theorem 2.1 can be written as:

$$h = f^+ + f^- - J^+ + J^- - \lambda$$
$$\lambda = \lambda^\uparrow + \lambda^\downarrow.$$

Corollary 3.2. *To reverse the coorientation of a basic front having Maslov index equal to μ, one has to make at least $|\mu|$ cusp crossings.*

Proof. The difference of the values of the invariant $p = p^\uparrow + p^\downarrow$ on two fronts with opposite coorientations is 2μ. At every cusp, crossing the invariant p changes by $+2$ or -2.

Remark 3.5. Unfortunately Theorem 3.1 does not allow us to prove the Arnold conjecture that one of the two curves of fig. 1a cannot be transformed into the other without crossing of the discriminant of the parallel coorientations self-tangencies K^+ if the cusps births are forbidden. They are indeed distinguished only by the values of the invariants f^+ and f^-. Different paths connecting the two curves in $\Omega_{1,0}$ are shown in fig. 25.

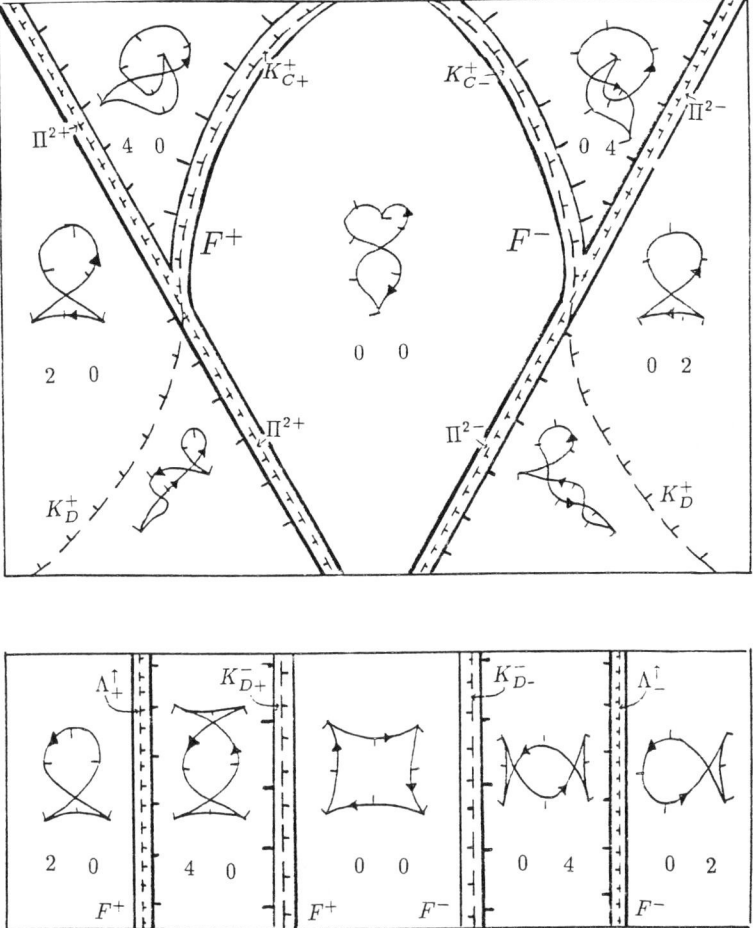

Figure 25. Examples of the F^+ and F^- discriminants. The values of f^+ and f^- are shown in the connected components of the complement

In table 8 a list of oriented and cooriented fronts in the oriented plane with the values of their 10 local additive invariants is presented. The action of RO on these fronts gives the representatives of *all* the front classes with at most two double points and at most two cusp points. Their invariants are obtained applying Lemma 3.3.

i = 0, μ = 0 (first row of fronts)

	1	2	3	4	5	6	7	8	9	10	11	12
$J^+\ J^-$	0 −1	0 1	0 −1	0 −1	0 −1	0 −1	−2 1	−2 −1	−2 −1	0 1	0 1	−2 −3
S_p	0	0	0	0	0	0	0	2	2	−2	−2	4
$f^+\ f^-$	2 0	0 0	4 0	4 0	4 0	4 0	0 0	2 0	2 0	2 0	2 0	4 0
$p^+\ p^-$	0 0	0 0	0 0	0 0	0 0	0 0	0 0	−2 0	2 0	−2 0	2 0	0 0
$\lambda^+\ \lambda^-$	0 0	2 0	2 0	2 0	2 0	2 0	2 0	2 0	2 0	2 0	2 0	2 0

i = 0, μ = 0 (continued) and **i = 0, μ = 2**

	1	2	3	4	5	6	7	8	9	(i=0, μ=2) 1	2	3
$J^+\ J^-$	−2 −3	−2 −3	−2 −3	−4 −1	2 1	−2 1	2 1	−4 −5	−4 −3	−1 −1	−1 1	−3 −1
S_p	4	4	4	4	−4	0	−4	8	6	2	0	4
$f^+\ f^-$	2 0	4 0	2 2	0 0	2 2	0 0	4 0	4 0	0 2	0 2	0 0	0 0
$p^+\ p^-$	0 0	0 0	0 0	4 0	0 0	0 0	0 0	0 0	−2 0	1 1	−1 1	1 −1
$\lambda^+\ \lambda^-$	2 0	2 0	2 0	2 0	2 0	2 0	2 0	2 0	2 0	1 1	1 1	1 1

i = 2, μ = 0 (first row of fronts)

	1	2	3	4	5	6	7	8	9	10	11	12
$J^+\ J^-$	−2 −3	−2 −1	−2 −3	−2 −3	−2 −3	−2 −3	−2 −3	−2 −1	−4 −3	−4 −3	−4 −3	−4 −3
S_p	4	4	4	4	4	4	4	2	6	6	6	6
$f^+\ f^-$	2 0	0 0	4 0	4 0	4 0	4 0	4 0	2 0	2 0	2 0	0 2	2 0
$p^+\ p^-$	0 0	0 0	0 0	0 0	0 0	0 0	0 0	−2 0	0 2	−2 0	−2 0	2 0
$\lambda^+\ \lambda^-$	0 0	2 0	0 2	2 0	2 0	2 0	2 0	2 0	0 2	2 0	2 0	2 0

i = 2, μ = 0 (second row of fronts)

	1	2	3	4	5	6	7	8	9	10	11	12
$J^+\ J^-$	0 −1	0 −1	−4 −5	−6 −3	−4 −5	−4 −1	−4 −1	−4 −5	−4 −5	−2 −3	−4 −3	−4 −5
S_p	0	0	8	8	8	4	4	8	8	8	8	8
$f^+\ f^-$	4 0	2 2	2 2	0 0	2 2	0 0	0 0	4 0	4 0	4 0	2 0	2 2
$p^+\ p^-$	0 0	0 0	0 0	−4 0	0 0	0 0	0 0	0 0	0 0	0 −2	0 0	0 0
$\lambda^+\ \lambda^-$	2 0	2 0	0 2	2 0	2 0	2 0	2 0	2 0	2 0	0 2	0 2	0 2

i = 2, μ = 2 and remaining **i = 2, μ = 0** fronts

	1	2	3	4	5	6	7	8	(i=2, μ=2) 1	2	3	4
$J^+\ J^-$	−4 −5	−6 −5	−6 −3	−6 −7	−4 −1	−6 −5	−6 −7	−4 −5	−3 −3	−3 −1	−5 −3	−3 −3
S_p	8	10	8	12	4	10	12	8	6	4	8	6
$f^+\ f^-$	4 0	2 0	0 0	4 0	0 0	2 0	4 0	2 2	0 2	0 0	0 0	2 0
$p^+\ p^-$	0 0	0 −2	0 −4	0 0	0 0	0 0	0 0	0 0	1 1	1 3	3 1	3 3
$\lambda^+\ \lambda^-$	0 2	0 2	0 2	0 2	0 2	2 0	2 0	2 0	1 1	1 1	1 1	1 1

i = 4, μ = 0

	1	2	3	4	5	6	7	8	9	10	11	12
$J^+\ J^-$	−6 −7	−6 7	−10 −7	−8 −9	−12 −13	−8 −9	−8 −7	−10 −11	−10 −9	−10 −11	−12 −9	−12 −11
S_p	12	12	16	16	24	16	14	20	18	20	20	22
$f^+\ f^-$	4 0	2 2	0 0	4 0	4 0	4 0	2 0	4 0	2 0	2 2	0 0	0 2
$p^+\ p^-$	0 0	0 0	0 0	0 0	0 0	−2 0	0 0	0 0	2 0	0 0	−4 0	−2 0
$\lambda^+\ \lambda^-$	2 0	2 0	2 0	2 0	2 0	2 0	2 0	2 0	2 0	2 0	2 0	2 0

Table 8a,b. The additive local invariants of fronts
with at most two double points and at most two cusp points

Row 1

$i=1$ $\mu=0$												$i=1$ $\mu=2$	
J^+ J^-	0 0	0 -2	0 -2	0 0	0 0	-2 -4	-2 -2	-2 0	-2 -2	-2 0		J^+ J^-	-1 0
Sp	0	0	0	0	0	4	4	2	4	2		Sp	2
f^+ f^-	0 0	4 0	2 0	2 0	4 0	2 0	2 0	0 0	2 0	0 0		f^+ f^-	0 0
$p^\uparrow$ $p^\downarrow$	0 0	0 0	0 0	0 0	0 0	0 0	-2 0	0 0	0 0	0 2		$p^\uparrow$ $p^\downarrow$	1 1
$\lambda^\uparrow$ $\lambda^\downarrow$	0 0	0 0	0 0	2 0	0 2	0 0	2 0	2 0	0 2	0 2		$\lambda^\uparrow$ $\lambda^\downarrow$	1 1

Row 2

J^+ J^-	-1 -2	-1 0	-1 -2	-1 -2	-3 -4	-1 -2	-1 -2	-1 -2	-1 0	-1 -2	-1 -2	-1 0	-5 -4
Sp	2	0	2	2	6	2	2	2	0	2	2	0	8
f^+ f^-	0 4	0 2	0 4	2 2	4 0	0 4	0 4	4 0	2 0	2 2	4 0	0 2	2 0
$p^\uparrow$ $p^\downarrow$	1 1	-1 1	-1 -1	-1 -1	1 1	1 1	3 3	3 3	3 3	1 1	1 1	1 3	1 -1
$\lambda^\uparrow$ $\lambda^\downarrow$	1 1	1 1	1 1	1 1	1 1	1 1	1 1	1 1	1 1	1 1	1 1	1 1	1 1

Row 3

J^+ J^-	-3 -2	-3 -4	-3 0	-3 -4	-3 -4	-3 -2	-3 -4	-3 0	-3 -2	-3 0	-3 -2	-3 0	-3 -2
Sp	4	6	2	6	6	4	6	2	4	2	4	2	4
f^+ f^-	0 2	4 0	0 0	4 0	0 4	0 2	2 2	0 0	0 2	0 0	0 2	0 0	2 0
$p^\uparrow$ $p^\downarrow$	1 1	-1 -1	1 1	1 1	1 1	-1 1	1 1	1 1	-1 3	3 3	1 1	1 1	3 1
$\lambda^\uparrow$ $\lambda^\downarrow$	1 1	1 1	1 1	1 1	1 1	1 1	1 1	1 1	1 1	1 1	1 1	1 1	1 1

Row 4

J^+ J^-	-5 -4	-7 -4	$i=3$ $\mu=0$						$i=3$ $\mu=2$				
			J^+ J^-	-4 -6	-4 -4	-6 -8	-6 -4	-6 -6	J^+ J^-	-7 -6	-7 -4	-7 -8	-7 -8
Sp	8	10	Sp	8	8	12	10	12	Sp	12	10	14	14
f^+ f^-	2 0	0 0	f^+ f^-	4 0	2 0	4 0	0 0	2 0	f^+ f^-	2 0	0 0	4 0	4 0
$p^\uparrow$ $p^\downarrow$	3 1	3 -1	$p^\uparrow$ $p^\downarrow$	0 0	0 0	0 0	0 0	-2 0	$p^\uparrow$ $p^\downarrow$	3 1	3 3	1 1	3 3
$\lambda^\uparrow$ $\lambda^\downarrow$	1 1	1 1	$\lambda^\uparrow$ $\lambda^\downarrow$	2 0	0 0	2 0	0 2	2 0	$\lambda^\uparrow$ $\lambda^\downarrow$	1 1	1 1	1 1	1 1

Row 5

J^+ J^-	-7 -8	-7 -8	-7 -6	-7 -8	-7 -4	-7 -6	-5 -6	-5 -4	-5 -6	-5 -6	-9 -8	-9 -8	-11 8
Sp	14	14	12	14	10	12	10	8	10	10	16	16	18
f^+ f^-	2 2	4 0	2 0	0 4	0 0	2 0	0 4	0 2	0 4	2 2	0 2	0 2	0 0
$p^\uparrow$ $p^\downarrow$	3 3	5 5	1 3	3 3	3 5	3 3	3 3	1 3	1 1	1 1	3 1	5 3	5 1
$\lambda^\uparrow$ $\lambda^\downarrow$	1 1	1 1	1 1	1 1	1 1	1 1	1 1	1 1	1 1	1 1	1 1	1 1	1 1

3.2 The RO-dependent 1-cocycles

We solve the coherence system at the strata of codimension 2 for the RO-dependent local strata in two steps. In the first one we give the

Proof of Lemma 3.1. We write the partial coherence systems for the reduced jumps of type ΔJ, ΔSt, $\Delta \pi$.

Stratum KK. For the reduced jumps at the strata of self-tangencies, the partial coherence system is

$$M_{(KK)} \nabla J = 0 \qquad (III)$$

where ∇J is the vector of the coefficients of the RO-dependent basic jumps defining the jump

$$\xi = \nabla J_{C+}^+ \cdot \Delta J_{C+}^+ + \ldots + \nabla J_{D-,i}^- \cdot \Delta J_{D-,i}^-,$$

namely

$$\nabla J = (\nabla J_{C+}^+, \nabla J_{C-}^+, \nabla J_{D,i+}^+, \nabla J_{D,i-}^+, \nabla J_{C,i}^-, \nabla J_{C,e}^-,$$
$$\nabla J_{D+,e}^-, \nabla J_{D+,i}^-, \nabla J_{D-,e}^-, \nabla J_{D-,i}^-)$$

and

$$M_{(KK)} = \begin{pmatrix} 0 & 0 & 1 & -1 & 0 & 0 & 0 & 0 & 0 & 0 \\ 0 & 0 & 0 & 0 & 1 & -1 & 0 & 0 & 0 & 0 \\ 0 & 0 & 0 & 0 & 0 & 0 & 1 & -1 & 0 & 0 \\ 0 & 0 & 0 & 0 & 0 & 0 & 0 & 0 & 1 & -1 \end{pmatrix}$$

There are 6 different cases involving the 10 local RO-dependent strata giving the coorientability conditions at the strata KK of codimension 2. They are shown in fig. 26. In (a) and (b) the coorientability condition is automatically satisfied, (c), (d), (e) and (f) give respectively the rows 1–4 of $M_{(KK)}$. The generators of the solution of the system (III) consist of the reduced jumps denoted 1)–6) in the statement of Lemma 3.1.

Remark. We remark that the reduced jumps ΔJ_C^- and ΔJ_D^+ are RO-independent, while the other four jumps are $R_{(ab)(bc)}$-invariants.

For the jumps at the strata of triple points, we have to solve the partial coherence system for the vector $\nabla(St)$ of the coefficients ot the 24 elementary RO-dependent jumps,

$$M_{(KT,T\Pi)} \nabla St = 0. \qquad (IV)$$

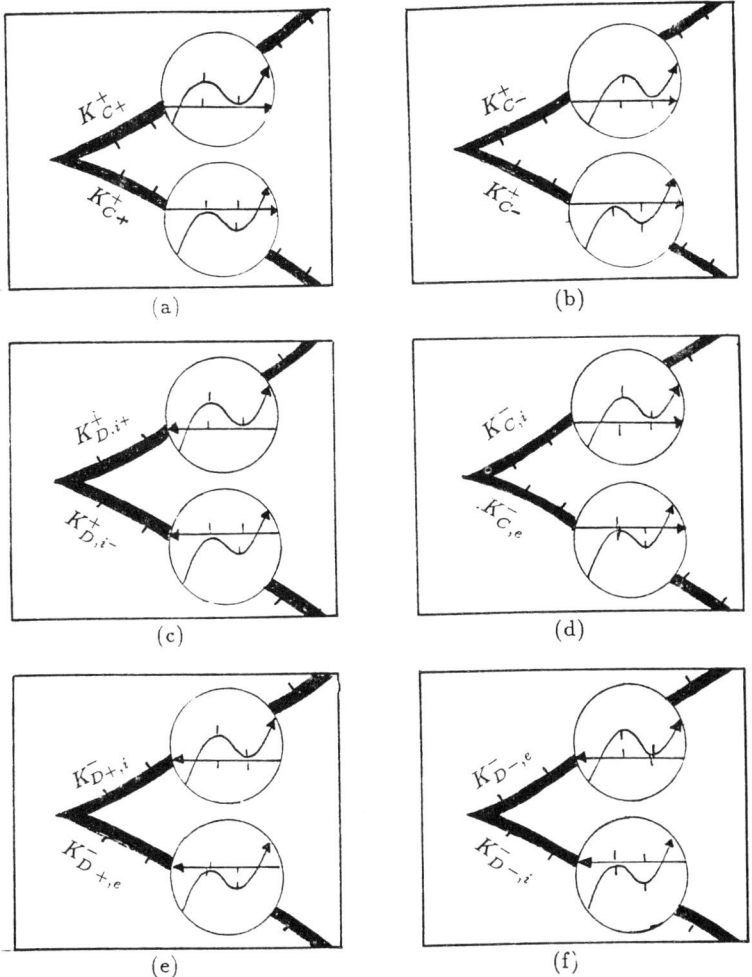

Figure 26. The RO-dependent strata KK

Indeed the coorientability conditions at the strata KT and $T\Pi$ involve nontrivially only the strata of type T. The coorientations of the two branches of these strata meeting at the strata of codimension 2 are compatible (see Lemma 2.1). Hence all equations of the system (IV) come from equations of the type $\nabla St(1) - \nabla St(2) = 0$, $\Delta St(1)$ and $\Delta St(2)$ being the elementary jumps of type ΔSt at the particular local strata of triple points consisting of the branches $T(1)$ and $T(2)$, respectively).

Stratum KT. We first solve the system given by the conditions at strata KT

$$M_{(KT)}\nabla St = \begin{pmatrix} M_A^I & M_B^I \\ M_A^{II} & M_B^{II} \end{pmatrix} \begin{pmatrix} \nabla St_A \\ \nabla St_B \end{pmatrix} = 0 \qquad (IV')$$

where

$$\begin{aligned}
\nabla St_A =(&\nabla St_{A\longrightarrow}^{+++}, \nabla St_{A\longleftarrow}^{+++}, \nabla St_{A\longrightarrow}^{---}, \nabla St_{A\longleftarrow}^{---}, \\
&\nabla St_{A\longrightarrow}^{++-}, \nabla St_{A\longleftarrow}^{++-}, \nabla St_{A\longrightarrow}^{-++}, \nabla St_{A\longleftarrow}^{-++}, \\
&\nabla St_{A\longrightarrow}^{--+}, \nabla St_{A\longleftarrow}^{--+}, \nabla St_{A\longrightarrow}^{+--}, \nabla St_{A\longleftarrow}^{+--}, \\
&\nabla St_{A\longrightarrow}^{+-+}, \nabla St_{A\longleftarrow}^{+-+}, \nabla St_{A\longrightarrow}^{-+-}, \nabla St_{A\longleftarrow}^{-+-}) \\
\nabla St_B =(&\nabla St_{B\longrightarrow}^{+++}, \nabla St_{B\longleftarrow}^{+++}, \nabla St_{B\longrightarrow}^{---}, \nabla St_{B\longleftarrow}^{---}, \\
&\nabla St_{B\longrightarrow}^{++-}, \nabla St_{B\longleftarrow}^{++-}, \nabla St_{B\longrightarrow}^{--+}, \nabla St_{B\longleftarrow}^{--+})
\end{aligned}$$

are the coefficient of the jumps $(\Delta St_{A\longrightarrow}^{+++}, \ldots, \Delta St_{B\longleftarrow}^{--+})$ respectively, M_B^I is the zero matrix, M_A^{II} is the identity matrix, while

$$M_A^I = \begin{pmatrix}
1 & -1 & 0 & 0 & 0 & 0 & 0 & 0 & 0 & 0 & 0 & 0 & 0 & 0 & 0 & 0 \\
0 & 0 & 1 & -1 & 0 & 0 & 0 & 0 & 0 & 0 & 0 & 0 & 0 & 0 & 0 & 0 \\
0 & 0 & 0 & 0 & 1 & 0 & 0 & 0 & 0 & 0 & 0 & 0 & 0 & -1 & 0 & 0 \\
0 & 0 & 0 & 0 & 0 & 1 & 0 & 0 & 0 & 0 & 0 & 0 & -1 & 0 & 0 & 0 \\
0 & 0 & 0 & 0 & 0 & 0 & 1 & -1 & 0 & 0 & 0 & 0 & 0 & 0 & 0 & 0 \\
0 & 0 & 0 & 0 & 0 & 0 & 0 & 0 & 1 & 0 & 0 & 0 & 0 & 0 & 0 & -1 \\
0 & 0 & 0 & 0 & 0 & 0 & 0 & 0 & 0 & 1 & 0 & 0 & 0 & 0 & -1 & 0 \\
0 & 0 & 0 & 0 & 0 & 0 & 0 & 0 & 0 & 0 & 1 & -1 & 0 & 0 & 0 & 0 \\
0 & 0 & 0 & 0 & 1 & -1 & 0 & 0 & 0 & 0 & 0 & 0 & 0 & 0 & 0 & 0 \\
0 & 0 & 0 & 0 & 0 & 0 & 0 & 0 & 1 & -1 & 0 & 0 & 0 & 0 & 0 & 0 \\
0 & 0 & 0 & 0 & 0 & 1 & 0 & 0 & 0 & 0 & 0 & 0 & 0 & -1 & 0 & 0 \\
0 & 0 & 0 & 0 & 0 & 0 & 1 & 0 & 0 & 0 & 0 & 0 & -1 & 0 & 0 & 0 \\
0 & 0 & 0 & 0 & 0 & 0 & 0 & 0 & 0 & 0 & 1 & 0 & 0 & 0 & 0 & -1 \\
0 & 0 & 0 & 0 & 0 & 0 & 0 & 0 & 0 & 0 & 0 & 1 & 0 & 0 & -1 & 0
\end{pmatrix}$$

$$M_B^{II} = \begin{pmatrix} -1 & 0 & 0 & 0 & 0 & 0 & 0 & 0 \\ 0 & -1 & 0 & 0 & 0 & 0 & 0 & 0 \\ 0 & 0 & -1 & 0 & 0 & 0 & 0 & 0 \\ 0 & 0 & 0 & -1 & 0 & 0 & 0 & 0 \\ 0 & 0 & 0 & 0 & -1 & 0 & 0 & 0 \\ 0 & 0 & 0 & 0 & 0 & -1 & 0 & 0 \\ 0 & 0 & 0 & 0 & -1 & 0 & 0 & 0 \\ 0 & 0 & 0 & 0 & 0 & -1 & 0 & 0 \\ 0 & 0 & 0 & 0 & 0 & 0 & -1 & 0 \\ 0 & 0 & 0 & 0 & 0 & 0 & 0 & -1 \\ 0 & 0 & 0 & 0 & 0 & 0 & -1 & 0 \\ 0 & 0 & 0 & 0 & 0 & 0 & 0 & -1 \\ 0 & 0 & 0 & 0 & -1 & 0 & 0 & 0 \\ 0 & 0 & 0 & 0 & 0 & -1 & 0 & 0 \\ 0 & 0 & 0 & 0 & 0 & 0 & -1 & 0 \\ 0 & 0 & 0 & 0 & 0 & 0 & 0 & -1 \end{pmatrix}$$

The equations (IV') come from different cases shown in fig. 27.

In cases (a) and (b) $T(1)$ and $T(2)$ are both of type T_A: the cyclic order is reversed because the second and the third segments are interchanged (a), or the first and the second ones (b). We have from (a)

$$\nabla St^{\alpha\beta\gamma}_{A\ \eta} - \nabla St^{\alpha\gamma\beta}_{A\ -\eta} = 0$$

and from (b)

$$\nabla St^{\alpha\beta\gamma}_{A\ \eta} - \nabla St^{\beta\alpha\gamma}_{A\ -\eta} = 0$$

where if η is the symbol "$\rightarrow$" then $-\eta$ is "$\leftarrow$" and vice-versa. These equations correspond to the first 14 rows of $M_{(KT)}$, i.e., to the rows of M_A^I.

In the case (c) $T(1)$ and $T(2)$ are of type T_A and T_B. The cyclic orders on both strata are the same. Hence we obtain the successive 16 rows of $M_{(KT)}$ (i.e. the rows of M_{AB}^{II}) in the form

$$\nabla St^{\alpha\beta\gamma}_{A\ \eta} - \nabla St^{\alpha\beta\gamma}_{B\ \eta} = 0$$

(where in the case T_B the three signs α, β, γ have to be considered up to cyclic permutations).

The space of the solutions of the system (IV') has dimension 4: we choose the *first reduced* jumps:

$$\Delta St^0 = \sum_{Z,\eta} \Delta St_{Z\ \eta}^{+++}$$

$$\Delta St^1 = \sum_{\eta,\sigma} \Delta St_{A\ \eta}^{\sigma(++-)} + \sum_{\eta} \Delta St_{B\ \eta}^{++-}$$

$$\Delta St^2 = \sum_{\eta,\sigma} \Delta St_{A\ \eta}^{\sigma(--+)} + \sum_{\eta} \Delta St_{B\ \eta}^{--+}$$

$$\Delta St^3 = \sum_{Z,\eta} \Delta St_{Z\ \eta}^{---}$$

where Z runs on the set (A, B) and σ runs through all the permutations of the three symbols.

Stratum $T\Pi$. For the jump $\Delta St = \sum_0^3 \nabla St^i \cdot \Delta St_i$ we write now the partial coherence system

$$M_{(T\Pi)} \nabla St = 0 \qquad\qquad (IV'')$$

where ∇St is the vector of the coeficients ∇St_i, and

$$M_{(T\Pi)} = \begin{pmatrix} 1 & -1 & 0 & 0 \\ 0 & 1 & -1 & 0 \\ 0 & 0 & 1 & -1 \end{pmatrix}$$

Indeed it is sufficient to note that to the strata $T(1)$ and $T(2)$ meeting at the stratum $T\Pi$ there always correspond different types of triple points. The types of jumps ΔSt^i are distinguished only by the number i of segments of sign $+$ which cross at the triple point. Two of these signs are fixed; the third, being the sign of one of two branches of a cusp, changes from $T(1)$ to $T(2)$ (see fig. 28). So there are 3 possibilities for the pair $(i, i+1)$ being $i = 0, 1, 2, 3$.

The matrix $M_{(T\Pi)}$ has rank equal to 3, so the solution we obtain for the reduced jumps at the strata of triple points has dimension 1. Therefore the reduced jump ΔSt is the sum of all the 24 local elementary RO-dependent jumps.

Stratum $\Pi\Pi$. For the reduced jumps at the strata of the cusp crossings, the partial coherence system is

$$M_{(\Pi\Pi)} \nabla \Pi = 0 \qquad\qquad (V)$$

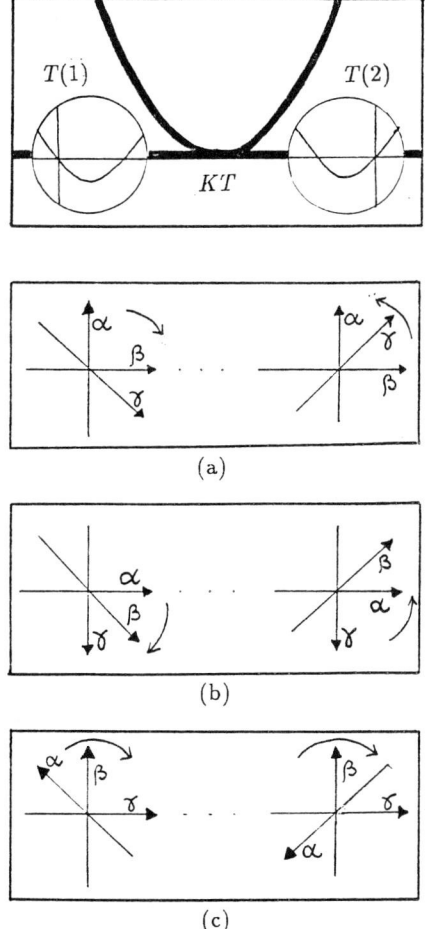

Figure 27. Local coherence at the strata KT

where

$$\nabla \pi = (\nabla \pi_{C\uparrow}^{-+}, \nabla \pi_{C\uparrow}^{--}, \nabla \pi_{C\uparrow}^{++}, \nabla \pi_{C\uparrow}^{+-}, \nabla \pi_{C\downarrow}^{-+}, \nabla \pi_{C\downarrow}^{--}, \nabla \pi_{C\downarrow}^{++}, \nabla \pi_{C\downarrow}^{+-},$$
$$\nabla \pi_{D\uparrow}^{-+}, \nabla \pi_{D\uparrow}^{--}, \nabla \pi_{D\uparrow}^{++}, \nabla \pi_{D\uparrow}^{+-}, \nabla \pi_{D\downarrow}^{-+}, \nabla \pi_{D\downarrow}^{--}, \nabla \pi_{D\downarrow}^{++}, \nabla \pi_{D\downarrow}^{+-})$$

is the vector of the coefficients of the corresponding basic jumps $(\Delta \pi_{C\uparrow}^{-+}, \dots, \Delta \pi_{D\downarrow}^{+-})$ and the linear operator $M_{(\text{III})}$ is defined by the

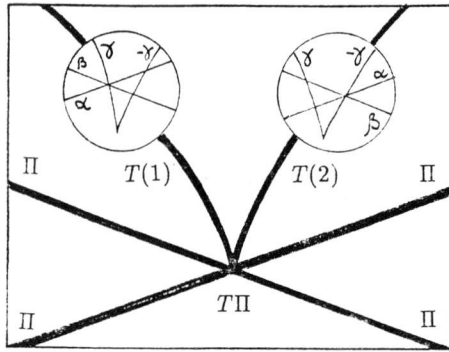

Figure 28. Local coherence at the stratum $T\Pi$

following matrix

$$\begin{pmatrix}
0 & 0 & 1 & -1 & 0 & 0 & 0 & 0 & 0 & 0 & 1 & -1 & 0 & 0 & 0 & 0 \\
1 & 0 & -1 & 0 & 0 & 0 & 0 & 0 & 0 & -1 & 0 & 1 & 0 & 0 & 0 & 0 \\
0 & -1 & 0 & 1 & 0 & 0 & 0 & 0 & 1 & 0 & -1 & 0 & 0 & 0 & 0 & 0 \\
-1 & 1 & 0 & 0 & 0 & 0 & 0 & 0 & -1 & 1 & 0 & 0 & 0 & 0 & 0 & 0 \\
0 & 0 & 0 & 0 & 0 & 0 & -1 & 1 & 0 & 0 & 0 & 0 & 0 & 0 & -1 & 1 \\
0 & 0 & 0 & 0 & 0 & 1 & 0 & -1 & 0 & 0 & 0 & 0 & -1 & 0 & 1 & 0 \\
0 & 0 & 0 & 0 & -1 & 0 & 1 & 0 & 0 & 0 & 0 & 0 & 0 & 1 & 0 & -1 \\
0 & 0 & 0 & 0 & 1 & -1 & 0 & 0 & 0 & 0 & 0 & 0 & 1 & -1 & 0 & 0 \\
0 & 0 & 1 & 0 & 0 & 0 & -1 & 0 & 0 & 0 & 0 & -1 & 0 & 0 & 0 & 1 \\
1 & 0 & 0 & 0 & 0 & 0 & 0 & -1 & 0 & -1 & 0 & 0 & 0 & 0 & 1 & 0 \\
0 & 0 & 0 & 1 & -1 & 0 & 0 & 0 & 0 & 0 & -1 & 0 & 0 & 1 & 0 & 0 \\
0 & 1 & 0 & 0 & 0 & -1 & 0 & 0 & -1 & 0 & 0 & 0 & 1 & 0 & 0 & 0 \\
0 & 0 & 0 & -1 & 0 & 0 & 0 & 1 & 0 & 0 & 1 & 0 & 0 & 0 & -1 & 0 \\
0 & 0 & -1 & 0 & 0 & 1 & 0 & 0 & 0 & 0 & 0 & 1 & -1 & 0 & 0 & 0 \\
0 & -1 & 0 & 0 & 0 & 0 & 1 & 0 & 1 & 0 & 0 & 0 & 0 & 0 & 0 & -1 \\
-1 & 0 & 0 & 0 & 1 & 0 & 0 & 0 & 0 & 1 & 0 & 0 & 0 & -1 & 0 & 0
\end{pmatrix}$$

The equation (V) is obtained at the strata $\Pi\Pi\Pi$. Consider first the case of two cusp points both of type *up*. There are 4 cases depending on the coorientations of the cusp branches. The two cases (a) and (b) of fig. 29 give respectively the rows 1 and 2 of the matrix $M_{(\Pi\Pi)}$. Reversing the coorientation of the front, the two signs of every elementatry jump are reversed while the other symbols remain unchanged. Then one obtains from the previous ones the cases (d) and (c), giving the rows (4) and (3) respectively.

　　Consider now the case when the two cusps are both of type *down*. It is obtained applying the involution b to the previous situations. Every

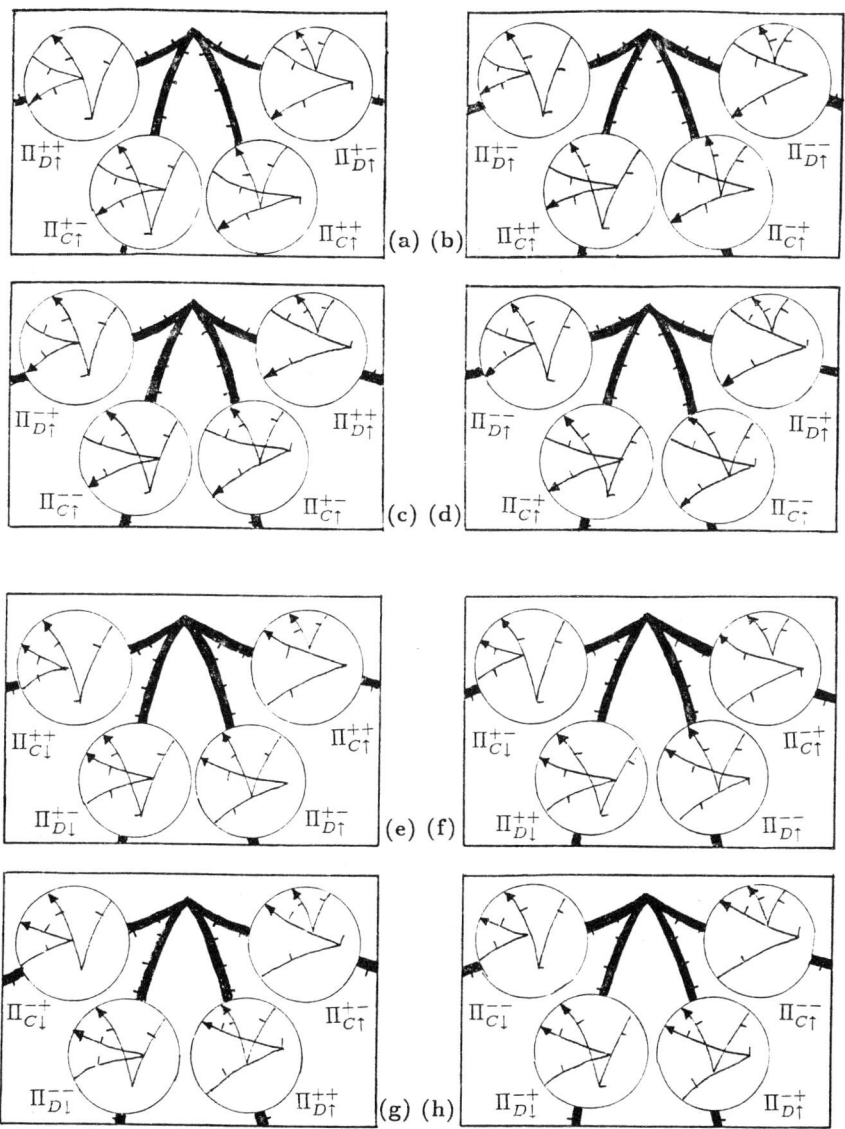

Figure 29. *RO*-dependent strata IIII

elementary jump changes in this way:

$$\Delta \pi_{Z\uparrow}^{\alpha \beta} \longrightarrow \Delta \pi_{Z\downarrow}^{\alpha \ -\beta}$$

so that we obtain rows 5–8 of $M_{(\text{IIII})}$ from rows 1–4 reversing the signs of β and the direction of the arrows.

Two of the four cases when the cusp points are one of type up and the other of type down are shown in fig. 29: (e) and (f) give respectively the rows 9 and 10 of $M_{(\text{IIII})}$. Applying the involution c to these two cases one obtains the cases (h) and (g) and the rows 12 and 11. The other 4 cases obtained applying the involution b to the four preceding cases (the stratum IIII is not invariant with respect to this involution), give the rows 13–16.

The action of the involution a on this stratum is equivalent to the action of the involution bc, so there are no other cases.

The rank of the matrix $M_{(\text{IIII})}$ is equal to 7, so that we choose the 9 reduced jumps (8)–(16) of type $\Delta\pi$ as defined in the statement of Lemma 3.1 in terms of the elementary RO-dependent jumps.

We are now ready to give the

Proof of Lemma 3.2. We have still to consider the strata $K\Pi$, $\Lambda\Lambda$ and $\Pi\Lambda$ of codimension 2.

Stratum $K\Pi$. The coorientations of the local strata of self-tangencies and of cusp crossings meeting at this stratum and their characters C and D have been discussed in the RO-independent case. In the RO-dependent case there are 16 distinguishable situations. Indeed, consider the cases (a) and (b) of fig. 30. They are distinguishable also in the RO-independent case. We note that in the two occurrences of the cusp crossing along the closed path linked with the stratum $K\Pi$, the types *up* or *down* of the cusps are the same as the signs of all segments. Therefore the two elementary jumps of type $\Delta\pi$ have coincident upper signs and arrows. In terms of the coefficients of the elementary jumps of type $\Delta\pi$ the cases shown in (a) and (b) give:

$$\nabla\pi_{C\uparrow}^{++} - \nabla J_{C+}^{+} - \nabla\pi_{D\uparrow}^{++} + \nabla J_{D}^{+} = 0$$
$$\nabla\pi_{C\uparrow}^{++} - \nabla J_{C}^{-} - \nabla\pi_{D\uparrow}^{++} + \nabla J_{D+}^{-} = 0.$$

So we obtain rows 1 and 2 of the matrix M^{or}, since in terms of the linearly independent reduced jumps

$$\nabla\pi_{C\uparrow}^{\alpha\beta} - \nabla\pi_{D\uparrow}^{\alpha\beta} = \nabla\pi_{\uparrow}^{\alpha\beta} - \nabla\pi_{\uparrow}^{\alpha-\beta} + \nabla\pi_{C}.$$

We find the other equations applying the RO-involutions to these two cases. We notice that such involutions do not change the upper signs of the jumps of type ΔJ, and do not change the characters C and D of the strata.

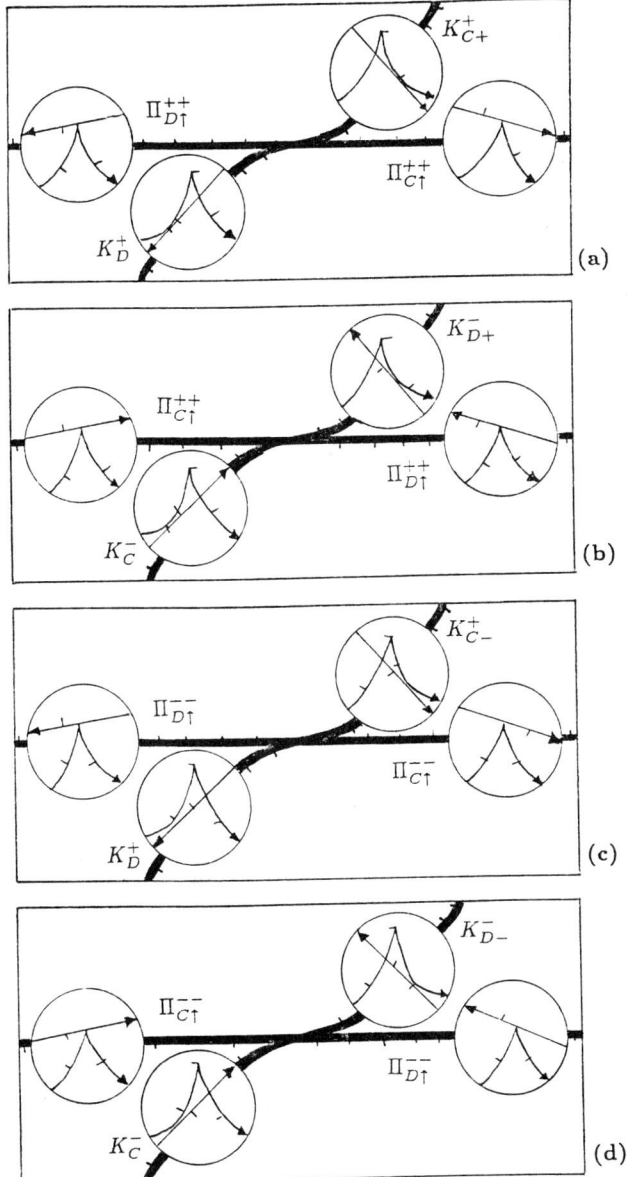

Figure 30. Four examples of RO-dependent strata $K\Pi$

For example the cases (c) and (d) in the figure, which give the rows 2 and 4 of the matrix M^{or}, are obtained applying the involution c to the two preceding cases. The result of this involution reversing all the signs of the segments is the following: for the jumps of type ΔJ the lower signs are reversed, while for the jumps of type $\Delta \pi$ the upper signs are reversed.

If one applies the involution b, again the lower signs of the jumps of type ΔJ are reversed, while only the second upper sign (the sign of the smooth segment crossing the cusp) is reversed in $\Delta\pi_\eta^{\alpha\beta}$. Moreover the character *up* becomes *down*:

$$\Delta\pi_{Z\uparrow}^{\alpha\beta} \longrightarrow \Delta\pi_{Z\downarrow}^{\alpha-\beta}. \qquad (Z = C, D)$$

The involution a reverses the lower signs of the jumps of type ΔJ and all the upper signs and the arrow of $\Delta\pi$. Therefore the involution ab leaves unchanged the jumps of type ΔJ and the arrows of the jumps $\Delta\pi$, and reverses only the first of the upper signs of $\Delta\pi$:

$$\Delta\pi_{Z\uparrow}^{\alpha\beta} \longrightarrow \Delta\pi_{Z\uparrow}^{-\alpha\beta}, \qquad (Z = C, D)$$

i.e., for the reduced jumps

$$\Delta\pi_\uparrow^{\alpha\beta} \longrightarrow \Delta\pi_\uparrow^{-\alpha\beta}.$$

In this way one obtains the rows 5–8 of M^{or} from the first 4 rows.

Moreover it follows that the involution ac only reverses the arrows of the reduced jumps at the cusp crossings. One obtains the rows 9–16 of the matrix M^{or} applying this involution to the rows 1–8.

Remark. We have chosen the closed path linked with the stratum $K\Pi$ such that the crossing of the local stratum of type Π_C is always positive (see fig. 30). Hence the value of the coefficient of $\nabla\pi_C$ is equal to 1 in every equation.

Stratum $\Lambda\Lambda$. The two strata of type Λ and those of type Π are all of type *up* or all of type *down*; moreover, they are cooriented always as shown in fig. 31.

The two strata Π can be only both of type C, as we remarked in the case of RO-independence (§2.3), where there is only one situation. In the RO-dependent case, we note that the situation is always invariant with repect to $R_{(ab)}$. Hence there are four distinguishable cases, shown in fig. 31. In the case (a) one has

$$\nabla\lambda_+^\uparrow + \nabla\pi_{C\uparrow}^{+-} - \nabla\pi_{C\uparrow}^{++} - \nabla\lambda_-^\uparrow = 0$$

We can reverse the orientation and the coorientation of the front, so obtaining the other 3 cases. Reversing the orientation, the character

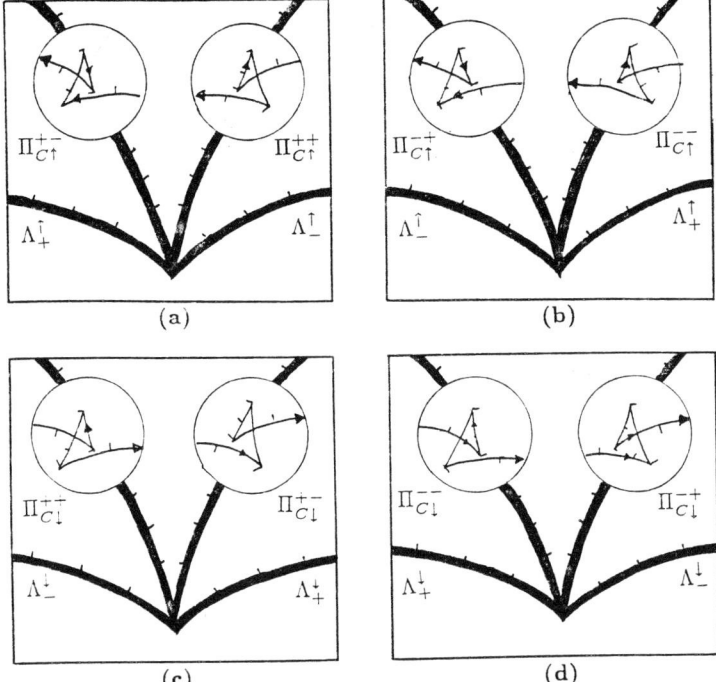

Figure 31. The RO-dependent strata $\Lambda\Lambda$

up becomes *down* and all the signs are reversed. Reversing the coorientation, only the signs are reversed while the arrows remain unchanged. One has always:

$$\nabla\lambda_\alpha^\eta + \nabla\pi_\eta^{\alpha\beta} - \nabla\pi_\eta^{\alpha\ -\beta} - \nabla\lambda_{-\alpha}^\eta = 0$$

Remark. The coefficient of $\nabla\pi_C$ is always trivially equal to zero.

The cases (a) and (b) in the figure give the rows 17 and 18 of M^{or}, while (c) and (d) give the rows 19 and 20.

Stratum $\Pi\Lambda$. The two local strata Λ separated by the stratum $\Pi\Lambda$ are of the same type and coherently cooriented. The two strata Π non coherently cooriented are always both of type *up* or both of type *down*, and one of type C and the other of type D. The coorientation of the stratum T is consistent with the coorientations of both strata Π (see §2.3). Consider the example of fig. 32.

The two strata of cusp crossings at the points (a) and (b) are $\Pi_{C\downarrow}^{++}$

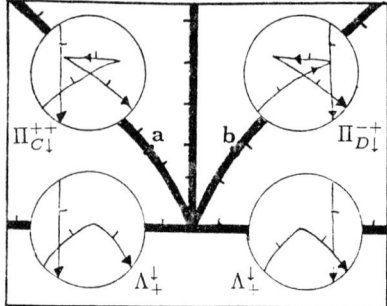

Figure 32. A RO-dependent stratum $\Pi\Lambda$

and $\Pi_{D\downarrow}^{-+}$. Hence the corresponding equation is:

$$\nabla\pi_{C\downarrow}^{++} - \nabla St + \nabla\pi_{D\downarrow}^{-+} = 0.$$

Changing the orientations and the coorientations of the segments one can obtain every local stratum in (a), say $\Pi_{Z\eta}^{\alpha,\beta}$. At the same time the stratum in (b) becomes $\Pi_{\overline{Z}\eta}^{-\alpha\beta}$, (where if $Z = C$ then $\overline{Z} = D$ and vice versa). Therefore for every pair of signs (α, β) and for every arrow (η) there is a situation where the nonzero coefficients of the reduced jumps on a closed path linked with the stratum $\Pi\Lambda$ satisfy

$$\nabla\pi_\eta^{\alpha\beta} - \nabla St + \nabla\pi_\eta^{-\alpha-\beta} + \nabla\pi_C = 0.$$

Taking into account the symmetry of the equation $(\alpha, \beta) \leftrightarrow (-\alpha, -\beta)$ we find 4 distinguishable cases. These four cases give the rows 21–24 of the matrix M^{or}.

To complete the proof of the lemma we have only to solve the linear sysem (II). The rank of the matrix M^{or} is equal to 11, so the dimension of the set of solutions is 9. The generators we choose satisfy (II) and are linearly independent.

3.3 The symmetries of the RO-dependent basic jumps

Proof of Lemma 3.3. To prove this lemma we have to study the action of the group RO on the reduced jumps (of which the basic jumps are linear combinations).

The reduced jumps ΔJ_{C+}^+, ΔJ_{D+}^- and $\Delta\pi^{2+}$ are transformed into ΔJ_{C-}^+, ΔJ_{D-}^- and $\Delta\pi^{2-}$ by the involutions a, b and c. Indeed these involutions reverse the signs of all segments of the front.

The reduced jump $\Delta\pi_\uparrow^{1+}$ becomes $\Delta\pi_\downarrow^{1+}$ under the action of the involution b. Indeed the character *up* is changed into *down* and viceversa while the sign of the cusp branch going away from the cusp point does not change under this involution. The involution c, reversing this sign, transforms $\Delta\pi^{1+}$ into $\Delta\pi^{1-}$. The involution a acts on $\Delta\pi$ as bc.

The reduced jump $\Delta\lambda_+^\uparrow$ is transformed into $\Delta\lambda_-^\downarrow$ by the involutions a and b. Indeed each of these involutions reverses the local orientation. The involution c, revrsing only the coorientation, transforms $\Delta\lambda_+^\uparrow$ into $\Delta\lambda_-^\uparrow$.

The symmetries of the basic jumps follow from the symmetries of the reduced jumps which constitute them.

3.4 The cohomological triviality of the RO-dependent 1-cocycles

Proof of Lemma 3.4. To prove this lemma, we have to verify that the value of every basic 1-cocycle on every closed path in $\Omega_{i,\mu}$ vanishes. We proceed exactly as in the proof of the corresponding lemma in the case of RO-independence, doing this verification for a representative of the unique homotopy class of noncontractile paths in the space of the marked L-immersions for every values of the indices i and μ.

In the case $\mu \neq 0$, there are noncontractile closed paths in $\Omega_{i,\mu}^0$ such that there are no crossing of the discriminant at all (see for example fig. 14).

In the case ($\mu = 0$, $i = 1$) consider the path shown in fig. 15, where all the 14 crossing of the discriminant are numbered and the corresponding RO-dependent elementary jumps are presented. The sum δ of the values of the basic RO-dependent jumps along this path can be easily calculated (the value of the jump at the event x being denoted by (x)):

$$\delta(f^+) = -(5) + (7) + (8) + (9) + 2(10) = 0;$$
$$\delta(f^-) = -2(1) + (2) - (3) - 2(4) + (11) + (13) - 2(14) = 0;$$
$$\delta(p^\uparrow) = -(3) + (5) = 0;$$
$$\delta(p^\downarrow) = (9) - (11) = 0;$$
$$\delta(\lambda^\uparrow) = (7) + (13) = 0;$$
$$\delta(\lambda^\downarrow) = (2) + (8) = 0.$$

In the case of $\Omega_{i,0}$ with $i \neq 1$, consider again the paths shown in fig. 16. The values of the basic 1-cochains on these closed paths are still vanishing if they are vanishing on the part of the path consisting of the bypassing of a loop by a pair of cusps of the same sign. The crossings of the discriminant and the corresponding elementary RO-dependent jumps are shown and numbered in fig. 17. We can thus calculate (with the same notation as before) the values of the 1-cocycles on the closed paths

$$\delta(f^+) = -(1) - (2) + (3) + (4) = 0;$$
$$\delta(f^-) = 0;$$
$$\delta(p^\uparrow) = (1) + (3) = 0;$$
$$\delta(p^\downarrow) = -(2) - (4) = 0;$$
$$\delta(\lambda^\uparrow) = 0;$$
$$\delta(\lambda^\downarrow) = 0.$$

For the paths in $\Omega_{\pm i, \pm \mu}$ one obtain the same results acting on the paths shown in fig. 17 by the involutions b and c and applying Lemma 3.3.

3.5. Definitions of the basic RO-dependent invariants

Proof of Theorem 3.1 I) Existence of the additive invariants.

Lemma 3.4 ensures that the basic RO-dependent jumps given by the solutions of the eq. (II), define RO-dependent local invariants in the connected components of the space of L-immersions up to additive constants which are still arbitrary. We shall now choose the values of these invariantss on the basic curves, making these invariants additive under the connected summation of fronts.

An additive RO-dependent invariant X has to verify:

i) $X(\theta) = 0$ on the four front classes $(\theta, a\theta, c\theta, ac\theta)$ of the circle (which are $R_{(ab)}$-invariant);

ii) $X(\alpha \# \beta) = X(\alpha) + X(\beta)$ for every generic points of α and β where the gluing takes place (when the orientations and the coorientations allow it);

iii) if $X(\gamma) = (f^+, f^-, p^\uparrow, p^\downarrow, \lambda^\uparrow, \lambda^\downarrow)$ are the values of the 6 invariants on a generic front γ, then the following symmetry conditions have to hold (by Lemma 3.3):

$X(\gamma):$	f^+	f^-	$p^\uparrow$	$p^\downarrow$	$\lambda^\uparrow$	$\lambda^\downarrow$
$X(a\gamma):$	f^-	f^+	$-p^\downarrow$	$-p^\uparrow$	$\lambda^\downarrow$	$\lambda^\uparrow$
$X(b\gamma):$	f^-	f^+	$p^\downarrow$	$p^\uparrow$	$\lambda^\downarrow$	$\lambda^\uparrow$
$X(c\gamma):$	f^-	f^+	$-p^\uparrow$	$-p^\downarrow$	$\lambda^\uparrow$	$\lambda^\downarrow$

In the case $\mu = 0$ we obtain the values of the invariants of $\omega_{i,0}$ $(i > 0)$ as $i + 1$ times the values of the invariants of $\omega_{0,0}$. Indeed $\omega_{i,0}$ is the connected sum of $i + 1$ identical copies of $\omega_{0,0}$. The values of the invariants on the curve $\omega_{0,0}$ are found by (i) (see fig.33).

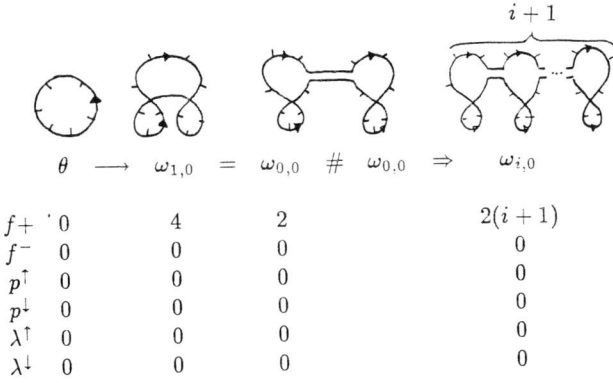

$$\theta \longrightarrow \omega_{1,0} = \omega_{0,0} \ \# \ \omega_{0,0} \Rightarrow \omega_{i,0}$$

	θ	$\omega_{1,0}$	$\omega_{0,0} \ \# \ \omega_{0,0}$	$\omega_{i,0}$
$f+$	0	4	2	$2(i+1)$
$f-$	0	0	0	0
$p^{\uparrow}$	0	0	0	0
$p^{\downarrow}$	0	0	0	0
$\lambda^{\uparrow}$	0	0	0	0
$\lambda^{\downarrow}$	0	0	0	0

Figure 33. The RO-dependent basic invariants of $\omega_{i,0}$

Remark 3.6. Every curve $\omega_{i,0}$ has an $R_{(ab)}$-invariant representative. The values of the invariants on the curves $b\omega_{i,0}$ and $c\omega_{i,0}$ are obtained applying (iii).

In the case of the curves $\omega_{1,\mu}$ we start from $\omega_{1,0}$ and transform it into the curve γ (see fig. 34) so that $\gamma = \alpha \# \beta$ where α and β are two heart curves, both of index 1 but with opposite coorientations, so that they have Maslov indices 2 and -2, respectively. ($\alpha = \omega_{1,2}$ and $c\alpha = \omega_{1,-2}$).

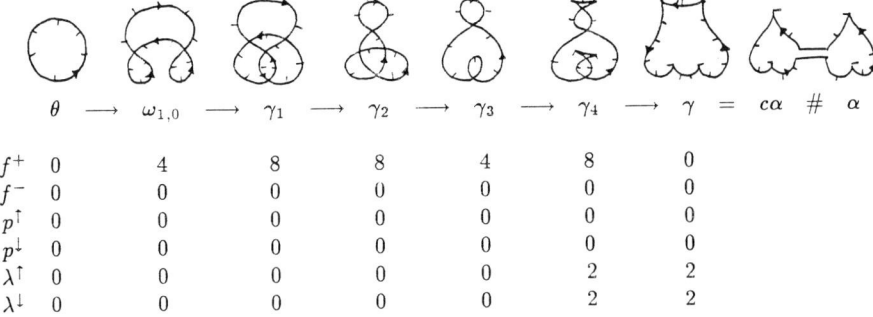

$$\theta \longrightarrow \omega_{1,0} \longrightarrow \gamma_1 \longrightarrow \gamma_2 \longrightarrow \gamma_3 \longrightarrow \gamma_4 \longrightarrow \gamma = c\alpha \ \# \ \alpha$$

	θ	$\omega_{1,0}$	γ_1	γ_2	γ_3	γ_4	$\gamma = c\alpha \# \alpha$
$f+$	0	4	8	8	4	8	0
$f-$	0	0	0	0	0	0	0
$p^{\uparrow}$	0	0	0	0	0	0	0
$p^{\downarrow}$	0	0	0	0	0	0	0
$\lambda^{\uparrow}$	0	0	0	0	0	2	2
$\lambda^{\downarrow}$	0	0	0	0	0	2	2

Figure 34. The curves α and $c\alpha$

We obtain:

1) $\quad f^+(\alpha) + f^+(c\alpha) = 0$
2) $\quad f^-(\alpha) + f^-(c\alpha) = 0$
3) $\quad p^\uparrow(\alpha) + p^\uparrow(c\alpha) = 0$
4) $\quad p^\downarrow(\alpha) + p^\downarrow(c\alpha) = 0$
5) $\quad \lambda^\uparrow(\alpha) + \lambda^\uparrow(c\alpha) = 2$
6) $\quad \lambda^\downarrow(\alpha) + \lambda^\downarrow(c\alpha) = 2$

From (1) (or from (2)) applying (iii) one has

$$f^+(\alpha) = -f^-(\alpha)$$

but α satisfies $abc\alpha = \alpha$, hence again from (iii)

$$f^+(abc\alpha) = f^-(\alpha) = f^+(\alpha).$$

Therefore f^+ and f^- are both vanishing on all the four front classes of the heart curve.

From (iii) $p^{\uparrow\downarrow}(c\alpha) = -p^{\uparrow\downarrow}(\alpha)$, thus eqq.(3) and (4) do not give any information.

The values of the invariants $\lambda^\uparrow$ and $\lambda^\downarrow$ do not change under the involution c. Hence equations (4) and (5) give:

$$\lambda^\uparrow(\alpha) = 1, \quad \lambda^\downarrow(\alpha) = 1.$$

We have still to find the value of the invariants $p^\uparrow$ and $p^\downarrow$ on the basic curve $\omega_{1,2}$.

Consider the two curves δ and ζ shown in fig. 35

One has:

$$\delta = \alpha \,\#\, \omega_{0,0} \,\#\, a\alpha,$$
$$\zeta = \alpha \,\#\, \alpha \,\#\, c\omega_{0,0}.$$

We can calculate all their invariants since

$$p^\eta(\delta) = p^\eta(\alpha) + p^\eta(\omega_{0,0}) + p^\eta(a\alpha) = 0 \qquad (\eta = \uparrow, \downarrow)$$

while the values of $p^\eta(\zeta)$ are found transforming δ into ζ. (See fig. 36)

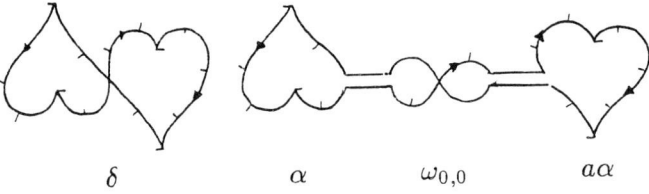

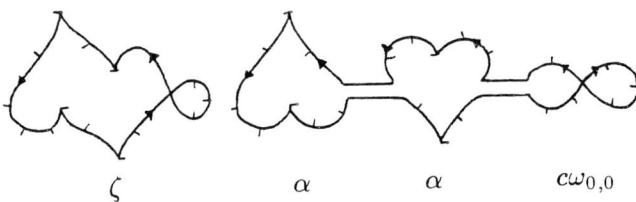

Figure 35. The two curves δ and ζ

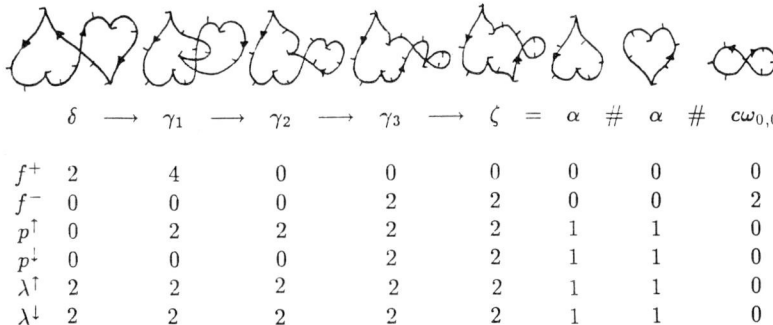

	δ	γ_1	γ_2	γ_3	$\zeta =$	α #	α #	$c\omega_{0,0}$
f^+	2	4	0	0	0	0	0	0
f^-	0	0	0	2	2	0	0	2
$p^\uparrow$	0	2	2	2	2	1	1	0
$p^\downarrow$	0	0	0	2	2	1	1	0
$\lambda^\uparrow$	2	2	2	2	2	1	1	0
$\lambda^\downarrow$	2	2	2	2	2	1	1	0

Figure 36. A generic path from δ to ζ

Hence we obtain from

$$p^\eta(\zeta) = 2 = 2p^\eta(\alpha) + p^\eta(c\omega) \qquad (\eta =\uparrow, \downarrow)$$

that $p^\uparrow(\alpha) = p^\downarrow(\alpha) = 1$.

Having the values of the basic invariants on the basic curve $\alpha = \omega_{1,2}$ we find, by connected sum, the values on the basic fronts $\omega_{1,\mu}$.

Remark 3.7. Every curves $\omega_{1,\mu}$ has a representative which is invariant under $R_{(abc)}$.

The values of the RO-dependent invariants on the curves $\omega_{i,\mu}$ are finally calculated as sums of the values of these invariants on the curves $\omega_{i,0}$ and $\omega_{-1,\mu}$.

To finish the proof of the theorem, we have to verify the condition (ii). As in the case of RO-independence, it suffices to verify it on the basic curves.

We can repeat step by step the proof of Theorem 2.1, taking this time into account all the orientations. This is rather long. Therefore first of all we observe the two following facts, whose proof is in §3.6: i) For a generic oriented and cooriented front in the oriented plane whose L-immersion belongs to $\Omega_{i,\mu}$ the invariant $\lambda^\uparrow$ ($\lambda^\downarrow$) (defined by its jump and its values on the basic curves $\omega_{i,\mu}$ shown in table 7) is equal to the number of the cusp points of the front of type *up* (of type *down* respectively); ii) the invariants f^+ (f^-) (defined by its jump and its values on the basic curves shown in table 7) is equal to twice to the number of the double points of the front such that the two crossing segments are both positive (both negative respectively).

The additivity of the invariants $\lambda^\uparrow$, $\lambda^\downarrow$, f^+ and f^- under the connected summation follows immediately from this observation.

To prove the additivity of the invariants $p^\uparrow$ and $p^\downarrow$ we proceed as in the proof of Theorem 2.1 (see §2.4).

The events of cusp crossing which take place in the transformation of a connected sum of basic curves into a basic curve, occur in the sets of perestroikas called P_1 and P_2. The variations of $p^\uparrow$ and $p^\downarrow$ at P_1 are both vanishing. The variations at P_2 are both equal to -2, in the case shown in fig. 37. (The variations of $p^\uparrow$ and $p^\downarrow$ in the case of different orientations is found applying the lemma 3.3).

We consider now the different cases of connected summation of basic curves depending on the positions of the gluing points on the curves. If the gluing point on a curve is of type (3) or (4) (see fig. 38), we can always transform the curve before the gluing so that this point, remaining always generic and at the boundary of the curve, becomes of type (1). We have only to transport cusps points beyond loops. It is achieved by the perestroikas P_1, which do not change the values of $p^\uparrow$ and $p^\downarrow$.

We remark that in the case of a point of type (3') (i.e., when the number of cusp points to be transported is odd), the resulting basic front (containing the gluing point at the position (1)) is obtained from the original front by the *abc* involution. This involution does not change the value of the invariants $p^\uparrow$ and $p^\downarrow$ (see Remark 3.4).

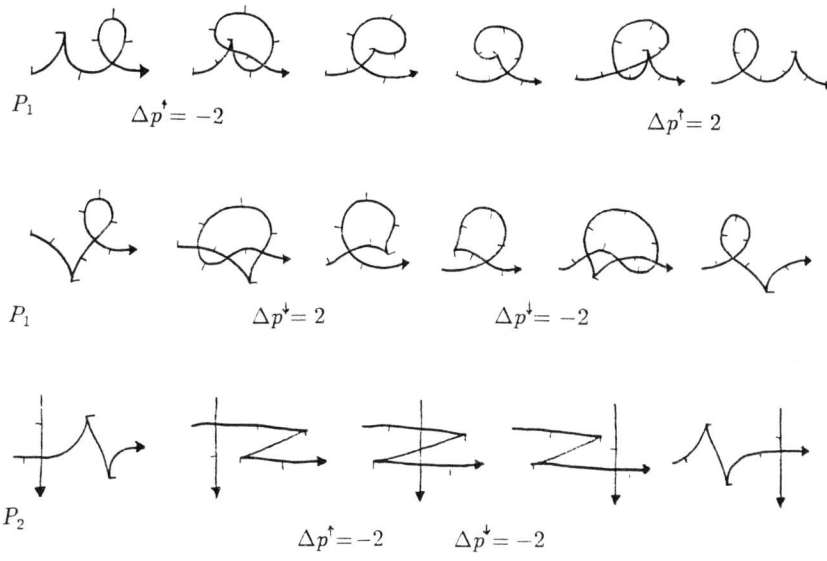

Figure 37. The P_1 and P_2 perestroikas

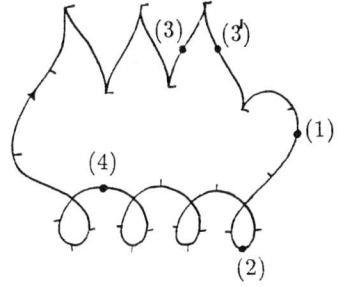

Figure 38. Different places of the gluing points

The perestroika P_2 (the only one changing the value of $p^{\uparrow\downarrow}$) occurs only when the gluing points are of type (1)-(2) (see fig. 24). In the case of fig. 24 the perestroikas P_2 are exactly of the type considered in fig. 37. In the transformation of the connected sum of the basic curves: $(i, j, \mu, \nu \geq 0; \; i \geq j)$

$$\omega_{i,\mu} \# \omega_{-j,\nu}$$

the perestroika P_2 occurs exactly $k = \nu/2$ times. Hence the total variation of $p^{\uparrow}$ and $p^{\downarrow}$ are both equal to $-2k$. This is in agreement with the values of $p^\uparrow$ and $p^\downarrow$ on the basic curves (see Remark 3.4):

$$p^\eta(\omega_{i,2m}) + p^\eta(\omega_{-j,2k}) = k - m$$

$$p^\eta(\omega_{i-j+1,2m+2k}) = -m - k.$$

where $\eta = (\uparrow, \downarrow)$. Hence the invariants $p^\uparrow$ and $p^\downarrow$ are additive under the connected summation of the basic fronts of type (1)-(2).

In all the other cases of transformation of the connected sum of two basic fronts into a basic front, when the perestroika P_2 does not enter, one verify easily that the value of the invariant $p^\uparrow$ and $p^\downarrow$ on the resulting basic curve is equal to the sum of the values of these invariants on the two original basic curves.

The proof of the existence of the additive invariants is thus completed.

II) Uniqueness of the additive invariants.

Lemma 3.6. *An additive invariant of oriented and cooriented fronts in the oriented plane is unambiguously defined by its jumps, provided that it is invariant under the reversal of the coorientation of the front. If it is not required to be invariant under the reversal of the coorientation of the front, it is defined by its jumps up to the addition of a multiple of the Maslov index.*

Proof. The difference of two invariants with equal jumps is a function locally constant. Hence it may be considered as a function on the set of the components $\Omega_{i,\mu}$ of the space of the legendrian immersions.

The connected sum defines two operations on this set. Indeed, there are four possible types of bridges connecting two oriented and cooriented fronts, one in the left half-plane, the other in the right half-plane (fig. 39).

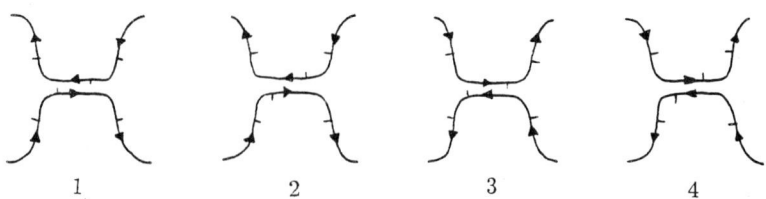

Figure 39. The four types of bridges connecting two fronts

Using the elementary perestroikas one is able to prepare on the left and on the right curves two pieces, ready for the joining with any of the four bridges (fig. 40).

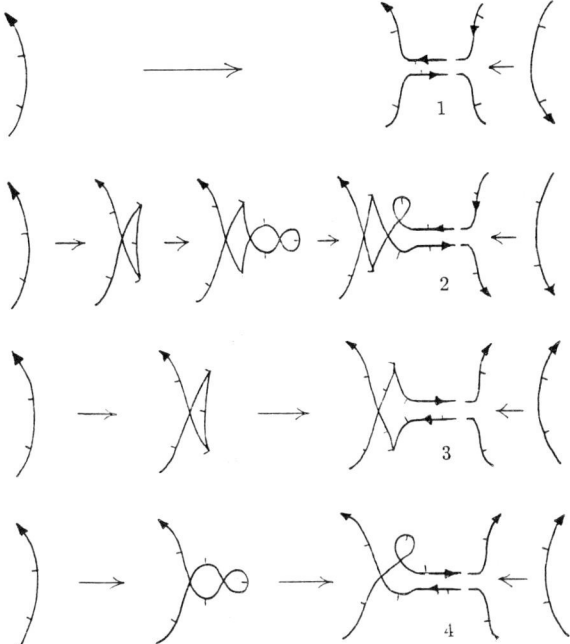

Figure 40. Preparation of the pieces ready to the joining

Therefore one can add two curves of any two classes Ω_{i_1,μ_1} and Ω_{i_2,μ_2} by any of the four bridges.

The Maslov index is additive under any of the four types of the connected summation. The index of the sum is $i_1 + i_2 + 1$ for the first two bridges (differing only by the coorientation) and $i_1 + i_2 - 1$ for the other pair of bridges (obtained from the first pair by the orientation reversal). Hence the connected summation defines two operations on the set of connected components $\mathbf{Z}^2 = \{i, k = \mu/2\}$:

$$A_+ : (i_1, k_1), (i_2, k_2) \mapsto (i_1 + i_2 + 1, k_1 + k_2)$$
$$A_- : (i_1, k_1), (i_2, k_2) \mapsto (i_1 + i_2 - 1, k_1 + k_2)$$

Both operations are commutative as operations on $\mathbf{Z}^2$.

The additivity of the difference of two additive invariants with equal jumps for the A_+ connected sum means that this function on the affine

$\mathbf{Z}^2$ is linear with respect to the abelian group structure in $\mathbf{Z}^2$, whose null element is $(i = -1, k = 0)$. All such functions have the form $f(i, k) = a(i + 1) + bk$. Similarly, the additivity with respect to the second kind of connected sum implies that $f(i, k) = c(i - 1) + dk$. Hence $f(i, k) = ek$, which proves the lemma.

Corollary 3.3. *The only RO-independent local additive invariants are the linear combinations of* $(J^+, J^-, Sp, h, \lambda)$.

Proof. Indeed, the linear combinations of the basic RO-independent jumps are the only possibilities for the RO-independent jumps (see the proof of Lemma 2.2) and the jumps define unambiguously the invariants by the above lemma.

3.6. Explicit formulas for some basic local invariant

Explicit formulas for the local invariants of the immersions of a circle in the plane and for the RO-independent invariants of the planar fronts have been found [5,6,7]. There are two simple formulas for the invariants f^+, f^-, $\lambda^\uparrow$ and $\lambda^\downarrow$ of a generic front γ.

Let $n^\uparrow$ and $n^\downarrow$ be the number of cusp points of γ of type *up* and of type *down* respectively.

Proposition 3.1.

$$n^\uparrow = \lambda^\uparrow, \quad n^\downarrow = \lambda^\downarrow$$

Proof. This is true when γ is the basic curve $\omega_{i,\mu}$ $(\mu = 2k)$ having k cusp points of type *up* and k cusp points of type *down*, and having $\lambda^\uparrow = \lambda^\downarrow = k$. At the birth of two cusp points, they are both of type *up* or both of type *down* at the corresponding increment by 2 of $\lambda^\uparrow$ or of $\lambda^\downarrow$. The character *up* or *down* of a cusp does not change along any path in $\Omega_{i,\mu}$.

Let n^+ (n^-) be the number of double points of γ such that both intersecting branches are positive (negative).

Proposition 3.2.

$$f^+ = 2n^+, \quad f^- = 2n^-$$

Proof. For the basic curve $\gamma = \omega_{i,\mu}$ (see table 7), the number of such double points is $n^+ = i + 1$ while $f^+ = 2(i + 1)$, $f^- = 0$. The numbers of double points of $a\gamma$, $b\gamma$ and $c\gamma$ are equal to $n^- = i+1$ while $f^+ = 0$, $f^- = 2(i+1)$.

The number n^+ increases by 2 at the positive crossing of the local stratum K_{C+}^+ and at the negative crossing of the local stratum K_{D+}^-, and remains unchanged at the crossing of the other local strata of self-tangencies. At the strata of cusp crossings, n^+ increases by 1 when $\Delta\pi^{2+} = 2$ and reamins unchanged in the other cases. Indeed at the cusp crossing where the smooth segment crossing the cusp is positive, $\Delta\pi^{2+} = 2$ and $\Delta n^+ = 1$ (since one of the branches into which the cusp point divides the cusped segment is always positive). If the smooth segment is negative, $\Delta\pi^{2+} = \Delta n^+ = 0$. Moreover, n^+ increases by 1 while f^+ increases by 2 at the positive crossings of the local strata $\Lambda_+^\uparrow$ and $\Lambda_+^\downarrow$. Hence $f^+ = 2n^+$ for any planar front. Similar arguments hold for the invariant n^-.

Remark 3.8. From the preceding observations it follows that the values of the invariants f^+, f^-, $\lambda^\uparrow$ and $\lambda^\downarrow$ cannot be negative.

Remark 3.9. From the equality (see Theorem 2.1 and Corollary 3.1):

$$f^+ + f^- - J^+ + J^- - \frac{1}{2}\lambda = n$$

one obtains the result of M. Polyak ([8]):

$$J^+ - J^- = n_+ - n_- - c$$

where n_+ is the number of double points of the front where the two intersecting branches are both positive or both negative, n_- is the number of double points where the two intersecting branches are one positive and the other negative and $2c$ is the number of the cusp points.

Indeed, $n_+ = \frac{1}{2}(f^+ + f^-)$, $n_+ + n_- = n$ and $c = \frac{1}{2}\lambda$.

3.7 Other types of invariance

The local basic additive invariants are functions constant on the front classes (on the orbits of the generic legendrian immersions of a circle to the space of the cooriented contact elements of the plane) under the action of the group of the orientation preserving diffeomorphisms of the source circle and of the target plane.

Some of the invariants we find are also invariant under the action of larger groups of diffeomorfisms of the plane and of the circle which are not necessarily preserving all the orientations. We obtain the following facts from Theorems 2.1 and 3.1.

Corollary 3.4. *The basic local additive invariants of the front classes of legendrian immersions of the circle in the space of the cooriented contact elements of the plane up to diffeomorphisms preserving only the orientation of the plane, only the orientation of the circle, or non preserving these orientations at all, are shown in the following table*

$\mathbf{R}^2$ and S^1	$\mu,$	$J^+,$	$J^-,$	$Sp,$	$f^+,$	$f^-,$	$p^\uparrow,$	$p^\downarrow,$	$\lambda^\uparrow,$	$\lambda^\downarrow$
$\mathbf{R}^2$	$\mu,$	$J^+,$	$J^-,$	$Sp,$	$(f^+ + f^-),$		$(p^\uparrow + p^\downarrow),$		$(\lambda^\uparrow + \lambda^\downarrow)$	
S^1		$J^+,$	$J^-,$	$Sp,$	$(f^+ + f^-),$		$(p^\uparrow - p^\downarrow),$		$(\lambda^\uparrow + \lambda^\downarrow)$	
$-$		$J^+,$	$J^-,$	$Sp,$	$(f^+ + f^-),$				$(\lambda^\uparrow + \lambda^\downarrow)$	

Remark. In the last three cases only the absolute values of the index is invariant.

Proof. Consider the symmetry types of the reduced elementary jumps which constitute the basic RO-dependent invariants (see §3.3). For example, an invariant of a front class defined up to diffeomorphisms preserving only the orientation of the plane has to be invariant with respect to the b involution (which reverses the orientation of the preimage circle), and so on. Of course the five RO-independent basic invariants are still invariants in the four cases.

Consider now a similar problem, when the front is defined as the projection to the plane of a legendrian immersion of a circle in the space of the *noncooriented* contact elements of the plane, with the same contact structure as in the cooriented case.

All the codimension 1 and 2 strata of the discriminant can be defined in this case, where the normal vector of the front is absent. In this case the answer depends on the parity of the the number of cusp points. This parity is invariant under all the perestroikas, the birth and the death of cusp points occurring in pairs.

We study only the case of an even number of cusp points. The index, i.e., the number of complete turns of the normal vector of the front, is thus an integer number. If the orientation of the preimage circle or that of the plane is not preserved, then only the absolute value of the index is invariant.

In terms of the front, the definition of the Maslov index as the difference betweeen the numbers of positive and negative cusp points is no longer possible.

The front of a legendrian immersion of a circle into the space of the noncooriented local elements of the plane with an even number of cusp points has two coorientations. The Maslov index of these two cooriented fronts are opposite. The absolute value of each of them is a well-defined (non additive) invariant of the noncooriented front.

Corollary 3.5. *The space of the local additive invariants of the non-cooriented fronts with an even number of cusp points has dimension 6 in the case where the plane and the front are oriented. The generators of this space are $(J^+, J^-, Sp, f^+ + f^-, \lambda^\uparrow, \lambda^\downarrow)$.*

The spaces of the local additive invariants of the noncooriented fronts with an even number of cusps which are invariant under the reversal i) of the orientation of the plane; ii) of the orientation of the front; iii) of both orientations, coincide and have dimension 5. Their generators are $(J^+, J^-, Sp, f^+ + f^-, \lambda^\uparrow + \lambda^\downarrow)$.

Proof. Acting by the coorientation reversal involution c on one of the two cooriented fronts $\tilde{\gamma}_1$ covering the given noncooriented front γ, one obtains the other cooriented covering front $\tilde{\gamma}_2$. The arithmetical mean of the values of a local additive invariant on the two cooriented fronts $\tilde{\gamma}_1$ and $\tilde{\gamma}_2 = c\tilde{\gamma}_2$ is a local additive invariant of the original noncooriented front. In fact this construction provides all the local additive invariants of the noncooriented fronts.

References

[1] V.I.Arnold, *Invarianty i perestroiki ploskikh frontov*, Trudy Math. Inst. Steklova (1993)

[2] V.I.Arnold, *Plane Curves, Their Invariants, Perestroikas and Classifications* Adv. in Sov. Math., vol.21, AMS, Providence (1994)

[3] V.I.Arnold, *Rutgers Lectures on plane curves and Wave fronts*, Preprint Ceremade 9143 (1994)

[4] F. Aicardi, *Remarks on the symmetries of planar fronts*, Revista Matematica de la Universidad Complutense de Madrid (1995)

[5] F. Aicardi, *Partial indices and linking numbers of planar fronts*, Preprint (1994)

[6] O.Viro, *First degree invariants of generic curves on surfaces*, Preprint Uppsala University U.U.D.M. 1994:21 (1994)

[7] A. Shumakovich, *Formulas for Strangeness of plane curves*, S.Pbg Math. Jour. (1995)

[8] M. Polyak, *Invariants of generic plane curves via Gauss diagrams,* Preprint Bonn (1994).

F. Aicardi
International Centre for Theoretical Physics
Strada Costiera, 11
I-34014 Trieste

Received March 1995

Crofton Densities, Symplectic Geometry and Hilbert's Fourth Problem

J. C. Alvarez   I.M. Gelfand   M. Smirnov

Abstract

We study the relation between the integral geometric and the symplectic construction of Desarguesian metrics on $\mathbf{R}^n$ and show that these constructions characterize all Desarguesian Finsler metrics.

1. Introduction

A *Desarguesian metric* on an open convex subset $\mathcal{O} \subset \mathbf{R}^n$ is defined by a distance function

$$d : \mathcal{O} \times \mathcal{O} \longrightarrow \mathbf{R}$$

with the usual properties:

D1. $d(x, y) \geq 0$
D2. $d(x, y) = 0 \Leftrightarrow x = y$
D3. $d(x, y) = d(y, x)$
D4. $d(x, z) \leq d(x, y) + d(y, z)$.

Together with the *Desargueness condition:*

D5. If x, y, and z are collinear points and y is between x and z, then

$$d(x, z) = d(x, y) + d(y, z) \ .$$

If $\gamma : [a, b] \to \mathcal{O}$ is a continuous curve we define its arclength with respect to a metric d by the following formula:

$$l(\gamma) := sup\{\sum_{i=0}^{k-1} d(\gamma(t_i), \gamma(t_{i+1})) :$$

$a = t_0 < \cdots < t_k = b$ is a partition of $[a, b]\} \ .$

Clearly, a metric d on $\mathcal{O}$ is Desarguesian if and only if the shortest curve between two points is the line segment joining them. The problem of finding all geometries for which the line segment is the shortest curve is known as *Hilbert's fourth problem*.

In this paper we present some constructions of Desarguesian metrics and study their degree of generality. More specifically, we give variants of the integral geometric construction of Busemann, the symplectic construction of Álvarez, show the beautiful relation between them, and prove that any Desarguesian Finsler metric may be obtained by either of these constructions.

The present paper is based on the work of the first author on the symplectic geometry of spaces of geodesics (see [A1] and [A2]) and the work on Crofton densities ([GS]) by the second and third authors. The relation between symplectic and integral geometry resulted in a clear and simple understanding of both the symplectic and the integral geometric constructions of Desarguesian metrics.

For the authors, a great part of the beauty of our approach to Hilbert's fourth problem is that it relates a number of concepts which are 2-dimensional in nature:

- A Desarguesian metric on $\mathbf{R}^n$ induces on every 2-dimensional affine subspace a Desarguesian metric.

- A subset of $\mathbf{R}^n$ is convex if and only if all its 2-dimensional sections are convex. A function on $\mathbf{R}^n$ defines a norm if and only if its restriction to every 2-dimensional subspace defines a norm.

- A symplectic form, as a 2-form, ultimately depends only on its restrictions to 2-dimensional submanifolds.

A word on the organization of the paper: in Section 2 we present some constructions of Desarguesian metrics. Section 3 relates Hilbert's fourth problem to the calculus of variations by discussing Finsler metrics and variational problems whose extremals are lines. In section 4 we show that the integral geometric construction presented in 2.4. characterizes all Desarguesian Finsler metrics. Finally, in Section 5 we relate the integral geometric construction and the symplectic construction proving that the symplectic construction also characterizes all Desarguesian Finsler metrics.

2. Constructions

In this section we review some constructions of Desarguesian metrics on $\mathbf{R}^n$. For the most part this constructions are well known, but we have modified the integral geometric construction of Busemann and the symplectic construction presented in [A] in order to underline the relation between the two.

2.1 Minkowskian Geometries

The simplest construction of Desarguesian metrics is due to Minkowski. In modern terminology, it amounts to giving a Banach norm on $\mathbf{R}^n$.

Given a bounded convex set $C \subset \mathbf{R}^n$ which is centrally symmetric with respect to the origin we define the distance d by the following procedure:

Draw the line segment joining a point x with $\vec{0}$ and denote the point of intersection with the boundary of C by x_0. We now define $d(x,0)$ by the equality $d(x,0)\overrightarrow{0x_0} = \overrightarrow{0x}$. If x and y are two points in $\mathbf{R}^n$ we define $d(x,y) = d(x-y,0)$.

The space $\mathbf{R}^n$ with such a metric structure is called a *Minkowskian space*. It is easy to see that a Minkowskian space is Desarguesian.

Note that Minkowskian geometries are characterized by their translation invariance.

2.2 Hilbert Geometries

In [H] Hilbert noted that the Cayley-Klein model of hyperbolic geometry could be modified to yield Desarguesian metrics on open bounded convex subset of $\mathbf{R}^n$.

Let $\mathcal{O} \subset \mathbf{R}^n$ be an open bounded convex set, and let x and y be two points in $\mathcal{O}$. Draw the line defined by x and y and mark the two intersection points of the line with the boundary of $\mathcal{O}$. Denote by B the intersection point such that y belongs to the segment xB, and denote the other intersection point by A. Now define

$$d(x,y) := ln\left(\frac{\|x-A\| \, \|y-B\|}{\|x-B\| \, \|y-A\|}\right) ,$$

where $\| \cdot \|$ denotes the Euclidean norm.

The multiplicativity of the cross ratio immediately implies that the Desargueness property is satisified. Unlike Minkowskian geometries, Hilbert geometries have no obvious symmetries. However, like Minkowskian geometries, the metric is not Riemannian unless the convex set used to define it is an ellipsoid. In this case we have the Cayley-Klein model of hyperbolic geometry.

2.3 2-D Case

It is to be noted that, roughly speaking, the construction of Minkowskian or Hilbert geometries in n-dimensional space depends on a function of $n-1$ variables (the choice of a convex hypersurface). We shall now give a geometric construction of Desarguesian metrics on $\mathbf{R}^2$ that depends on a function of 2 variables.

Consider an area form on the cylinder $\{(x, y, z) \in \mathbf{R}^3 : x^2 + y^2 = 1\}$ which is invariant under the antipodal map. We associate to this area form a metric on $\mathbf{R}^2$ as follows:

Identify $\mathbf{R}^2$ with $\mathbf{RP}^2 \setminus \mathbf{RP}^1$ or, more precisely, with the set of planes in $\mathbf{R}^3$ which pass through the origin but do not contain the z-axis. If $\mathbf{x}$ and $\mathbf{y}$ are two such planes we consider their intersection with the cylinder. This intersection consists of two ellipses which divide the cylinder into four connected components, two bounded and two unbounded. As a consequence of the invariance under the antipodal map, the bounded components have equal areas. We denote the common area by $a(\mathbf{x}, \mathbf{y})$ and define the distance between $\mathbf{x}$ and $\mathbf{y}$ by

$$d(\mathbf{x}, \mathbf{y}) := \frac{1}{2} a(\mathbf{x}, \mathbf{y}) \ .$$

If $\mathbf{x}, \mathbf{y}$ and $\mathbf{z}$ are three collinear points in $\mathbf{R}^2$, then the planes they represent intersect at a line. In the cylinder we have three ellipses that intersect in two antipodal points. It is easy to see that if $\mathbf{y}$ is between $\mathbf{x}$ and $\mathbf{z}$, then $a(\mathbf{x}, \mathbf{z}) = a(\mathbf{x}, \mathbf{y}) + a(\mathbf{y}, \mathbf{z})$.

The reader may enjoy checking that if the area form on the cylinder is the one induced by its embedding into $\mathbf{R}^3$, then the resulting metric on the plane is the Euclidean metric. Note also that if the area form is independent of the z-coordinate on the cylinder the resulting metric is Minkowskian.

2.4 Integral Geometric Construction

Let $\mathbf{H}$ denote the set of all affine hyperplanes in $\mathbf{R}^n$. The set $\mathbf{H}$ inherits a manifold structure from its natural identification with the quotient of $S^{n-1} \times \mathbf{R}$ by the involution $(\xi, p) \mapsto (-\xi, -p)$. The manifold $S^{n-1} \times \mathbf{R}$ is itself identified with the set $\tilde{\mathbf{H}}$ of all oriented affine hyperplanes in $\mathbf{R}^n$.

Let σ be a smooth signed measure on $\mathbf{H}$. Given two points $x, y \in \mathbf{R}^n$ we denote by $\mathbf{H}_{xy}$ the set of hyperplanes which intersect the segment xy. Let us define $d(x, y) := \sigma(\mathbf{H}_{xy})$.

We would like to find conditions on σ which assure that d defines a metric on $\mathbf{R}^n$. It is easy to see that such a metric would necessarily be Desarguesian.

Quasi-positive measures

Definition 1 *A smooth signed measure σ on* **H** *is said to be quasi-positive if it satisfies the following condition: If U is a Borel subset of the set of lines lying on a 2-dimensional affine subspace $P \subset \mathbf{R}^n$ and $\hat{U}$ is the set of all hyperplanes that contain a line in U, then $\sigma_P(U) := \sigma(\hat{U}) > 0$.*

Theorem 1 *Let σ be a smooth quasi-positive measure on* **H**, *the function*

$$d : \mathbf{R}^n \times \mathbf{R}^n \longrightarrow [0, \infty)$$

defined by $d(x, y) := \sigma(\mathbf{H}_{xy})$ defines a Desarguesian metric on $\mathbf{R}^n$.

The Desarguesian metrics obtained by this construction will be called σ-*metrics*.

Let us elaborate somewhat on the quasi-positivity condition:

Let $P \subset \mathbf{R}^n$ be a 2-dimensional affine subspace, and let $\mathbf{H}_P \subset \mathbf{H}$ denote the set of hyperplanes intersecting P on a line. If $L(P)$ denotes the set of lines lying on P, then we have a natural projection

$$\pi_P : \mathbf{H}_P \longrightarrow L(P) .$$

The push forward of the measure σ under the map π_P is what we have called σ_P. For every 2-dimensional affine subspace P a quasi-positive measure induces an area form on $L(P)$. This reduces the construction of Desarguesian metrics to the 2-dimensional construction.

When the measure σ is invariant under the action of the translation group on **H** it can be identified with a measure on the projective space $\mathbf{RP}^{n-1}$. In this case the σ-metric is Minkowskian and is, therefore, defined by a convex subset of $\mathbf{R}^n$. The explicit relation between quasi-positivity and convexity will be presented in section 4.2.

2.5 Symplectic Construction

Let $\tilde{L}(\mathbf{R}^n)$ denote the set of all oriented lines in $\mathbf{R}^n$. The set $\tilde{L}(\mathbf{R}^n)$ inherits a manifold structure from its natural identification with the tangent bundle of the $(n\text{-}1)$-dimensional sphere, TS^{n-1}:

An oriented line is specified uniquely by its direction vector and the *foot vector* defined by the point on the line closest to $\vec{0}$. The direction vector lies in the unit sphere and the foot vector is orthogonal to it. Together they define a point in TS^{n-1}.

Let ω be a differential 2-form on $\tilde{L}(\mathbf{R}^n)$. Let x, y be two points in $\mathbf{R}^n$ and choose a 2-dimensional affine subspace $P \subset \mathbf{R}^n$ containing x and y. Denote by $\tilde{L}(P)$ the set of all oriented lines lying on P and let

$$i_P : \tilde{L}(P) \longrightarrow \tilde{L}(\mathbf{R}^n)$$

be the natural inclusion.

Set

$$d(x, y; P) := \int_{\tilde{L}(P)_{xy}} |i_P^* \omega| \, ,$$

where $\tilde{L}(P)_{xy} \subset \tilde{L}(P)$ is the set of all oriented lines on P intersecting the segment xy.

We would like to find conditions on ω which assure that the quantity $d(x, y; P)$ is independent of P and defines a metric on $\mathbf{R}^n$. It is easy to see that such metric would necessarily be Desarguesian.

Admissible Symplectic Forms.   Before defining admissible symplectic forms let is quickly review the definitions of symplectic form and Lagrangian submanifold.

Definition 2 *A differential 2-form ω on a smooth manifold M is said to be symplectic if it is closed and at every point $m \in M$ the bilinear form*

$$\omega_m : T_m M \times T_m M \longrightarrow \mathbf{R}$$

is nondegenerate.

Definition 3 *Let M be a smooth manifold and let ω be a symplectic form on M. A smooth submanifold $\Gamma \subset M$ is said to be Lagrangian if ω restricted to Γ is identically zero, and the dimension of Γ is half that of M.*

Let $\Gamma_x \subset \tilde{L}(\mathbf{R}^n)$ denote the set of all oriented lines passing through a point $x \in \mathbf{R}^n$. Clearly the Γ_x are smooth embedded spheres in $\tilde{L}(\mathbf{R}^n)$.

Definition 4 *A symplectic form ω on $\tilde{L}(\mathbf{R}^n)$ is said to be admissible if for each point $x \in \mathbf{R}^n$ the submanifold Γ_x is Lagrangian, and if $a^* \omega = -\omega$, where $a : \tilde{L}(\mathbf{R}^n) \to \tilde{L}(\mathbf{R}^n)$ is the involution defined by changing the orientation of the lines.*

Theorem 2 *Let ω be an admissible symplectic form on $\tilde{L}(\mathbf{R}^n)$. The function*

$$d(x, y; P) := \int_{\tilde{L}(P)_{xy}} |i_P^* \omega|$$

is independent of P and defines a Desarguesian metric on $\mathbf{R}^n$.

Distance functions on $\mathbf{R}^n$ obtained by this construction will be called ω-*metrics*.

Note that both the symplectic construction and the integral geometric construction, as presented in the previous section, reduce the problem of finding Desarguesian metrics on $\mathbf{R}^n$ to the problem of finding a *compatible* choice of positive measures on the family of submanifolds $L(P) \subset L(\mathbf{R}^n)$ parametrized 2-dimensional affine subspaces of $\mathbf{R}^n$. For the integral geometric construction the measures are the pushforwards σ_P, for the symplectic construction they are the absolute values of the pullback forms $i_P^*\omega$.

3. Desarguesian Finsler Metrics

The constructions of Desarguesian metrics presented in the previous sections all yield continuous metrics (i.e., the topology induced on $\mathbf{R}^n$ by these metrics is the standard). The smoothness requirements on the quasi-positive measures and admissible symplectic forms further restricts the class of metrics that can be obtained by these constructions. The main restriction is that these metrics be defined "infinitesimally." For example, the Euclidean length of a curve γ on $\mathbf{R}^n$ may be defined by the integral

$$S(\gamma) := \int_\gamma \|\dot{\gamma}(t)\| dt \ .$$

The Euclidean distance between two points $x, y \in \mathbf{R}^n$ is now recovered through the formula

$$d(x,y) := inf\{S(\gamma) : \gamma \text{ joins } x \text{ and } y\} \ .$$

One may now try to abstract the properties of the integrand $\|\dot{\gamma}(t)\|$ that garantee that d defines a metric on $\mathbf{R}^n$ and then one may try to classify all Desarguesian metrics of this form. In this section we study these metrics and give the precise statements of our main results.

3.1 Finsler Metrics and Even 1-Densities

Let $\varphi : \mathbf{R}^n \times (\mathbf{R}^n \setminus 0) \to \mathbf{R}$ be a smooth function. If γ is a smooth parametrized curve on $\mathbf{R}^n$ we define its action by the integral

$$S_\varphi(\gamma) := \int_\gamma \varphi(x, \dot{x}) dt \ .$$

To make the integral invariant under reparametrizations of γ we require that the function φ satisfy

$$\varphi(x, \lambda\dot{x}) = |\lambda|\varphi(x, \dot{x}) \ ,$$

for any nonzero real number λ. Such a function will be called an *even 1-density*.

We propose constructing metrics of the form

$$d(x,y) := inf\{S_\varphi(\gamma) : \gamma \text{ joins } x \text{ and } y\} \ ,$$

where φ is an even 1-density. It is well known that the following are sufficient conditions on an 1-density φ for d to define a metric on $\mathbf{R}^n$:

- $\varphi(x, \dot{x}) > 0$, for any $\dot{x} \neq 0$

- The matrix of second derivatives of the squared function φ^2 with
  respect to the the the $\dot{x}$'s

$$\left(\frac{\partial^2 \varphi^2}{\partial \dot{x}_i \partial \dot{x}_j} \right)_{i,j}$$

is positive definite.

The metric d obtained from a 1-density φ satisfying the above conditions is said to be a *Finsler metric*. We will also call the 1-density φ a
Finsler metric.

We now state our main results.

Theorem 3 *A Finsler metric on $\mathbf{R}^n$ is Desarguesian if and only if it is a
σ-metric.*

Theorem 4 *A Finsler metric on $\mathbf{R}^n$ is Desarguesian if and only if it is
an ω-metric.*

3.2 Hamel's Theorem

When an even 1-density φ is not Finsler we may no longer speak of its
geodesics as the curves which locally minimize length, but we may speak of
its geodesics as *extremals*. These extremals are the solutions of the system
of Euler-Lagrange differential equations:

$$\frac{d}{dt} \left(\frac{\partial \varphi}{\partial \dot{x}_i} \right) - \frac{\partial \varphi}{\partial x_i} = 0 \ , \ i = 1, ..., n \ .$$

Hamel, a student of Hilbert, was the first to study the Euler-Lagrange
equations of a 1-density whose extremals are straight lines. We state his
result with some modifications.

Theorem 5 *An even 1-density φ satisfies the equations*

$$\frac{\partial^2 \varphi}{\partial x_k \partial \dot{x}_j} = \frac{\partial^2 \varphi}{\partial x_j \partial \dot{x}_k}, \ k, j = 1, ..., n, \ \dot{x} \neq 0$$

if and only if straight lines are geodesics.

The similarity between this system of equations and the equations for a closed 1-form on $\mathbf{R}^n$ was noted and clarified in [GS]. In the k-dimensional case [GS] similar equations can be written and these equations similar to equations for closed differential k-forms. Other analogues of differential forms and De Rham complex were studied by M. Baranov and A.S.Shvarts [BS].

We may now pose the problem of characterizing all even 1-densities on $\mathbf{R}^n$ whose extremals are straight lines. This problem may be taken as the first step in the integral geometric solution of Hilbert's fourth problem.

4. Crofton Densities

In this section we present the main ideas that go into the proof of theorem 3. We start by giving an integral geometric characterization of even 1-densities whose extremals are straight lines.

Most of what follows has been taken from [GS].

If γ is a curve on $\mathbf{R}^n$ and $\zeta \in \mathbf{H}$ is a hyperplane, we define $Crof_\gamma(\zeta)$ to be the number of points in $\gamma \cap \zeta$. If γ is a regular curve, the set of hyperplanes for which $Crof_\gamma(\zeta)$ is infinite has measure zero.

Definition 5 (GS) *An even 1-density φ in $\mathbf{R}^n$ is called a Crofton 1-density if there exists a signed measure $\rho(\zeta)d\zeta$ on $\mathbf{H}$ such that for every smooth curve γ in $\mathbf{R}^n$*

$$\int_\gamma \varphi = \int_\mathbf{H} Crof_\gamma(\zeta)\rho(\zeta)d\zeta \ .$$

The measure ρ is called the dual measure of the Crofton density φ.

Theorem 6 (GS) *A smooth even 1-density φ is a Crofton density if and only if its extremals are straight lines.*

Theorem 3 follows from the preceding theorem and the following characterization of those Crofton densities which are Finsler:

Theorem 7 *A Crofton density is Finsler if and only if its dual measure is quasi-positive.*

4.1 Sketch of the Proof of Theorem 6

In what follows it will be advantageous, from a notational point of view, to work with the set of oriented hyperplanes $\tilde{\mathbf{H}}$. The reason is that $\tilde{\mathbf{H}}$ is an orientable manifold and we may work with volume forms instead of smooth measures. Moreover, using the natural identification of $\tilde{\mathbf{H}}$ with $\mathbf{R} \times S^{n-1}$, we may write any differential n-form η as $\nu(p,\xi)dp\wedge\Omega$, where ν is a function on $\mathbf{R} \times S^{n-1}$. Differential n-forms which are even, or odd, with respect to the involution $(p,\xi) \mapsto (-p,-\xi)$ induce smooth signed measures on $\mathbf{H}$. We will only consider such forms. An n-form η on $\tilde{\mathbf{H}}$ will be said to be *quasi-positive* if it induces a quasi-positive measure on $\mathbf{H}$.

Our point of departure is an interesting and classical integral representation of smooth homogeneous functions on $\mathbf{R}^n \setminus \{0\}$.

Theorem 8 *Let $\varphi : \mathbf{R}^n \setminus \{0\} \to \mathbf{R}$ be a smooth function homogeneous of degree one. There exists a unique smooth even function μ on S^{n-1} such that*

$$\varphi(v) = \int_{\xi \in S^{n-1}} |\xi \cdot v| \mu(\xi)\Omega .$$

From this theorem we obtain an integral representation of even 1-densities.

Proposition 1 *If $\varphi(x,v)$ is an even 1-density on $\mathbf{R}^n$, then there exists a unique smooth function $\mu(x,\xi)$ on $\mathbf{R}^n \times S^{n-1}$ which is even in ξ, and such that*

$$\varphi(x,v) = \int_{\xi \in S^{n-1}} |\xi \cdot v| \mu(x,\xi)\Omega .$$

In the context of the integral representation of even 1-densitiesm, the representation of Crofton densities is as follows:

Proposition 2 (GS) *A smooth even 1-density φ is a Crofton density if and only if there exists a smooth even function $\nu(p,\xi)$ on $\mathbf{R} \times S^{n-1}$ such that*

$$\varphi(x,v) := \int_{\xi \in S^{n-1}} |\xi \cdot v| \nu(\xi \cdot x, \xi)\Omega ,$$

where Ω is the standard volume form on the unit sphere in $\mathbf{R}^n$.

Finally, the proof of theorem 6 follows from Hamel's theorem and the following technical result:

Theorem 9 *The 1-density $\varphi(x,v) = \int_{\xi \in S^{n-1}} |\xi \cdot v| \mu(x,\xi)\Omega$ satisfies the equations*

$$\frac{\partial^2 \varphi}{\partial x_k \partial v_j} = \frac{\partial^2 \varphi}{\partial x_j \partial v_k}, \quad k,j = 1,...,n, \ v \neq 0$$

if and only if there exists a smooth function

$$\nu : R \times S^{n-1} \longrightarrow R$$

such that $\mu(x,\xi) = \nu(\xi \cdot x, \xi)$.

4.2 Proof of theorem 7

As in the proof of Theorem 6, we start by studying the representation of even 1-densities which are invariant under translations. In this case a Finsler metric is just a Banach norm with certain smoothness requirements. The following is the precise definition of the norms we will work with:

Definition 6 *A smooth function* $\varphi : \mathbf{R}^n \setminus \{0\} \to \mathbf{R}$ *homogeneous of degree one is said to be a* norm function *if*

- $\varphi(v) > 0$, *for any* $v \neq 0$

- *The matrix of second derivatives of the squared function* φ^2

$$\left(\frac{\partial^2 \varphi^2}{\partial v_i \partial v_j} \right)_{i,j}$$

is positive definite.

The strong convexity conditions required by the the positive-definiteness of the Hessian assure that not only the norm, but also the dual norm is smooth.

The problem of characterizing the functions $\mu : S^{n-1} \to \mathbf{R}$ such that

$$\varphi(v) := \int_{\xi \in S^{n-1}} |\xi \cdot v| \mu(\xi) \Omega \ .$$

is a norm function is known as Funk's problem. The notion of quasi-positivity allows us to give the following solution :

Theorem 10 *Let* $\mu : S^{n-1} \to R$ *be a smooth even function. The function*

$$\varphi := \int_{\xi \in S^{n-1}} |\xi \cdot v| \mu(\xi) \Omega$$

is a norm function if and only if $\mu(\xi)dp \wedge \Omega$ *is a quasi-positive top order form on* $\tilde{\mathbf{H}}$.

This theorem follows simply from the remark that a homogeneous function of degree one is a norm function if and only if the restriction to any 2-dimensional vector subspace is a norm function, and the following result:

Proposition 3 *Let* $\phi : \mathbf{R}^2 \to \mathbf{R}$ *be a smooth function homogeneous of degree one, and let* $\mu : S^1 \to \mathbf{R}$ *be a smooth even function such that*

$$\phi(v_1, v_2) = \int_0^{2\pi} |cos(\theta)v_1 + sin(\theta)v_2| \mu(\theta) d\theta \ .$$

The function ϕ *is a norm function if and only if* $\mu > 0$.

The characterization of those even 1-densities which are Finsler is now straightforward.

Proposition 4 *An even 1-density $\varphi(x,v)$ is a Finsler metric if and only if*

$$\varphi(x,v) = \int_{\xi \in S^{n-1}} |\xi \cdot v| \mu(x,\xi) \Omega \ ,$$

where for each point $x \in \mathbf{R}^n$ the form $\mu(x,\xi)dp \wedge \Omega$ is a quasi-positive n-form on $\tilde{\mathbf{H}}$.

Theorem 7 follows directly form this result and the characterization of Crofton densities given by proposition 2.

5. Integral Geometric Construction
of Admissible Symplectic Forms

In this section we show how to construct admissible symplectic forms on $\tilde{L}(\mathbf{R}^n)$ from quasi-positive top order forms on $\tilde{\mathbf{H}}$.

Let $\tilde{\mathbf{H}}$ be the space of oriented hyperplanes on $\mathbf{R}^n$ and let η be a top order form on $\tilde{\mathbf{H}}$. The form η induces a 2-form $\mathcal{R}(\eta)$ on the space of oriented lines by the following procedure:

Let $A = \{(l,h) \in \tilde{L}(\mathbf{R}^n) \times \tilde{\mathbf{H}} : l \subset h\}$ and let

$$\tilde{L}(\mathbf{R}^n) \xleftarrow{\ \pi_1\ } A \xrightarrow{\ \pi_2\ } \tilde{\mathbf{H}}$$

be the natural projections.

If $l \in \tilde{L}(\mathbf{R}^n)$ is an oriented line and $w_1, w_2 \in T_l \tilde{L}(\mathbf{R}^n)$ are two tangent vectors over l, we choose at each hyperplane $h \supset l$ two tangent vectors $\hat{w}_1, \hat{w}_2 \in T_h \tilde{\mathbf{H}}$ such that w_i and $\hat{w}_i$ have a common preimage in $T_{(l,h)}A$.

The contracted form $\iota_{\hat{w}_1 \wedge \hat{w}_2} \eta_h$ vdoes not depend on the choice of $\hat{w}_1, \hat{w}_2$. We define

$$\mathcal{R}(\eta)_l(w_1, w_2) := \int_{h \supset l} \iota_{\hat{w}_1 \wedge \hat{w}_2} \eta_h \ .$$

Roughly speaking, the 2-form $\mathcal{R}(\eta)$ at the line l is the integral of the form η over the codimension 2 submanifold $\{h \in \tilde{\mathbf{H}} : h \supset l\}$.

It is easy to see that the form $\mathcal{R}(\eta)$ is necessarily closed.

Theorem 11 *A symplectic form ω in $\tilde{L}(\mathbf{R}^n)$ is admissible if and only if there exists a quasi-positive n-form η on $\tilde{\mathbf{H}}$ such that $\omega = \mathcal{R}(\eta)$.*

We may now use this characterization of admissible symplectic forms to prove theorem 4. Indeed, it is easy to see that if η is a quasi-positive n-form on $\tilde{\mathbf{H}}$ which induces a quasi-positive measure σ on $\mathbf{H}$ and an admissible symplectic form $\omega := \mathcal{R}(\eta)$ on $\tilde{L}(\mathbf{R}^n)$, then the pushforward measure σ_P and the measure $|i_P{}^*\omega|$ coincide on each 2-dimensional affine subspace $P \subset \mathbf{R}^n$. It then follows that the σ-metric defined by σ coincides with the ω-metric defined by $\omega = \mathcal{R}(\eta)$.

6. Crofton densities and Desarguesian Finsler metrics as even differentials of functions and Hilbert transforms of closed 1-forms

In this section we demonstrate that there is another way to construct 1-densities whose extremals are straight lines.

We recall that an *even 1-density* is a function $\varphi(x, v)$ of a point $x \in \mathbf{R}^n$ and a tangent vector v at the point x such that $\varphi(x, \lambda v) = |\lambda| \varphi(x, v)$ for every $\lambda \in \mathbf{R}$.

An *odd 1-density* is a function $\theta(x, v)$ of a point $x \in \mathbf{R}^n$ and a tangent vector v such that $\theta(x, \lambda v) = \lambda \theta(x, v)$ for every $\lambda \in \mathbf{R}$. Differential 1-forms give a specific example of odd 1-densities, they are linear in v.

Definition 7 *The 1-density* $(H\varphi)(x, v)$

$$(H\varphi)(x, v) = P.V. \frac{1}{\pi} \int\limits_{-\infty}^{+\infty} \frac{\varphi(x - vt, v)}{t} dt =$$

$$= \frac{1}{\pi} \int\limits_{0}^{+\infty} \frac{\varphi(x + vt, v) - \varphi(x - vt, v)}{t} dt$$

is called the Hilbert transform *of a 1-density* φ.

Here we suppose that φ satisfy growth conditions for example it is rapidly decreasing. This means that $\varphi(x - vt, v)$ is a rapidly decreasing function of t for any fixed v and x). So its Hilbert transform exists.

Theorem 12 *The Hilbert transform of an even 1-density is an odd 1-density. The Hilbert transform of an odd 1-density is an even 1-density. The Hilbert transform of a closed 1-form with rapidly decreasing coefficients is the Crofton 1-density. The Hilbert transform of a rapidly decreasing Crofton 1-density* φ *is a closed 1-form.*

Let f be a function in $\mathbf{R}^n$. Its differential df depends on a point x from $\mathbf{R}^n$ and a tangent vector $dx = v$ from $T_x\mathbf{R}^n$ can be written as

$$(df)(x, v) = \int\limits_{-\infty}^{+\infty} \delta'(t) f(x - vt) dt.$$

Definition 8 *The* nonlocal (or even) differential *of a function f is*

$$(d^o f)(x,v) = \int_{-\infty}^{+\infty} \frac{1}{t^2} f(x-vt) dt = \int_0^{+\infty} \frac{f(x+vt)+f(x-vt)-2f(x)}{t^2} dt.$$

We suppose that f satisfies some growth conditions so that this integral converges at infinity.

Let $f(x)$ be a rapidly decreasing function in the Schwartz space $S(\mathbf{R}^n)$. Suppose that f is the Radon transform of a function $F(\alpha)$:
$$f(x_1,\ldots,x_n) =$$

$$= \int_{-\infty}^{+\infty} F\big(\alpha_1,\ldots,\alpha_{n-1},\ x_n - (\alpha_1 x_1 + \cdots + \alpha_{n-1}x_{n-1})\big) d\alpha_1 \ldots d\alpha_{n-1}.$$

Let
$$\varphi(x,v) = (d^o f)(x,v) = \int_{-\infty}^{+\infty} \frac{1}{t^2}\, f(x-vt) dt$$

be an even differential of the function F. Then

(1) $\varphi(x,v)$ is a Crofton density in $\mathbf{R}^n$.

(2) Let us denote by $D_{\alpha_n}^o F(\alpha_1,\ldots,\alpha_n)$, an "even partial derivative" of F

$$D_{\alpha_n}^o F(\alpha_1,\ldots,\alpha_n) = \int \frac{1}{t^2} F(\alpha_1,\ldots,\alpha_n - t) dt.$$

Then the $D^o F$ is the dual function for φ: $\mu(\alpha_1,\ldots,\alpha_n) = D_{\alpha_n}^o F(\alpha_1,\ldots,\alpha_n)$.

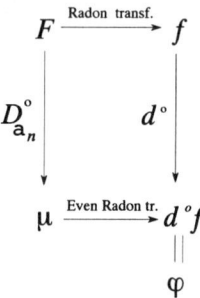

Theorem 13 *Suppose f is a rapidly decreasing function, $\omega = df$ is its differential and $\varphi = \dfrac{1}{\pi} d^o f$ is its nonlocal differential (which is the Crofton 1-density). Then $(H\varphi)(x,v) = \omega(x,v)$ and $(H\omega)(x,v) = \varphi(x,v)$. So the Hilbert transform maps even differentials into odd ones and vice versa.*

We can summarize properties of Crofton 1-densities on the following diagram:

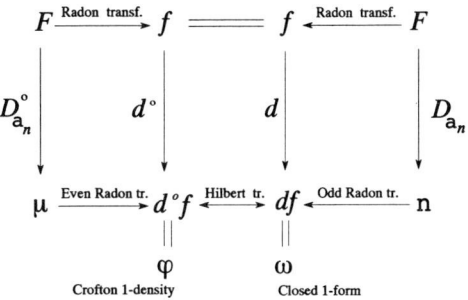

Acknowledgments. The authors are happy to acknowledge helpful conversations with A.B. Goncharov, E. Grinberg, D. Kazhdan, and S. Gelfand. The authors also thank C. Duran for his help in making this paper more readable.

References

[A1]   Álvarez, J.C. *The Symplectic Geometry of Spaces of Geodesics* . Ph.D. Thesis, Rutgers University, 1995.

[A2]   Álvarez, J.C. *A Symplectic Construction of Desarguesian Metrics on* $\mathbf{R}^n$. Preprint.

[AGS] Àlvarez, J.C., Gelfand, I.M., Smirnov, M. *Crofton Densities, Symplectic Geometry, and Hilbert's Fourth Problem*. In preparation.

[B]   Busemann, H. *Geometries in which Planes Minimize Area*. Annali Mat. Pure Appl. (IV) 55, 1961.

[G]   Gelfand, I.M., Gindikin, S.G., Graev, M.I. *Integral Geometry in Affine and Projective Spaces*. Journal of Soviet Mathematics, 1982.

[GS]   Gelfand, I.M., Smirnov, M. *Lagrangians satisfying Crofton Formulas, Radon Transforms and Nonlocal Differentials*. Advances in Mathematics 109, 1994.

[H]   Hamel, H. *Über die Geometrieen, in denen die Geraden Kúrzesten sind*. Math. Ann. 57, 1903.

[P]   Pogorelov, A.V. *Hilbert's Fourth Problem*. Translated by R. Silverman. Scripta Series in Mathematics, Winston and Sons, 1979.

[Sz]   Szabó, Z. *Hilbert's Fourth Problem. I.* Advances in Mathematics
        59, 1986.

J. C. Alvarez   I. M. Gelfand   M. Smirnov
Dept. of Math.   Dept. of Math.   Dept. of Math
Rutgers University   Rutgers University   Columbia University
New Brunswick, NJ 08903 New Brunswick, NJ 08903 New York, NY 10027

Received October 1995

Projective convex curves

S. Anisov

Abstract

We deal with convex curves and surfaces in real projective space. Questions concerning the affinity of convex curves, the flattening points and the loss of convexity are regarded. Finally, we prove that there are no convex surfaces except curves and hypersurfaces.

1. Main Definitions

Definition. (See [1]). A smooth embedding $\Gamma : S^1 \to \mathbb{R}P^n$ is said to be a *convex curve* if every hyperplane intersects $\Gamma(\varphi)$ no more than n times according to multiplicities. The parameter φ on S^1 is defined to be mod 2π.

Since n-multiple crossings are allowed (but not $(n+1)$-multiple ones), we must suppose that the map Γ is smooth of class C^n.

Suppose also that $0 \neq \frac{d}{d\varphi}\Gamma(\varphi)$ (i.e., $D_\varphi \Gamma$). Otherwise a transversal intersection can incidentally be regarded as a multiple one.

Example. $\Gamma(\varphi) = (1 : t : t^2 : \ldots : t^{n+1})$, where $t = \tan\frac{\varphi}{2}$, is a smooth closed convex curve in $\mathbb{R}P^n$; $\Gamma(\varphi) = (\sin\psi : \cos\psi : \sin 3\psi : \cos 3\psi : \ldots : \sin(2n+1)\psi : \cos(2n+1)\psi)$, where $\psi = \frac{\varphi}{2}$, is a smooth closed convex curve in $\mathbb{R}P^{2n+1}$.

Let $A_1, \ldots, A_k$ be different points of $\Gamma(\varphi)$ and $m_1, \ldots, m_k$ integers such that $m_1 + \ldots + m_k = n$. Then there exists a hyperplane that crosses Γ at A_i, $i = 1, \ldots, k$ with multiplicities m_i. If the curve Γ is convex, then such a hyperplane is unique and, of course, has no other intersections with Γ.

Fix an affine chart such that $\Gamma(\varphi_0) = 0$ and consider vectors $v_i = (\Gamma(\varphi))^{(n)}|_{\varphi=\varphi_0}$, $i = 1, \ldots, n$. If they are linearly dependent, then the point $\Gamma(\varphi_0)$ (or, for the sake of simplicity, the point φ_0) is called a *flattening point* of the curve Γ. Otherwise, the point φ_0 is called *regular*.

Definition. An *osculating hyperplane* $\Pi(\varphi_0)$ at the point φ_0 is any hyperplane having intersection of order no less than n with the curve Γ at $\Gamma(\varphi_0)$.

1991 *Mathematics Subject Classification.* 26B25, 52A01, 52A20.

Key words and phrases. Disconjugate differential equation, Real projective space, Smooth curve.

The author was supported in part by ISF Grant MHN 000 and the Grant *Pro-Mathematica* of the French Mathematical Society

Remark. 1. Osculating hyperplanes are well-defined.

2. The osculating hyperplane $\Pi(\varphi_0)$ at a regular point φ_0 is the span of the vectors $v_1, \ldots, v_{n-1}$. The intersection's order is equal to n in this case.

3. The osculating hyperplane at an isolated flattening point is the limit of the osculating hyperplanes at nearby points

Let $\mu(\varphi_0)$ be the multiplicity of the intersection of $\Pi(\varphi_0)$ and $\Gamma(\varphi)$ at the point φ_0. It is clear that $\mu(\varphi) = n$ at any regular point φ. If $\mu(\varphi) = n + k$, then k is said to be a flattening *order* of Γ at the point φ. Evidently, a convex curve has no flattening points and it crosses $\Pi(\varphi)$ at $\Gamma(\varphi)$ alone.

Question (V.I. Arnold, [1]). Is the converse true?

Remark. Another facet of a convex curve is the Chebysheff system of functions (cf. [1], [2]), considered up to transformations $f_i(\varphi) \mapsto F(\varphi) f_i(\varphi)$, where $F(\varphi) > 0$, and linear transformations $f_i \mapsto a_i^j f_j$, $i, j = 1, 2, \ldots, n+1$, $\det(a_i^j) \neq 0$. For example, the affinity of a convex curve (cf. below) is equivalent to the existence of a representative $(1 : f_1 : \ldots : f_n)$ in the corresponding class of Chebysheff systems. Only the segment I and the circle S^1 can be equipped with a Chebysheff system, cf. [2]. The case of a segment was explored by M.Z.Shapiro. See his Ph. D. thesis [3], where convex curves appeared as fundamental solution systems of disconjugate differential equations

2. How Loss of Convexity Occurs

The answer is furnished by the following

Theorem 1. *Let $\Gamma_t(\varphi)$ be continuous (with respect to t) family of C^{n+1}-smooth closed curves in $\mathbb{R}P^n$. Suppose all the curves Γ_t are convex whenever $t < 0$, but the curve $\Gamma(\varphi) = \Gamma_0(\varphi)$ is not convex. Then there exists a flattening point of order 2 or greater on $\Gamma(\varphi)$. Moreover, every hyperplane that crosses $\Gamma(\varphi)$ more than n times (according to multiplicities) is an osculating hyperplane at a certain flattening point of order no less than 2.*

Proof. The proof is by induction over the dimension of $\mathbb{R}P^n$. For $n = 2$ this fact is well known, but let us start from $n = 1$. For $n = 1$ the statement is: "Suppose C^2-smooth maps $\Gamma_t(\varphi) : S^1 \to \mathbb{R}P^1$ are diffeomorphisms whenever $t < 0$ and $\Gamma_0(\varphi)$ is not a diffeomorphism. Then there exists a point φ_0 such that $\Gamma'(\varphi_0) = \Gamma''(\varphi_0) = 0$, but $\varphi_1 \neq \varphi_2$ implies $\Gamma(\varphi_1) \neq \Gamma(\varphi_2)$. Moreover, $\Gamma'(\varphi) = 0$ implies $\Gamma''(\varphi) = 0$." This fact is evident and its proof is omitted.

We now need the following

Construction. Put $A = \Gamma(\varphi_0)$ or without loss of generality $A = \Gamma(0)$. All the straight lines incident to A form a projective space $\mathbb{R}P^{n-1}$. So, the central projection $\pi_A \colon \mathbb{R}P^n \to \mathbb{R}P^{n-1}$ with the center at A is defined. Denote $\gamma(\varphi) = \pi_A(\Gamma(\varphi))$ if $\varphi \neq 0 \bmod 2\pi$ and let $\gamma(0)$ be the projection on $\mathbb{R}P^{n-1}$ of the tangent line of $\Gamma(\varphi)$ at the point A.

It is clear that $\gamma(\varphi)$ is a C^n-smooth closed curve in $\mathbb{R}P^{n-1}$; if $\varphi \neq 0$, the curve $\gamma(\varphi)$ is C^{n+1}-smooth. Furthermore, let β be a hyperplane in $\mathbb{R}P^{n-1}$ and $B = \pi_A^{-1}(\beta)$ a hyperplane in $\mathbb{R}P^n$ passing through A. By $k(\varphi)$ we denote the multiplicity of the intersection of β and γ at the point $\gamma(\varphi)$ and by $K(\varphi)$ we denote the multiplicity of the intersection of B and Γ at $\Gamma(\varphi)$. Then $k(\varphi) = K(\varphi)$ for all $\varphi \neq 0 \bmod 2\pi$ and $K(0) = k(0) + 1$. We obtain immediately

Lemma. *If the curve $\Gamma(\varphi)$ is convex, then the curve $\gamma(\varphi)$ is also convex.*

Suppose the theorem is proved for $\mathbb{R}P^1$, $\mathbb{R}P^2$, ... , $\mathbb{R}P^{n-1}$; let us prove it for $\mathbb{R}P^n$. Assume the converse. Then the curve $\Gamma_0(\varphi)$ is not convex and in $\mathbb{R}P^n$ there is a hyperplane B that intersects $\Gamma(\varphi)$ more than n times (according to multiplicities), but there are at least two different points of intersection. Recall also that all the curves $\Gamma_t(\varphi)$, $t < 0$, are convex. It follows from here that there is a point A of multiple intersections; we can assume that $A = \Gamma(0)$. In fact, if all the intersections are transversal, then all the curves $\Gamma_t(\varphi)$ with small $|t|$ are not convex because they intersect B more than n times.

Now consider the central projection $\pi_A \colon \mathbb{R}P^n \to \mathbb{R}P^{n-1}$. We can suppose $\Gamma_t(0) \equiv \Gamma_0(0) = A$. Put $\gamma_t(\varphi) = \pi_A(\Gamma_t(\varphi))$. We know that all the curves $\gamma_t(\varphi)$, $t < 0$, are convex, but the curve $\gamma_0(\varphi)$ is not convex and the hyperplane $\beta = \pi_A(B)$ (it is really a hyperplane because $A \in B$) intersects γ more than $n-1$ times (according to multiplicities). Finally, there are at least two different points of intersection: $\gamma(0)$ and the projections π_A of the other intersections of B and Γ. This contradicts the induction hypothesis. The proof is finished.

3. Affinity of convex curves in $\mathbb{R}P^{2n}$

Definition. A subset M of a projective space $\mathbb{P}$ is called *affine* if there exists a hyperplane Π that has no intersections with M. So, if the hyperplane Π is infinite, then the whole set M lies in the finite part of $\mathbb{P}$.

Theorem 2. *Every convex curve $\Gamma(\varphi)$ in an even-dimensional projective space $\mathbb{R}P^{2n}$ is affine.*

Proof. First recall that $\Gamma \in C^{2n}(S^1, \mathbb{R}P^{2n})$. Let A be an arbitrary point on Γ, to be definite, $A = \Gamma(0)$. Put $\Pi = \Pi(0)$ (see §1). By the previous construction, Π crosses Γ at the point A only and the order of this crossing equals $2n$. Fix an affine chart such that the equation of Π is $x_{2n} = 0$ and A is the origin. Consider the C^{2n}-smooth function $x_{2n}(\Gamma(\varphi))$. It has a zero of order $2n$ (but not $2n+1$) at the point $\varphi = 0$. So,

$$\frac{d^{2n}}{d\varphi^{2n}} x_{2n}(\Gamma(\varphi))|_{\varphi=0} \neq 0$$

and near the origin A the curve Γ lies at the one side of Π. We can suppose $x_{2n}(\Gamma(\varphi)) \geq 0$ whenever $\varphi \in (-\delta, \delta)$. Draw the hyperplane $\Pi_\varepsilon \colon x_{2n} = -\varepsilon$, where $\varepsilon > 0$. If ε is small enough, then Γ and Π_ε do not intersect: there are no intersections at the points $\Gamma(\varphi)$, $\varphi \in (-\delta, \delta)$ by the construction and at other points of Γ as well by upper semi-continuity of the crossing's order. The theorem is proved.

For a given convex curve $\Gamma \subset \mathbb{R}P^{2n}$ denote by $M(\Gamma) \subset (\mathbb{R}P^{2n})^*$ the set of hyperplanes having no intersections with Γ.

Theorem 3. *The set $M(\Gamma)$ is contractible.*

Proof. Fix an affine chart containing the whole curve Γ. The existence of such a chart is provided by Theorem 2. Suppose the origin lies on Γ. Then every hyperplane Π of the set M has the equation

$$\sum a_i x_i = d , \quad where \quad \sum a_i^2 = 1 \quad and \quad d > 0$$

($d \neq 0$ because the origin does not belong to Π since it lies on Γ). Let $\Pi(t)$ be the hyperplane defined by the equation

$$\sum a_i x_i = d \left(1 + \tan \frac{\pi t}{2} \right).$$

This formula prescribes a contraction of the set M to its single point (the infinite hyperplane). In fact, the substitution $t = 1$ yields the equation $\sum a_i x_i = \infty$ for the hyperplane $\Pi(1)$ and the described contraction is continuous on $M \times [0, 1]$. So, the theorem is proved.

4. Odd-dimensional case

Let us consider a convex curve Γ in $\mathbb{R}P^{2n+1}$ now. Choose $2n+1$ different points on Γ and draw a hyperplane Π through chosen points. Γ crosses Π $2n + 1 \equiv 1 \mod 2$ times. So, $[\Gamma]$ and $[0]$ are two different elements of $\pi_1(\mathbb{R}P^{2n+1})$. Since affine curves are homotopic to zero, any convex curve in $\mathbb{R}P^{2n+1}$ cannot be affine.

Theorem 4. *There exist hyperplanes crossing the curve Γ only once. Such a hyperplane can pass through an arbitrary point of Γ.*

Proof. The case $n = 0$ is trivial. Assume that $n > 0$ and A is arbitrary point on Γ. Put $\gamma = \pi_A(\Gamma) \subset \mathbb{R}P^{2n}$. The curve γ is convex by Lemma from §2. It follows from Theorem 2 that there exists a hyperplane $\beta \subset \mathbb{R}P^{2n}$ having no intersections with γ. Then the hyperplane $B = \pi_A^{-1}(\beta) \subset \mathbb{R}P^{2n+1}$ crosses Γ transversally at point A and does not cross Γ at other points. Q. E. D.

For a convex curve $\Gamma \subset \mathbb{R}P^{2n+1}$ we denote by $M(\Gamma) \subset (\mathbb{R}P^{2n+1})^*$ the set of hyperplanes that cross the curve Γ only once (transversally, of course).

Theorem 5. *The set $M(\Gamma)$ is contractible to S^1; this circle is not homotopic to 0 in $\mathbb{P} = (\mathbb{R}P^{2n+1})^*$.*

Proof. The first idea is to note that the image of the natural map $\Phi : M \to S^1$ defined by the rule $\Phi(\Pi) = \Gamma^{-1}(\Pi \cap \Gamma(\varphi))$ is S^1 and the preimage of each point φ is contractible by Theorem 3. But this observation leads us to weak homotopy equivalence only.

To prove the proclaimed theorem, let us fix a certain Riemannian metric on $\mathbb{P}$. Put $M_\varphi = \Phi^{-1}(\varphi)$. Evidently $M_\varphi \subset \mathbb{R}P^{2n} \subset \mathbb{P}$.

Lemma 1. *M_φ is convex in any affine chart that contains the whole set M_φ.*

Lemma 2. *There exists a continuous "section" $s : S^1 \to M$ such that $\Phi \circ s = id_{S^1}$.*

Both proofs are straightforward. Note also that Lemma 1 gives another way to prove Theorem 3.

Let $\rho(x)$ be the distance between x and $s(\Phi(x))$ along the segment $[x, s(\Phi(x))] = I_x$ and let $x_t \in I_x$ be the point of I_x such that $\rho(x_t) = (1-t)\rho(x)$. Put $F(x,t) = x_t$. Then $F(x,t)$ defines the required contraction. The second statement of the theorem follows from the fact that for any point $A \in \Gamma$ the intersection index of the circle $s(\Phi)$ and the dual hyperplane $A^* \in \mathbb{P}$ is equal to $1 \bmod 2$.

5. Generalization: convex surfaces

Definition. A smooth k-dimensional surface Γ^k embedded in $\mathbb{R}P^n$ is called *convex* if every plane of codimension k intersects Γ no more than $n - k + 1$ times.

Example. Every point is 0-dimensional convex surface. Every domain in $\mathbb{R}P^n$ is n-dimensional convex surface. Convex curves (cf. above) are 1-dimensional convex surfaces. Spheres in $\mathbb{R}^n \subset \mathbb{R}P^n$ are $(n-1)$-dimensional convex surfaces.

For hypersurfaces our definition is the same as that of the classical one. So it can be proved that every convex hypersurface in $\mathbb{R}P^n$ is affine and is homeomorphic to S^{n-1} (cf. [4], for example).

Theorem 6. *There are no convex surfaces Γ^k in $\mathbb{R}P^n$ if $k = 2, 3, \ldots, n-2$.*

Definition. A smooth k-dimensional ($k > 0$) surface $\Gamma^k \subset \mathbb{R}P^n$ is called *almost convex* if there exists a unique plane $\Pi^{k-1} \subset \Gamma^k$ and every plane B^{n-k} intersects Γ at most $n-k+1$ times whenever $\dim(B^{n-k} \cap \Pi^{k-1}) < 1$.

Lemma. *Almost convex surfaces of codimension 1 do not exist.*

Proof. The case $k = 1$ is evident: any point $A \in \Gamma$ is a plane of dimension $k - 1$. So, suppose $k > 1$. Let A be an arbitrary point in $\Gamma^k \setminus \Pi^{k-1}$. Since $k > 1$, the tangent plane $T_A\Gamma$ and Π^{k-1} have non-empty intersection. If $A' \in T_A\Gamma \cap \Pi^{k-1}$, then the straight line AA' intersects Γ at least three times: twice at A and once more at A'. So the surface Γ^k is not an almost convex surface.

Proof. (of the theorem) Consider the central projection $\pi_A : \mathbb{R}P^n \to \mathbb{R}P^{n-1}$, where $A \in \Gamma$. Put $\gamma = \pi_A(\Gamma)$. It is easy to prove that $\gamma \subset \mathbb{R}P^{n-1}$ is an almost convex surface of dimension k and codimension $n - k - 1$. It contains the plane $\Pi^{k-1} = \pi_A(A)$. Next, consider the central projection $\pi_a : \mathbb{R}P^{n-1} \to \mathbb{R}P^{n-2}$, where $a \in \Pi^{k-1} \subset \gamma$. Then $\pi_a(\gamma) \subset \mathbb{R}P^{n-2}$ is an almost convex surface of dimension k and codimension $n - k - 2$, and so forth. Finally, we get an almost convex hypersurface in $\mathbb{R}P^{k+1}$. This contradicts the previous lemma. The theorem is proved.

Remark. We have obtained in the proof of Theorem 6 that there are almost no convex surfaces of any codimension.

The author is grateful to V. I. Arnold, V. A. Vassiliev, A. G. Khovanskii and M. Z. Shapiro for useful discussions.

References

[1] V. I. Arnold, On the Number of Flattening Points on Space Curve, preprint of the Institute Mittag Leffler, Djurisholm, Sweden, 1994/5, page 13

[2] M. G. Krein and A. A. Nudelman, *Markov's Momenta Problems and Extremal Problems*, Nauka, Moscow, 1973 (Russian), page 373

[3] M. Z. Shapiro, Disconjugate Differential Equations, Ph.D. Thesis, Moscow State University, Moscow, 1991, in Russian, pg. 121

[4] V. I. Arnold, Ramified Covering $\mathbb{C}P^2 \to S^4$, Hyperbolicity and Projective Topology, *Siberian Math. Journal* **29**:5 (1988), 3446–47 (in Russian)

Chair of Differential Geometry and its Applications
Department of Mathematics and Mechanics
Moscow State University
119899 Moscow, Russia
email: anisov@sch57.mcn.msk.su

Received March 1995

Topological classification of real trigonometric polynomials and cyclic serpents polyhedron

V. Arnold

The goal of this paper is the study of the manifold of the real trigonometric polynomials of degree n having the maximal possible number $(2n)$ of real critical points (M-polynomials).

We shall find the number of the connected components of the space of the generic M-polynomials, having $2n$ different critical values. We construct a polyhedral model of the manifold of M-polynomials and a real algebraic diffeomorphism sending the manifold of M-polynomials onto the interior of a convex polyhedral cone over the product of two simplices of dimension $n-1$ and of a line. Those polynomials, which are not generic, are sent onto some diagonal hyperplanes of this polyhedral cone. This diffeomorphism can be continued as a homeomorphism up to the boundary of the cone (and defines the diffeomorphisms on the interior parts of the boundary faces of all dimensions).

1. The space of cyclic serpents

Definition. A *cyclic serpent* of order $2n$ is a sequence of $2n$ real numbers, verifying the inequalities

$$c_0 \leq c_1 \geq c_2 \leq .... \geq c_{2n} = c_0.$$

The serpent is *proper*, if all the inequalities are strict, and *generic*, if all the $2n$ numbers are different.

The space of the cyclic serpents of order $2n$ is a polyhedral convex cone in $\mathbb{R}^{2n}$. The hyperplane $c_1 + ... + c_{2n} = 0$ intersects it along the strictly convex *cone of traceless cyclic serpents*. The cone of all cyclic serpents is the product of the cone of the traceless cyclic serpents with $\mathbb{R}$.

Definition. The intersection of the cone of the traceless cyclic serpents of order $2n$ with the hyperplane $\mathbb{R}^{2n-2}$, where $\Sigma(-1)^{i+1}c_i = a > 0$, is called the *cyclic serpent polyhedron P_n*.

It is a compact convex polyhedron in $\mathbb{R}^{2n-2}$.

Example. P_2 is a square in the plane. The interior of a square is diffeomorphic to the domain, bounded by an astroid (which is the caustic of an ellipse). The diagonals of the square are sent by a natural diffeomorphism to the Maxwell stratum (consisting of two segments joining the opposite cusps of the astroid).

The diffeomorphism of the serpent polyhedron onto the space of trigonometric polynomials that we shall construct below is a higher-dimensional generalization of this combinatorial description of the astroid caustic and of its Maxwell stratum.

Theorem 1. *The cyclic serpent polyhedron P_n is a convex polyhedron in $\mathbb{R}^{2n-2}$, having $2n$ faces and n^2 vertices : the direct product of two simplices in $\mathbb{R}^{n-1}$.*

Proof. Let $c_{2k+1} - c_{2k} = a_k$, $c_{2k+1} - c_{2k+2} = b_k$ $(k = 0, \ldots, n-1)$. The equation of the hyperplane $\mathbb{R}^{2n-2}$ implies the conditions $\Sigma a_k = a$, $\Sigma b_k = a$. The cyclic serpent definition implies the inequalities $a_k \geq 0$, $b_k \geq 0$. Any solution of these equations and inequalities defines exactly one traceless cyclic serpent belonging to P_n. Hence P_n is the product of the simplices $\Sigma a_k = a$, $a_k \geq 0$ and $\Sigma b_k = a$, $b_k \geq 0$ in the hyperplanes of dimension $n-1$ defined by the equations $\Sigma a_k = a$, $\Sigma b_k = a$ in the n-dimensional a-space and b-space. The $2n$ faces are defined by the equations $a_k = 0$, $b_k = 0$.

Now we describe the diagonal hyperplanes of the product of two simplices, defined by the conditions $c_i = c_j$ in the original coordinate system in the serpent space.

Write the cyclical word $A_1 B_1 A_2 B_2 ... A_n B_n (A_1 B_1 ...)$. Decompose it into two connected arcs. Each decomposition defines a diagonal hyperplane in the product of two simplices with vertices $(A_1 A_2 ... A_n)$ and $(B_1 B_2 ... B_n)$.

Definition. The *principal diagonal* of P_n is the hyperplane containing the vertices (A_i, B_j) where A_i and B_j belong to the same arc. (Here A_k is the vertice of the simplex in $\mathbb{R}^{n-1}$ described above, defined by $a_k = a$, B_k - by $b_k = a$).

The number of such diagonals is equal to $n(2n-1)$ and they are defined by the equations $c_i = c_j$ in the original coordinates.

Theorem 2. *The principal diagonal hyperplanes subdivide the polyhedron P_n into the simplicial domains. The number of these domains is equal to nT_n, where the tangens numbers $T_n = 1, 2, 16, 272, ...(n = 1, 2, 3, 4, ...)$ are defined by the exponential generating function*

$$\sum_{n=1}^{\infty} T_n \, x^{2n-1} \Big/ (2n-1)! \; = \; \tan x.$$

Example. For $n = 2$ we get $2T_2 = 4$. Two diagonals cut a square into 4 triangles.

Proof. It suffices to study the components of the space of all cyclic serpents (the trace zero and scaling restrictions being irrelevant). The principal diagonal hyperplanes are defined by the equations $c_i = c_j$. Their complement is the space of the generic serpents.

Construct a mapping from the space of generic cyclic traceless serpents of order $2n$ onto the space of the generic ordinary serpents of order $2n - 1$. Delete the smallest of the numbers c_i (say, c_{2k}). Rename the remaining numbers, denoting them by

$$b_i = c_{2k+i(\mathrm{mod}\,2n)}, \quad i = 1, \ldots, 2n - 1.$$

The sequence $b_1 > b_2 < b_3 > \ldots < b_{2n-1}$ is an ordinary serpent on the ray $(c_{2k}, +\infty)$, diffeomorphic to $\mathbb{R}$.

We have thus constructed a locally diffeomorphic mapping from the space of the (virtually nongeneric) traceless cyclic serpents of order $2n$ onto the space of the (virtually nongeneric) ordinary serpents of order $2n - 1$ in a neighbourhood of every point c, at which the minimum of the c_i-s is attained only once. This diffeomorphism sends the generic cyclic serpents to the generic ordinary ones. It follows that the components of the manifold of the generic traceless cyclic serpents are simplicial cones (as well as the components of the space of the ordinary serpents). The components of the complement to the principal diagonal in P_n are hence open simplices.

The number of such simplicial cones (simplices) is equal to the product of the number of the components of the space of the ordinary generic serpents of order $2n - 1$ (which is equal to T_n, see for example [1]) and of the number n of the possible values of $k = 1, \ldots, n$, defining the minimum $c_{2k} = \min c_i$.

2. The space of real trigonometric M-polynomials

We normalize the trigonometric polynomials of degree n fixing the highest term and removing the constant term,

$$f(t) = \cos nt + \sum_{0 < k < n} (a_k \cos kt + b_k \sin kt). \tag{1}$$

We consider the space of these polynomials as the real linear space $\mathbb{R}^{2n-2}$ with coordinates (a_k, b_k).

Definition. An *M-polynomial* is a trigonometric polynomial of degree n having no complex critical points (the $2n$ real critical points might be degenerate). It is *proper*, if all the $2n$ critical points are different (and hence nondegenerate). It is *generic*, if all the $2n$ critical values are different.

The M-polynomials (1) form a real semialgebraic set in the space $\mathbb{R}^{2n-2}$. The main result of the present paper is

Theorem 3. *The set of M-polynomials (1) in $\mathbb{R}^{2n-2}$ is homeomorphic to the cyclic serpent polyhedron P_n. There exists a semialgebraic homeomorphism, inducing the diffeomorphism of the space of the proper M-polynomials on the interior domain of P_n, and sending the nongeneric M-polynomials onto the principal diagonal hyperplanes of the polyhedron P_n.*

Corollary. *The connected components of the space of the generic M-polynomials (1) are contractible and their number is equal to nT_n.*

Example. For $n = 2$ the manifold of the proper M-polynomials (1) is the domain of the (a_1, b_1)-plane, bounded by an astroid. The generic M-polynomials form four triangles, into which the domain is cut by the two segments, joining the opposite cusp points. This decomposition is diffeomorphic to the decomposition of the open square P_2 into four triangles by its diagonals.

Proof. The complex version of the required mapping is constructed in [2] - it is the $L3$ (Lyashko-Looijenga-Laurent) mapping. The real version for the ordinary polynomials case is described in [3] as the real version of the $L2$ (Lyashko-Looijenga) mapping. The polyhedral model, similar to our P_n, is in this case the Springer cone.

To adapt these constructions to the trigonometric case one uses the following fact (extending the Vacoulenko construction described for the ordinary polynomials in [3]).

Lemma. *Every cyclic serpent of order $2n$ is, up to an additive constant and to the multiplication by a positive number, the sequence of the consecutive critical values of a real trigonometric M-polynomial (1).*

This polynomial is defined by the serpent uniquely, up to the shift of the argument t by a multiple of $2\pi /n$.

We shall prove this for the proper serpents, the proof in the nonproper case being similar.

Proof. Take $2n$ copies X_j of the (extended) complex plane of c ($j = 0, \ldots, 2n$; $X_{2n} = X_0$). Cut X_{2k} along the real axis from c_{2k} to $-\infty$ and from c_{2k+1} to $+\infty$, and X_{2k+1} from c_{2k+1} to $+\infty$ and from c_{2k+2} to $-\infty$. Glue the opposite sides of the common cut $(c_{j+1}, (-1)^j \infty)$ on X_j and on $X_{j+1}(j = 0, 1, \ldots 2n - 1)$.

The Riemann surface X thus constructed is a sphere (according to the Riemann-Hurwitz formula). The point $c = \infty$ is covered by two points of the surface X, each with multiplicity n.

The Riemann theorem on the unicity of the complex structure on the sphere implies that the projection of the surface X onto the c-plane can be written as a Laurent polynomial

$$c = f(z) = A_n z^{-n} + ... + A_n z^n$$

where z is a coordinate on X, such that the real axis on X (the fixed points set of the antihomomorphic involution, covering the complex conjugation of c) is $|z| = 1$.

The reality condition implies that $f(1/\bar{z}) = \bar{f}(z)$. Thus f is a real trigonometric polynomial of t where $z = e^{it}$. Choosing the origin of t we can reduce the highest degree term to $a \cos nt$, $a > 0$.

We have thus constructed the required polynomial. Its unicity follows from the fact, that the topological type of the covering is unambiguously defined by the serpent (see [2]). Hence all the real trigonometric polynomials defining the same serpent may be obtained one from the other by an automorphism of the sphere X.

The automorphisms, sending $z^n + z^{-n}$ to itself, are $z \mapsto \epsilon z$ and $z \mapsto \epsilon/z$, where $\epsilon^n = 1$. The second type automorphism (say with $\epsilon = 1$) would transform the serpent $c_0 \leq c_1 \geq c_2 \leq ...$ into $c_0 \leq c_{2n-1} \geq c_{2n-2} \leq ....$ We deduce that $c_k = c_{2n-k}$ and hence f is even and invariant under the automorphism. Thus only the translations of t by $2\pi k/n$ remain, and the Lemma follows.

To prove the theorem, we kill the constant term in the polynomial (adding a constant to the critical values). Multiplying f by a positive number , we reduce the coefficient a of $\cos nt$ to one. We have thus constructed a diffeomorphism of every connected component of the manifold of the generic M-polynomials (1) onto one of the simplices into which the polyhedron P_n is cut by the principal diagonals.

This simplex is defined by the generic cyclical serpent of the critical values only up to the action of the cyclic group of order n, since the polynomial does not know which of its local minima is c_0.

The component of the space of the generic M-polynomials (1) is defined by the cyclic serpent also only up to the action of the cyclic group (of the shifts of the argument t by the multiples of $2\pi/n$).

Both ambiguities are, however, cocordant.

Choose any pair (component, simplex). This choice defines an $\mathbb{Z}_n$-equivariant diffeomorphism of the manifold of the proper M-polynomials (1) onto the interior domain of the cyclic serpent polyhedron P_n (this diffeomorphism is the real version of $L3$).

Indeed, consider a generic path, connecting a generic M-polynomial (1) with its version, shifted by $2\pi k/n$ in the space of proper M-polynomials. This path is covered by a well-defined generic path in the proper serpent space, connecting the serpent of the critical values of the original M-polynomial to the shifted serpent.

A generic path, connecting a generic serpent with its shifted version in the space of proper serpents, defines unambiguously a generic path in the space of the proper M-polynomials (1), connecting the original generic M-polynomial (1) with its shifted version.

Hence we obtain a homeomorphism of the space of the proper M-polynomials (1) to the interior domain of the principal cyclic polyhedron P_n. This homeomorphism is a diffeomorphism, since, according to [2], the critical values are the local coordinates in the space of the proper M-polynomials (1).

References

[1] Arnold V.I. *Enumeration of snakes and combinatorics of Bernoulli, Euler and Springer numbers of Coxeter groups.* Russian Mathematical Surveys, **47**:1 (1992), 1-51.

[2] Arnold V.I. *Topological classification of complex trigonometric polynomials and combinatorics of graphs with equal numbers of vertices and edges.* Functional Analysis and its Applications, **30** : 1 (1996), 1-18.

[3] Arnold V.I. *Springer numbers and morsification spaces*, Journal of Algebraic Geometry, **1** (1992), 197-214.

CEREMADE, URA CNRS N° 749
Université Paris-Dauphine
Place du Maréchal de Lattre de Tassigny
75775 PARIS CEDEX 16 (FRANCE)

and

Steklov Mathematical Institute
42, Vavilova Street,
117966 MOSCOW, GSP-1 , RUSSIA

Received April 1, 1996

Singularities of short linear waves on the plane

Ilia A. Bogaevski

ABSTRACT. The subject of the paper is the geometrical optics of short linear waves on plane. We describe perestroikas of momentary fronts and scattering of rays when the light hypersurface has generic conical singularities. They appear if the waves propagate in a nonhomogeneous anisotropic medium and are controlled by a hyperbolic variational principle.

The geometrical optics of short linear waves on plane is described by some hypersurface in the contact space of the projectivised cotangent bundle of space-time. This hypersurface is called *light*. Projections of Legendre submanifolds of the light hypersurface define in space-time surfaces called *big* fronts and describing the wave propagation – their sections with isochrones are *momentary* fronts evolving in time. Big fronts consist of *rays* which are projections of characteristics of the light hypersurface. Usually the light hypersurface is *evolutionary*. It means the rays are transversal to the isochrones. In this case the evolution of a given momentary front is completely defined by some *initial condition* which is an integral (with respect to the contact structure) curve on the light hypersurface. Namely, the corresponding Legendre submanifold is the extension of the initial condition along characteristics of the light hypersurface and the projection of the initial condition is the given front.

A momentary front evolving in time can experience *perestroikas* (= bifurcations, metamorphoses). In case of a smooth evolutionary light hypersurface and a typical initial condition all perestroikas of momentary fronts are well known even if their dimensions does not exceed four (see, for example, [1], [2], [3]).

However, if the oscillations in the waves have more than one degrees of freedom, on the light hypersurface there can be conical singularities ([4]). Let the initial condition be smooth and do not pass through any singularity of the light hypersurface. Then it is called *regular*. But since some characteristics can pass through singularities of the light hypersurface, even in that case the extended Legendre submanifold can became singular too. Its singularities generate new singularities of big fronts and new perestroikas of momentary fronts. They are found in book [4] in case of a typical evolutionary light hypersurface having only conical singularities and a typical regular initial condition. However, the proposed there description of the

This research is partially supported by the National Science Foundation (grant MSD000) and the Russian Foundation for Basic Researches (grant 94-01-01203)

corresponding singularities of big fronts and perestroikas of momentary fronts is not correct.

Our results imply there are two types of the new perestroikas of momentary fronts. These types we call *refraction* and *reflection* of waves and show in Fig. 1.

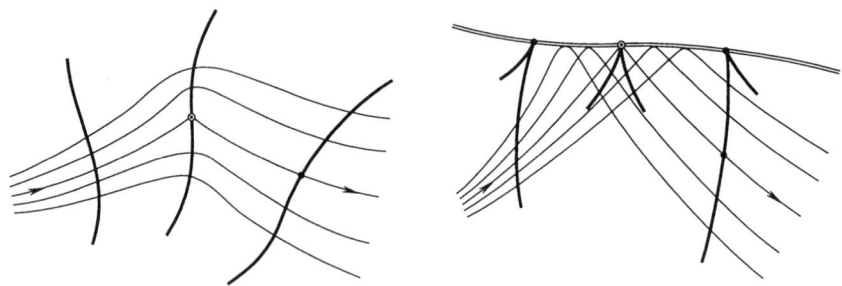

Figure 1. Refraction and reflection of waves.
Perestroikas of fronts and scattering of rays.

In the both cases one ray consists of two branches of intersecting smooth curves. These branches are the incoming ray and the outgoing one. The other rays are smooth and scatter. Namely, if ε is the distance between a fixed point of the incoming ray and a smooth ray, the distance between the same ray and a fixed point of the outgoing ray is equal to $\mu\varepsilon \ln \varepsilon + o(\varepsilon \ln \varepsilon)$ as $\varepsilon \to 0$ and $\mu < 0$. At the perestroika instant the momentary front consists of two tangent branches having different curvatures at the point of their tangency. Each of them can be given in suitable smooth coordinates by the local equation $y = x^2/\ln x + o(x^2/\ln x)$ as $x \to 0$. After the perestroika instant the momentary front acquires a discontinuity of the third derivative. The second derivative proves to be continuous but not to satisfy the Hölder condition. The discontinuity propagates along the outgoing ray and in its neighborhood the momentary front can be given in suitable smooth coordinates by the equation $y = x^2/\ln|x| + o(x^2/\ln|x|)$ as $x \to 0$.

In the case of reflection the cusp of the momentary front propagates along the *caustic*. It is shown in Fig. 1 by the double line. When the momentary front experiences a perestroika, the caustic has a discontinuity of the second derivative. At this point the caustic's curvature is infinite and changes a sign. The caustic can be given by the local equation $y = (\lambda - \operatorname{sign} x)x/\ln|x| + o(x/\ln|x|)$ as $x \to 0$ and $\lambda > 1$.

1. Singularities of Light Hypersurfaces

A singular point of a light hypersurface

$$\Sigma^4 \subset PT^*\mathbb{R}^{2+1}$$

in the contact space of the projectivised cotangent bundle of space-time $\mathbb{R}^{2+1}$ is called *conical* if in its neighborhood Σ^4 has the local equation $u^2 + v^2 = w^2$ in suitable smooth coordinates. There are two types of generic conical points of light hypersurfaces with respect to the contact structure in $PT^*\mathbb{R}^{2+1}$ (see [4], [5]). Namely, a conical singularity of the light hypersurface Σ^4 is *hyperbolic* if it has the local equation $P_1 Q_1 + P_2^2 = 0$ in suitable formal coordinates (P_1, P_2, Q_1, Q_2, U) such that the contact structure is given by the equation

$$dU = \frac{1}{2}(P dQ - Q dP).$$

If Σ^4 is given by the local equation

$$P_1^2 + Q_1^2 = P_2^2$$

then its conical singularity is called *elliptic*.

A characteristic of Σ^4 does not pass through any its elliptic singularity. On the contrary, there exist a couple of characteristics passing through each hyperbolic singularity of Σ^4. Their equations in the above coordinates are $P_1 = P_2 = Q_2 = U = 0$ and $P_2 = Q_1 = Q_2 = U = 0$.

It turns out there are three types of generic hyperbolic singularities of light hypersurfaces with respect to the natural projection

$$\pi : PT^*\mathbb{R}^{2+1} \to \mathbb{R}^{2+1}.$$

Namely, let us say these types are *refraction*, *reflection*, and *pseudoreflection* of characteristics. They can be defined if the two following conditions of regularity hold at the considered hyperbolic singularity.

1) The projection of the surface of all hyperbolic singularities into space-time has a full rank.

2) The projections of the directions of the two characteristics passing through our hyperbolic singularity are different.

Let $H^2 \subset \Sigma^4$ be the surface of all hyperbolic singularities of the light hypersurface, $N^4 = \pi^{-1}(\pi(H^2))$, and $\widetilde{H}^3 \subset \Sigma^4$ be the union of all characteristics passing through H^2. Locally $\widetilde{H}^3 = \widetilde{H}_P^3 \cup \widetilde{H}_Q^3$ and $H^2 = \widetilde{H}_P^3 \cap \widetilde{H}_Q^3$, where the local components $\widetilde{H}_P^3$ and $\widetilde{H}_Q^3$ are given by the equations

$$P_1 = P_2 = 0 \quad (\widetilde{H}_P^3) \qquad \text{and} \qquad P_2 = Q_1 = 0 \quad (\widetilde{H}_Q^3).$$

All submanifolds Σ^4, N^4, $\widetilde{H}_P^3$, and $\widetilde{H}_Q^3$ contain H^2. The *refraction*, *reflection* and *pseudoreflection* of characteristics are all different variants of the arrangement of these submanifolds providing N^4 is tangent to none of the submanifolds Σ^4, $\widetilde{H}_P^3$, and $\widetilde{H}_Q^3$ at the considered point. These variants are shown in Fig. 2.

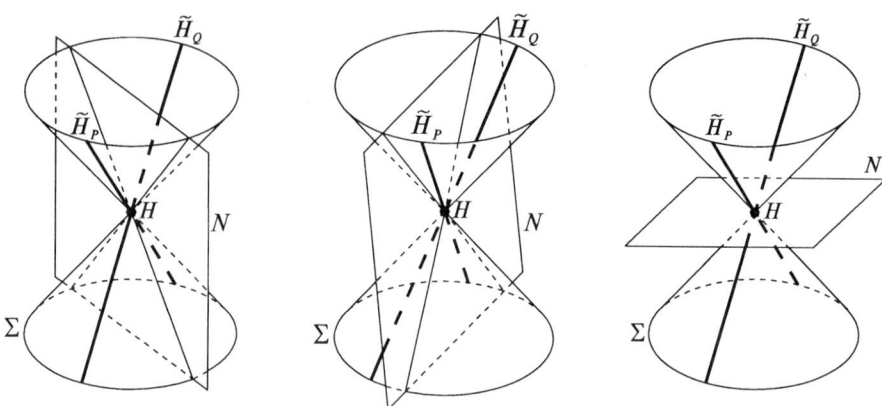

Figure 2. Points of refraction, reflection, and pseudoreflection of characteristics.

In space-time there exist smooth coordinates (y, z) such that the projection $\pi : (P, Q, U) \mapsto (y, z)$ is given by the asymptotic formulas

$$y_1 = aP_1 + bP_2 - Q_1 + \dots ,$$
$$y_2 = bP_1 + cP_2 - Q_2 + \dots ,$$
$$z = P_1 y_1/2 + P_2 y_2/2 - U + \dots ,$$

where the dots denote the terms of higher degrees providing $\deg P = \deg Q = 1$ and $\deg U = 2$. Then

$a < 0$ and $b \neq 0$ in the case of refraction of characteristics at the origin of the coordinates (P, Q, U);

$0 < 4a < b^2$ in the case of reflection of characteristics;

$0 < b^2 < 4a$ in the case of pseudoreflection of characteristics.

2. Singularities of Big Fronts

Let $L^1 \subset \Sigma^4$ be a regular initial condition. Its nontrivial intersection at isolated points with the submanifold $\tilde{H}^3 \subset \Sigma^4$ (which consists of all characteristics passing through the surface $H^2 \subset \Sigma^4$ of the hyperbolic singularities) is a typical phenomenon. Let our initial condition be transversal to $\tilde{H}^3$. Then the extended Legendre submanifold $\tilde{L}^2$ can intersect the surface H^2 at isolated points. The singularities of $\tilde{L}^2$ at these points are described by the parametric normal form

$$P_1 = A^2, \quad P_2 = AB, \quad Q_1 = -B^2, \quad Q_2 = -2AB \ln A^2, \quad U = -A^2 B^2/2$$

found by V. I. Arnold in [4] (A, B are the parameters). The normal forms of the contact structure and the light hypersurface remain the same as before.

By substituting these equations into the asymptotic formulas for the projection π from the previous section we obtain the local asymptotic parametrization of the big front $\pi(\ \widetilde{L}^2)$

$$y_1 = aA^2 + bAB + B^2 + \dots,$$
$$y_2 = bA^2 + cAB + 2AB \ln A^2 + \dots,$$
$$z = aA^4/2 + bA^3B + (1 + c/2)A^2B^2 + A^2B^2 \ln A^2 + \dots.$$

By the change of parameters $A \mapsto \alpha A$, $B \mapsto \beta B$ one can achieve $a = \pm 1$, $b > 0$, $c = 0$. Then $a = -1$ in the case of refraction of characteristics, $a = 1$ and $b > 2$ in the case of reflection of characteristics, $a = 1$ and $b < 2$ in the case of pseudoreflection of characteristics. The singularities of big fronts and the rays on their in these three cases are shown in Fig. 3.

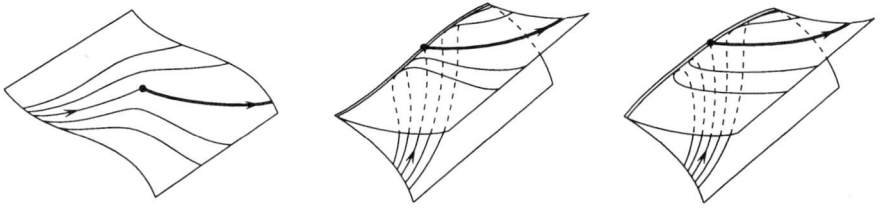

Figure 3. Refraction, reflection, and pseudoreflection of rays. Singularities of big fronts

3. Perestroikas of Momentary Fronts

If our light hypersurface is evolutionary then the pseudoreflection of characteristics is not realized. The rest singularities of big fronts describe the refraction and reflection of momentary fronts. Namely, the momentary fronts are the sections of the big front with the isochrones and transversal to the rays.

References

1. V. I. Arnold, *Wave front evolution and equivariant Morse lemma*, Comm. Pure Appl. Math. **29** (1976), no. 6, 557–582.
2. V. I. Arnold, *Catastrophe theory*, Springer-Verlag, Berlin, 1992.
3. I. A. Bogaevski, *Perestroikas of fronts in evolutionary families*, Proc. Steklov Inst. Math. **209** (1995) (to appear).
4. V. I. Arnold, *Singularities of caustics and wave fronts*, Math. Appl. (Soviet Ser.), vol. 62, Kluwer Acad. Publ., Dordrecht, 1990.
5. V. I. Arnold, *On the interior scattering of waves, defined by hyperbolic variational principles*, J. Geom. Phys. **5** (1988), no. 4.

Department for Higher Mathematics
Moscow State Forestry University,
141001, Mytischi-1, Moscow region, RUSSIA

Received March 1995

New generalizations of
Poincaré's geometric theorem

Yu. V. Chekanov[1]

Poincaré's geometric theorem claims that an area-preserving diffeomorphism of an annulus which shifts the boundary circles at opposite directions has at least two fixed points. The present paper consists of two parts. In the first one, we show that such a diffeomorphism has more than just two fixed points provided the shift of the boundaries is large enough. In the second part, we prove symplectic fixed point theorems which can be viewed as generalizations of Poincaré's geometric theorem to higher dimensions.

1.

Consider the annulus $A = [a_+, a_-] \times S^1$, where $S^1 = \mathbf{R}/\mathbf{Z}$, endowed with an area form ω. Let g be an area-preserving diffeomorphism of A.

Denote by $\tilde{g}$ the diffeomorphism of $\tilde{A} = [a_-, a_+] \times \mathbf{R}$ which covers g. Define the shift functions $\varphi_\pm \colon \mathbf{R} \to \mathbf{R}$ by

$$\tilde{g}(a_\pm, y) = (a_\pm, y + \varphi_\pm(y)).$$

Poincaré's geometric theorem (actually proved by Birkhoff [1]) claims that if $\varphi_+ > 0$ and $\varphi_- < 0$ then g has at least two fixed points.

We prove the following stronger result.

Theorem 1.1. *If $\varphi_+ > N_+$ and $\varphi_- < N_-$, where $N_\pm \in \mathbf{Z}$, then g has at least $2(1 + N_+ - N_-)$ distinct fixed points.*

In the middle of 1960s V. I. Arnold demonstrated that Poincaré's geometric theorem is a consequence of the fact that any area-preserving diffeomorphism of a 2-torus which is a time 1 map of a (non-autonomous) Hamiltonian flow has at least three fixed points. He formulated a conjecture estimating from below the number of fixed points of a Hamiltonian symplectomorphism of an arbitrary symplectic manifold [1,2]. Later C. Conley and E. Zehnder proved the Arnold conjecture for standard symplectic tori [4].

We show Theorem 1.1 to follow in a quite elementary way from the Conley–Zehnder theorem for the 2-torus. The proof of Theorem 1.1 is based on the following

---

[1] The work has been partially supported by the ISF grant MSD000 and the INTAS grant 4373.

Lemma 1.2. *If $\varphi_+ > 0$ and $\varphi_- < 0$ then the projection $\pi: \tilde{A} \to A$ sends the set of fixed points of $\tilde{g}$ to a set containing at least two points.*

Proof 1°. By the Moser theorem, we can assume the area form ω to coincide with the standard area form $dx \wedge dy$ on the annulus

$$A = \{(x, y) \mid x \in [a_-, a_+], \ y \in \mathbf{R}/\mathbf{Z}\}.$$

Choose a smooth family $\{g^t\}$, $t \in [0, 1]$ of ω-preserving diffeomorphisms of A in such a way that, for some covering family $\{\tilde{g}\}$ of diffeomorphisms of $\tilde{A}$, which preserve the form $\tilde{\omega} = \pi^*(\omega)$, the following holds:

$$\tilde{g}^0 = \mathrm{id}, \ \tilde{g}^1 = \tilde{g}, \ \tilde{g}^t(a_\pm, y) = (a_\pm, y + t\varphi_\pm(y)).$$

(To do this we first construct such an isotopy in the class of all diffeomorphisms, then apply the 1-parametric version of the Moser theorem.)

2°. Let $\{h^t\}$, $t \in [0, 1]$ be the family of Hamiltonians on $\tilde{A}$ generating the flow $\{\tilde{g}^t\}$. It means that $dh^t = i_{v_t}\omega$, where $v_t = \partial \tilde{g}^t/\partial t$. The form dh is clearly invariant under the action of the covering transformation group $\mathbf{Z}$ on $\tilde{A}$. Since the vector field v_t is tangent to $\partial \tilde{A}$, the restriction of h^t to either component of $\partial \tilde{A}$ is a constant function for any $t \in [0, 1]$. Therefore the Hamiltonians h^t are $\mathbf{Z}$-invariant.

3°. Since the functions h^t are defined up to additive constants, we can assume them to be positive. Let $\beta_\pm^\varepsilon$, $\varepsilon > 0$, be smooth monotone functions on $[a_-, a_+]$ such that $\beta_+^\varepsilon = 0$ on $[a_-, a_+ - 2\varepsilon]$, $\beta_+^\varepsilon = 1$ on $[a_+ - \varepsilon, a_+]$, $\beta_-^\varepsilon = 0$ on $[a_- + 2\varepsilon, a_+]$, $\beta^\varepsilon = 1$ on $[a_-, a_- + \varepsilon]$. Define a family of functions on $\tilde{A}$ by

$$\tilde{h}_\varepsilon^t(x, y) = \beta_+^\varepsilon(x)(a_+ - x) + \beta_-^\varepsilon(x)(x - a_-) + (1 - \beta_+^\varepsilon(x) - \beta_-^\varepsilon(x))h^t(x, y).$$

Since the Hamiltonians $\tilde{h}_\varepsilon^t$ are $\mathbf{Z}$-invariant, we have $\tilde{h}_\varepsilon^t = h_\varepsilon^t \circ \pi$ for some family $\{h_\varepsilon^t\}$, $t \in [0, 1]$, of functions on A. Consider the flow $\{g_\varepsilon^t\}$ on A associated with the family of Hamiltonians $\{h_\varepsilon^t\}$, and the flow $\{\tilde{g}_\varepsilon^t\}$ on $\tilde{A}$ associated with the family of Hamiltonians $\{\tilde{h}_\varepsilon^t\}$.

For ε small enough, the fixed points of $\tilde{g}_\varepsilon^1$ are exactly the fixed points of $\tilde{g}$. To complete the proof of Lemma 1.2, it suffices to show that there exist at least two points $z \in A$ such that $g_\varepsilon^1(z) = z$ and the loop $\{g_\varepsilon^t(z)\}$, $t \in [0, 1]$, is contractible because these points are exactly the images of fixed points of $\tilde{g}_\varepsilon^1$ under the projection π.

4°. Consider the torus

$$T = \{(x, y) \mid x \in \mathbf{R}/(2a_+ - 2a_-)\mathbf{Z}, \ y \in \mathbf{R}/\mathbf{Z}\},$$

equipped with the area form $dx \wedge dy$ as a double of the annulus A. More precisely, define $I_{1,2}: A \to T$ by

$$I_1(x, y) = ([x], y), \quad I_2(x, y) = ([2a_+ - x], y),$$

where $[r]$ denotes the class of r in $\mathbf{R}/(2a_+ - 2a_-)\mathbf{Z}$, then $T = I_1(A) \cup I_2(A)$.

Define the flow $\{G_\varepsilon^t\}$, $t \in [0, 1]$, on T by doubling the flow $\{g_\varepsilon^t\}$ on A. More formally,

$$G_\varepsilon^t(z) = I_j \circ g_\varepsilon^t \circ I_j^{-1}(z) \text{ when } z \in I_j(A).$$

This flow is defined by the family of Hamiltonians $\{H_\varepsilon^t\}$, where $H_\varepsilon^t(z) = h_\varepsilon^t \circ I_1^{-1}(z)$ when $z \in I_1(A)$ and $H_\varepsilon^t(z) = -h_\varepsilon^t \circ I_2^{-1}(z)$ when $z \in I_2(A)$.

Let N denote the number of points $z \in T$ such that $G_\varepsilon^1(z) = z$ and the loop $\{G_\varepsilon^t(z)\}$, $t \in [0, 1]$, is contractible. By the Conley–Zehnder theorem, $N \geq 3$ [4]. Hence there exist $N/2 \geq 2$ points $z \in A$ such that $g_\varepsilon^1(z) = z$ and the loop $\{g_\varepsilon^t(z)\}$, $t \in [0, 1]$, is contractible. This completes the proof. ∎

Proof of Theorem 1.1 Using the action of the covering transformation group $\mathbf{Z}$ on $\tilde{A}$, one easily constructs $1 + N_+ - N_-$ different diffeomorphisms of $\tilde{A}$ which cover the diffeomorphism g and satisfy the hypothesis of Lemma 1.2. Their fixed points project to A disjointly. Hence the theorem follows from the lemma. ∎

2.

A. Floer introduced in [5,6] an elliptic version of infinite-dimensional Morse theory and used it to prove the Arnold conjecture for monotone symplectic manifolds. We use the same technique to obtain fixed points theorems for certain symplectic manifolds with boundaries. These results can be viewed as generalizations of Poincaré's geometric theorem and Theorem 1.1.

Let g be a Riemannian metric on a manifold M. Define the function $L_g: T^*M \to \mathbf{R}$ to measure the lengths of covectors in the metric g^* dual to g:

$$L_g(p) = |p| = \sqrt{g^*(p, p)} \text{ for } p \in T_q^*M.$$

Denote $A_g(M) = \{x \in T^*M \mid L_g(x) \leq 1\} \subset T^*M$.

Fix a flat metric g on $T^n = \mathbf{R}^n/\mathbf{Z}^n$ and let $H: [0, 1] \times A_g(T^n) \to \mathbf{R}$ be a time depending Hamiltonian on $A_g(T^n)$ coinciding with L_g in a neighbourhood of $\partial A_g(T^n)$. Define G to be the time 1 map of the Hamiltonian flow associated with H. Let $N(g)$ denote the number of elements of $\pi_1(T^n)$ represented by closed geodesics of length ≤ 1 in the metric g (constant loops are assumed to be geodesics throughout).

Theorem 2.1. *Symplectomorphism G has at least $2^n N(g)$ fixed points provided all of them are non-degenerate.*

Consider the case $n = 1$ to describe the relation between this result and Theorem 1.1. For $n = 1$, $A_g(T^n)$ is just the annulus. The metric g has form $c\,(dq) \otimes (dq)$, where q is the natural coordinate on $T^1 = \mathbf{R}/\mathbf{Z}$. Assume $0 \leq N_0 < 1/c < N_0 + 1$ for some $N_0 \in \mathbf{Z}$, then $N(g) = 2N_0 + 1$. One easily checks that G satisfies the hypothesis of Theorem 1.1 with $N_+ = (N(g) - 1)/2$, $N_- = (1 - N(g))/2$. Hence the case $n = 1$ of Theorem 2.1 is a corollary of Theorem 1.1.

Proof $1°$. The function L_g on T^*T^n defines outside the zero section the Hamiltonian system describing the velocity 1 motion on T^n:

$$\dot{p} = 0, \quad \dot{q} = i(p)/|p|, \tag{2.1}$$

where $i : (\mathbf{R}^n)^* \to \mathbf{R}^n$ is the isomorphism defined by g). The projection $T^*T^n \to T^n$ sends time t trajectories of this system to length t geodesics. If there exists a length 1 closed geodesics on T^n, then there are infinitely many trajectories of (2.1) projecting to this geodesics which lie in ∂A_g. Then G has infinitely many fixed points and the claim of Theorem 2.1 is trivially satisfied. So we assume that there are no such closed geodesics. Extend H to a function on $[0, 1] \times T^*T^n$ by letting $H(t, \cdot) = L_g$ outside $A_g(T_n)$. Assume now that H is actually defined on $(S^1 = \mathbf{R}/\mathbf{Z}) \times T^*T^n$ instead of $[0, 1] \times T^*T^n$ (one can check that the general case follows from this one). The 1-periodic solutions of

$$\dot{x}(t) = \operatorname{sgrad} H(t, (x(t))) \tag{2.2}$$

(sgrad denotes Hamiltonian vector field) are in one-to-one correspondence (given by $x \mapsto x(0)$) with the fixed points of G. Our goal is to estimate from below the number of such solutions. The desired estimate will be obtained by means of the appropriate version of Floer homology.

$2°$. Fix a connected component of the loop space $C^\infty(S^1, T^n)$ containing a closed geodesics of length < 1 (there exist $N(g)$ such components). Denote by Ω the corresponding connected component of $C^\infty(S^1, T^*T^n)$.

Define $S(\Omega, H)$ to be the set of 1-periodic solutions of (2.2) belonging to Ω. To prove the theorem, it suffices to show that $S(\Omega, H)$ contains at least 2^n loops if they all are non-degenerate.

Consider the symplectic action functional $\mathcal{A}_H \colon C^\infty(S^1, T^*T^n) \to \mathbf{R}$,

$$\mathcal{A}_H(x) = \int_x p\,dq - \int_0^1 H(t, (x(t))\,dt.$$

The critical set of its restriction to Ω is $S(\Omega, H)$. Endow T^*T^n with the complex structure J,

$$J(\partial/\partial q_i) = \partial/\partial p_i, \quad J(\partial/\partial p_i) = -\partial/\partial q_i.$$

The symplectic form $\omega = dp \wedge dq$ and the complex structure J define the metric $\omega(J\cdot, \cdot)$ on T^*T^n and the metric

$$\int_0^1 \omega(J\cdot(t), \cdot(t))\,dt$$

on $C^\infty(S^1, T^*T^n)$.

The (anti) gradient flow equation for the functional $\mathcal{A}_H$ reads

$$\frac{du(s)}{ds}(t) = -J\frac{\partial u(s)(t)}{\partial t} - \nabla H^t(u(s)(t)). \tag{2.3}$$

For $x_-, x_+ \in S(\Omega, H)$, define $\mathcal{M}(x_-, x_+)$ to be the set of gradient trajectories (solutions of (2.3)) $u \colon \mathbf{R} \to \Omega$ satisfying the asymptotic conditions $\lim_{s \to \pm\infty} u(s) = x_\pm$.

One can prove that the collection of the points $u(s)(t)$ over all $u \in \mathcal{M}(x_-, x_+)$, $x_\pm \in S(\Omega, H)$, $s \in \mathbf{R}$, $t \in S^1$ form a compact subset of T^*T^n (for a similar case, the proof has been carried out in [8]). Therefore, in view of the fact that ω is exact and hence no bubbling can occur, we can immediately apply the results established by Floer for the case of closed symplectic manifolds. This yields the following. (1) To any non-degenerate $x \in S(\Omega, H)$, one can attach the integer $\mathrm{ind}(x)$ called the (relative) Maslov–Conley–Zehnder index.

These indices are defined up to a uniform additive constant. Assuming all critical loops to be non-degenerate, we get a grading (defined up to a shift) on the chain space $CF_*(\Omega, H)$ spanned over $\mathbf{Z}_2$ by the points of $S(\Omega, H)$. (2) After a generic perturbation of H, any set $\mathcal{M}(x_-, x_+)$ is a manifold of dimension $\mathrm{ind}(x_-) - \mathrm{ind}(x_+)$. (3) The mapping $\partial \colon CF_*(\Omega, H) \to CF_{*-1}(\Omega, H)$ given by

$$\partial(x) = \sum_{\substack{y \in S(\Omega, H), \\ \mathrm{ind}(y) = \mathrm{ind}(x) - 1}} \#\big(\hat{\mathcal{M}}(x, y)\big)\,y,$$

where $x \in S(\Omega, H)$, $\hat{\mathcal{M}}(x, y) = \mathcal{M}(x, y)/\mathbf{R}$ ($\mathbf{R}$ acts by translations in s variable), is a differential. The resulting homology $HF_*(\Omega, H)$ does not depend on the choice of (generic) H as long as $H(t, \cdot)$ coincides with L_g outside a compact (or differs by a C^2-small function).

$3°$. To prove Theorem 2.1, it suffices to show that $\dim HF_*(\Omega, H) = 2^n$ (since this implies $\dim CF_*(\Omega, H) \geq 2^n$). We construct a special H to symplify the computation of the homology. Let λ be the length of the longest geodesics of length < 1. Fix a smooth function σ on $[0, +\infty[$ such that $\sigma(r) = r^2/2$ for $r \leq 1$, $\sigma(r) = r$ for r large enough, and $\sigma'(r) > \lambda$ for $r > 1$. The Hamiltonian system associated with the function $H(t, p, q) = \sigma(|p|)$ reads

$$\dot{p} = 0, \quad \dot{q} = \sigma'(|p|)\, i(p)/|p| \,.$$

Its time t trajectory starting at the point (p, q) projects to a length $\sigma'(|p|)t$ geodesics starting at the point q. The definition of σ implies the critical set $S(\Omega, H)$ to be an n-dimensional torus consisting of trajectories

$$t \mapsto (p_\Omega, q + ti(p)),$$

where $p_\Omega \in (\mathbf{R}^n)^*$ depends on Ω, and q is an arbitrary point of T^n. One can check that, for any $x \in S(\Omega, H)$, $\ker d\mathcal{A}_H(x) = T_x S(\Omega, H)$ and hence the restriction of $\mathcal{A}_H$ to Ω is a Morse–Bott function whose critical set $S(\Omega, H)$ is diffeomorphic to T^n. Having in mind the analogy with the finite-dimensional (or Palais–Smale) Morse–Bott theory, we conclude that $HF_*(\Omega, H) \cong H_*(T^n, \mathbf{Z}_2)$ (after a suitable choice of grading). Unfortunately, the Morse–Bott version of Floer homology has not been explicitly developed yet. So, to be more formal, we have to perturb H to make $\mathcal{A}_H$ a genuine Morse function.

$4°$. First, we prove that the homologies $HF_*(\Omega, H)$ are isomorphic for all Ω. Then we carry out the computation of $HF_*(\Omega, H)$ for the case where $\Omega = \Omega_0$ is the component comprising constant loops, H is some time-independent function.

Consider the mapping

$$X_\Omega \colon S^1 \times T^*T^n \to S^1 \times T^*T^n, \quad (t, p, q) \mapsto (t, p + p_\Omega, q + ti(p_\Omega)).$$

For any $t \in S^1$, $X_\Omega(t, \cdot)$ is a symplectomorphism. The trajectories of the Hamiltonian system associated with H are sent by X_Ω to the trajectories of the Hamiltonian system associated with H^Ω defined by

$$H^\Omega(t, p, q) = H(t, p - p_\Omega, q - ti(p_\Omega)) + b_\Omega(p), \tag{2.4}$$

where $b_\Omega(p) = g^*(p, p_\Omega) = \langle p, i(p_\Omega) \rangle$. Since $X_\Omega(t, \cdot)$ preserves J, the mapping X_Ω yields a one-to-one correspondence between the gradient trajectories (2.3) for $\mathcal{A}_H$ and that for $\mathcal{A}_{H^\Omega}$. Let $H_\varepsilon(t, p, q) = \sigma(|p|) + \varepsilon\varphi(t, p, q)$, where $\varepsilon \neq 0$, φ is a generic C^2-bounded smooth function.

Consider the gradient trajectories for the restriction of $\mathcal{A}_{H_\varepsilon}$ to Ω. The weak compactness argument [6,7] shows that, for ε small enough, their images are contained in a neighbourhood of $T_\Omega = \{p = p_\Omega\}$.

Define

$$\check{H}_\varepsilon(t, p, q) = \sigma(|p|) + \varepsilon\varphi(t, p - p_\Omega, q + ti(p_\Omega)).$$

Since in a neighbourhood of T_Ω the Hamiltonians H_ε^Ω (obtained by applying (2.4) to H_ε) and $\check{H}_\varepsilon$ differs by a constant, X_Ω yields (for ε small enough) the one-to-one correspondence between the gradient trajectories for the restriction of $\mathcal{A}_{H_\varepsilon}$ to Ω_0 and the gradient trajectories for the restriction of $\mathcal{A}_{\check{H}_\varepsilon}$ to Ω. Hence (for ε small enough) X_Ω induces an isomorphism between $HF_*(\Omega_0, H_\varepsilon)$ and $HF_*(\Omega, H'_\varepsilon)$.

Choose now H_ε to be time-independent:

$$H_\varepsilon(t, p, q) = F_\varepsilon(p, q) = \sigma(|p|) + \varepsilon f(q),$$

where f is a generic Morse function on T^n. One easily checks that, for ε small enough, $S(\Omega_0, H_\varepsilon)$ consists of constant loops:

$$S(\Omega_0, H_\varepsilon) = \{x \mid x(t) = (0, q), \, df(q) = 0\}$$

As in [6,7,9], one shows that there is a natural one-to-one correspondence between the gradient trajectories (in the loop space of T^*T^n) connecting the critical points of the restriction of $\mathcal{A}_{H_\varepsilon}$ to Ω_0 and the gradient trajectories (on T^*T^n) connecting the critical points of F_ε. Though T^*T^n is not a compact manifold (as the manifolds considered in [6]), F_ε is a proper Morse function on T^*T^n and hence we get an isomorphism $HF_*(\Omega_0, H_\varepsilon) \cong (T^*T^n, \mathbf{Z}_2)$. We get $\dim HF_*(\Omega_0, H_\varepsilon) = 2^n$, which completes the proof of Theorem 2.1. ∎

Consider now the case of an arbitrary manifold M and an arbitrary metric g on M. Provided there is no length 1 closed geodesics, one easily constructs the homology $HF_*(\Omega, H)$ for any connected component Ω of the loop space of T^*M. The number of the periodic trajectories of the Hamiltonian system coinciding with (2.1) outside a compact can be estimated from below by the dimension of $HF_*(\Omega, H)$ (in the non-degenerate case).

The problem is to compute the homology. In some particular cases, this can be easily done. For example, the claim of Theorem 2.1 remains true while g is deformed in such a way that no length 1 closed geodesics appear. Another result of this kind is the following.

Theorem 2.2. *Let g be a metric on a closed manifold M. Assume that each non-constant contractible geodesics has length > 1, and there exist N elements of $\pi_1(M)$ represented by exactly one closed geodesics of length < 1. If $\{H^t\}$, $t \in [0,1]$, is a family of Hamiltonians on $A_g(T^n)$ which coincide with L_g in a neighbourhood of $\partial A_g(M)$ then the corresponding time 1 flow map has at least $\dim H_*(M, \mathbf{Z}_2) + 2N$ fixed points provided they are non-degenerate.*

Note that the hypothesis of this theorem holds for a negative sectional curvature metric as well as for a sufficiently small metric.

Proof of this theorem essentially follows the lines of the proof of Theorem 2.1.

For $H = \sigma(|p|)$, the functional $\mathcal{A}_H$ on Ω is a Morse–Bott function whose critical set is diffeomorphic to M when Ω consists of contractible loops, and to S^1 when Ω contains a non-constant geodesics of length < 1. This yields the desired estimate.   ∎

In the general case, it seems natural to conjecture that

$$\bigoplus_\Omega HF_*(\Omega, H) = H_*(M, \mathbf{Z}_2) \oplus H_*(S^1, \mathbf{Z}_2) \otimes H_*(E_1, E_0, \mathbf{Z}_2),$$

where E_1 is the space of loops on M having length ≤ 1, and E_0 is the subspace of constant loops.

References

[1] Birkhoff G. D., Proof of Poincaré's geometric theorem, *Trans. AMS* **14** (1913), 14–22

[2] Arnold V. I., Sur une propriété topologique des application globalement canoniques de la mécanique classique, *C. R. Acad. Sci. Paris* **261** (1965), 3719–3722

[3] Arnold V. I., Mathematical methods of classical mechanics (Appendix 9), *Nauka* 1974; Engl. transl., Springer Heidelberg, New York, 1978

[4] Conley C., Zehnder E, The Birkhoff–Lewis fixed point theorem and a conjecture of V. I. Arnold, *Invent. Math.* **73** (1983), 33–49

[5] Floer A., Morse theory for Lagrangian intersections, *J. Diff. Geom* **28** (1988), 513–547

[6] Floer A., Symplectic fixed points and holomorphic spheres. *Comm. Math. Phys.* **120** (1989), 575–611

[7] Floer A., Witten's complex and infinite dimensional Morse theory, *J. Diff. Geom* **30** (1989), 207–221

[8] Floer A., Hofer. H., Symplectic homology I, *Math. Zeit.* **215** (1994), 37–88

[9] Salamon D., Zehnder. E., Morse theory for periodic solutions of Hamiltonian systems and the Maslov index. *Comm. Pure Appl. Math.* **XLV** (1992), 1303–1360

Independent University of Moscow
Moscow, Russia

Received November 1995

Explicit formulas
for Arnold's generic curve invariants

S. Chmutov and S. Duzhin[1]

Abstract

We review the explicit formulas for Arnold's generic curve invariants due to Viro, Shumakovich and Polyak and add some remarks concerning the invariants of spherical curves and curves immersed in arbitrary orientable surfaces.

1. Statement of the problem

A generic curve is a smooth immersion of the circle into the plane whose only singularities are transversal double points. Up to a diffeomorphism of the plane, how many generic closed curves are there?

One can immediately invent a host of invariants for generic plane curves, for example, the total number of double points, the Whitney index or the (unordered) sequence of edge numbers of the regions into which the curve divides the plane.

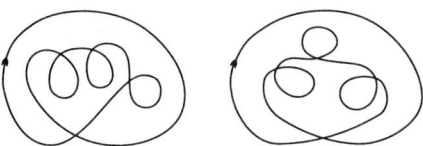

Figure 1: Two different curves with an equal number of double points ($= 4$), equal Whitney numbers ($= -3$) and equal sequences of edge numbers $(6, 6, 1, 1, 1, 1)$.

The problem we will discuss is the study of invariants and the classification of plane curves and also curves immersed in orientable surfaces. Although this problem has a venerable history (it was first studied by Gauss), the most basic invariants of plane curves, J^+, J^- and St, were only discovered in 1993 by V. I. Arnold who used V. Vassiliev's approach to topological invariants through discriminants and singularity theory.

[1]Both authors acknowledge financial support from the International Science Foundation and the Russian Foundation for Fundamental Research.

The discriminant is the set of all nongeneric curves viewed as a singular hypersurface in the space of all immersed curves. It divides the whole space into a number of connected components. Two generic curves are equivalent if and only if they belong to the same connected component, i.e., there exists a path in the space of generic curves connecting one curve with another.

From this viewpoint, the role of basic invariants should be played by functions which are constant on each connected component and have a pre-scribed behavior when the path intersects the discriminant. In the general position such a path meets the discriminant only at a finite number of its generic points. A generic point of the discriminant corresponds to a curve with only one generic singular point belonging to one of the three categories: a direct self-tangency, an inverse self-tangency or a triple point.

Arnold's seminal work [A1] gives an axiomatic description of the three invariants J^+, J^-, St which correspond to these types of singularities and prove their existence from the general properties of Serre fibrations. In 1994, two sets of explicit combinatorial formulas for the Arnold invariants were found by O. Viro [V], A. Shumakovich [Sh] and M. Polyak [Po].

In the present text, we will give an overview of the work by Viro, Shumakovich and Polyak and add several observations on our own part that might shed more light on the meaning of these formulas and emphasize their relation with other viewpoints on the invariant theory of generic curves.

Since we are interested in plane curves as well as in the curves immersed in surfaces, we will treat only closed curves — observing that, in the plane, a parallel and in many respects simpler theory exists for *long curves* ([Tab]).

Now we state the problem in a more precise manner.

Let $\gamma : S^1 \to \mathbb{R}^2$ be a smooth immersion of the circle into the plane whose image is a smooth submanifold of $\mathbb{R}^2$ everywhere except for a finite number of double points with a transversal intersection. The problem is to study the invariants and find a classification of such immersions ("generic curves") with respect to an arbitrary diffeomorphism of the circle S^1 and an arbitrary diffeomorphism of the plane $\mathbb{R}^2$.

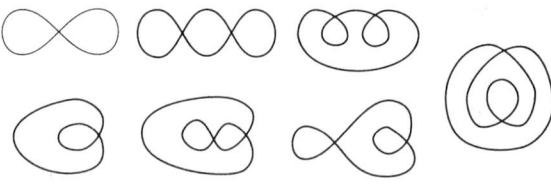

Figure 2: Plane curves with 1 and 2 double points.

Example 1. There are 2 nonequivalent plane curves with 1 double point and 5 nonequivalent plane curves with 2 double points (Fig. 2).

There are three nonequivalent variations of this problem: one can consider only orientation-preserving diffeomorphisms of the circle, or only orientation-preserving diffeomorphisms of the plane, or both (see [A1]). In this paper, we will only consider the totally nonoriented case.

One can also study a similar problem for immersions of the circle in arbitrary 2-surfaces. For example, there exists a one spherical curve with 1 double point, two spherical curves with 2 double points and one curve with 2 double points on the torus which is not spherical (see Fig. 3).

Figure 3: Spherical and toric curves with 1 and 2 double points.

Note that although in the case of spherical curves a complete invariant (Gauss diagram) is well-known, the study of the invariants which have a special behavior with respect to the discriminant is interesting in itself. Of course, Arnold's basic invariants for spherical curves can be expressed through the Gauss diagram, but these expressions are quite nontrivial (Polyak's formulas modified for spherical case, see last section).

2. Gauss diagrams

In this section, we will describe the classical construction of a Gauss diagram which gives an important combinatorial invariant of plane curves as well as of curves immersed in closed orientable surfaces.

Let $\gamma : S^1 \to \mathbb{R}^2$ be a generic curve with n double points. Denote the double points by n letters $a_1, a_2, \ldots, a_n$ and mark the preimages of these points on the circle by the corresponding letters with exponents $+1$ or -1 according to the following rule.

The natural counterclockwise orientation of the circle S^1 induces a direction for travelling along the curve. A passage of the double point is said to be *positive* if the traveller sees another branch of the curve as going from left to right.

Definition 1 *A Gauss diagram of order n is a circle with n distinguished ordered pairs of distinct points, considered up to an arbitrary diffeomorphism of the circle and up to the operation of a simultaneous reversing of order in each pair.*

Gauss diagrams are in one-to-one correspondence with *signed Gauss words*, i.e., elements of the quotient set

$$W(n)/S_n \times D_{2n} \times \mathbb{Z}_2,$$

where $W(n)$ is the set of all permutations of $2n$ letters $a_1, a_1^{-1}, \ldots, a_n, a_n^{-1}$ in which the symmetric group S_n acts by permutations of subscripts, the dihedral group D_{2n} acts by cyclic shifts and reflections, and the nontrivial element of the group $\mathbb{Z}_2$ changes the exponents of all letters.

A Gauss diagram can be depicted as a chord diagram with oriented chords. The arrow on a chord goes from the positive to the negative passage of the double point. The simultaneous reversing of the orientations of all chords corresponds to an orientation-reversing diffeomorphism of the plane.

The curves of Figure 1 have different Gauss diagrams:

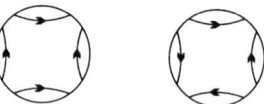

Figure 4: Gauss diagrams of the curves of Figure 1.

The construction of the Gauss diagram $G(\Gamma)$ is valid not only for a plane curve Γ, but also for a generic immersion of the circle into an arbitrary orientable surface.

There are three different Gauss diagrams of order 2 (see Fig. 5). Among them, two first diagrams correspond to plane curves, while the last one can only be realized by a curve in the torus (and hence in an arbitrary surface of genus higher than one).

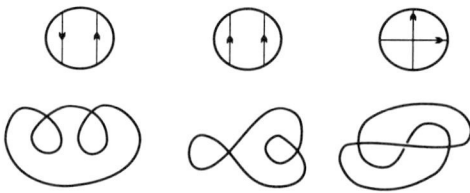

Figure 5: Curves corresponding to Gauss diagrams of order 2.

In general, any arbitrary Gauss diagram corresponds to a curve immersed in a surface of a sufficiently high genus. This was proved by J. Scott Carter [C] who closely followed L. Heffter's [H] classical construction of a closed oriented surface containing a given graph with a prescribed cyclical order of edges at every vertex.

Theorem 1 (J. Scott Carter [C]) *For any Gauss diagram G there is a minimal number g (the genus of G) such that G is the Gauss diagram of a certain generic curve in the closed orientable surface of genus g. This curve is defined uniquely up to a diffeomorphism of the surface. In other words,* the Gauss diagram constitutes a complete invariant for those curves immersed in a surface that cannot be immersed in a surface of smaller genus. In particular, the Gauss diagram of a plane curve is the complete invariant of its spherical class. Moreover, if two immersions of S^1 in two different surfaces Q_1, Q_2 have the same Gauss diagram, then one can remove some handles from either or both of the surfaces Q_1, Q_2 so that the resulting immersions will be equivalent.

The minimal surface of a Gauss diagram can be obtained by attaching disks to the curve viewed as a 1-skeleton in such a way that in the double points these pieces are patched together consistently with the cyclic order of the 4 branches of the curve. This can always be done in a unique way; an example is shown in Fig. 6.

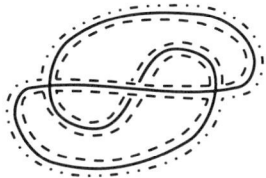

Figure 6: The Carter's surface of this curve is a torus, which can be obtained by attaching disks along each of the two dotted lines.

If n is the order of the Gauss diagram (the number of double points) and c is the number of faces attached, then the Euler characteristic of Carter's curface is $2 - 2g = n - 2n + c$, hence $g = (2 + n - c)/2 \leq (n + 1)/2$.

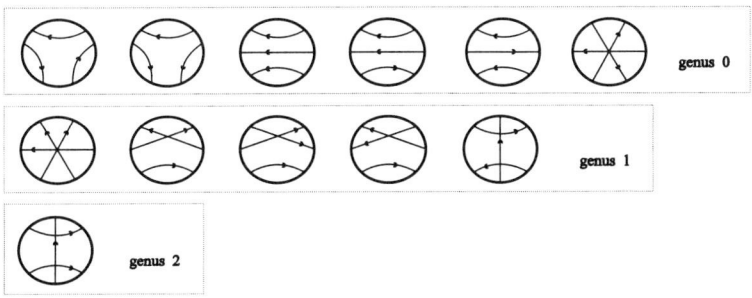

Figure 7: Gauss diagrams of order 3. Out of the total number of 12 different Gauss diagrams of order three, 6 diagrams have genus zero, 5 diagrams are of genus one and 1 diagram of genus two.

3. Arnold's basic invariants

Let X be the space of all smooth immersions of the circle S^1 in the plane $\mathbb{R}^2$. Elements of X will be referred to as *curves*.

According to a classical theorem due to H. Whitney [Wh], the space X is made up of a countable number of connected components each containing all curves with the same value i of *rotation number*. The rotation number, or Whitney index, of a curve is the degree of the mapping $S^1 \to S^1$ which sends every point in the direction of the tangent vector at this point, i.e., the total number of revolutions made by the tangent vector when the point traverses the curve once in the positive direction. Since we identify the two immersions that differ by a mirror symmetry of the plane, we will suppose the Whitney index to be nonnegative.

Choose a standard curve K_i for every nonnegative value of i:

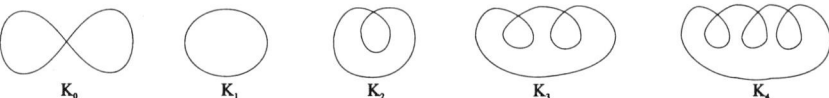

$$K_0 \qquad K_1 \qquad K_2 \qquad K_3 \qquad K_4$$

Figure 8: Standard curves.

Denote by Σ^+ the set of all curves that have a point of direct self-tangency, i. e. a point which is visited at least twice with tangent vectors having the same direction (Fig. 9a). In a similar way, let Σ^- be the set of all curves that have a point of inverse self-tangency (Fig. 9b) and Σ^{St}, the set of all curves having at least one triple point.

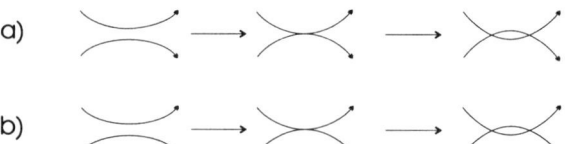

Figure 9: Direct and inverse self-tangency.

The discriminant is the union of the three varieties Σ^+, Σ^- and Σ^{St}. Set
$$G = X \setminus (\Sigma^+ \cup \Sigma^- \cup \Sigma^{St}).$$

An invariant of generic curves is a function on the space G which is constant on connected components, i.e., whose value may only change when we move through the discriminant hypersurface. Therefore the question initially raised by V. Arnold in [A1] is quite natural: are there any invariants of generic curves which have *constant* jumps on the discriminant? To state the main theorem of [A1], we need to introduce a coorientation of each of the three components, Σ^+, Σ^- and Σ^{St}, of the discriminant.

If two curves are close to a generic point of Σ^+ and lie on either side of this hypersurface, then they are identical everywhere but in a small disk, where one of them has two additional double points. We say that this curve is on the *positive* side of Σ^+.

The same definition also holds for the coorientation of the component Σ^-.

The fact that the triple point component is also coorientable is not so evident, and the corresponding definition is more tricky. We explain this construction in Fig. 10. It takes some imagination to understand that a triple point can be resolved in only two essentially different ways, i.e., that the neighborhood of a curve with one triple point in the space of generic curves consists of two connected components. Either of these two components is given a sign $+$ or $-$ according to the following rule:

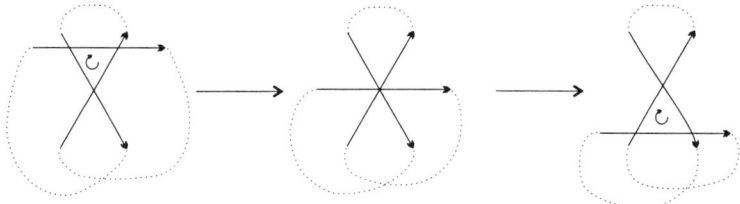

Figure 10: Coorientation of Σ^{St}.

Consider the new-born triangle that has replaced the triple point. Choose an arbitrary direction on the curve. It supplies each of the three sides of the triangle with a certain direction. There is also a cyclic order in the set of the three sides defined by the order in which the sides are passed when moving along the curve in the choosen direction. This cyclic order induces another direction on each of the three sides. The *sign* of the new-born triangle is equal to

$$(-1)^{\text{the number of sides on which the two directions are the same}}.$$

This sign does not depend on the choice of the direction on the curve.

Definition 2 *The strangeness St is an invariant which jumps by 1 only at a positive crossing of the cooriented stratum Σ^{St}.*

J^+ *is an invariant which jumps by 2 only at a positive crossing of the cooriented stratum Σ^+.*

J^- *is an invariant which jumps by -2 only at a positive crossing of the cooriented stratum Σ^-.*

Theorem 2 (V. Arnold [A1]) *For any sequence* $\mathbf{k} = (k_0, k_1, k_2, \ldots)$
there are unique invariants $St_{\mathbf{k}}$, $J_{\mathbf{k}}^+$, $J_{\mathbf{k}}^-$ *with normalization*

$$St_{\mathbf{k}}(K_i) = k_i, \qquad J_{\mathbf{k}}^+(K_i) = k_i, \qquad J_{\mathbf{k}}^-(K_i) = k_i.$$

Different choices of the normalizing sequence $\mathbf{k}$ may lead to additional remarkable properties of the invariants $St_{\mathbf{k}}$, $J_{\mathbf{k}}^+$, $J_{\mathbf{k}}^-$. In [A1] Arnold used the following normalization:

for St: $k_0 = 0$, $k_i = i - 1$ $(i = 1, 2, 3, \ldots)$;
for J^+: $k_0 = 0$, $k_i = -2(i - 1)$ $(i = 1, 2, 3, \ldots)$;
for J^-: $k_0 = -1$, $k_i = -3(i - 1)$ $(i = 1, 2, 3, \ldots)$

— and proved that it is the only normalization which leads to invariants, additive under the connected sum of curves. Below for the invariants with this normalization we will omit the subscript $\mathbf{k}$.

Another choice of normalization $\mathbf{k}$ gives invariants of spherical classes of curves (see Sec. 6).

To calculate the invariants of a curve by definition, we connect the given curve with one of standard curves K_i by a smooth path in the space of immersions that meets the discriminant only in its generic points. The value of the invariant on the standard curve is known. Accumulating all jumps of the invariant during deformation, we obtain the value of the invariant on the initial curve.

Example 2. As an example, consider the plane curve

$$\Gamma = $$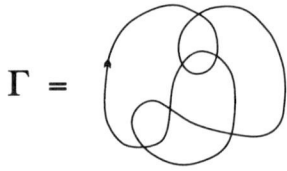

Figure 11: An example curve.

Figure 12 shows a deformation of this curve to K_1 and the corresponding jumps of the basic invariants.

It follows that $St_{\mathbf{k}}(\Gamma) = k_1 + 1$, $J_{\mathbf{k}}^+(\Gamma) = k_1 + 2$, $J_{\mathbf{k}}^-(\Gamma) = k_1 - 4$.
In particular, with the normalization we have adopted

$$St(\Gamma) = 1, \quad J^+(\Gamma) = 2, \quad J^-(\Gamma) = -4.$$

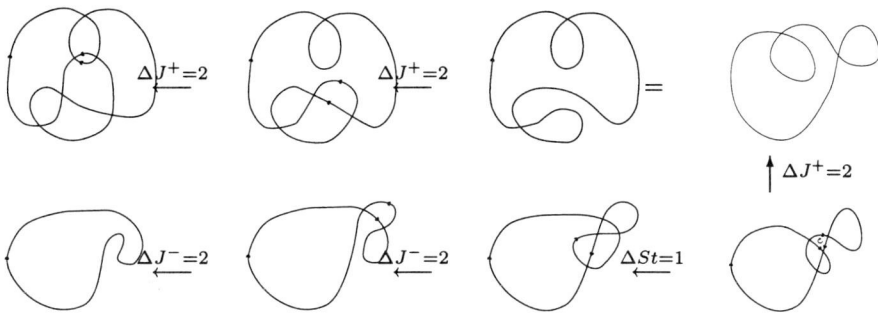

Figure 12: Calculation of the invariants.

4. Formulas of Viro and Shumakovich

Choose an arbitrary orientation of the curve $\Gamma \subset \mathbb{R}^2$. Pick a connected component C of $\mathbb{R}^2 \setminus \Gamma$ and a point $x \in C$.

Definition 3 *The index* $\mathrm{ind}_\Gamma(C)$ *of the region* C *is the algebraic number of turns made by the radius-vector from* x *to a point travelling along* Γ.

Evidently, $\mathrm{ind}_\Gamma(C)$ does not depend on the choice of $x \in C$ but does depend on the orientation of Γ.

The index of the exterior component is zero. If C_1 and C_2 are two adjacent components and C_1 lies on the left of Γ while C_2 lies on the right, then $\mathrm{ind}_\Gamma(C_2) = \mathrm{ind}_\Gamma(C_1) + 1$. We arrive at the *Alexander rule*[2] for the indices of four connected components in the vicinity of a double point of Γ (see Fig. 13).

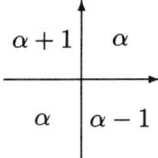

Figure 13: Alexander numbering.

Now consider the smoothing of a double point determined by the orientation (Fig. 14).

---

[2] J. Alexander [Al] was the first to notice that any generic plane curve admits such a distribution of indices.

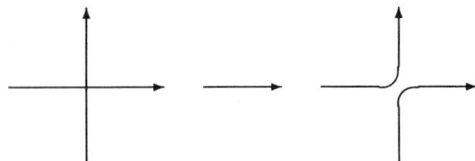

Figure 14: Smoothing.

Let $\tilde{\Gamma}$ be obtained from Γ by smoothing all double points of Γ. It is a union of circles. Each smoothing joins two local components of the complement of Γ with the same index into one component of the complement of $\tilde{\Gamma}$. Therefore a connected component $\tilde{C}$ of $\mathbb{R}^2 \setminus \tilde{\Gamma}$ inherits the index $\text{ind}_{\tilde{\Gamma}}(\tilde{C})$.

Theorem 3 (O. Viro [V])

$$J^+(\Gamma) = 1 + n - \sum_{\tilde{C}} \text{ind}_{\tilde{\Gamma}}^2(\tilde{C}) \chi(\tilde{C});$$

$$J^-(\Gamma) = 1 - \sum_{\tilde{C}} \text{ind}_{\tilde{\Gamma}}^2(\tilde{C}) \chi(\tilde{C}),$$

where n is the number of double points of Γ and χ is the Euler characteristic.

This theorem can be proved by checking that the right hand side expressions give correct values for the standard curves K_i and verifying that their jumps under the three elementary perestroikas conform to the definition of the invariants. The same remark also holds for the formula of Shumakovich (Theorem 4) and formula of Polyak (Theorem 5).

Example 3. Figure 15 shows the result of smoothing all the double points of our example curve.

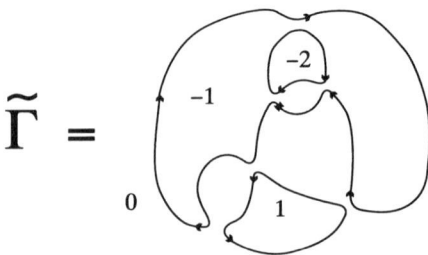

Figure 15: Smoothed curve.

Numbers in this picture indicate the indices $\operatorname{ind}_{\tilde{\Gamma}}(\tilde{C})$ for all components $\tilde{C}$. So,

$$
\begin{aligned}
J^+(\Gamma) &= 1 + 6 - (1 + 4) = 2, \\
J^-(\Gamma) &= 1 - (1 + 4) = -4.
\end{aligned}
$$

A. Shumakovich [Sh] gave a formula for the strangeness in similar terms.

Fix a base point $*$ on the (oriented) curve Γ. Then each double point p of Γ obtains a sign $s(p) = +1$ or -1 depending on whether the first *passage* of this point starting from the base point $*$ is positive or negative. (We have defined the sign of the passage earlier in Sec. 2). The signs of double points depend on the choice of the base point $*$.

Definition 4 *The index* $\operatorname{ind}_{\Gamma}(p)$ *of a double point* p *of* Γ *is the arithmetic mean of indices of the four components of the complement to* Γ *near* p. *(If one component approaches* p *twice from two opposite directions, then its index is counted twice.)*

Theorem 4 (A. Shumakovich [Sh]) *If the base point* $*$ *lies on the exterior contour of* Γ, *then*

$$
St(\Gamma) = \sum_p s(p) \operatorname{ind}_{\Gamma}(p).
$$

Example 4. We determine the signs and indices of all double points of our curve Γ:

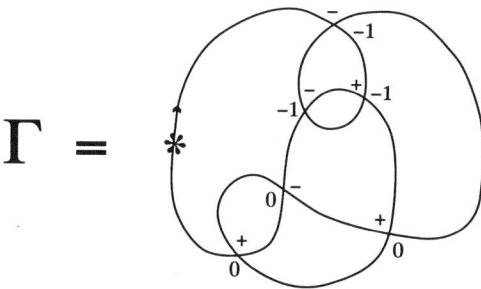

Figure 16: Indices and signs of double points of the example curve.

— and find that $St(\Gamma) = 1 - 1 + 1 = 1$.

5. Formulas of Polyak

M. Polyak in [Po] gave explicit formulas for Arnold's invariants in terms of *based Gauss diagrams*, i. e. Gauss diagrams with a fixed base point $*$, or more exactly, with a distinguished arc on the outer circle of the diagram.

The choice of the base point defines a sign $s(c)$ of every chord c according to the rule: $s(c) = s(p)$ where p is the double point that corresponds to the chord c and $s(p)$ is the sign introduced in the previous section. In other words, the chord is positive with respect to the chosen base point $*$, if, starting from $*$ and moving counterclockwise, its endpoint which is passed first corresponds to a positive passage of the double point.

Let A be a *based chord diagram*. This is an object that differs from a based Gauss diagram in that the chords of A are non-oriented.

A based chord diagram A defines a rational-valued function $G \mapsto \langle A, G \rangle$ on the set of based Gauss diagrams. In the definition that follows a *subdiagram* of type A means a based Gauss subdiagram of G which becomes A after forgetting the orientation of all its chords.

Definition 5

$$\langle A, G \rangle = \sum_{G_A} \prod_{c \in G_A} s(c),$$

where the sum is taken over all subdiagrams G_A of type A and the product is taken over all chords of G_A.

It turns out that, in the space of all generic curves with a fixed number of double points and a fixed Whitney index, Arnold's basic invariants J^+, J^-, St can be expressed in terms of the three functions $\langle A, \cdot \rangle$, where A is a based chord diagram of order two.

Theorem 5 (M. Polyak [Po]) *Let G be a based Gauss diagram of a generic curve Γ. Then*

$$J^+(\Gamma) = \left\langle \vcenter{\hbox{}}, G \right\rangle - \left\langle \vcenter{\hbox{}}, G \right\rangle - 3\left\langle \vcenter{\hbox{}}, G \right\rangle - \frac{n + \mathrm{ind}^2\Gamma - 1}{2};$$

$$J^-(\Gamma) = \left\langle \vcenter{\hbox{}}, G \right\rangle - \left\langle \vcenter{\hbox{}}, G \right\rangle - 3\left\langle \vcenter{\hbox{}}, G \right\rangle - \frac{3n + \mathrm{ind}^2\Gamma - 1}{2};$$

$$St(\Gamma) = -\frac{1}{2}\left\langle \vcenter{\hbox{}}, G \right\rangle + \frac{1}{2}\left\langle \vcenter{\hbox{}}, G \right\rangle + \frac{1}{2}\left\langle \vcenter{\hbox{}}, G \right\rangle$$
$$+ \frac{n + \mathrm{ind}^2\Gamma - 1}{4},$$

where n is the number of double points of Γ and $\mathrm{ind}\Gamma$ is the rotation number of Γ.

Example 5.

For our test curve

Figure 17: Example curve with a base point.

we have indΓ $= -1$, $n = 6$ and the based Gauss diagram looks as follows:

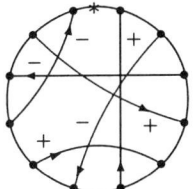

Figure 18: Based Gauss diagram.

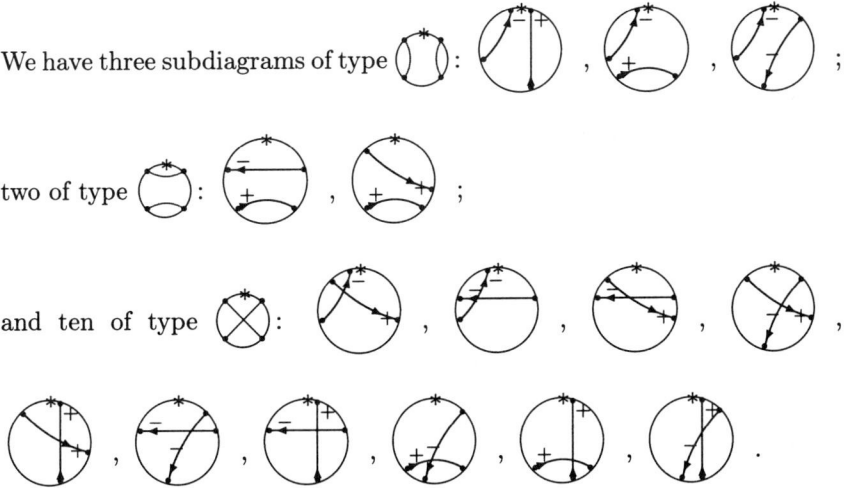

Hence,

$$\langle \bigcirc\!\!\bigcirc, G \rangle \ = \ -1 - 1 + 1 = -1,$$

$$\langle \ominus, G \rangle \ = \ -1 + 1 = 0,$$

$$\langle \otimes, G \rangle \ = \ -1 + 1 - 1 - 1 + 1 + 1 - 1 - 1 + 1 - 1 = -2.$$

and

$$J^+(\Gamma) \ = \ -1 + 6 - \frac{6 + 1 - 1}{2} = 2;$$

$$J^-(\Gamma) \ = \ -1 + 6 - \frac{18 + 1 - 1}{2} = -4;$$

$$St(\Gamma) \ = \ (1/2) - 1 + \frac{6 + 1 - 1}{4} = 1,$$

6. Arnold's inariants for spherical curves

After adding one point "at infinity" the plane $\mathbb{R}^2$ becomes the sphere S^2. Hence, a plane curve Γ can always be considered as a spherical curve. Since under this process different plane curves may become equivalent spherical curves, the natural question is: what plane curve invariants in fact represent invariants of spherical curves?

In a recent paper [A3] Arnold noted that $SJ^+(\Gamma) = J^+(\Gamma) + \frac{1}{2}\mathrm{ind}^2\Gamma$, where $\mathrm{ind}\Gamma$ is the rotation number of Γ, is a spherical invariant. Similarly $SJ^-(\Gamma) = J^-(\Gamma) + \frac{1}{2}\mathrm{ind}^2\Gamma$ and $SSt(\Gamma) = St(\Gamma) - \frac{1}{4}\mathrm{ind}^2\Gamma$ are also spherical invariants. Since the rotation number $\mathrm{ind}\,\Gamma$ does not change under local perestroikas that consist in passing through Σ^+, Σ^-, Σ^{St}, the spherical invariants SJ^+, SJ^-, SSt can be considered as usual plane curve invariants $J_{\mathbf{k}}^+$, $J_{\mathbf{k}}^-$, $St_{\mathbf{k}}$ with a special normalization:

for SSt: $\quad k_0 = 0, \quad k_i = -\dfrac{(i-2)^2}{4} \quad (i = 1, 2, 3, \ldots);$

for SJ^+: $\quad k_0 = 0, \quad k_i = \dfrac{(i-2)^2}{2} \quad (i = 1, 2, 3, \ldots);$

for SJ^-: $\quad k_0 = -1, \quad k_i = \dfrac{(i-3)^2 - 3}{2} \quad (i = 1, 2, 3, \ldots).$

Example 6. The spherical invariants of our curve Γ (Fig. 11) are $SSt(\Gamma) = 3/4$; $SJ^+(\Gamma) = 5/2$; $SJ^-(\Gamma) = -7/2$.

A natural complete invariant of spherical curves is its Gauss diagram. So each spherical invariant in principle can be expressed in terms of Gauss diagram. Explicit formulas for SJ^+, SJ^-, SSt can be obtained from Polyak's formulas by excluding the base point information. The following numbers take part in the answer.

Definition 6 *Let* *be a Gauss subdiagram of a Gauss diagram G. The endpoints of its two chords divide the circle into four arcs. Let a be the number of endpoints of chords of G that belong to the upper arc, i. e. lie between the two arrowheads of the oriented chords, and b, c, d are the numbers of endpoints on the other three arcs in the counterclockwise order. Then the weight* w *of the Gauss subdiagram is the number* w $=$ $-a + b - c + d$.

Theorem 6

$$SJ^+(\Gamma) = -\frac{n-1}{2} - \#\,\,\circleddash + \#\,\,\circleddash + \frac{3n}{2}\sum w$$

$$SJ^-(\Gamma) = -\frac{3n-1}{2} - \#\,\,\circleddash + \#\,\,\circleddash + \frac{3n}{2}\sum w$$

$$SSt(\Gamma) = \frac{n-1}{4} + \frac{1}{2}\#\,\,\circleddash - 1/2\#\,\,\circleddash - \frac{n}{4}\sum w$$

where $\#\,\,\circleddash$ and $\#\,\,\circleddash$ are the numbers of Gauss subdiagrams of Γ of the indicated form.

Example 7. It is easy to see from the example of Sec. 5 that for our favourite curve Γ:

$$\#\,\,\circleddash = 3, \qquad \#\,\,\circleddash = 2$$

and weights of the ten Gauss subdiagrams of type are equal respectively to

$$0, \quad 4, \quad 8, \quad 0, \quad 4, \quad -4, \quad 0, \quad 4, \quad 0, \quad 8.$$

So, $\sum w = 24$ and

$$SJ^+(\Gamma) = -\frac{5}{2} - 3 + 2 + \frac{3}{12} \cdot 24 = 5/2;$$

$$SJ^-(\Gamma) = -\frac{17}{2} - 3 + 2 + \frac{3}{12} \cdot 24 = -7/2;$$
$$SSt(\Gamma) = \frac{5}{4} + \frac{3}{2} - \frac{2}{2} - \frac{1}{24} \cdot 24 = 3/4.$$

References

[A1] V. I. Arnold, *Plane curves. Their invariants, perestroikas and classifications*, Advances in Soviet Math. **21** (1995), 00–00.

[A2] V. I. Arnold, *Topological invariants of plane curves and caustics*, University Lecture Series, Vol. 5, AMS, Providence, 1994.

[A3] V. I. Arnold, *Geometry of spherical curves and algebra of quaternions*, (in Russian), Russian Math. Surveys **50** (1995), no. 1, 3–68.

[Al] J. Alexander, *Topological invariants of knots and links*, Trans. Amer. Math. Soc. **30** (1928), 275–306.

[C] J. Scott Carter, *Classifying immersed curves*, Proc. Amer. Math. Soc. **111** (1991), no. 1, 281–287.

[H] L. Heffter, *Über das Problem der Nachbargebiete*, Math. Annalen **38** (1891), 477–508.

[Po] M. Polyak, *Invariants of plane curves and Legendrian fronts via Gauss diagrams*, preprint, 1994.

[Sh] A. Shumakovich, *Explicit formulas for the strangeness of a plane curve*, preprint (in Russian), 1994.

[Tab] S. Tabachnikov, *Invariants of smooth triple point free plane curves*, preprint, University of Arkansas, 1994.

[V] O. Viro, *First degree invariants of generic curves on surfaces*, preprint, Uppsala University, 1994.

[Wh] H. Whitney, *On regular closed curves in the plane*, Compositio Math. **4** (1937), 276–284.

Program Systems Institute,
Pereslavl-Zalessky, 152140, Russia
E-mail: chmutov@math.botik.yaroslavl.su,
        duzhin@math.botik.yaroslavl.su.

Nonlinear integrable equations
and nonlinear Fourier transform

A. S. Fokas, I. M. Gelfand, M. V. Zyskin

Introduction

In this paper we study nonlocal functionals whose kernels are homogeneous generalized functions. We also use such functionals to solve the Korteweg-de Vries (KdV), the nonlinear Schrödinger (NLS) and the Davey-Stewartson (DS) equations.

The solution of certain integrable equations in terms of formal power series was obtained in [4], [5]. In these papers the solution was expressed in a formal power series involving scattering data. In this paper in addition to developing techniques for multiplying and inverting nonlocal functionals we also:

(a) Give the correct version of these series by giving meaning to the relevant kernels, see (2.10) and (3.18)).

(b) We invert these series to obtain scattering data in terms of initial data.

(c) Prove the convergence of these series.

(d) We extend these results to equations in two space dimensions.

1. Nonlocal analytic functionals withhomogeneous kernels

The calculus of local functionals was developed by Gelfand and Dikii [4]. Local functionals of one function $u(x)$ can be written as multiple integrals using the kernels given by the δ-function and its derivatives. For example,

$$\int u^2(x)dx = \int u(x_1)u(x_2)\delta(x_1 - x_2) \; dx_1dx_2 \; ,$$

$$\int (u')^3 dx = \int u(x_1)u(x_2)u(x_3)\delta'(x - x_1)\delta'(x - x_2)\delta'(x - x_3) \; dxdx_1dx_2dx_3 \; ,$$

$$\int u^2(u')^3 dx = \int u(x_1)u(x_2)u(x_3)u(x_4)u(x_5)\delta(x - x_1)\delta(x - x_2)\delta'(x - x_3)$$
$$\delta'(x - x_4)\delta'(x - x_5) \; dxdx_1dx_2\ldots dx_5 \; .$$

and so on.

Nonlocal analytic functionals are those functionals whose kernels involve homogeneous generalized functions.

In the case of a real variable a basis in the space of homogeneous generalized functions is [1]

$$\frac{x_+^{\lambda-1}}{\Gamma(\lambda)}, \quad \frac{x_-^{\lambda-1}}{\Gamma(\lambda)}, \quad (x+i0)^\lambda.$$

In the case of a complex variable a basis in the space of homogeneous generalized functions is

$$z^s \bar{z}^{s+n} \quad n = 0, \pm 1, \pm 2, \ldots$$

$$\delta - function \quad and \quad its \quad derivatives$$

Remarks.

1)   Only those functionals which make sense in the framework of generalized functions are allowed. For example, the functionals

$$\int u(k_1)u(k_2)\frac{1}{k_1+k_2+i0}\delta(k_1+k_2)\,dk_1 dk_2,$$

and

$$\int u(k_1)u(k_2)\frac{1}{k_1+k_2+i0}\frac{1}{k_1+k_2-i0}\,dk_1 dk_2,$$

are not allowed, while the functional

$$\int u^2(k_1)u(k_2)\frac{1}{(k_1+k_2)^2}dk_1 dk_2 :=$$

$$\frac{1}{2}\int_0^\infty dk \int_{-\infty}^\infty dq \frac{1}{k^2}\left(u^2\left(\frac{q+k}{2}\right)u\left(\frac{q-k}{2}\right)\right.$$

$$\left. +u^2\left(\frac{q-k}{2}\right)u\left(\frac{q+k}{2}\right) - 2u^3(q)\right)$$

is allowed. (see [1] for details).

2)   Local functionals are a particular case of functionals with homogeneous kernels, for example,

$$\int u^2(x)dx =$$

$$= \int u(x_1)u(x_2)\delta(x_1-x_2)\,dx_1 dx_2 = \int u(x_1)u(x_2)\frac{(x_1-x_2)_+^{\lambda-1}}{\Gamma(\lambda)}\bigg|_{\lambda=0}dx_1 dx_2.$$

The product of two nonlocal analytic functionals is also a nonlocal analytic functional whose kernel is the direct product of the kernels of the two starting functionals. For example,

$$\left(\int u(x_1)u(x_2)\frac{(x_1-x_2)^\lambda}{\Gamma(\lambda+1)}\right) \cdot \left(\int u(y_1)u(y_2)u(y_3)\frac{(y_1-y_2)^{\mu_1}}{\Gamma(\mu_1+1)}\frac{y_3^{\mu_2}}{\Gamma(\mu_2+1)}\right)$$

$$= \int u(x_1)u(x_2)u(x_3)u(x_4)u(x_5)\frac{(x_1-x_2)^\lambda}{\Gamma(\lambda+1)}\frac{(x_3-x_4)^{\mu_1}}{\Gamma(\mu_1+1)}\frac{x_5^{\mu_2}}{\Gamma(\mu_2+1)}\,dx_1\ldots dx_5$$

There are certain relations in the algebra of nonlocal analytic functionals.

Examples.

1)

$$\left(\int u_1(x_1)u_2(x_2)\Theta(1-x_1)\Theta(x_1-x_2)\Theta(x_2)\, dx_1 dx_2 \right) \cdot$$

$$\left(\int v_1(y_1)v_2(y_2)v_3(y_3)\Theta(1-y_1)\Theta(y_1-y_2)\Theta(y_2-y_3)\Theta(y_3)\, dy_1 dy_2 dy_3 \right)$$

$$= \int_0^1 u_1(x_1)u_2(x_2)v_1(x_3)v_2(x_4)v_3(x_5)\Theta_{12}\Theta_{34}\Theta_{45}\, dx_1 \ldots dx_5$$

$$= \int_0^1 u_1(x_1)u_2(x_2)v_1(x_3)v_2(x_4)v_3(x_5)\, (\Theta_{12}\Theta_{23}\Theta_{34}\Theta_{45} + \Theta_{13}\Theta_{32}\Theta_{24}\Theta_{45}$$

$$+\Theta_{13}\Theta_{34}\Theta_{42}\Theta_{25} + \Theta_{13}\Theta_{34}\Theta_{45}\Theta_{52} + \Theta_{31}\Theta_{12}\Theta_{24}\Theta_{45}$$

$$+\Theta_{31}\Theta_{14}\Theta_{42}\Theta_{25} + \Theta_{31}\Theta_{14}\Theta_{45}\Theta_{52} + \Theta_{34}\Theta_{41}\Theta_{12}\Theta_{25}$$

$$+\Theta_{34}\Theta_{41}\Theta_{15}\Theta_{52} + \Theta_{34}\Theta_{45}\Theta_{51}\Theta_{12})\, dx_1 \ldots dx_5$$

where $\Theta(x) = \dfrac{x_+^0}{\Gamma(1)} = \begin{cases} 1, & x \geq 0 \\ 0, & x < 0 \end{cases}$, and $\Theta_{ij} := \Theta(x_i - x_j)$

We have multiplied two functionals, with kernels of degree 0, and with integration domain given by the simplexes, $0 < x_1 < x_2 < 1$ and $0 < y_1 < y_2 < y_3 < 1$,respectively. For the product functional, the integration domain is not a simplex, but we can write it as sum of functionals, such that the integration domain of each functional is a simplex. The simplexes are given by all possible orderings of the letters x_1, x_2, y_1, y_2, y_3 such that $x_1 < x_2$ and $y_1 < y_2 < y_3$ for all the orderings (all shuffles of $(x_1 x_2)(y_1 y_2 y_3)$).

Remark. The functional $\int u_1(x_1)u_2(x_2)\Theta(y-x_1)\Theta(x_1-x_2)\Theta(x_2)\, dx_1 dx_2$ with $u_2(x) = \frac{1}{1-x}$, $u_1(x) = \frac{1}{x}$ is the dilogarithm $Li_2(y)$.

Example 2.

$$\int u_1(x_1)u_2(x_2)u_3(x_3)u_4(x_4)u_5(x_5)\Theta_{12}\Theta_{32}\Theta_{34}\Theta_{54}\, dx_1 \ldots dx_5$$

$$- \left(\int u_1(x_1)u_2(x_2)u_3(x_3)u_4(x_4)u_5(x_5)\Theta_{12}\Theta_{32}\, dx_1 \ldots dx_3 \right)$$

$$\left(\int u_4(x_4)u_5(x_5)\, \Theta_{54}dx_4 = \int u_1(x_1)u_2(x_2)\ldots u_5(x_5)\Theta_{12}\Theta_{23}\Theta_{34}\Theta_{45}\, dx_1 \ldots dx_5 \right.$$

Example 3.

$$\left(\int u_1(k_1)u_2(k_2)\frac{1}{k_1+i0}\delta(k_1+k_2)\frac{dk_1dk_2}{(2\pi i)^2} \right)$$

$$\cdot \left(\int v_1(q_1)v_2(q_2)v_3(q_3)\frac{1}{(q_1+i0)(q_1+q_2+i0)}\delta(q_1+q_2+q_3)\cdot\frac{dq_1dq_2dq_3}{(2\pi i)^3} \right)$$

$$= \int u_1(k_1)u_2(k_2)v_1(k_3)v_2(k_4)v_3(k_5)\frac{\delta(k_1+k_2)\delta(k_1+k_2+k_3)}{(k_1+i0)(k_3+i0)(k_3+k_4+i0)}\frac{dk_1\ldots dk_5}{(2\pi i)^5}$$

$$= \int u_1(k_1)u_2(k_2)v_1(k_3)v_2(k_4)v_3(k_5)\delta(k_1+k_2+k_3+k_4+k_5)$$

$$\cdot \left(p(1,2,3,4)+p(1,3,2,4)+p(1,3,4,2)+p(1,3,4,5)+p(3,1,2,4) \right.$$

$$\left. +p(3,1,4,2)+p(3,1,4,5)+p(3,4,1,2)+p(3,4,1,5)+p(3,4,5,1) \right)\frac{dk_1\ldots dk_5}{(2\pi i)^4}$$

where

$$p(m_1,m_2,m_3,m_4)$$

$$= \frac{1}{(k_{m_1}+i0)(k_{m_1}+k_{m_2}+i0)(k_{m_1}+k_{m_2}+k_{m_3}+i0)(k_{m_1}+k_{m_2}+k_{m_3}+k_{m_4}+i0)}$$

Nonlocal analytic functionals appear naturally in nonlinear integrable equations. For example, in the KdV equation the transformation from the potential to scattering data and the inverse transformation are given by nonlocal analytic functionals. We will write these functionals using the inverse scattering formalism [3]. Alternatively, one could start directly from the nonlinear equation.

2. Nonlocal functionals for the KdV equation

Let $u(x)$ be a C^∞ real-valued function of a real variable x, with the fast decrease as $x \to \pm\infty$.

We construct the following functionals of u :

$$a(k) = 1 + \sum_{n=1}^{\infty}(-)^n \int u(x_2)u(x_4)\ldots u(x_{2n})\Theta_{12}\Theta_{23}\Theta_{34}\cdot\ldots\cdot\Theta_{2n-1,2n} \tag{1}$$

$$\delta(x-x_1+x_2-\ldots+x_{2n})\exp(2ikx)\,dxdx_1dx_2\ldots dx_{2n},$$

$$b(k) = \frac{1}{2i(k+i0)}\sum_{n=0}^{\infty}(-)^{n+1}\int u(x_1)u(x_3)\cdot\ldots\cdot u(x_{2n+1})\Theta_{12}\Theta_{23}\cdot\ldots\cdot\Theta_{2n,2n+1}$$

$$\delta(-x+x_1-x_2+x_3-\ldots+x_{2n+1})\exp(-2ikx)\,dxdx_1\ldots dx_{2n},$$

$$\tag{2}$$

$$\Psi(k,x) = 1 + \sum_{n=1}^{\infty} (-)^n \int u(x_2)u(x_4) \cdot \ldots \cdot u(x_{2n})\Theta(x_1 - x)\Theta_{21}\Theta_{32} \cdot \ldots \cdot \Theta_{2n,2n-1}$$

$$\delta(x_0 - x_1 + x_2 - \ldots + x_{2n})\exp(-2ikx_0)\ dx_0 dx_1 \ldots dx_{2n},$$

$$(3)$$

$$\Phi(k,x) = 1 + \sum_{n=1}^{\infty} (-)^n \int u(x_2)u(x_4) \cdot \ldots \cdot u(x_{2n})\Theta(x - x_1)\Theta_{12}\Theta_{23} \cdot \ldots \cdot \Theta_{2n-1,2n}$$

$$\delta(-x_0 - x_1 + x_2 - \ldots + x_{2n})\exp(-2ikx_0)\ dx_0 dx_1 \ldots dx_{2n}.$$

$$(4)$$

Remark. In the series, say, for $\Psi(k,x)$ we can integrate over x_{2m+1}, $m = 0, 1, \ldots$, to get another formula for $\Psi(k,x)$:

$$\Psi(k,x) = 1 + \sum_{n=1}^{\infty} \frac{(-)^n}{(2ik)^n} \int u(x_1)u(x_2) \ldots u(x_{n-1})u(x_n)$$

$$\Theta(x_1 - x)\Theta_{21}\Theta_{32}\Theta_{43} \cdot \ldots \cdot \Theta_{n,n-1}$$

$$\cdot (e^{-2ik(x-x_1)} - 1)(e^{-2ik(x_1-x_2)} - 1) \ldots (e^{-2ik(x_{n-1}-x_n)} - 1)\ dx_1 \ldots dx_n.$$

We define $S(k)$ by

$$S(k) = \frac{b(k)}{a(k)}.$$

$$(5)$$

If $|1 - a(k)| < 1$, $S(k)$ can be written as

$$S(k) = -\frac{1}{2i(k+i0)} \sum_{n=0}^{\infty} \int u(x_1)u(x_3) \ldots u(x_{2n+1})$$

$$\Theta_{21}\Theta_{23}\Theta_{43}\Theta_{45} \ldots \Theta_{2n,2n-1}\Theta_{2n,2n+1}$$

$$(6)$$

$$\delta(-x + x_1 - x_2 + x_3 - \ldots + x_{2n+1})\exp(-2ikx)\ dx dx_1 dx_2 \ldots dx_{2n+1}$$

Convergence.
The series (1)–(4) converge for $k \neq 0$. They also converge at $k = 0$ if the moments of function $u(x)$, that is, $\int u(x)dx$, $\int xu(x)dx$, $\int x^2 u(x)dx$, $\ldots$ are small. The series (5) is convergent if (1) and (2) are convergent, and $|1 - a(k)| < 1$.

Indeed, let $k \neq 0$. Then

$$|\Phi(k,x)| \leq 1 + \sum_{n=1}^{\infty} \frac{1}{|k|^n} \int |u(x_1)u(x_2)\ldots u(x_n)|$$

$$\sin k(x-x_1)\sin k(x_1-x_2)\ldots\sin k(x_{n-1}-x_n)$$

$$\Theta(x-x_1)\Theta_{12}\Theta_{23}\Theta_{34}\cdot\ldots\cdot\Theta_{n-1,n}\ dx_1 dx_2\ldots dx_n$$

$$\leq 1 + \sum_{n=1}^{\infty} \frac{1}{n!}\frac{(\int|u(x)|dx)^n}{k^n}$$

For all k, including $k = 0$,

$$|\Phi(k,x)| \leq 1 + \sum_{n=1}^{\infty}\int|u(x_1)u(x_2)\ldots u(x_n)(x-x_1)(x_1-x_2)\ldots(x_{n-1}-x_n)|$$

$$\cdot\Theta(x-x_1)\Theta_{12}\cdot\ldots\cdot\Theta_{n-1,n}\ dx_1 dx_2\ldots dx_n$$

If the moments of function u are small, the series is convergent for $k = 0$ as well. We will consider two operations on functionals (1)–(5): multiplication and inversion. These operations are infinite-dimensional analogues of multiplication of functions and the inverse function. There could be some relations among the products of functionals.

Example 1.

Let us show that $a(k)a(-k) - b(k)b(-k) = 1$ for all $k \neq 0$. Indeed,

$$a(k)a(-k) = 1 + \sum_{n=1}^{\infty}\sum_{m=0}^{n}(-)^n\int u(x_2)u(x_4)\ldots u(x_{2n})\Theta_{12}\Theta_{23}\ldots\Theta_{2m-1,2m}$$

$$\cdot\Theta_{2m+2,2m+1}\cdot\Theta_{2m+3,2m+2}\ldots\Theta_{2n,2n-1}e^{2ikx_0}$$

$$\delta(-x_0+x_1-x_2+x_3-x_4+\ldots-x_{2n})\ dx_0 dx_1 dx_2\ldots dx_{2n},$$

$$b(k)b(-k) = \sum_{n=1}^{\infty}\sum_{m=1}^{n-1}(-)^n\int u(x_2)u(x_4)\ldots u(x_{2n})\Theta_{12}\Theta_{23}\cdot\ldots\cdot\Theta_{2m,2m+1}$$

$$\Theta_{2m+3,2m+2}\cdot\Theta_{2m+4,2m+3}\cdot\ldots\cdot\Theta_{2n,2n-1}e^{2ikx_0}$$

$$\delta(-x_0+x_1-x_2+x_3-x_4+\ldots-x_{2n})\ dx_0 dx_1 dx_2\ldots dx_{2n},$$

$$a(k)a(-k)-b(k)b(-k)-1 = \sum_{n=1}^{\infty}(-)^n\int u(x_2)u(x_4)\ldots u(x_{2n})$$

$$\exp(2ik(x_1-x_2+\ldots-x_{2n}))\Theta_{23}\Theta_{34}\cdot\Theta_{45}\cdot\ldots\cdot\Theta_{2n-1,2n}$$

$$dx_1\ldots dx_{2n} = 0,\quad k \neq 0.$$

From ( ) one can see that $\overline{a(k)} = a(-k), \overline{b(k)} = b(-k)$; therefore, $|a(k)| \geq 1$ and $|S(k)| = \left|\dfrac{b(k)}{a(k)}\right| = \dfrac{|b(k)|}{\sqrt{1 + |b(k)|^2}} \leq 1, \quad k \neq 0$.

Example 2.

Let us show that $\Psi(-k, y)a(k) + \Psi(k, y)b(k)e^{2iky} = \Phi(k, y)$. Indeed,

$$\Psi(-k, y)a(k) = 1 + \sum_{n=1}^{\infty}(-)^n \int (u(x_2)u(x_4)\ldots u(x_{2n}))$$
$$\cdot (\Theta_{12}\Theta_{23}\Theta_{34} \cdot \ldots \cdot \Theta_{2n-1,2n}$$

$$+ \sum_{m=1}^{n-1}(\Theta(x_1 - y)\Theta_{21}\Theta_{32} \cdot \ldots \cdot \Theta_{2m,2m-1}\Theta_{2m+1,2m+2}$$
$$\cdot \Theta_{2m+2,2m+3} \cdot \Theta_{2n-1,2n}) + \Theta(x_1 - y)\Theta_{21}\Theta_{32} \cdot \ldots \cdot \Theta_{2n,2n-1})$$

$$e^{-2ikx_0}\delta(x_0 + x_1 - x_2 + \ldots - x_{2n})\, dx_0 \ldots dx_{2n},$$

also,

$$\Psi(k, y)b(k)e^{2iky} = \sum_{n=1}^{\infty}(-)^{n+1} \int u(x_2)u(x_4) \cdot \ldots \cdot u(x_{2n})$$
$$\cdot (\Theta(x_1 - y)\Theta_{23}\Theta_{34}\Theta_{45} \cdot \ldots \cdot \Theta_{2n-1,2n}$$

$$+ \sum_{m=1}^{n-2}\Theta(x_1 - y)\Theta_{21}\Theta_{32} \cdot \ldots \cdot \Theta_{2m+1,2m} \cdot \Theta_{2m+2,2m+3}$$
$$\cdot \Theta_{2m+3,2m+4} \cdot \ldots \cdot \Theta_{2n-1,2n} + \Theta(x_1 - y)\Theta_{21}\Theta_{32} \cdot \ldots \cdot \Theta_{2n-1,2n})$$

$$e^{-2ikx_0}\delta(x_0 + x_1 - x_2 + \ldots - x_{2n})\, dx_0 dx_1 \ldots dx_{2n},$$

and

$$\Psi(-k, y)a(k) + \Psi(k, y)b(k)e^{2iky} = 1 + \sum_{n=1}^{\infty}(-)^n \int u(x_2)u(x_4) \cdot \ldots \cdot u(x_{2n})$$
$$\Theta(y - x_1)\Theta_{12}\Theta_{23} \cdot \ldots \cdot \Theta_{2n-1,2n}e^{-2ikx_0}$$

$$\delta(x + x_1 - x_2 - \ldots + x_{2n-1} - x_{2n})\, dx_0 dx_1 \ldots dx_{2n} = \Phi(k, y).$$

Example 3.

Let us take $S(k)$,given by the formal series (6), and $a(k)$, $b(k)$, given by the convergent series (1), (2). We can prove the following relation for the series in u(x):

$$a(k)S(k) = b(k).$$

The computation is similar to that of the Examples 1 and 2.

Inversion.

Let us define $S(x) = \int S(k)e^{2ikx}\dfrac{dk}{\pi}$. We suppose that $S(x)$ is a fast decreasing function as $x \to +\infty$. Formula (6) can be written as

$$\frac{d}{dx}S(x) = -\left(u(x) + \int u(x_1)u(x_3)\Theta_{21}\Theta_{23}\delta(x - x_1 + x_2 - x_3)dx_1dx_2dx_3\right.$$

$$+ \ldots \int u(x_1)u(x_3)\ldots u(x_{2n+1})\Theta_{21}\Theta_{23}\Theta_{43}\Theta_{45}\ldots\Theta_{2n,2n-1}\Theta_{2n,2n+1}$$

$$\left.\delta(x - x_1 + x_2 - x_3 + \ldots - x_4)\,dx_1dx_2\ldots dx_{2n+1} + \ldots\right)$$

(here the right-hand side is a formal series in $u(x)$).

We can invert it, that is, we can express $u(x)$ in terms of $S(x)$ by the formal series $u(x) = \displaystyle\sum_{k=1}^{\infty} W_k(x)$, where $W_k(x)$ is a nonlocal analytic functional of degree k in $S(x)$ ($S(x)$ has degree one). The functionals $W_k(x)$ are determined recursively:

$$W_1(x) = -\frac{d}{dx}S(x),$$

$$W_2(x) = -\int \frac{d}{dx_1}S(x_1)\frac{d}{dx_3}S(x_3)\Theta_{21}\Theta_{23}\delta(x - x_1 + x_2 - x_3)\,dx_1dx_2dx_3$$

$$= -\int S(x_1)S(x_3)\left(\delta(x_2 - x_1)\delta(x_2 - x_3)\delta(x - x_1 + x_2 - x_3)\right.$$

$$+\delta(x_2 - x_1)\Theta_{23}\delta'(x - x_1 + x_2 - x_3)$$

$$+\Theta_{21}\delta(x_2 - x_3)\delta'(x - x_1 + x_2 - x_3)$$

$$\left.+\Theta_{23}\Theta_{23}\delta''(x - x_1 + x_2 - x_3)\right)\,dx_1dx_2dx_3$$

$$= -\int S(x_1)S(x_3)\delta(x_2 - x_1)\delta(x_2 - x_3)\delta(x - x_1 + x_2 - x_3)\,dx_1dx_2dx_3$$

$$= -S^2(x) = \frac{d}{dx}\int S(x_1)S(x_2)\Theta(x_1 - x)\delta(x_1 - x_2)\,dx_1dx_2,$$

(we integrated by parts),

$$W_n(x) = -\sum_{k=2}^{n}\sum_{\substack{m_1,m_2\ldots m_k \geq 1 \\ m_1+m_2+\ldots+m_k=n}}\int W_{m_1}(x_1)W_{m_2}(x_3)\ldots W_{m_k}(x_{2k+1})$$

$$\Theta_{21}\Theta_{23}\cdot\Theta_{43}\Theta_{45}\cdot\ldots\cdot\Theta_{2k,2k-1}\Theta_{2k,2k+1}$$

$$\delta(x - x_1 + x_2 - \ldots + x_{2k} - x_{2k+1})\,dx_1dx_2\ldots dx_{2k+1}.$$

Lemma. *Consider the functionals of $S(x)$, given by the formal series*

$$\tilde{u}(x) = -\frac{d}{dx}S(x) + \frac{d}{dx}\int S(x_1)S(x_2)\Theta(x_1 - x)\delta(x_1 - x_2)\, dx_1 dx_2$$

$$-\frac{d}{dx}\sum_{n=1}^{\infty}\int S(x_1)S(x_2)\dots S(x_{2n+1})\Theta(-x_1 + x_2)$$

$$\Theta(-x_1 + x_2 - x_3 + x_4)\dots\Theta(-x_1 + x_2 - \dots + x_{2n})\Theta(-x_{2n+1} + x_{2n}) \tag{7}$$

$$\Theta(-x_{2n+1} + x_{2n} - x_{2n-1} + x_{2n-2})\dots\Theta(-x_{2n+1} + x_{2n} - \dots + x_2)$$

$$\delta(x - x_1 + x_2 - x_3 + \dots - x_{2n+1})\, dx_1 dx_2 \dots dx_{2n+1}$$

$$+\frac{d}{dx}\sum_{n=2}^{\infty}\int S(x_1)S(x_2)\dots S(x_{2n})$$

$$\Theta(-x_1 + x_2)\Theta(-x_1 + x_2 - x_3 + x_4)\dots\Theta(-x_1 + x_2 - \dots + x_{2n-2})$$

$$\Theta(x_1 - x)\Theta(x_1 - x_2 + x_3 - x)\dots\Theta(x_1 - x_2 + x_3 - \dots + x_{2n-1} - x)$$

$$\delta(x - x_2 + x_3 - x_4 + \dots - x_{2n})\, dx_1 \dots dx_{2n}.$$

$$\tilde{\Psi}(x, y) = \delta(x) - \sum_{n=0}^{\infty}\Theta(-x)\int S(x_1)S(x_3)\dots S(x_{2n+1})$$

$$\Theta(-x_1 + x_2)\Theta(-x_1 + x_2 - x_3 + x_4)\dots\Theta(-x_1 + x_2 - \dots + x_{2n})$$

$$\Theta(x_1 - y)\Theta(x_1 - x_2 + x_3 - y)\dots\Theta(x_1 - x_2 + x_3 - \dots + x_{2n+1} - y) \tag{8}$$

$$\delta(x - y + x_1 - x_2 + x_3 - \dots + x_{2n+1})\, dx_1 \dots dx_{2n+1}$$

$$+\sum_{n=1}^{\infty}\Theta(-x)\int S(x_1)S(x_2)\dots S(x_{2n})$$

$$\Theta(-x_1 + x_2)\Theta(-x_1 + x_2 - x_3 + x_4)\Theta(-x_1 + x_2 - \dots + x_{2n-2})$$

$$\Theta(x_1 - y)\Theta(x_1 - x_2 + x_3 - y)\dots\Theta(x_1 - x_2 + x_3 - \dots + x_{2n-1} - y)$$

$$\delta(x - x_1 + x_2 - \dots + x_{2n})\, dx_1 dx_2 \dots dx_{2n}.$$

We can substitute in these series $S(x)$ as a functional of $\{u(x)\}$, given by the formal series

$$S(x) = \int \Theta(x_1 - x)u(x_1)dx_1 + \sum_{n=1}^{\infty}\int u(x_1)u(x_3)\dots u(x_{2n+1})$$

$$\Theta_{21}\Theta_{23}\Theta_{43}\Theta_{45}\cdot\ldots\cdot\Theta_{2n,2n-1}\Theta_{2n,2n+1} \tag{9}$$

$$\cdot\delta(x_0 - x_1 + x_2 - x_3 + \dots - x_{2n+1})\, dx_0 dx_1 \dots dx_{2n+1}.$$

As a result of this substitution, we will have $\tilde{u}(x)$ and $\tilde{\Psi}(x,y)$ given by formal series in $\{u(x)\}$. Moreover, $\tilde{u}(x) = u(x)$,

$$\tilde{\Psi}(x,y) = \Psi(x,y) := \int \Psi(k,y)e^{2ikx}\frac{dk}{\pi}, \text{ where } \Psi(k,y) \text{ is given by (3).}$$

Proof

We will prove the lemma by induction in the degree of $\{u(x)\}$.

1) In the first order in $\{u(x)\}$

$$S_{(1)}(x) = \int \Theta(x_1 - x)u(x_1) \, dx_1,$$

$$\tilde{u}(x)_1 = -\frac{d}{dx}S_{(1)}(x) = u(x),$$

$$\tilde{\Psi}_{(1)}(x,y) = -\Theta(-x)\int S_{(1)}(x_1)\delta(x - y + x_1) \, dx_1$$

$$= -\Theta(-x)\int \Theta(x_2 - x_1)u(x_2)\delta(x - y + x_1) \, dx_1 dx_2$$

$$= -\int \Theta(x_1 - y)\Theta(x_2 - x_1)u(x_2)\delta(x - y + x_1) \, dx_1 dx_2$$

$$-\int \Theta(x_1 - y)\Theta(x_2 - x_1)u(x_2)\delta(x - x_1 + x_2)$$

$$\cdot dx_1 dx_2 = \Psi_{(1)}(x,y).$$

(In the last step we have made the change of variables $x_1 \to y + x_2 - x_1$).

2) Suppose that we have proved that

$$\tilde{\Psi}(x,y) = \delta(x) + \sum_{n=1}^{N}(-)^n \int u(x_2)u(x_4)\ldots u(x_{2n})\Theta(x_1 - y)\Theta_{21}\Theta_{32}\ldots\Theta_{2n,2n-1}$$

$$\delta(x - x_1 + x_2 - \ldots + x_{2n}) \, dx_1 dx_2 \ldots dx_{2n} + O(u^{N+1}).$$

From the definition of $\tilde{\Psi}(x,y)$

$$\tilde{\Psi}(x,y) = \delta(x) - \Theta(-x)\int \tilde{\Psi}(x_1,y)S(y - x - x_1) \, dx.$$

But $S(x)$ is a series in $\{u(x)\}$ with terms of degree ≥ 1 in u, therefore, if we know $\tilde{\Psi}(x,y)$ as functionals of $\{u(x)\}$ up to degree n, we can compute it in the next order $(n+1)$.

Notice that

$$(-)^n \int u(x_2)u(x_4)\ldots u(x_{2n})\Theta(x_1 - y)\Theta_{21}\Theta_{32}\Theta_{43}\ldots\Theta_{2n,2n-1}$$

$$\delta(x - x_1 + x_2 - \ldots x_{2n})\, dx_1 dx_2 \ldots + dx_{2n}$$

$$= (-)^n \int u(x_2)u(x_4)\ldots u(x_{2n})\Theta(x_1 - y)\Theta_{21}\Theta_{32}\Theta_{43}\ldots\Theta_{2n,2n-1}$$

$$\delta(x - y + x_1 - x_2 + x_3 - \ldots + x_{2n-1})\, dx_1 dx_2 \ldots dx_{2n}$$

$$= (-)^{n+1}\frac{d}{dx}\int u(x_2)u(x_4)\ldots u(x_{2n})\Theta(x_1 - y)\Theta_{21}\Theta_{32}\Theta_{43}\ldots\Theta_{2n,2n-1}\Theta_{2n+1,2n}$$

$$\cdot\delta\left(x - y + x_1 - x_2 + x_3 - \ldots + x_{2n-1} - x_{2n} + x_{2n+1}\right)\, dx_1 dx_2 dx_3 \ldots dx_{2n} dx_{2n+1}$$

(In the first step we have used the change of variables $x_1 \to x_2 + y - x_1, x_3 \to x_2 + x_4 - x_3, \ldots, x_{2n-1} \to x_{2n-2} + x_{2n} - x_{2n-1}$).

Also, $\delta(x) = -\dfrac{d}{dx}\displaystyle\int \Theta(x_1 - y)\delta(x - y + x_1)dx_1$, and

$$\tilde{\Psi}_{N+1}(x, y) = -\sum_{m=0}^{N}\Theta(-x)\int \tilde{\Psi}_{(m)}(x_0, y)S_{(N+1-m)}(y - x - x_0)\, dx_0$$

$$= -\sum_{m=0}^{N}\Theta(-x)(-)^{m+1}\int \frac{d}{dx_0}\left(u(x_2)u(x_4)\ldots u(x_{2m})\right.$$

$$\Theta(x, y_1)\Theta_{21}\Theta_{32}\Theta_{43}\ldots\Theta_{2m,2m-1}\Theta_{2m+1,2m}$$

$$\cdot\delta\left(x_0 - y + x_1 - x_2 + x_3 - \ldots + x_{2m+1}\right))$$

$$\cdot S_{(N+1-m)}(y - x - x_0)dx_0 dx_1 dx_2 \ldots dx_{2m+1}$$

$$= \Theta(-x)\int u(x_2)u(x_4)\ldots u(x_{2n+2})$$

$$\left(\sum_{m=0}^{N}(-)^{m+1}\Theta(x_1 - y)\Theta_{21}\Theta_{32}\ldots\Theta_{2m,2m-1}\Theta_{2m+1,2m}\cdot(\Theta_{2m+2,2m+1} + \Theta_{2m+1,2m+2})\right.$$

$$\left.(\Theta_{2m+3,2m+2}\Theta_{2m+3,2m+4}\cdot\ldots\cdot\Theta_{2N+1,2N}\Theta_{2N+1,2N+2})\right)$$

$$\cdot\delta(-x + x_1 - x_2 + x_3 - \ldots + x_{2N+1} - x_{2N+2})\, dx_1 dx_2 \ldots dx_{2N+2}$$

$$= (-)^{N+1}\Theta(-x)\int u(x_2)u(x_4)\ldots u(x_{2N+2})$$

$$(\Theta(x_1 - y_1)\Theta_{21}\Theta_{32}\Theta_{43}\cdot\ldots\cdot\Theta_{2N+2,2N+1} - (-)^{N+1}$$

$$\cdot\Theta(x_1 - y)\Theta_{12}\Theta_{32}\Theta_{34}\Theta_{54}\cdot\ldots\cdot\Theta_{2N+1,2N}\Theta_{2N+1,2N+2})$$

$$\delta(-x + x_1 - x_2 + x_3 - \ldots + x_{2N+1} - x_{2N+2})\, dx_1 dx_2 \ldots x_{2N+2}$$

The second term in the last expression is zero, because the volume of the integration domain vanishes ; the first term coincides with the term of degree $(N + 1)$ in $\Psi(x, y)$, see (3).

To prove the formula for $\tilde{u}(x)$ we use the relation $\tilde{u}(x) = -\dfrac{d}{dx} \displaystyle\int$ $S(x - x_1)\tilde{\Psi}(x_1, x)dx_1$, which follows from the definition of $\tilde{u}$ and $\tilde{\Psi}$. We know both $S(x)$ and $\tilde{\Psi}(x, y)$ as functionals in $\{u(x)\}$. The calculation of the same type as above gives that only the first order term in $\{u(x)\}$ is not zero:

$$\tilde{u}(x) = \frac{d}{dx} \sum_{n=1}^{\infty} (-)^n \int u(x_2)u(x_4)\ldots u(x_{2n})\delta(-x_1 + x_2 - \ldots + x_{2n})$$

$$\cdot (\Theta(x_1 - x)\Theta_{21}\Theta_{32}\Theta_{43} \cdot \ldots \cdot \Theta_{2n,2n-1} - (-)^n$$

$$\Theta(x_1 - x)\Theta_{12}\Theta_{32}\Theta_{34}\Theta_{54} \cdot \ldots \cdot \Theta_{2n-1,2n-2}\Theta_{2n-1,2n})$$

$$dx_1 dx_2 \ldots dx_{2n} = u(x)$$

The formula (7) can be written as follows:

$$u(x) = 4 \sum_{n=1}^{\infty} \int \frac{S(k_1)S(k_2)\ldots S(k_n)(k_1 + k_2 + \ldots + k_n)}{(k_1 + k_2 + i0)(k_2 + k_3 + i0)\ldots(k_{n-1} + k_n + i0)}$$

$$\exp(2ikx)\delta(k - k_1 - k_2 - \ldots - k_n)\frac{dk dk_1 dk_2 \ldots dk_n}{(2\pi i)^n} \tag{10}$$

Any polynomial in $u(x)$ and its derivative can be written as the functional of the same type, but with some polynomial in $\{k_i\}$ in the numerator. This property is similar to the usual Fourier transform of the linear function of u and its derivatives . Therefore it is natural to call transformation (10) the Nonlinear Fourier transform of u :

$$u^{(d_1)}u^{(d_2)}\ldots u^{(d_m)} = 4^m(2i)^{d_1+d_2+\ldots+d_m}$$

$$\sum_{n=m}^{\infty} \int \frac{S(k_1)S(k_2)\ldots S(k_n)}{(k_1 + k_2 + i0)(k_2 + k_3 + i0)\ldots(k_{n-1} + k_n + i0)}$$

$$\exp(2ikx) \cdot \delta(k - k_1 - k_2 - \ldots - k_n) \cdot \mathrm{Sym}_{d_1,d_2\ldots d_m}$$

$$\sum_{1 \le p_1 < p_2 < \ldots < p_{m-1} < n} (k_1 + k_2 + \ldots + k_{p_1})^{d_1+1}(k_{p_1} + k_{p_1+1})$$

$$(k_{p_1+1} + k_{p_1+2} + \ldots + k_{p_2})^{d_2+1}(k_{p_2} + k_{p_2+1})(k_{p_2+1} + k_{p_2+2} + \ldots + k_{p_3})^{d_3+1}$$

$$\cdot(k_{p_3} + k_{p_3+1})(k_{p_{m-1}+1} + k_{p_{m-1}+2} + \ldots + k_n)^{d_m+1}\frac{dk dk_1 dk_2 \ldots dk_n}{(2\pi i)^n}.$$

$$\tag{11}$$

Examples.

$$6uu_x = 32i \sum_{n=2}^{\infty} \int \frac{S(k_1)S(k_2)S(k_3)\dots S(k_n)}{(k_1 + k_2 + i0)(k_2 + k_3 + i0)\dots(k_{n-1} + k_n + i0)}$$

$$\exp(2ikx) \cdot \delta(k - k_1 - k_2 - \dots - k_n)$$

$$\cdot k \cdot ((k_1 + k_2 + \dots + k_n)^3 - (k_1{}^3 + k_2{}^3 + \dots + k_n{}^3)) \frac{dk\, dk_1\, dk_2 \dots dk_n}{(2\pi i)^n}$$

$$u_{xxx} + 6uu_x = -32i \sum_{n=1}^{\infty} \int \frac{S(k_1)S(k_2)S(k_3)\dots S(k_n)}{(k_1 + k_2 + i0)(k_2 + k_3 + i0)\dots(k_{n-1} + k_n + i0)}$$

$$\exp(2ikx) \cdot \delta(k - k_1 - k_2 - \dots - k_n)$$

$$\cdot k \cdot (k_1{}^3 + k_2{}^3 + \dots + k_n{}^3)) \frac{dk\, dk_1 dk_2 \dots dk_n}{(2\pi i)^n}$$

$$(12)$$

We see that for the differential polynomial $u_{xxx} + 6uu_x$ the corresponding polynomial in $\{k\}$ after the nonlinear Fourier transform is $(k_1{}^3 + k_2{}^3 + \dots + k_n{}^3)$, n= 1,2, ... We can use this fact to solve some nonlinear equation. Let $S(k, t) = S_0(k)e^{8ik^3 t}$, and $u(x, t)$ is defined as a functional of $S(k, t)$ by (10) (with $S(k, t)$ instead of $S(k)$). Such $u(x, t)$ solves the KdV equation

$$-\frac{d}{dt}u(x, t) = \frac{\partial^3}{\partial x^3}u(x, t) + 6u\frac{\partial u}{\partial x}.$$

Indeed, for such $u(x, t)$ the right-hand side of the equation is expressed in terms of $S(k, t)$ by (12), and it coincides with $-\frac{d}{dt}u(x, t)$.

Let us find polynomials in u and derivatives of u such that after the Nonlinear Fourier Transform the corresponding polynomial in k is given by $\sum_i k_i{}^{2l-1}$,

$l = 1, 2, \dots$ .In order to do this, let us first compute the resolvent:

$$\Psi(k, x) = 1 + \sum_{n=1}^{\infty} \int \frac{S(k_1)S(k_2)S(k_3)\dots S(k_n)}{(k + k_1 + i0)(k_2 + k_3 + i0)\dots(k_{n-1} + k_n + i0)}$$

$$\exp 2i(k_1 + \dots + k_n)x \, \frac{dk_1 \dots dk_n}{(2\pi i)^n},$$

$$R(k, x) = \Psi(k, x)\Psi(-k, x - 0)$$

$$= \frac{1}{+} \sum_{n=1}^{\infty} \int S(k_1)S(k_2)\dots S(k_n)\exp(2i(k_1 + \dots + k_n)x)\cdot$$

$$\left(\sum_{m=1}^{n-1} \right.$$

$$\frac{1}{(k_1+k_2+i0)(k_2+k_3+i0)\dots(k_{m-1}+k_m+i0)(k_m+k+i0)(k_{m+1}-k+i0)(k_{m+1}+k_{m+2}+i0)\dots(k_{n-1}+k_n+i0)}$$

$$+ \frac{1}{(k+k_1+i0)(k_1+k_2+i0)\dots(k_{n-1}+k_n+i0)}$$

$$+ \left. \frac{1}{(-k+k_1+i0)(k_1+k_2+i0)\dots(k_{n-1}+k_n+i0)} \right) \frac{dk_1\dots dk_n}{(2\pi i)^n}$$

$$= 1 + \sum_{n=1}^{\infty} \int \frac{S(k_1)S(k_2)\dots S(k_n)\exp(2i(k_1 + \dots k_n)x)}{(k_1+k_2+i0)(k_2+k_3+i0)\dots(k_{n-1}+k_n+i0)}$$

$$\left(\sum_{m=1}^{n-1} \left(\frac{1}{k_m+k+i0} + \frac{1}{k_{m+1}-k+i0} \right) + \frac{1}{k+k_1+i0} + \frac{1}{-k+k_1+i0} \right) \frac{dk_1\dots dk_n}{(2\pi i)^n}$$

$$\sim 1 - 2\sum_{n=1}^{\infty} \int \frac{S(k_1)S(k_2)\dots S(k_n)}{(k_1+k_2+i0)(k_2+k_3+i0)\dots(k_{n-1}+k_n+i0)} \exp 2i(k_1 + \dots + k_n)x$$

$$\sum_{l=1}^{\infty} \frac{1}{k^{2l}}(k_1^{2l-1} + k_2^{2l-1} + \dots + k_n^{2l-1})\cdot\frac{dk_1\dots dk_n}{(2\pi i)^n}.$$

Lemma.
1)

$$I_1 = \frac{1}{4}\int u(x)dx$$

$$= \sum_{n=1}^{\infty} \int \frac{S(k_1)S(k_2)\dots S(k_n)k\exp(2ikx)}{(k_1+k_2+i0)(k_2+k_3+i0)\dots(k_{n-1}+k_n+i0)}$$

$$\delta(-k+k_1+\dots+k_n)\frac{dkdk_1\dots dk_n}{(2\pi i)^n}\ dx$$

$$I_3 = -\frac{1}{16} \int (3u^2 + u'')$$

$$= \pi \sum_{n=1}^{\infty} \int \frac{S(k_1) \dots S(k_n)(k_1{}^3 + k_2{}^3 + \dots + k_n{}^3)}{(k_1 + k_2 + i0)(k_2 + k_3 + i0) \dots (k_{n-1} + k_n + i0)}$$

$$\delta(k_1 + \dots + k_n) \frac{dk_1 \dots dk_n}{(2\pi i)^n}$$

$$I_5 = \frac{1}{64} \int (10u^3 + 10uu'' + 5(u')^2 + u^{1v})$$

$$= \pi \sum_{n=1}^{\infty} \int \frac{S(k_1) \dots S(k_n)(k_1{}^5 + k_2{}^5 + \dots + k_n{}^5)}{(k_1 + k_2 + i0)(k_2 + k_3 + i0) \dots (k_{n-1} + k_n + i0)}$$

$$\delta(k_1 + \dots + k_n) \frac{dk_1 \dots dk_n}{(2\pi i)^n}$$

$$\dots$$

2) *let* $S(k,t) = S(k)e^{ik^{2l+1}t}, l = 0, 1, 2, \dots$, *and*

$$I_{l,m} = -\frac{1}{2} \sum_{n=1}^{\infty} \int \frac{S(k_1,t)S(k_2,t) \dots S(k_n,t)}{(k_1 + k_2 + i0)(k_2 + k_3 + i0) \dots (k_{n-1} + k_n + i0)}$$

$$(k_1{}^{2m+1} + k_2{}^{2m+1} + \dots + k_n{}^{2m+1})\delta(k_1 + \dots + k_n) \frac{dk_1 \dots dk_n}{(2\pi i)^{n-1}}$$

Such functionals $I_{l,m}$ *are conserved:* $\frac{d}{dt} I_{l,m} = 0$.

Theorem 2.1 (Solution of the *KdV* equation).
Consider the Cauchy problem for the *KdV* equation

$$-\frac{\partial}{\partial t}u(x,t) = \frac{\partial^3}{\partial x^3}u(x,t) + 6u\frac{\partial u}{\partial x}, t \geq 0 \quad u(x,0) = u(x) \qquad (13)$$

Solution in formal power series.

$$u(x,t) = 4 \sum_{n=1}^{\infty} \int \frac{S(k_1,t)S(k_2,t) \dots S(k_n,t)(k_1 + \dots + k_n)}{(k_1 + k_2 + i0)(k_2 + k_3 + i0) \dots (k_{n-1} + k_n + i0)}$$

$$\exp(2i(k_1 + \dots + k_n)x) \frac{dk_1 \dots dk_n}{(2\pi i)^n}$$

where $S(k,t) = S(k)\exp(8ik^3 t)$

$$S(k) = -\frac{1}{2i(k + i0)} \sum_{n=0}^{\infty} \int u(x_1)u(x_3) \dots u(x_{2n+1})$$

$$\Theta_{21}\Theta_{23}\Theta_{43}\Theta_{45} \cdot \dots \cdot \Theta_{2n,2n-1}\Theta_{2n,2n+1}$$

$$\delta(-x + x_1 - x_2 + x_3 - \dots + x_{2n+1})\exp(-2ikx) \, dx dx_1 \dots dx_{2n+1}$$

$$u(x) \equiv u(x,0)$$

is the solution of the Cauchy problem.

Solution in convergent series.

1) Starting from $u(x)$ define $S(k)$, $\{i\kappa_n\}_{n=1}^N$, $\{c_n\}_{n=1}^N$ as follows: $S(k) = \dfrac{b(k)}{a(k)}$, $k \in \mathbb{R}$, where

$$a(k) = 1 + \sum_{n=1}^{\infty} (-)^n \int u(x_2)u(x_4)\ldots u(x_{2n}) \exp(+2ikx_0)$$

$$\delta(x_0 - x_1 + x_2 - \ldots + x_{2n})\Theta_{12}\Theta_{23}\ldots\Theta_{2n-1,2n}$$

$$dx_0 dx_1 \ldots dx_{2n}, \quad k \in \mathbb{C}^+$$

$$b(k) = \frac{1}{2i(k+i0)} \sum_{n=0}^{\infty} (-)^{n+1} \int u(x_1)u(x_3)\ldots u(x_{2n+1}) \exp(-2ikx_0)$$

$$\delta(-x_0 + x_1 - x_2 + x_3 - \ldots + x_{2n+1})\Theta_{12}\Theta_{23}\ldots\Theta_{2n,2n+1}$$

$$dx_0 dx_1 \ldots dx_{2n+1}$$

$S(k) = \bar{S}(-k); |S(k)| \leq 1, k \neq 0$

$\{i\kappa_n\}_{n=1}^N, \kappa_n \in \mathbb{R}^+$ is the set of zeroes of the function $a(k)$, $k \in \mathbb{C}^+$

$$c_n = \frac{\tilde{c}_n}{\frac{\partial}{\partial k} a(k)\,|_{k=i\kappa_n}}, \text{ where}$$

$$\tilde{c}_n = \frac{e^{\kappa_n x}\Phi(i\kappa_n, x)}{\Psi(i\kappa_n, x)} = \frac{\Phi(i\kappa_n, 0)}{\Psi(i\kappa_n, 0)} \text{ (the ratio does not depend on } x\text{)}$$

$\Phi(k, x)$ and $\Psi(k, x)$ defined in (3), (4).

Define

$$S(x, t) = \int S(k)\exp(2ikx + 8ik^3 t)\frac{dk}{\pi}$$

$$c_n(t) = c_n \exp(8\kappa_n^3 t)$$

For the class of initial data $u(x)$ such that the function $S(x)$ is of fast decrease as $x \to +\infty$, the solution could be written as follows:

Define the Fredholm determinant and the 1st minor to be

$$D_{x,t} =$$

$$1 + \sum_{n=1}^{\infty} \frac{1}{n!} \int_{-\infty}^{\infty} \Theta(-y_1)\Theta(-y_2)\dots\Theta(-y_n)\dots \det_{ij}[S(x - y_i - y_j, t)]dy_1 dy_2 \dots dy_n$$

$$D_{x,t}\binom{y}{y_0} =$$

$$\Theta(-y)S(x - y - y_0, t) + \sum_{n=1}^{\infty} \frac{1}{n!}\Theta(-y)\int \Theta(-y_1)\Theta(-y_2)\dots\Theta(-y_n)$$

$$\cdot \begin{vmatrix} S(x - y - y_0, t) & S(x - y - y_1, t) & \dots & S(x - y - y_n, t) \\ S(x - y_1 - y_0, t) & S(x - y_1 - y_1, t) & \dots & S(x - y_1 - y_n, t) \\ & \dots & & \\ S(x - y_n - y_0, t) & S(x - y_n - y_1, t) & \dots & S(x - y_n - y_n, t) \end{vmatrix} \cdot dy_1 dy_2 \dots dy_n$$

$$u(x, t) = -\frac{\partial}{\partial x}\int S(x - y, t)\left(\delta(y) - \frac{D_{x,t}\binom{y}{0}}{D_{x,t}}\right)dy \quad -2\frac{\partial^2}{\partial x^2}\ln\det A(x, t)$$

$$A(x, t)_{mn} = \delta_{mn} - \frac{ic_n(t)e^{-(\kappa_n + \kappa_m)x}}{(\kappa_n + \kappa_m)}$$

$$+2ic_n(t)e^{-(\kappa_n + \kappa_m)x}\int e^{2(\kappa_m y_0 + \kappa_n y_1)}\frac{D_{x,t}\binom{y_0}{y_1}}{D_{x,t}}\Theta(-y_1)dy_0 dy_1.$$

This solution is defined for the class of initial data such that the Fredholm determinant and 1st minor are convergent. For convergence it is enough to have $S(x)$ of fast decrease as $x \to +\infty$. It can be proved that the Fredholm determinant is not zero.

3. Nonlocal transformations for the nonlinear Schrodinger equation
(defocusing case)

Let $q(x)$ be a C^{∞} complex-valued function of a real variable x, with fast decrease as $x \to \pm\infty$ (Schwarz class).

From $q(x)$ construct the following series:

$$a(x) = \delta(x) + \sum_{n=1}^{\infty}\int_{-\infty}^{\infty}\bar{q}(x_1)q(x_2)\bar{q}(x_3)\dots q(x_{2n})\Theta_{21}\Theta_{32}\Theta_{43}\dots\Theta_{2n, 2n-1}$$

$$\delta(x + x_1 - x_2 + x_3 - \dots + x_{2n})\, dx_1 dx_2 \dots dx_{2n}$$

$$(1)$$

$$b(x) = \sum_{n=0}^{\infty} \int_{-\infty}^{\infty} q(x_1)\bar{q}(x_2)q(x_3)\ldots\bar{q}(x_{2n})q(x_{2n+1})\Theta_{21}\Theta_{32}\Theta_{43}\cdot\ldots\cdot\Theta_{2n+1,2n}$$
$$\delta(x - x_1 + x_2 - \ldots + x_{2n} - x_{2n+1})\,dx_1 dx_2 \ldots dx_{2n+1}$$

(2)

$$\Phi_1(x,y) = \delta(x) + \sum_{n=1}^{\infty} \int_{-\infty}^{\infty} q(x_1)\bar{q}(x_2)q(x_3)\ldots\bar{q}(x_{2n})\Theta(y-x_1)\Theta_{12}\Theta_{23}\ldots\Theta_{2n+1,2n}$$
$$\cdot\delta(x - x_1 + x_2 - x_3 + \ldots - x_{2n-1} + x_{2n})\,dx_1 \ldots dx_{2n}$$

(3)

$$\Phi_2(x,y) = \sum_{n=0}^{\infty} \int_{-\infty}^{\infty} \bar{q}(x_1)q(x_2)\bar{q}(x_3)\ldots q(x_{2n})\bar{q}(x_{2n+1})\Theta(y-x_1)\Theta_{12}\Theta_{23}\ldots\Theta_{2n,2n+1}$$
$$\cdot\delta(x - y + x_1 - x_2 + \ldots - x_{2n} + x_{2n+1})\,dx_1 \ldots dx_{2n+1}$$

(4)

$$\Psi_1(x,y) = -\sum_{n=0}^{\infty} \int_{-\infty}^{\infty} q(x_1)\bar{q}(x_2)q(x_3)\ldots\bar{q}(x_{2n})q(x_{2n+1})\Theta(x_1-y)\Theta_{21}\Theta_{32}\ldots\Theta_{2n+1,2n}$$
$$\delta(x + y - x_1 + x_2 - x_3 + \ldots - x_{2n+1})\,dx_1 \ldots dx_{2n+1}$$

(5)

$$\Psi_2(x,y) = \delta(x) + \sum_{n=1}^{\infty} \int_{-\infty}^{\infty} \bar{q}(x_1)q(x_2)\bar{q}(x_3)\ldots q(x_{2n})\Theta(x_1-y)\Theta_{21}\Theta_{32}\ldots\Theta_{2n,2n-1}$$
$$\delta(x + x_1 - x_2 + x_3 - \ldots + x_{2n-1} - x_{2n})\,dx_1 \ldots dx_{2n}$$

(6)

The integration domain, say for nth term in $\Psi_1(x,y)$, is the intersection of the region $y \leq x_1 \leq x_2 \leq \ldots \leq x_{2n+1}$ with the hyperplane $x + y - x_1 + x_2 - x_3 + \ldots - x_{2n+1} = 0$.

Lemma. *The series (1)–(6) are convergent.*

Consider, for example, the series for $\Phi_1(x,y)$. Let $Q = \max|q(x)|$.

$$\left| \int q(x_1)\bar{q}(x_2)q(x_3)\ldots\bar{q}(x_{2n})\Theta(y-x_1)\Theta_{12}\Theta_{23}\cdot\ldots\cdot\Theta_{2n-1,2n} \right.$$
$$\delta(x - x_1 + x_2 - \ldots + x_{2n})\,dx_1 \ldots dx_{2n}$$

$$\leq Q \cdot \int |q(x_1)||q(x_2)|\ldots|q(x_{2n-1})|\Theta_{12}\Theta_{23}\ldots\Theta_{2n-1,2n}\,dx_1 \ldots dx_{2n-1}$$

$$= \frac{Q}{(2n-1)!} \left(\int_{-\infty}^{\infty} |q(x)|dx \right)^{2n-1}.$$

Define $S(x) = \int \frac{b(k)}{a(k)} e^{2ikx} \frac{dk}{\pi}$, where

$$b(k) = \int b(x) e^{-2ikx} \, dx, \quad a(k) = \int a(x) e^{-2ikx} \, dx \qquad (7)$$

Lemma.

1).For $q(x)$ such that $|1 - a(k)| < 1$ $S(x)$ is given by the convergent series

$$S(x) = \sum_{n=0}^{\infty} (-)^n \int q(x_1)\bar{q}(x_2)q(x_3)\ldots\bar{q}(x_{2n})q(x_{2n+1})$$

$$\Theta_{12}\Theta_{32}\Theta_{34}\Theta_{54} \cdot \ldots \cdot \Theta_{2n-1,2n}\Theta_{2n+1,2n} \qquad (8)$$

$$\cdot \delta(x - x_1 + x_2 - \ldots + x_{2n} - x_{2n+1}) \, dx_1 dx_2 \ldots dx_{2n+1}$$

2).Consider $S(x)$,given by the formal series (8). We can prove the following relation:

$$\int S(x_1)a(x - x_1)dx_1 = b(x).$$

Here $a(x)$ and $b(x)$ are functionals in $q(x), \bar{q}(x)$, given by the series (1) and (2).

Indeed, let us collect all the terms of the same order in $q, \bar{q}$ in the convolution of the series (1) and (8)

$$\int S(y)a(x - y)dy = \int q(x_1)\delta(y - x_1)\delta(x - y)dx_1 dy + \int q(x_1)\bar{q}(x_2)q(x_3)$$

$$(-\Theta_{12}\Theta_{32}\delta(y - x_1 + x_2 - x_3)\delta(x - y)+$$

$$\Theta_{32}\delta(y - x_1)\delta(x - y + x_2 - x_3) \, dx_1 dx_2 dx_3 dy$$

$$+ \ldots + \int q(x_1)\bar{q}(x_2)\ldots q(x_{2n+1})(-)^n$$

$$(\Theta_{12}\Theta_{32}\Theta_{34}\Theta_{54}\ldots\Theta_{2n-1,2n-2}\Theta_{2n-1,2n}\Theta_{2n+1,2n}$$

$$+ \sum_{m=0}^{n-1} (-)^{m-n}\Theta_{12}\Theta_{32}\Theta_{34}\Theta_{54}\ldots\Theta_{2m-1,2m}\Theta_{2m+1,2m}$$

$$(\Theta_{2m+2,2m+1} + \Theta_{2m+1,2m+2})\Theta_{2m+3,2m+2}$$

$$\cdot \Theta_{2m+4,2m+3} \cdot \ldots \cdot \Theta_{2n+1,2n}) \, \delta(x - x_1 + x_2 - x_3 + \ldots - x_{2n+1}) \, dx_1 \ldots dx_{2n}$$

$$= \sum_{n=0}^{\infty} \int q(x_1)\bar{q}(x_2)\ldots q(x_{2n+1})\Theta_{21}\Theta_{32}\Theta_{43} \cdot \ldots \cdot \Theta_{2n+1,2n}$$

$$\delta(x - x_1 + x_2 - x_3 + \ldots - x_{2n+1}) \, dx_1 \ldots dx_{2n}$$

$$= b(x)$$

(We used $\Theta_{ij} + \Theta_{ji} = 1$).

The convolutions of functionals (1)–(8) are again the functionals of the same type. There are certain relations for the convolutions:

$$\int \bar{a}(x_1)a(x + x_1)dx_1 = \delta(x) + \int \bar{b}(x_1)b(x + x_1)\,dx_1 \qquad (9)$$

$$\int \left(b(x_1)\Phi_2(x - x_1 + y, y) - a(x_1)\bar{\Phi}_1(-x + x_1, y)\right)dx_1 = -\Psi_2(x, y) \qquad (10)$$

$$\int \left(b(x_1)\Phi_1(x - x_1 + y, y) - a(x_1)\bar{\Phi}_2(-x + x_1, y)\right)dx_1 = -\Psi_1(x, y) \qquad (11)$$

Let us prove (10). We have to collect all the terms of the same degree in q, $\bar{q}$ in the left-hand side and to compare with the right-hand side.

$$\int (\Phi_2(x - s + y, y)b(s) - \bar{\Phi}_1(-x + s, y)a(s))\,ds$$

$$= -\delta(x) + \int \bar{q}(x_1)q(x_2)\left(\Theta(y - x_1)\delta(x - s + x_1)\delta(s - x_2) - \Theta(y - x_1)\Theta_{12}\right.$$

$$\delta(x - s + x_1 - x_2)d(s) \cdot -\Theta_{21}\delta(x - s)\delta(s + x_1 - x_2))\,dx_1dx_2ds + \ldots$$

$$+ \int \bar{q}(x_1)q(x_2)\bar{q}(x_3)\ldots q(x_{2n})$$

$$\cdot \left(\sum_{m=0}^{n-1} \Theta(y - x_1)\Theta_{12}\Theta_{23}\Theta_{34} \cdot \ldots \cdot \Theta_{2m,2m+1}(\Theta_{2m+1,2m+2} + \Theta_{2m+2,2m+1})\right.$$

$$\Theta_{2m+3,2m+2}\Theta_{2m+4,2m+3} \cdot \ldots \cdot \Theta_{2n,2n-1} - \Theta_{21}\Theta_{32} \cdot \ldots \cdot \Theta_{2n,2n-1}$$

$$- \sum_{m=1} \Theta(y - x_1)\Theta_{12}\Theta_{23} \cdot \ldots \cdot \Theta_{2m-1,2m}(\Theta_{2m,2m+1} + \Theta_{2m+1,2m})$$

$$\Theta_{2m+2,2m+1} \cdot \Theta_{2m+3,2m+2} \cdot \ldots \cdot \Theta_{2n,2n-1}$$

$$- \Theta(y - x_1)\Theta_{12}\Theta_{23} \cdot \ldots \cdot \Theta_{2n,2n-1}\right)\delta(x + x_1 - x_2 + \ldots - x_{2n})\,dx_1\ldots dx_{2n}$$

$$= -\delta(x) - \sum_{n=1}^{\infty} \int \bar{q}(x_1)q(x_2)\bar{q}(x_3)\ldots q(x_{2n})\Theta(x_1 - y)\Theta_{21}\Theta_{32} \cdot \ldots \cdot \Theta_{2n,2n-1}$$

$$\delta(x + x_1 - x_2 + \ldots - x_{2n})dx_1\ldots dx_{2n} = -\Psi_2(x, y)$$

The proof of the other relations is similar.

In addition to convolution of two functionals, there is another operation for our functionals , namely inversion.It is the infinite-dimensional analoque of the inverse function. Consider the series (8)

$$S(x) = \sum_{i=0}^{\infty} q_{(2i+1)}(x) :=$$

$$= q(x) - \int q(x_1)\bar{q}(x_2)q(x_3)\Theta_{12}\Theta_{32}\delta(x - x_1 + x_2 - x_3)$$

$$+ \int q(x_1)\bar{q}(x_2)q(x_3)\bar{q}(x_4)q(x_5)\Theta_{12}\Theta_{32}\Theta_{34}\Theta_{54}$$

$$\delta(x - x_1 + x_2 - x_3 + x_4 - x_5) - \dots$$

$S(x)$ is a formal series , its nth term to is a nonlocal analytic functional of $q(x)$, $\bar{q}(x)$ of degree $(2n + 1)$ in $q, \bar{q}$. It can be inverted, namely, $q(x)$ can be expressed in terms of $S(x)$:

$$q(x) = S_{(1)}(x) + S_{(3)}(x) + S_{(5)}(x) + \dots$$

where $S_m(x)$ is a nonlocal analytic functional of $S(x), \bar{S}(x)$ of degree m in $S, \bar{S}$.

$$S_{(1)}(x) = S(x)$$

$$S_{(3)}(x) = \int S(x_1)\bar{S}(x_2)S(x_3)\delta(x - x_1 + x_2 - x_3)\Theta_{12}\Theta_{32} \ dx_1 dx_2 dx_3$$

$$S_{(5)}(x) = -\int S(x_1)\bar{S}(x_2)S(x_3)\bar{S}(x_4)S(x_5)\Theta_{12}\Theta_{32}\Theta_{34}\Theta_{54}$$

$$\delta(x - x_1 + x_2 - x_3 + x_4 - x_5) \ dx_1 \dots dx_5$$

$$+ \int \left(S_{(3)}^{(3)}(x_1)\bar{S}(x_2)S(x_3) + S(x_1) \right.$$

$$\bar{S}_{(3)}^{(3)}(x_2)S(x_3) + S(x_1)\bar{S}(x_2)S_{(3)}^{(3)}(x_3) \Big)$$

$$\Theta_{12}\Theta_{32}\delta(x - x_1 + x_2 - x_3) \ dx_1 dx_2 dx_3$$

$$= \int S(x_1)\bar{S}(x_2)S(x_3)\bar{S}(x_4)S(x_5)\delta(x - x_1 + x_2 - x_3 + x_4 - x_5)$$

$$(-\Theta_{12}\Theta_{32}\Theta_{34}\Theta_{54}$$

$$+ \Theta_{12}\Theta_{32} \quad \Theta_{54}\Theta(x_1 - x_2 + x_3 - x_4)$$

$$\Theta_{23}\Theta_{43} \quad \Theta(x_1 - x_2 + x_3 - x_4)\Theta(x_5 - x_4 + x_3 - x_2)$$

$$+ \Theta_{12} \quad \Theta_{34}\Theta_{54}\Theta(x_5 - x_4 + x_3 - x_2)) \, dx_1 \ldots dx_5$$

$$= \int S(x_1)\bar{S}(x_1)S(x_3)\bar{S}(x_4)S(x_5)\delta(x - x_1 + x_2 - x_3 + x_4 - x_5)$$

$$\cdot \Theta_{12}\Theta(x_1 - x_2 + x_3 - x_4)\Theta_{54}\Theta(x_5 - x_4 + x_3 - x_2)$$

$$(-\Theta_{32}\Theta_{34} + \Theta_{32} + \Theta_{23}\Theta_{43} + \Theta_{34}) \, dx_1 \ldots dx_5$$

$$= \int S(x_1)\bar{S}(x_2)S(x_3)\bar{S}(x_4)S(x_5)\Theta_{12}\Theta(x_1 - x_2 + x_3 - x_4)\Theta_{54}$$

$$\cdot \Theta(x_5 - x_4 + x_3 - x_2)\delta(x - x_1 + \ldots - x_5) \, dx_1 \ldots dx_5$$

(Indeed, $-\Theta_{32}\Theta_{34} + \Theta_{32} + \Theta_{23}\Theta_{43} + \Theta_{34} = -\Theta_{32}\Theta_{34} + \Theta_{32}(\Theta_{34} + \Theta_{43}) + \Theta_{23}\Theta_{43} + \Theta_{34} = (\Theta_{23} + \Theta_{32})\Theta_{43} + \Theta_{34} = \Theta_{43} + \Theta_{34} = 1$.

Lemma. *Consider the nonlocal analytic functionals of $\{S(x)\}$, given by formal series*

$$\tilde{q}(x) = S(x) + \sum_{n=1}^{\infty} \int \left(S(x_1)\bar{S}(x_2)S(x_3) \ldots \bar{S}(x_{2n})S(x_{2n+1}) \right.$$

$$\Theta(x_1 - x_2)\Theta(x_1 - x_2 + x_3 - x_4) \ldots \Theta(x_1 - x_2 + \ldots - x_{2n})$$

$$\Theta(x_{2n+1} - x_{2n})\Theta(x_{2n+1} - x_{2n} + x_{2n-1} - x_{2n-2}) \ldots \Theta(x_{2n+1} - x_{2n} + \ldots - x_2)$$

$$\delta(x - x_1 + x_2 - \ldots + x_{2n} - x_{2n+1}) \, dx_1 \ldots dx_{2n+1}$$

$$\tag{12}$$

$$\tilde{\Phi}_1(x, y) = \delta(x) + \sum_{n=1}^{\infty} \int \left(S(x_1)\bar{S}(x_2)S(x_3) \ldots \bar{S}(x_{2n}) \right.$$

$$\cdot \Theta(y - x_1)\Theta(y - x_1 + x_2 - x_3) \ldots \Theta(y - x_1 + \ldots - x_{2n-1})$$

$$\cdot \Theta(x_1 - x_2)\Theta(x_1 - x_2 + x_3 - x_4) \cdot \ldots \cdot \Theta(x_1 - x_2 + \ldots - x_{2n})$$

$$\delta(x - x_1 + x_2 - x_3 + \ldots - x_{2n+1} + x_{2n}) \, dx_1 \ldots dx_{2n}$$

$$\tag{13}$$

$$\tilde{\Phi}_2(x, y) = \sum_{n=0}^{\infty} \int \bar{S}(x_1)S(x_2)\bar{S}(x_3) \ldots S(x_{2n})\bar{S}(x_{2n+1})$$

$$\cdot \Theta(y - x_1) \cdot \Theta(y - x_1 + x_2 - x_3) \cdot \ldots \cdot \Theta(y - x_1 + x_2 - x_3 + \ldots - x_{2n-1})$$

$$\cdot \Theta(x_1 - x_2)\Theta(x_1 - x_2 + x_3 - x_4) \cdot \ldots \cdot \Theta(x_1 - x_2 + \ldots - x_{2n})$$

$$\delta(x - y + x_1 - x_2 + \ldots - x_{2n} + x_{2n+1}) \, dx_1 \ldots dx_{2n+1}$$

$$\tag{14}$$

Let us substitute in these series $S(x)$ as formal series in $\{q(x)\}$, (8). The *result of such substitution would be formal series in $\{q(x)\}$, and , moreover,*

$$\tilde{q}(x) = q(x)$$

$$\tilde{\Phi}_1(x,y) = \Phi_1(x,y)$$

$$\tilde{\Phi}_2(x,y) = \Phi_2(x,y)$$

as formal series in $q(x)$, with $\Phi_1(x,y)$ and $\Phi_2(x,y)$ given by (3) , (4).

The series (9) could be used to get the solution of a nonlinear equation. In order to see this, let us rewrite (9) in terms of $S(k) = \int_{-\infty}^{\infty} S(x)e^{-2ikx}dx$ and $\bar{S} = (k)\int_{-\infty}^{\infty} \bar{S}(x)e^{2ikx}dx$:

$$q(x) = 2\sum_{n=0}^{\infty} \int_{-\infty}^{\infty} \frac{S(k_1)\bar{S}(k_2)\ldots S(k_{2n+1})}{(k_2 - k_1 + i0)(k_3 - k_2 - i0)(k_4 - k_3 + i0)\ldots(k_{2n+1} - k_{2n} - i0)}$$
$$\cdot \exp(2i(k_1 - k_2 + k_3 - \ldots + k_{2n+1})x) \frac{dk_1 \ldots dk_{2n+1}}{(2\pi)^{2n+1}}$$

(15)

The kernels $\dfrac{1}{k \mp i0} = 2i \int_{-\infty}^{\infty} \Theta(x)exp(\mp 2ikx)$ appeared as the Fourier transform of the Heaviside function kernels.

The series (15) has the following property: a polynomial in $q(x)$, $\bar{q}(x)$ and their derivatives can also be written in the form (15) but with some polynomial in $\{k\}$ in the numerator:

$$\frac{d^n q(x)}{dx^n} = 2 \cdot (2i)^n \sum_{n=0}^{\infty}$$
$$\int \frac{S(k_1)\bar{S}(k_2)\ldots S(k_{2n+1})(k_1 - k_2 + \ldots - k_{2n} + k_{2n+1})^n}{(k_2 - k_1 + i0)(k_3 - k_2 - i0)(k_4 - k_3 + i0)\ldots(k_{2n+1} - k_{2n} - i0)}$$
$$\cdot \exp(2i(k_1 - k_2 + k_3 - \ldots + k_{2n+1})x) \frac{dk_1 \ldots dk_{2n+1}}{(2\pi)^{2n+1}}$$

(16)

$$q(x)\bar{q}(x) = 4\sum_{n=1}^{\infty}$$

$$\int \frac{S(k_1)\bar{S}(k_2)S(k_3)\ldots\bar{S}(k_{2n})(-k_1 + k_2 - \ldots + k_{2n})}{(k_2 - k_1 + i0)(k_3 - k_2 - i0)(k_4 - k_3 + i0)\ldots(k_{2n-1} - k_{2n-2} - i0)(k_{2n} - k_{2n-1} + i0)}$$

$$\cdot \exp(2i(k_1 - k_2 + \ldots - k_{2n})x)\,\frac{dk_1\ldots dk_{2n}}{(2\pi)^{2n}}$$

$$q(x)\bar{q}(x)q(x) = 4\sum_{n=1}^{\infty}$$

$$\int \frac{S(k_1)\bar{S}(k_2)S(k_3)\ldots\bar{S}(k_{2n})S(k_{2n+1})}{(k_2 - k_1 + i0)(k_3 - k_2 - i0)\ldots(k_{2n} - k_{2n-1} - i0)(k_{2n+1} - k_{2n} + i0)}$$

$$\cdot \left(-(k_1 - k_2 + \ldots - k_{2n+1})^2 - \sum_{p=1}^{n} k_{2p}^{\,2} + \sum_{p=0}^{n} k_{2p+1}^{\,2}\right)$$

$$\exp(2i(k_1 - k_2 + \ldots - k_{2n} + k_{2n+1}))\,\frac{dk_1\ldots dk_{2n+1}}{(2\pi)^{2n+1}}$$

$$q^{m+1}(x)\bar{q}^{m}(x) = 2^{2m+1}\sum_{n=m}^{\infty}$$

$$\int \frac{S(k_1)\bar{S}(k_2)S(k_3)\ldots\bar{S}(k_{2n})S(k_{2n+1})}{(k_2 - k_1 - i0)(k_3 - k_2 + i0)\ldots(k_{2n} - k_{2n-1} - i0)(k_{2n+1} - k_{2n} + i0)}$$

$$\sum_{1 \le p_1 < p_2 < \ldots < p_m \le n} (-k_1 + k_2 - \ldots + k_{2p_1})$$

$$(k_{2p_1+1} - k_{2p_1})(-k_{2p_1+1} + k_{2p_1+2} - \ldots + k_{2p_2})(k_{2p_2+1} - k_{2p_2}) \cdot \ldots$$

$$\cdot(-k_{2p_m+1} + k_{2p_m+2} - \ldots + k_{2p_m})(k_{2p_m+1} - k_{2p_m})\,\frac{dk_1\ldots dk_{2n+1}}{(2\pi)^{2n+1}}$$

$$(17)$$

The fact that both differentiation of $q(x)$ and nonlinearity under the "nonlinear Fourier transformation" (15) has the same effect, namely, some polynomial in $\{k\}$ appears in the numerator, can be used to solve nonlinear equations.

Theorem 3.1 (solution of the NLS Equation).
Consider the Cauchy problem for the NLS equation:

$$i\frac{\partial}{\partial t}q(x,t) + \frac{\partial^2}{\partial x^2}q(x,t) - 2\lambda|q|^2 q = 0, \quad \lambda = \begin{cases} 1, & \text{defocusing case} \\ -1, & \text{focusing case} \end{cases} \quad (18)$$

$$q(x,0) = q(x)$$

(Solution in formal series).
The solution of Cauchy problem is given by

$$q(x,t) = 2\sum_{n=0}^{\infty} \lambda^n \int \frac{S(k_1,t)\bar{S}(k_2,t)\dots S(k_{2n+1},t)}{(k_2-k_1+i0)(k_3-k_2-i0)\dots(k_4-k_3+i0)(k_{2n+1}-k_{2n}-i0)}$$

$$\exp(2i(k_1-k_2+\dots-k_{2n}+k_{2n+1}))$$

$$\frac{dk_1\dots dk_{2n+1}}{(2\pi)^{2n+1}}$$

$$\tag{19}$$

$$S(k,t) = e^{-4ik^2t}S(k)$$

$$S(k) = \sum_{n=0}^{\infty}(-)^n\lambda^n \int_{-\infty}^{\infty} q(x_1)\bar{q}(x_2)q(x_3)\dots\bar{q}(x_{2n})q(x_{2n+1})$$

$$\Theta_{12}\Theta_{32}\Theta_{34}\Theta_{54}\dots\Theta_{2n-1,2n}\Theta_{2n+1,2n}$$

$$\exp(-2ikx)\delta(x-x_1+x_2-\dots+x_{2n}-x_{2n+1})\,dxdx_1dx_2\dots dx_{2n+1}$$

$$\tag{20}$$

(Solution in convergent series, defocusing case $\lambda = 1$).

$$q(x,t) = \int S(x-y,t)\left(\delta(y) - \frac{D_{x,t}\begin{pmatrix}y\\0\end{pmatrix}}{D_{x,t}}\right)dy$$

$$S(x,t) = \int \frac{b(k)}{a(k)}\exp(2ikx-4ik^2t)\frac{dk}{\pi}, \quad b(k) \text{ and } a(k) \text{ are defined by (1) (2)}$$

$$D_{x,t}\begin{pmatrix}y\\y_0\end{pmatrix} = -\Theta(-y)K_{x,t}(y,y_0) + \sum_{n=1}^{\infty}\frac{(-)^{n+1}}{n!}$$

$$\Theta(-y)\int\Theta(-y_1)\Theta(-y_2)\dots\Theta(-y_n)$$

$$\begin{vmatrix} K_{x,t}(y,y_0) & K_{x,t}(y,y_1) & \cdots & K_{x,t}(y,y_n) \\ K_{x,t}(y_1,y_0) & K_{x,t}(y_1,y_1) & \cdots & K_{x,t}(y_1,y_n) \\ & \cdots & & \\ K_{x,t}(y_n,y_0) & K_{x,t}(y_n,y_1) & \cdots & K_{x,t}(y_n,y_n) \end{vmatrix} dy_1\dots dy_n$$

$$D_{x,t} = 1 + \sum_{n=1}^{\infty}\frac{(-)^n}{n!}\int\Theta(-y_1)\Theta(-y_2)\dots\Theta(-y_n)\,|K_{x,t}(y_i,y_j)|_{i,j}\,dy_1\dots dy_n$$

$$K_{x,t}(y_1,y_2) = \int\Theta(-y)\bar{S}(x-y_1-y)S(x-y-y_2,t)\,dy$$

The solution is defined for the class of initial data $q(x)$ such that the Fredholm determinant $D_{x,t}$ and the first minor $D_{x,t}\binom{y}{y_0}$ are convergent. It can be proved that the determinant is not zero.

Proof (formal series).

1) We compute $\dfrac{\partial^2}{\partial x^2}q(x,t)$ and $|q|^2q$ in the same way as before, see (16) ;

$$i\frac{\partial}{\partial t}q(x,t) + \frac{\partial^2}{\partial x^2}q(x,t) - 2|q|^2q$$

$$= 2\int(i\frac{\partial}{\partial t} - 4k_1{}^2)S(k,t)e^{2ik_1x}\,\frac{dk_1}{2\pi}$$

$$+\sum_{n=1}^{\infty}2\lambda^n(i\frac{\partial}{\partial t} - 4(k_1{}^2 - k_2{}^2 + k_3{}^2 - \ldots + k_{2n+1}{}^2))$$

$$\frac{S(k_1,t)\bar{S}(k_2,t)\ldots S(k_{2n+1},t)}{(k_2 - k_1 + i0)(k_3 - k_2 - i0)\ldots(k_{2n+1} - k_{2n} - i0)}$$

$$\cdot\exp(2i(k_1 - k_2 + k_3 - \ldots + k_{2n+1})x)\,\frac{dk_1\ldots dk_{2n+1}}{(2\pi)^{2n+1}} = 0$$

for $S(k,t) = e^{-4ik^2t}S(k,0)$.

2) The substitution of $S(k,0) = S(k)$ as series in $\{q(x)\}$ (see (19)) into (18) gives $q(x,0) = q(x)$ (see Lemma) .

4. Nonlinear transformations for Davey-Stewardson equation

Let $q(x,y)$ be a complex-valued C^{∞} function on the plane $\mathbb{R}^2$, with fast decrease as $|x|^2 + |y|^2 \to \infty$.

We construct the following nonlocal analytic functionals of $\{q\}$: [4]

$$\alpha(k,\bar{k}) = \sum_{n=0}^{\infty}$$

$$\int \frac{q(z_1,\bar{z}_1)\bar{q}(z_2,\bar{z}_2)q(z_3,\bar{z}_3)\ldots\bar{q}(z_{2n},\bar{z}_{2n})q(z_{2n+1},\bar{z}_{2n+1})}{(\bar{z}_1 - \bar{z}_2)(z_2 - z_3)(\bar{z}_3 - \bar{z}_4)\ldots(z_{2k} - z_{2k+1})} \tag{1}$$

$$\cdot\exp\left(\bar{k}(\bar{z}_1 - \bar{z}_2 + \bar{z}_3 - \ldots + \bar{z}_{2k+1}) - k(z_1 - z_2 + z_3 - \ldots + z_{2k+1})\right)$$

$$\frac{d^2z_1d^2z_2\ldots d^2z_{2n+1}}{(2\pi)^{2n+1}}$$

$$\mu_1(k, \bar{k}; z, \bar{z}) = 1 + \sum_{n=1}^{\infty}$$

$$\int \frac{q(z_1, \bar{z}_1)\bar{q}(z_2, \bar{z}_2)\dots\bar{q}(z_{2n-1}, \bar{z}_{2n-1})\bar{q}(z_{2n}, \bar{z}_{2n})}{(z - z_1)(\bar{z}_1 - \bar{z}_2)(z_2 - z_3)\dots(\bar{z}_{2n-1} - \bar{z}_{2n})}$$

$$\cdot \exp\left(\bar{k}(\bar{z}_1 - \bar{z}_2 + \bar{z}_3 - \dots + \bar{z}_{2n+1} - \bar{z}_{2n}) - k(z_1 - z_2 + z_3 - \dots + z_{2n+1} - z_{2n})\right)$$

$$\frac{d^2 z_1 \dots d^2 z_{2n}}{(2\pi)^{2n}}$$

$$(2)$$

$$\mu_2(k, \bar{k}; z, \bar{z}) = \sum_{n=0}^{\infty}$$

$$\int \frac{\bar{q}(z_1, \bar{z}_1)q(z_2, \bar{z}_2)\dots\bar{q}(z_{2n+1}, \bar{z}_{2n+1})}{(\bar{z} - \bar{z}_1)(z_1 - z_2)(\bar{z}_2 - \bar{z}_3)\dots(z_{2n-1} - z_{2n})(\bar{z}_{2n} - \bar{z}_{2n+1})}$$

$$\cdot \exp\left(\bar{k}(\bar{z} - \bar{z}_1 + \bar{z}_2 - \dots + \bar{z}_{2n} - \bar{z}_{2n+1}) - k(z - z_1 + z_2 - \dots + z_{2n} - z_{2n+1})\right)$$

$$\frac{d^2 z_1 \dots d^2 z_{2n}}{(2\pi)^{2n+1}}$$

$$(3)$$

where $z = x_1 + ix_2$, $\bar{z} = x_1 - ix_2$, $k = k_1 + ik_2$, $\bar{k} = k_1 - ik_2$, $d^2z := \frac{i}{2}dzd\bar{z}$.

The series are convergent for some class of functions q. We will not investigate convergence; we will work with these functionals as with formal series. Each term of these series is an integral, involving homogeneous generalized functions $\frac{1}{z}$ as kernels ([1]).

There are the following relations for the functionals (1)–(3):

$$\frac{\partial \mu_1}{\partial \bar{z}}(k, \bar{k}, z, \bar{z}) = \frac{1}{2}q(z, \bar{z})\mu_2(k, \bar{k}, z, \bar{z}).$$

$$\frac{\partial \mu_2}{\partial \bar{z}}(k, \bar{k}, z, \bar{z}) = k\mu_2(k, \bar{k}, z, \bar{z}) + \frac{1}{2}q(z, \bar{z})\mu_1(k, \bar{k}, z, \bar{z}).$$

$$\frac{\partial \mu_1}{\partial \bar{k}} = e^{\bar{k}\bar{z} - kz}\bar{\alpha}\bar{\mu}_2.$$

$$\frac{\partial \mu_2}{\partial \bar{k}} = e^{\bar{k}\bar{z} - kz}\bar{\alpha}\bar{\mu}_1.$$

The series (1) could be inverted, that is, $q(z, \bar{z})$ could be written as a functional of $\{\alpha(k, \bar{k})\}$:

$$
\alpha(k, \bar{k}) = \int q(z_1, \bar{z}_1) \exp(\bar{k}\bar{z}_1 - kz_1) \frac{d^2 z_1}{2\pi} + \int \left(\frac{q(z_1, \bar{z}_1)\bar{q}(z_2, \bar{z}_2)q(z_3, \bar{z}_3)}{(z_1 - z_2)(\bar{z}_2 - \bar{z}_3)} \right.
$$

$$
\left. \exp(\bar{k}(\bar{z}_1 - \bar{z}_2 + \bar{z}_3) - k(z_1 - z_2 + z_3)) \right) \frac{d^2 z_1 d^2 z_2 d^2 z_3}{(2\pi)^3} + \ldots
$$

$$
q(z, \bar{z}) = \alpha_{(1)}(z, \bar{z}) + \alpha_{(3)}(z, \bar{z}) + \ldots
$$

$$
\alpha_{(1)}(z, \bar{z}) = -2 \int \alpha(k_1, \bar{k}_1) \exp(-\bar{k}, \bar{z} + k\bar{z}) \frac{d^2 k}{\pi}
$$

$$
\alpha_{(3)}(z, \bar{z}) = -\frac{2}{\pi^7} \int \frac{\alpha(k_1, \bar{k}_1)\bar{\alpha}(k_2, \bar{k}_2)\alpha(k_3, \bar{k}_3)}{(z_1 - z_2)(\bar{z}_2 - \bar{z}_3)}
$$

$$
\exp\left((-\bar{k}_1 \bar{z}_1 + \bar{k}_2 \bar{z}_2 - \bar{k}_3 z_3 + \bar{k}(\bar{z}_1 - \bar{z}_2 + \bar{z}_3) - \bar{k}\bar{z})\right.
$$

$$
d^2 z_1 d^2 z_2 d^2 z_3 d^2 k_1 d^2 k_2 d^2 k_3 d^2 k
$$

$$
= -\frac{2}{\pi^3} \int \frac{\alpha(k_1, \bar{k}_1)\bar{\alpha}(k_2, \bar{k}_2)\bar{\alpha}(k_3, \bar{k}_3)}{(\bar{k} - \bar{k}_1)(k_2 - k_1)} \delta(k - k_1 + k_2 - k_3)
$$

$$
\exp(-\bar{k}\bar{z} + kz) d^2 k_1 d^2 k_2 d^2 k_3 d^2 k_4
$$

$$
= -2 \int \frac{\alpha(k_1, \bar{k}_1)\bar{\alpha}(k_2, \bar{k}_2)\alpha(k_3, \bar{k}_3)}{(k_2 - k_1)(\bar{k}_3 - \bar{k}_2)}
$$

$$
\exp(\bar{z}(-\bar{k}_1 + \bar{k}_2 - \bar{k}_3) - z(-k_1 + k_2 - k_3)) \frac{d^2 k_1 d^2 k_2 d^2 k_3}{\pi^3}
$$

Lemma.   *Consider the following functionals of $\{\alpha(k, \bar{k})\}$:*

$$
\tilde{\mu}_1(k, \bar{k}, z, \bar{z}) = 1 + \sum_{n=1}^{\infty}
$$

$$
\int \frac{\alpha(k_1, \bar{k}_1)\bar{\alpha}(k_2, \bar{k}_2) \ldots \alpha(k_{2n-1}, k_{2n-1})\bar{\alpha}(k_{2n}, k_{2n})}{(\bar{k}_2 - \bar{k}_1)(k_3 - k_2)(\bar{k}_4 - \bar{k}_3) \ldots (k_{2n-1}, k_{2n-2})(\bar{k}_{2n} - \bar{k}_{2n-1})(k - k_{2n})}
$$

$$
\cdot \exp(\bar{z}(-\bar{k}_1 + \bar{k}_2 - \bar{k}_3 + \ldots - \bar{k}_{2n-1} + \bar{k}_{2n})
$$

$$
-z(-k_1 + k_2 - k_3 + \ldots - k_{2n-1} + k_{2n})) \frac{d^2 k_1 \ldots d^2 k_n}{\pi^{2n}},
$$

$$
(4)
$$

$$\tilde{\mu}_2(k, \bar{k}, z, \bar{z}) = \sum_{n=0}^{\infty}$$

$$\int \frac{\bar{\alpha}(k_1, \bar{k}_1)\alpha(k_2, \bar{k}_2)\dots\bar{\alpha}(k_{2n-1}, \bar{k}_{2n-1})\alpha(k_{2n}, k_{2n})\bar{\alpha}(k_{2n+1}, \bar{k}_{2n+1})}{(k_2 - k_1)(\bar{k}_3 - \bar{k}_2)(k_4 - k_3)\dots(\bar{k}_{2n-1} - \bar{k}_{2n-2})(k_{2n} - k_{2n-1})(\bar{k}_{2n+1} - \bar{k}_{2n})(k - k_{2n+1})}$$

$$\exp(\bar{z}(\bar{k}_1 - \bar{k}_2 + \dots + \bar{k}_{2n+1}) - z(k_1 - k_2 + k_3 - \dots + k_{2n-1})),$$

$$\frac{d^2 k_1 \dots d^2 k_{2n+1}}{\pi^{2n+1}}$$

$$(5)$$

$$\tilde{q}(z, \bar{z}) = -2\sum_{n=0}^{\infty}$$

$$\int \frac{\alpha(k_1, k_1)\bar{\alpha}(k_2, k_2)\dots\alpha(k_{2n+1}, \bar{k}_{2n+1})}{(k_2 - k_1)(\bar{k}_3 - \bar{k}_2)(k_4 - k_3)\dots(\bar{k}_{2n+1} - \bar{k}_{2n})}$$

$$\cdot \exp(\bar{z}(-\bar{k}_1 + \bar{k}_2 - \dots + \bar{k}_{2n} - \bar{k}_{2n+1}) - z(-k_1 + k_2 - \dots + k_{2n} - k_{2n+1}))$$

$$\frac{d^2 k_1 \dots d^2 k_{2n+1}}{\pi^{2n+1}}.$$

$$(6)$$

We can substitute in (4)–(6) $\alpha(k, \bar{k})$ as a functional in $q(z, \bar{z})$, given by (1). As a result of this substitution we will obtain functionals in $\{q(z, \bar{z})\}$,given by formal series. Moreover,

$$\tilde{\mu}_1(k, \bar{k}, z, \bar{z}) = \tilde{\mu}_1(k, \bar{k}, z, \bar{z})$$

$$\tilde{\mu}_2(k, \bar{k}, z, \bar{z}) = \tilde{\mu}_2(k, \bar{k}, z, \bar{z})$$

$$\tilde{q}(z, \bar{z}) = q(z, \bar{z})$$

Theorem 4.1 (Solution of the Davey-Stewartson equation-II)
Consider the DS equation

$$i\frac{\partial}{\partial t}q(z, \bar{z}, t) = -(\partial^2 + \bar{\partial}^2)q(z, \bar{z}, t)$$

$$+\frac{1}{2}q(z, \bar{z}, t)(\bar{\partial}^{-1}\partial + \partial^{-1}\bar{\partial})(|q(z, \bar{z}, t)|^2)$$

$$q(z, \bar{z}, 0) = q(z, \bar{z})$$

(here $\partial = \dfrac{\partial}{\partial z}$, $\bar{\partial} = \dfrac{\partial}{\partial \bar{z}}$, $\bar{\partial}^{-1}f(z, \bar{z}) = \dfrac{1}{\pi}\displaystyle\int \dfrac{f(z^1, \bar{z}^1)}{z - z^1}\, d^2 z^1.$

The solution is

$$q(z,\bar{z},t) = -2 \sum_{n=0}^{\infty}$$

$$\int \frac{\alpha(k_1,\bar{k}_1,t)\bar{\alpha}(k_2,k_2,t)\ldots\alpha(k_{2n+1},\bar{k}_{2n+1},t)}{(k_1-k_2)(\bar{k}_2-\bar{k}_3)(k_3-k_4)\ldots(\bar{k}_{2n}-\bar{k}_{2n+1})}$$

$$\cdot \exp(\bar{z}(-\bar{k}_1+\bar{k}_2-\ldots+\bar{k}_{2n}-\bar{k}_{2n+1})-z(-k_1+k_2-\ldots+k_{2n}-k_{2n+1}))$$

$$\frac{d^2k_1\ldots d^2k_n}{\pi^{2n+1}}$$

(7)

$$\alpha(k,\bar{k}) = \sum_{n=0}^{\infty}$$

$$\int \frac{q(z_1,\bar{z}_1)\bar{q}(z_2,\bar{z}_2)q(z_3,\bar{z}_3)\ldots\bar{q}(z_{2n},\bar{z}_{2n})q(z_{2n+1},\bar{z}_{2n+1})}{(\bar{z}_1-\bar{z}_2)(z_2-z_3)(\bar{z}_3-\bar{z}_4)\ldots(z_{2k}-z_{2k+1})}$$

$$\cdot \exp(\bar{k}(-\bar{z}_1-\bar{z}_2+\bar{z}_3-\ldots+\bar{z}_{2k+1}-k(z_1+z_2+z_3-\ldots+z_{2n+1}))$$

(8)

$$\frac{d^2z_1 d^2z_2\ldots d^2z_{2n+1}}{(2\pi)^{2n+1}}$$

$$\alpha(k,\bar{k},t) = \alpha(k,\bar{k})e^{i(k^2+\bar{k}^2)t}.$$

Proof.

1) Let us compute $q\bar{\partial}^{-1}\partial(|q|^2)$ for $q(z,\bar{z},t)$ given by (6):

$$q\bar{\partial}^{-1}\partial(|q|^2) = -8 \sum_{n=1}^{\infty}$$

$$\int \frac{\alpha(k_1,\bar{k}_1,t)\bar{\alpha}(k_2,\bar{k}_2,t)\ldots\alpha(k_{2n+1},\bar{k}_{2n+1},t)}{(k_1-k_2)(\bar{k}_2-\bar{k}_3)(k_3-k_4)\ldots(\bar{k}_{2n}-\bar{k}_{2n+1})}$$

$$\cdot \exp(\bar{z}(-\bar{k}_1+\bar{k}_2-\ldots+\bar{k}_{2n}-\bar{k}_{2n+1})-z(-k_1+k_2-\ldots+k_{2n}-k_{2n+1}))$$

$$\sum_{\substack{m_1,m_2\geq 0 \\ m_1+m_2+1\leq n}} (k_{2m_1+1}-k_{2m_1+2})(\bar{k}_{2m_1+2m_2+2}-\bar{k}_{2m_1+2m_2+3})$$

$$\frac{k_{2m_1+2}-k_{2m_1+3}+\ldots-k_{2n+1}}{\bar{k}_{2m_1+2}-\bar{k}_{2m_1+3}+\ldots-\bar{k}_{2n+1}}$$

$$\frac{d^2k_1\ldots d^2k_n}{\pi^{2n+1}}$$

In the sum over m_1, m_2 let us sum over m_2 first, then the second multiplier and the denumerator cancels:

$$\sum_{\substack{m_1,m_2\geq 0 \\ m_1+m_2+1\leq n}} (k_{2m_1+1} - k_{2m_1+2})(\bar{k}_{2m_1+2m_2+2} - \bar{k}_{2m_1+2m_2+3})$$

$$\frac{k_{2m_1+2} - k_{2m_1+3} + \ldots - k_{2n+1}}{\bar{k}_{2m_1+2} - \bar{k}_{2m_1+3} + \ldots - \bar{k}_{2n+1}}$$

$$= \sum_{m_1\geq 0}^{n-1} (k_{2m_1+1} - k_{2m_1+2})(\bar{k}_{2m_1+2} - \bar{k}_{2m_1+3}\bar{k}_{2m_1+4} - \bar{k}_{2m_1+5} + \ldots + \bar{k}_{2n} - \bar{k}_{2n+1})$$

$$\cdot \frac{(k_{2m_1+2} - k_{2m_1+3} + \ldots - k_{2n+1})}{\bar{k}_{2m_1+2} - \bar{k}_{2m_1+3} + \ldots - \bar{k}_{2n+1}}$$

$$= \sum_{m_1=0}^{n-1} (k_{2m_1+1} - k_{2m_1+2})(k_{2m_1+2} - k_{2m_1+3} + \ldots - k_{2n+1})$$

$$= \frac{1}{2}\left(-(k_1 - k_2 + k_3 - \ldots + k_{2n+1})^2 - \sum_{p=1}^{n} k_{2p}{}^2 + \sum_{p=0}^{n} k_{2p+1}{}^2\right)$$

Let us substitute $q(z, \bar{z}, t)$ as a functional of $\{\alpha(k, \bar{k}, t)\}$ in the equation

$$\int (i\partial_t + k_1{}^2 + \bar{k}_1^2)\alpha(k_1, k_1, t)\exp(-\bar{z}_1, \bar{k}_1 + z_1 k_1)\frac{d^2 k_1}{\pi}$$

$$+ \sum_{n=1}^{\infty}\int\left(i\partial_t + \sum_{p=0}^{n}(k_{2p+1}^2 + \bar{k}_{2p+1}^2) - \sum_{p=1}^{n}(k_{2p}{}^2 + \bar{k}_{2p}^2)\right)$$

$$\frac{\alpha(k, \bar{k}, t)\bar{\alpha}(k_2, \bar{k}_2, t)\ldots\alpha(k_{2n+1}, \bar{k}_{2n+1}, t)}{(k_1 - k_2)(\bar{k}_2 - \bar{k}_3)(k_3 - k_4)\ldots(\bar{k}_{2n} - \bar{k}_{2n+1})}$$

$$\exp(\bar{z}(-\bar{k}_1 + \bar{k}_2 - \ldots + \bar{k}_{2n} - \bar{k}_{2n+1}) - z(-k_1 + k_2 - \ldots + k_{2n} - k_{2n+1}))$$

$$\frac{d^2 k_1 \ldots d^2 k_{2n+1}}{\pi^{2n+1}} = 0$$

if $\alpha(k, \bar{k}, t) = \alpha(k, \bar{k})e^{i(k^2+\bar{k}^2)t}$

2) Substitution of $\alpha(k, \bar{k}, 0) = \alpha(k, \bar{k})$, where $\alpha(k, \bar{k})$ is given by (8), into (7) gives $q(z, \bar{z}, 0) = q(z, \bar{z})$.

References

[1] I.M.Gelfand, G.E.Shilov. *Generalized functions* , v.1, Academic Press 1964.

[2] I.M.Gelfand, L.A.Dikii,*Russ. Math. Surv.* **30** (5),77-113,1975.

[3] A.S.Fokas, I.M.Gelfand,*Lett. Math. Phys*, **32**,189-210,1994.

[4] R.Rosales,*Stud. Appl. Math.* **59**,117-151,1978.

[5] S.Oishi. *Proc.Soc.Japan* **47**(3),1037-1038,1979.

A. S. Fokas
Dept. Math.
Imperial College of Science
Huxley Building
London SW7 2BZ, UK

I. M. Gelfand
Dept. Math.
Rutgers University
Busch Hill Center
New Brunswick, NJ 08903

M. V. Zyskin
Dept. of Math.
Courant Institute
New York University
New York, NY 10012

March 1995

Elliptic solutions of the Yang-Baxter equation and modular hypergeometric functions

Igor B. Frenkel and Vladimir G. Turaev

Introduction

Various results in algebra, analysis, and geometry can be generalized by replacing the ordinary numbers (integer, real or complex) by their trigonometric analogues. For $x \in \mathbb{C}$, the trigonometric number $[x]_h \in \mathbb{C}$ is defined by

$$(0.a) \qquad [x]_h = \frac{\sin(\pi h x)}{\sin(\pi h)}$$

where $h \in \mathbb{C}\backslash\mathbb{Z}$ is a fixed parameter. It is clear that $\lim_{h\to 0}[x]_h = x$; thus, $[x]_h$ may be viewed as a one-parameter deformation of x. The trigonometric numbers are not additive: generally speaking $[x+y]_h \neq [x]_h + [y]_h$. However, they satisfy a kind of additivity of "second order": for any $x, y, z \in \mathbb{C}$,

$$(0.b) \qquad [x + z]_h [x - z]_h = [x + y]_h [x - y]_h + [y + z]_h [y - z]_h.$$

Many identities between ordinary numbers can be proved using only the additivity of second order and therefore allow a trigonometric deformation. It is well known that trigonometric functions admit one parameter generalizations - elliptic functions. This suggests the notion of an elliptic number $[x]_{h,\tau} \in \mathbb{C}$ defined for any $x, h, \tau \in \mathbb{C}$ with $h \notin \mathbb{Z} + \mathbb{Z}\tau$, $\mathrm{Im}(\tau) > 0$ by

$$(0.c) \qquad [x] = [x]_{h,\tau} = \frac{\theta_{11}(hx, \tau)}{\theta_{11}(h, \tau)}$$

where θ_{11} is an odd Jacobi elliptic theta function (cf. Section 1.1). It follows from definitions that $\lim_{\mathrm{Im}(\tau)\to+\infty}[x]_{h,\tau} = [x]_h$. Therefore, we can view the elliptic numbers as one-parameter deformations of the trigonometric numbers and as two-parameter deformations of the ordinary numbers. The elliptic numbers do not satisfy formula (0.b) but satisfy the following additivity of the "fourth order"

$$[x+z][x-z][y+w][y-w] = [x+y][x-y][z+w][z-w]+[x+w][x-w][y+z][y-z].$$

This is a version of Riemann's theta identity. For trigonometric numbers, it follows from (0.b).

The identities that can be proven using only the additivity of the fourth order are much more exceptional than those that can be deduced from (0.b). An important class of such remarkable identities appear in statistical mechanics, specifically, in the theory of elliptic solutions of the Yang-Baxter equation with spectral parameter. The elliptic solutions were first constructed by R. Baxter [Ba], who deduced the Yang-Baxter relation from Riemann's theta identity. Baxter's solutions were substantially generalized in [DJMO]. We call the solutions of [Ba] and [DJMO] elliptic $6j$-symbols. They can be viewed as far reaching generalizations of the classical $6j$-symbols introduced by G. Racah and E. Wigner in the early 1940's (see [BL] for a survey). Besides the 6 integral parameters (as in the Racah-Wigner theory), the elliptic $6j$-symbols depend also on three complex parameters: h, τ, and the spectral parameter $x \in \mathbb{C}$. The elliptic $6j$-symbols are the main object of our study.

Converging h, τ, x to appropriate limits we can degenerate the elliptic $6j$-symbols into less powerful but better known functions. In the limit $\mathrm{Im}(\tau) \to +\infty$, the elliptic $6j$-symbols degenerate into the trigonometric $6j$-symbols studied in our previous paper [FT]. In the limit $h \to 0$, the trigonometric $6j$-symbols yield so-called rational $6j$-symbols. The elliptic, trigonometric, and rational $6j$-symbols are solutions of the Yang-Baxter equation with spectral parameter. Finally, we can eliminate the spectral parameter x by converging it to infinity and obtain in this way solutions of the Yang-Baxter equation without spectral parameter. In the limit $x \to +\infty$, the rational and trigonometric $6j$-symbols degenerate into, respectively, the classical Racah-Wigner $6j$-symbols and their q-deformations known as quantum $6j$-symbols (see [KR]). Observe that the spectral parameter of the elliptic $6j$-symbols lies on the elliptic curve determined by τ. Therefore, there is no direct way to consider its infinite limit and to define elliptic $6j$-symbols without spectral parameter.

We establish in this paper several new results concerning the elliptic $6j$-symbols. First, we prove the tetrahedral symmetry of the elliptic $6j$-symbols analogous to the well known symmetry of the classical and quantum $6j$-symbols and generalizing our results in the trigonometric case [FT]. This symmetry yields an example of a deep identity between trigonometric numbers that generalize to elliptic numbers. Secondly, we establish a relation between the elliptic $6j$-symbols and the trigonometric $6j$-symbols constructed in [FT] using a graphical calculus.

Thirdly, we describe the transformation properties of the elliptic $6j$-symbols under the natural action of the modular group $SL_2(\mathbb{Z})$ in the space of parameters. It is interesting to note that the naive deformation of quantum $6j$-symbols obtained by replacing trigonometric numbers by elliptic numbers in the standard explicit formulas does not give the elliptic $6j$-symbols and breaks all non-trivial identities (and also lacks the modular invariance).

An important aspect of the theory of $6j$-symbols is their relationships with hypergeometric series. The quantum $6j$-symbols can be expressed as balanced basic hypergeometric series $_4\phi_3$, see [KR]. By [GR], they can also be rewritten as very-well-poised series $_8\phi_7$. It was established in [FT] that the trigonometric $6j$-symbols (with spectral parameter) are balanced very-well-poised series $_{10}\phi_9$. Our study of elliptic $6j$-symbols naturally leads to a generalization of basic hypergeometric functions obtained via replacing trigonometric numbers by the elliptic numbers. We restrict our attention to balanced very-well-poised elliptic series. This is exactly the class of "elliptic" hypergeometric functions with nice modular properties! We call these elliptic series modular hypergeometric functions.

The hypergeometric series $_{10}\phi_9$ yields the simplest balanced very-well-poised basic hypergeometric series that are not reduced to a monomial. The corresponding modular hypergeometric functions turn out to be a disguised form of the elliptic $6j$-symbols. The tetrahedral symmetry of the elliptic $6j$-symbols implies an analogue of the Bailey transform for these modular hypergeometric functions. We show also that the Jackson summation formula for balanced very-well-poised series $_8\phi_7$ extends to modular hypergeometric functions. It is natural to conjecture that other known identities involving balanced very-well-poised basic hypergeometric series admit generalizations to the modular hypergeometric functions.

Although $6j$-symbols with spectral parameters were first constructed in the context of statistical mechanics, they are deeply related to the representation theory. The rational $6j$-symbols arise in the finite-dimensional representation theory of the Yangian $Y(\mathfrak{sl}_2)$. The trigonometric $6j$-symbols arise in the finite-dimensional representation theory of the quantum group $U_q(\hat{\mathfrak{sl}}_2)$. It is
interesting to find a representation theoretic description of the elliptic $6j$-symbols. Since the fundamental work of E. Sklyanin, there was a considerable progress in this matter, however, the existing interpretations require ad hoc ingredients. In this paper, we do not discuss this problem, although it was one of our motivations.

This paper consists of 5 sections. In Section 1, we recall the definition of the elliptic $6j$-symbols following [Ba] and [DJKMO]. In Section 2, we give an explicit formula for the elliptic $6j$-symbols revealing their tetrahedral symmetry. In Section 3, we relate the elliptic $6j$-symbols and the trigonometric $6j$-symbols constructed in [FT] using the graphical calculus. In Section 4, we describe the behavior of the elliptic $6j$-symbols under the modular transformations. In Section 5, we introduce modular hypergeometric functions and prove the analogues of the Jackson summation and the Bailey transform.

1. Elliptic 6j-symbols

1.1. Theta function. Recall the definition of the theta function θ_{11} (see [Mu, Section 5]). This is an analytic function in two variables $x \in \mathbb{C}, \tau \in \mathbb{C}$ with $\mathrm{Im}(\tau) > 0$ defined by

$$\theta_{11}(x, \tau) = \sum_{n \in \mathbb{Z}} \exp(\pi\sqrt{-1}(n + \tfrac{1}{2})^2\tau + 2\pi\sqrt{-1}(n + \tfrac{1}{2})(x + \tfrac{1}{2})).$$

For a fixed τ, the zeros of θ_{11} form the lattice $\mathbb{Z} + \mathbb{Z}\tau$.

Recall the well known transformation properties of θ_{11}:

(1.1.a) $$\theta_{11}(x + 1, \tau) = -\theta_{11}(x, \tau),$$

(1.1.b) $$\theta_{11}(x + \tau, \tau) = -\exp(-\pi\sqrt{-1}\tau - 2\pi\sqrt{-1}x)\,\theta_{11}(x, \tau),$$

(1.1.c) $$\theta_{11}(x, \tau + 1) = \exp(\pi\sqrt{-1}/4)\,\theta_{11}(x, \tau),$$

(1.1.d) $$\theta_{11}(\tfrac{x}{\tau}, -\tfrac{1}{\tau}) = -(-\sqrt{-1}\tau)^{1/2} \exp(\tfrac{\pi\sqrt{-1}x^2}{\tau})\,\theta_{11}(x, \tau)$$

where the square root of $-\sqrt{-1}\tau$ is chosen so that its real part is positive.

The function θ_{11} admits an infinite product expansion given by

$$\theta_{11}(x, \tau) = -2\exp(\pi\sqrt{-1}\tau/4)\sin(\pi x)\prod_{m \geq 1}(1 - p^m)(1 - 2p^m\cos(x) + p^{2m})$$

where $p = \exp(2\pi\sqrt{-1}\tau)$ (see [Mu, Section 14]). It follows from this formula that

(1.1.e) $$\lim_{\mathrm{Im}(\tau)\to+\infty} \exp(-\pi\sqrt{-1}\tau/4)\,\theta_{11}(x, \tau) = -2\sin(\pi x).$$

We shall systematically use the following notation. Fix complex numbers h, τ with $h \notin \mathbb{Z} + \mathbb{Z}\tau$ and $\mathrm{Im}(\tau) > 0$. For a complex number x, set

(1.1.f) $$[x] = [x]_{h,\tau} = \frac{\theta_{11}(hx, \tau)}{\theta_{11}(h, \tau)} \in \mathbb{C}.$$

Observe that $[-x] = -[x]$ and $[1] = 1$.

Unless explicitly stated to the contrary, we shall assume that $h \notin \mathbb{Q} + \mathbb{Q}\tau$. This assumption ensures that $[n] \neq 0$ for any integer $n \neq 0$.

1.2. Baxter's elliptic 6*j*-symbols. R. Baxter [Ba] introduced elliptic 6*j*-symbols as solutions to the so-called Yang-Baxter equation. Baxter's 6*j*-symbol is a complex-valued function of x, h, τ depending on 4 integer parameters i, j, k, l and denoted here by

$$(1.2.a) \qquad \begin{bmatrix} i & j & 1 \\ k & l & 1 \end{bmatrix} (x, h, \tau).$$

We shall usually fix h, τ and omit them from the notation for the 6*j*-symbol.

The number (1.2.a) is defined as follows. It is equal to 0 unless $i, j, k, l \geq 0$ and

$$|i - j| = |j - k| = |k - l| = |l - i| = 1.$$

Under these conditions, we have $i = l \pm 1, k = l \pm 1$, and $j = l$ or $j = l \pm 2$. Set

$$\begin{bmatrix} l \pm 1 & l \pm 2 & 1 \\ l \pm 1 & l & 1 \end{bmatrix} (x) = [x + 1],$$

$$\begin{bmatrix} l \mp 1 & l & 1 \\ l \pm 1 & l & 1 \end{bmatrix} (x) = \frac{[l + 1 \pm 1][x]}{[l + 1]},$$

$$\begin{bmatrix} l \pm 1 & l & 1 \\ l \pm 1 & l & 1 \end{bmatrix} (x) = \frac{[l + 1 \mp x]}{[l + 1]}.$$

Since $h \notin \mathbb{Q} + \mathbb{Q}\tau$, all the denominators on the right hand side are non-zero.

Note that the 6*j*-symbol (1.2.a) is an elliptic monomial, i.e., a product

$$\pm \prod_{r \geq 2} [r]^{a_r} \prod_{s \in \mathbb{Z}} [x + s]^{b_s}$$

with integer (possibly negative) a_r, b_s equal to zero except for a finite set of r, s. (The factor [1] may also occur but we can ignore it since $[1] = 1$.) The *degree* $\sum_s b_s$ of the monomials defining Baxter's 6*j*-symbols is equal to 1.

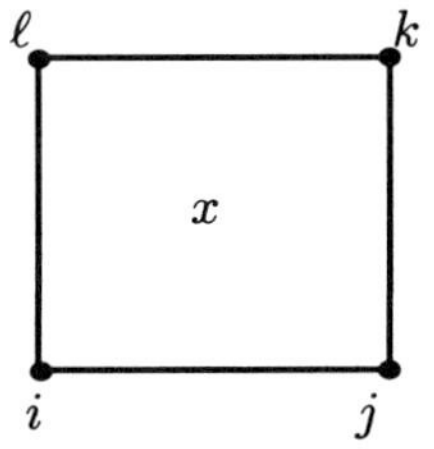

Figure 1.1

It will be convenient to represent the 6*j*-symbol (1.2.a) graphically by a square with the parameter x sitting inside and the numbers i, j, k, l attached to the vertices in the anticlockwise direction beginning in the bottom left corner, see Figure 1.1.

It is easy to check that under the opposite choice of direction we get an equal $6j$-symbol, see Figure 1.2.

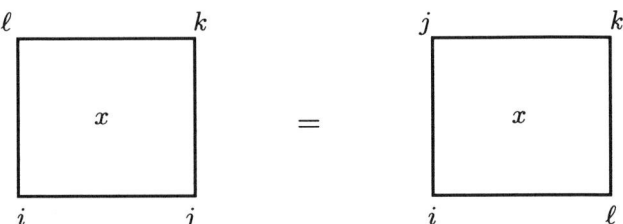

Figure 1.2

Our notation differs from the one in [DJKMO] where the square brackets are defined by $[x] = \theta_{11}(hx, \tau)$ and the $6j$-symbol (1.2.a) is denoted $W_{11}(i, j, k, l \mid x)$. The complex parameter ξ used in [DJKMO] is set here to be 1. For this value of ξ, the number $W_{11}(i, j, k, l \mid x)$ is defined in [DJKMO] by the same formulas as above, for any $i, j, k, l \in \mathbb{Z}$ with $l \neq -1$. We shall consider only non-negative i, j, k, l; this will allow us to consider tetrahedral symmetries of $6j$-symbols involving, in particular, permutations of l with i, j, k. Our notation is meant to emphasize this symmetry.

1.2.1. Theorem. ([Ba], [DJKMO]) *(i) For any $i, j, k, l, e, f \geq 0$ and $x, y \in \mathbb{C}$,*

$$(1.2.b) \qquad \sum_g \begin{bmatrix} i & j & 1 \\ g & f & 1 \end{bmatrix}(x) \begin{bmatrix} f & g & 1 \\ l & e & 1 \end{bmatrix}(x+y) \begin{bmatrix} g & j & 1 \\ k & l & 1 \end{bmatrix}(y) =$$

$$= \sum_g \begin{bmatrix} f & i & 1 \\ g & e & 1 \end{bmatrix}(y) \begin{bmatrix} i & j & 1 \\ k & g & 1 \end{bmatrix}(x+y) \begin{bmatrix} g & k & 1 \\ l & e & 1 \end{bmatrix}(x).$$

(ii) For any $i, j, k, l \geq 0$ and $x \in \mathbb{C}$, the sum

$$(1.2.c) \qquad \sum_g \begin{bmatrix} i & g & 1 \\ e & l & 1 \end{bmatrix}(x+1) \begin{bmatrix} g & j & 1 \\ k & e & 1 \end{bmatrix}(x)$$

does not depend on the choice of e such that $|e - k| = |e - l| = 1$.

Formula (1.2.b) is called the Yang-Baxter equation. Its graphical interpretation is given in Figure 1.3 where in both parts we multiply the $6j$-symbols associated with the three (distorted) squares and sum over the integer variable $g = 0, 1, 2, ...$ assigned to the vertex drawn in bold. Formula (1.2.c) plays a key role in the fusion procedure, see Section 1.3. Note that the sums in (1.2.b), (1.2.c) are finite, since their terms are equal to 0 except for a finite number of values of g.

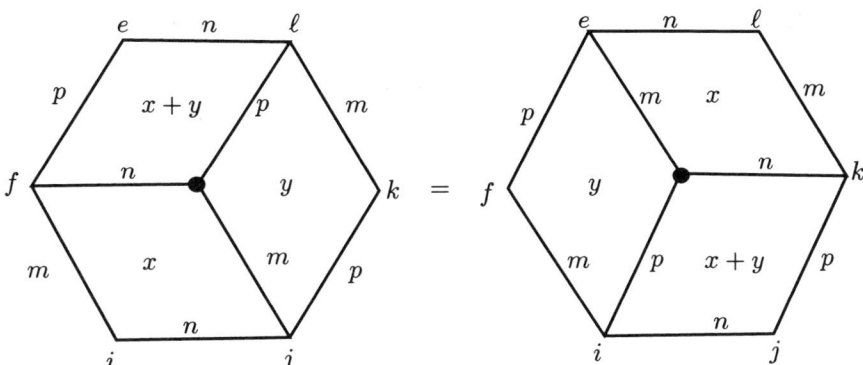

Figure 1.3

Proof of Theorem. Formula (1.2.b) follows from the formula

$$\sum_g W_{11}(i,j,g,f \,|\, x)\, W_{11}(f,g,l,e \,|\, x+y)\, W_{11}(g,j,k,l \,|\, y) =$$

$$= \sum_g W_{11}(f,i,g,e \,|\, y)\, W_{11}(i,j,k,g \,|\, x+y)\, W_{11}(g,k,l,e \,|\, x)$$

established in [Ba]. In the last formula g may take negative values but the corresponding terms are equal to 0. Indeed, the g-th term on the left hand side is 0 for $g \leq -2$ since in this case $|j - g| \neq 1$. The term corresponding to $g = -1$ might be non-zero only if $i = e = k = 1$ and $j = l = f = 0$. However, in this case it also equals 0 because $W_{11}(i,j,g,f \,|\, x) = W_{11}(1,0,-1,0 \,|\, y) = 0$. A similar argument applies to the right hand side.

For $k \geq 1$, the claim (ii) of the theorem follows from [DJKMO, Lemma 2.1.1 (i)]. For $k = 0$, there is nothing to prove since there is only one choice $e = 1$ for e.

1.3. General elliptic $6j$-symbols (see [DJKMO]). The general elliptic $6j$-symbols are complex-valued functions of x, h, τ

(1.3.a) $$\begin{bmatrix} i & j & n \\ k & l & m \end{bmatrix} (x, h, \tau)$$

depending on six non-negative integer parameters i, j, k, l, m, n with $m, n \geq 1$. We shall usually fix h, τ and omit them from the notation for the $6j$-symbol.

The number (1.3.a) is defined as follows. It is equal to 0 unless $i, j, k, l, m, n \geq 0$ and

$$i - j, k - l \in \{-n, -n+2, -n+4, ..., n-2, n\},$$

$$i - l, j - k \in \{-m, -m+2, -m+4, ..., m-2, m\}.$$

If these conditions are satisfied, then one can find a pair of sequences of non-negative integers $\alpha_{0,m} = l, \alpha_{1,m}, ..., \alpha_{n,m} = k$ and $\alpha_{n,m} = k$, $\alpha_{n,m-1}, ..., \alpha_{n,0} = j$ such that the neighboring terms differ by ± 1. Fix such sequences and consider their extension to an integer $(n+1) \times (m+1)$-matrix $A = (\alpha_{s,t})_{s,t}$ with $s = 0, 1, ..., n$, $t = 0, 1, ..., m$ such that $\alpha_{0,0} = i$ and $|\alpha_{s,t} - \alpha_{s+1,t}| = |\alpha_{s,t} - \alpha_{s,t+1}| = 1$ for any s, t. Set

$$R_{s,t} = \frac{m+n}{2} - s - t - 1$$

and

$$(1.3.b) \quad \begin{bmatrix} i & j & n \\ k & l & m \end{bmatrix}(x) = \sum_A \prod_{s=0}^{n-1} \prod_{t=0}^{m-1} \begin{bmatrix} \alpha_{s,t} & \alpha_{s+1,t} & 1 \\ \alpha_{s,t+1} & \alpha_{s+1,t+1} & 1 \end{bmatrix}(x + R_{s,t})$$

where the sum is taken over all matrices A as above (with values i, j, k, l in the corners and fixed last row and column). A graphical expression for this sum is given in Figure 1.4 where we multiply the $6j$-symbols associated with the mn unit squares and sum over the integer variables assigned to the vertices drawn in bold. The $6j$-symbol (1.3.a) is graphically presented by a square with the parameter x inside, the numbers i, j, k, l attached to the vertices and the numbers m, n attached to the edges.

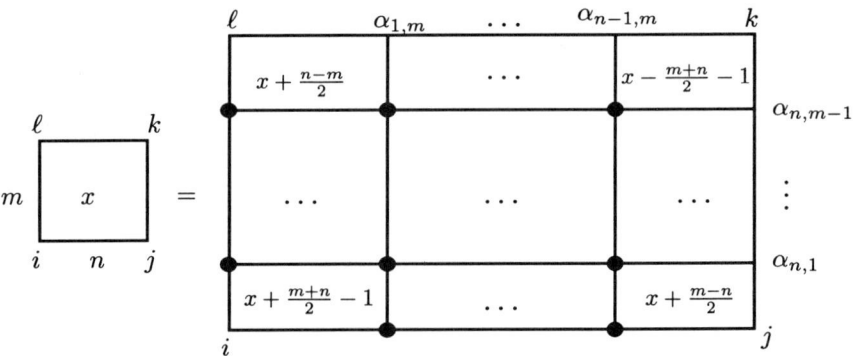

Figure 1.4

Observe that there is only a finite number of matrices A contributing non-zero to the right hand side of (1.3.b). Repeated use of Theorem 1.2.1 (ii) shows that the right hand side of (1.3.b) does not depend on the choice of the sequence $\alpha_{0,m} = l, \alpha_{1,m}, ..., \alpha_{n,m} = k$. Similarly, a version of Theorem 1.2.1 (ii) obtained by transposition as in Figure 1.2 shows that the right

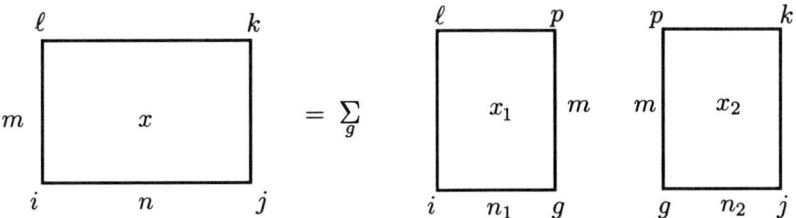

Figure 1.5

hand side of (1.3.b) does not depend on the choice of the sequence $\alpha_{n,m} = k, \alpha_{n,m-1}, ..., \alpha_{n,0} = j$. Therefore the $6j$-symbol (1.3.a) is well defined.

We shall say that a triple of non-negative integers i, j, k is admissible if $|i - j| \le k \le |i + j|$ and $i + j + k$ is even. A 6-tuple of non-negative integers (i, j, k, l, m, n) is said to be admissible if the triples (i, j, n), (k, l, n), $(i, l, m), (j, k, m)$ are admissible. By definition, the $6j$-symbol (1.3.a) may be non-zero only when the tuple (i, j, k, l, m, n) is admissible.

The next lemma follows directly from definitions.

1.3.1. Lemma. *Let (i, j, k, l, m, n) be an admissible tuple of integers with $m, n \ge 1$. If $n = n_1 + n_2$ with positive $n_1, n_2 \in \mathbb{Z}$ then for any non-negative $p \in \mathbb{Z}$ such that the triples $(l, n_1, p), (p, n_2, k)$ are admissible and for any $x \in \mathbb{C}$,*

$$(1.3.c) \qquad \begin{bmatrix} i & j & n \\ k & l & m \end{bmatrix}(x) = \sum_g \begin{bmatrix} i & g & n_1 \\ p & l & m \end{bmatrix}(x+\tfrac{n_2}{2}) \begin{bmatrix} g & j & n_2 \\ k & p & m \end{bmatrix}(x-\tfrac{n_1}{2}).$$

If $m = m_1 + m_2$ with positive $m_1, m_2 \in \mathbb{Z}$ then for any non-negative $p \in \mathbb{Z}$ such that the triples $(j, m_1, p), (p, m_2, k)$ are admissible and for any $x \in \mathbb{C}$,
(1.3.d)
$$\begin{bmatrix} i & j & n \\ k & l & m \end{bmatrix}(x) = \sum_g \begin{bmatrix} i & j & n \\ p & g & m_1 \end{bmatrix}(x + \tfrac{m_2}{2}) \begin{bmatrix} g & p & n \\ k & l & m_2 \end{bmatrix}(x - \tfrac{m_1}{2}).$$

A graphical expression corresponding to formula (1.3.c) is given in Figure 1.5.

The symmetry in Figure 1.2 generalizes to the symmetry

$$(1.3.e) \qquad \begin{bmatrix} i & j & n \\ k & l & m \end{bmatrix}(x) = \begin{bmatrix} i & l & m \\ k & j & n \end{bmatrix}(x).$$

Equality (1.2.b) generalizes to the following Yang-Baxter relation easily proven by induction.

1.3.2. Theorem ([DJKMO]).

$$(1.3.\text{f}) \quad \sum_g \begin{bmatrix} i & j & n \\ g & f & m \end{bmatrix}(x) \begin{bmatrix} f & g & n \\ l & e & p \end{bmatrix}(x+y) \begin{bmatrix} g & j & m \\ k & l & p \end{bmatrix}(y) =$$

$$= \sum_g \begin{bmatrix} f & i & m \\ g & e & p \end{bmatrix}(y) \begin{bmatrix} i & j & n \\ k & g & p \end{bmatrix}(x+y) \begin{bmatrix} g & k & n \\ l & e & m \end{bmatrix}(x).$$

For completeness, we mention another important identity, the so-called orthonormality relation. We need the following notation: for $x \in \mathbb{C}$ and non-negative integers m, n, set

$$(1.3.\text{g}) \qquad \Pi_{m,n}(x) = \prod_{s=0}^{n-1} \prod_{t=0}^{m-1} [x + \frac{m+n}{2} - s - t - 1].$$

This is an elliptic monomial of degree mn. Note that

$$\Pi_{m,n}(x) = \Pi_{n,m}(x) = (-1)^{mn}\Pi_{m,n}(-x) = (-1)^{mn}\Pi_{n,m}(-x).$$

We have

$$\sum_g \begin{bmatrix} i & j & n \\ g & l & m \end{bmatrix}(x) \begin{bmatrix} g & j & m \\ i' & l & n \end{bmatrix}(-x) = (-1)^{mn}\Pi_{m,n}(x+1)\,\Pi_{m,n}(x-1)\,\delta^i_{i'}.$$

1.4. Remarks. 1. The assumption $h \notin \mathbb{Q}+\mathbb{Q}\tau$ ensures that all $6j$-symbols (1.3.a) are well defined. For each particular $6j$-symbol (1.3.a) to be well defined, it suffices to assume that $Nh \notin \mathbb{Z} + \mathbb{Z}\tau$ for $N = 1, 2, ..., i + m + n$.

2. In [DJKMO] the $6j$-symbol (1.3.a) is denoted by $W'_{nm}(i, j, k, l \,|\, x + (m - n)/2)$.

2. Computation of the elliptic $6j$-symbols

In this section we shall give an explicit formula for the elliptic $6j$-symbols (Theorem 2.2). This formula reveals their tetrahedral symmetry. We first introduce the necessary notation.

2.1. Notation. For an admissible triple of non-negative integers i, j, k, set

$$(2.1.\text{a}) \quad \Theta(i, j, k) = (-1)^{(i+j+k)/2} \frac{[\frac{i+j+k}{2} + 1]!\,[\frac{i+j-k}{2}]!\,[\frac{i+k-j}{2}]!\,[\frac{j+k-i}{2}]!}{[i]!\,[j]!\,[k]!}.$$

Here for a positive integer a we have $[a]! = [1][2]...[a]$ and for $a = 0$ we have $[a]! = 1$. Note that $\Theta(i, j, k)$ is an elliptic monomial of degree 0. (The corresponding trigonometric monomial naturally appears as the bracket function of the so-called θ-net, cf. Section 3, [KL], [FT].)

2.2. Theorem. *Let (i, j, k, l, m, n) be an admissible 6-tuple of integers. Set*

$$\alpha = \frac{j + m - k}{2}, \quad \beta = \frac{k + m - j}{2}, \quad \gamma = \frac{j + k - m}{2}$$

and

$$\sigma = \alpha + \beta + \gamma = \frac{j + k + m}{2}.$$

For $x \in \mathbb{C}$, set

(2.2a)

$$\begin{bmatrix} i & j & n \\ k & l & m \end{bmatrix}'(x) =$$

$$\sum_g \left(\frac{[g+1]([\frac{g+n-\gamma}{2}]! \, [\frac{g+i-\alpha}{2}]! \, [\frac{g+l-\beta}{2}]!)^2}{([g]!)^3 \Theta(g, n, \gamma) \Theta(g, i, \alpha) \Theta(g, l, \beta)} \prod_{s=0}^{\sigma} [x - \frac{g + \sigma}{2} + s][x + \frac{g - \sigma}{2} + 1 + s] \right)$$

where g runs over non-negative integers such that the triples $(g, n, \gamma), (g, i, \alpha)$, and (g, l, β) are admissible. Then

(2.2.b)

$$\begin{bmatrix} i & j & n \\ k & l & m \end{bmatrix}(x) = (-1)^{\frac{i+k+m+n}{2}} \frac{[i+1] \, [\frac{i+j-n}{2}]! \, [\frac{k+l+n}{2} + 1]! \, [\frac{k+n-l}{2}]!}{[i]! \, [l]! \, [n]! \, [\frac{i+n-j}{2}]! \, [\frac{j+k-m}{2}]!} \times$$

$$\times \Pi_{m,n-1}(x + \frac{1}{2}) \prod_{r=-\frac{i+l}{2}}^{\frac{k-i}{2}} [x + r]^{-1} \prod_{r=\frac{l-j}{2}+1}^{\frac{k+i}{2}+1} [x + r]^{-1} \begin{bmatrix} i & j & n \\ k & l & m \end{bmatrix}'(x).$$

The proof of this theorem is given in Section 2.4. It is based on the following lemma whose proof is postponed to Section 3.

2.3. Lemma.

(2.3.a)

$$\begin{bmatrix} i & k+m & n \\ k & l & m \end{bmatrix}(x) =$$

$$= (-1)^{\frac{i+l+m}{2}} \frac{[i+1] \, [\frac{i+k+m-n}{2}]! \, [\frac{l+n-k}{2}]! \, ([\frac{i+l-m}{2}]!)^2}{[\frac{k+l-n}{2}]! \, [\frac{i+n-k-m}{2}]! \, [i]! \, [l]! \, \Theta(i, l, m)} \times$$

$$\times \Pi_{m,n-1}(x + \frac{1}{2}) \prod_{r=1}^{\frac{i+m-l}{2}} [x + 1 + \frac{k+m-l}{2} - r] \prod_{r=1}^{\frac{l+m-i}{2}} [x + 2 + \frac{k+l+m}{2} - r],$$

(2.3.b)

$$\begin{bmatrix} i & j & n \\ j+m & l & m \end{bmatrix}(x) =$$

$$= (-1)^{\frac{i+l-m}{2}} \frac{[\frac{i+j-n}{2}]! \, [\frac{i+n-j}{2}]! \, ([\frac{i+l-m}{2}]!)^2 \, [j+m]! \, [i+1] \Theta(l,n,j+m)}{[\frac{j+m+l-n}{2}]! \, [\frac{l+n-j-m}{2}]! \, ([i]!)^2 \, [j]! \Theta(i,j,n) \Theta(i,l,m)} \times$$

$$\times \Pi_{m,n-1}(x+\frac{1}{2}) \prod_{r=1}^{\frac{l+m-i}{2}} [x+r-\frac{j+m-i}{2}] \prod_{r=1}^{\frac{i+m-l}{2}} [x+r-\frac{i+j+m}{2}-1].$$

2.4. *Proof of Theorem 2.2.* Set in formula (1.3.d) $p = \gamma, m_1 = \alpha, m_2 = \beta$. We have

$$\begin{bmatrix} i & j & n \\ k & l & m \end{bmatrix}(x) = \sum_{g} \begin{bmatrix} i & j & n \\ \gamma & g & \alpha \end{bmatrix}(x+\beta/2) \begin{bmatrix} g & \gamma & n \\ k & l & \beta \end{bmatrix}(x-\alpha/2).$$

Note that $j = \gamma + \alpha$ and $k = \gamma + \beta$ so that the $6j$-symbols on the right hand side are the ones computed in Lemma 2.3. We obtain

$$\begin{bmatrix} i & j & n \\ k & l & m \end{bmatrix}(x) = \sum_{g} (-1)^{\frac{g+\alpha+i}{2}} \frac{[i+1] \, [\frac{i+j-n}{2}]! \, [\frac{g+n-\gamma}{2}]! \, ([\frac{i+g-\alpha}{2}]!)^2}{[\frac{\gamma+g-n}{2}]! \, [\frac{i+n-j}{2}]! \, [i]! \, [g]! \, \Theta(i,g,\alpha)} \times$$

$$\times \Pi_{\alpha,n-1}(x+\frac{\beta+1}{2}) \prod_{r=1}^{\frac{i+\alpha-g}{2}} [x+1+\frac{\sigma-g}{2}-r] \prod_{r=1}^{\frac{g+\alpha-i}{2}} [x+2+\frac{\sigma+g}{2}-r] \times$$

$$\times (-1)^{\frac{l+g-\beta}{2}} \frac{[\frac{g+\gamma-n}{2}]! \, [\frac{g+n-\gamma}{2}]! \, ([\frac{g+l-\beta}{2}]!)^2 \, [k]! \, [g+1] \Theta(l,n,k)}{[\frac{k+l-n}{2}]! \, [\frac{l+n-k}{2}]! \, ([g]!)^2 \, [\gamma]! \Theta(g,\gamma,n) \Theta(g,l,\beta)} \times$$

$$\times \Pi_{\beta,n-1}(x+\frac{1-\alpha}{2}) \prod_{r=1}^{\frac{l+\beta-g}{2}} [x+r-\frac{\sigma-g}{2}] \prod_{r=1}^{\frac{g+\beta-l}{2}} [x+r-\frac{g+\sigma}{2}-1].$$

To simplify this expression we observe that

$$\Pi_{\alpha,n-1}(x+\frac{\beta+1}{2}) \, \Pi_{\beta,n-1}(x+\frac{1-\alpha}{2}) = \Pi_{\alpha+\beta,n-1}(x+\frac{1}{2}) = \Pi_{m,n-1}(x+\frac{1}{2})$$

and that the product of the other expressions involving x is equal to

$$\prod_{r=-\frac{i+l}{2}}^{\frac{k-i}{2}} [x+r]^{-1} \prod_{r=\frac{l-i}{2}+1}^{\frac{k+i}{2}+1} [x+r]^{-1} \prod_{s=0}^{\sigma} [x-\frac{g+\sigma}{2}+s][x+\frac{g-\sigma}{2}+1+s].$$

We aslo replace $\Theta(k,l,n)$ using its definition. This gives formula (2.2.b). (To compute the sign one should use that $g = \beta - l(\mathrm{mod}\,2)$.)

2.5. Symmetries of the elliptic $6j$-symbols. Theorem 2.2 shows that the expression (2.2.a) yields the the elliptic $6j$-symbol (1.3.a), at least up to multiplication by a monomial. It is easy to see that the $6j$-symbol (2.2.a) is invariant under 6 permutations generated by the following two

$$(2.5.a) \qquad \begin{bmatrix} i & j & n \\ k & l & m \end{bmatrix}'(x) = \begin{bmatrix} l & k & n \\ j & i & m \end{bmatrix}'(x) = \begin{bmatrix} n & j & i \\ m & l & k \end{bmatrix}'(x).$$

As in [FT, Section 8.10], it is natural to associate with the $6j$-symbol (2.2.a) the geometric picture consisting of a triangle whose center is connected to the vertices and endowed with $x \in \mathbb{C}$, the six edges of the resulting graph are labelled by i, j, k, l, m, n as in Figure 2.1. The symmetries in (2.5.a) are induced by the automorphisms of this graph preserving its central vertex.

Together with (1.3.e) the symmetries (2.5.a) generate the full symmetric group on 4 elements. Thus, the elliptic $6j$-symbols considered up to multiplication by monomials have the full tetrahedral symmetry.

It is easy to deduce from definitions and the identity $[-x] = -[x]$ that the $6j$-symbol (2.2.a) is invariant also under the transformation $x \mapsto -1-x$.

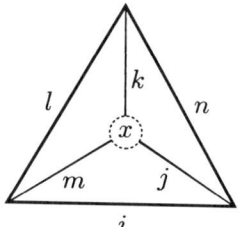

Figure 2.1

3. Comparison with the trigonometric $6j$-symbols

3.1. The trigonometric $6j$-symbols. In [FT, Section 6.3] we defined the trigonometric $6j$-symbol

$$(3.1.a) \qquad \left\{ \begin{matrix} i & j & n \\ k & l & m \end{matrix} \right\}_h (x) \in \mathbb{C}$$

where (i, j, k, l, m, n) is an admissible tuple of integers and $h \in \mathbb{C} \backslash \mathbb{Z}, x \in \mathbb{C}$. The definition of the trigonometric $6j$-symbol given in [FT] relies on the ideas of the theory of skein classes of knots. In Section 3 we assume that the reader is acquainted with this definition. For completeness, we reproduce in Section 3.4 the explicit formulas for the trigonometric $6j$-symbol obtained in [FT]. Note that the $6j$-symbol (3.1.a) is computed via the function

$$(3.1.b) \qquad [x]_h = \frac{\sin(\pi h x)}{\sin(\pi h)} = \frac{\exp(\pi \sqrt{-1} h x) - \exp(-\pi \sqrt{-1} h x)}{\exp(\pi \sqrt{-1} h) - \exp(-\pi \sqrt{-1} h)}.$$

In the notation of [FT] we have $q = \exp(\pi\sqrt{-1}h)$, $q^x = \exp(\pi\sqrt{-1}hx)$, and $[x]_h = (q^x - q^{-x})/(q - q^{-1})$. Here we shall not use this notation.

It follows from (1.1.e) and (1.1.f) that

$$(3.1.c) \qquad \lim_{\mathrm{Im}(\tau)\to+\infty} [x]_{h,\tau} = \lim_{\mathrm{Im}(\tau)\to+\infty} \frac{\theta_{11}(hx,\tau)}{\theta_{11}(h,\tau)} = \frac{\sin(\pi hx)}{\sin(\pi h)} = [x]_h.$$

3.2. Theorem. *Let (i,j,k,l,m,n) be an admissible tuple of integers and $x \in \mathbb{C}$. Then*

$$(3.2.a) \qquad \lim_{\mathrm{Im}(\tau)\to+\infty} \begin{bmatrix} i & j & n \\ k & l & m \end{bmatrix}(x,h,\tau) =$$

$$= (-1)^{mn+i} \frac{[i+1]_h\,[k]_h!\,[\frac{i+j-n}{2}]_h!\,[\frac{i+l-m}{2}]_h!}{[i]_h!\,[\frac{k+l-n}{2}]_h!\,[\frac{k+j-m}{2}]_h!\,\theta(i,j,n)\,\theta(i,l,m)} \left\{\begin{matrix} i & j & n \\ k & l & m \end{matrix}\right\}_h (-x)$$

where $\theta(i,j,k)$ is defined by the formula (2.1.a) with the elliptic square brackets [...] replaced everywhere with the trigonometric brackets [...]$_h$.

Here for a positive integer a we have $[a]_h! = [1]_h[2]_h...[a]_h$ and for $a = 0$ we have $[a]_h! = 1$.

Note that the $6j$-symbol on the left hand side of (3.2.a) is well defined if $h \notin \mathbb{Q}$ and $\mathrm{Im}(\tau)$ is big enough, say, $\mathrm{Im}(\tau) > (i+m+n)\,|\mathrm{Im}(h)|$, cf. Remark 1.4.

Proof of Theorem. It follows from formula (3.1.c) that in the case $m = n = 1$ the limit

$$\lim_{\mathrm{Im}(\tau)\to+\infty} \begin{bmatrix} i & j & n \\ k & l & m \end{bmatrix}(x,h,\tau)$$

exists and is given by the same formulas as in Section 1.2 with the only difference that the brackets [...] should be replaced everywhere with [...]$_h$. It follows from this fact and the definition of the elliptic $6j$-symbols that the limit on the left hand side of (3.2.a) exists for all m, n. (This limit does not depend on $\mathrm{Re}(\tau)$.)

Let us compute the right hand side of (3.2.a) for $m = n = 1$. We can assume that $|i - j| = |j - k| = |k - l| = |l - i| = 1$. By definition of the trigonometric $6j$-symbol, we have the equalities in Figure 3.1. The right hand side equals

$$[x]_h\,\delta_l^j\,\frac{\theta(k,l,1)\,\theta(i,j,1)}{(-1)^l[l+1]_h} + [x-1]_h\,\delta_k^i\,\frac{\theta(k,l,1)\,\theta(i,j,1)}{(-1)^i[i+1]_h}.$$

It is easy to compute that for $j = i + 1$,

(3.2.b) $\qquad \theta(i, j, 1) = \theta(j, i, 1) = (-1)^{i+1}[i + 2]_h.$

Now, a direct computation yields that

$$\begin{Bmatrix} l \pm 1 & l \pm 2 & 1 \\ l \pm 1 & l & 1 \end{Bmatrix}_h (x) = (-1)^{l+1\pm 1}[l + 2 \pm 1]_h[x - 1]_h,$$

$$\begin{Bmatrix} l \mp 1 & l & 1 \\ l \pm 1 & l & 1 \end{Bmatrix}_h (x) = (-1)^{l+1}[l + 2]_h[x]_h,$$

$$\begin{Bmatrix} l \pm 1 & l & 1 \\ l \pm 1 & l & 1 \end{Bmatrix}_h (x) = (-1)^l \frac{[l + 3/2 \pm 1/2]_h}{[l + 1/2 \pm 1/2]_h}[l + 1 \pm x]_h.$$

Substituting these expressions and (3.2.b) in the right hand side of (3.2.a) and observing that $[1]_h = 1$ we obtain precisely the formulas of Section 1.2 for the Baxter $6j$-symbol with the brackets [...] replaced everywhere by [...]$_h$. This proves (3.2.a) in the case $m = n = 1$.

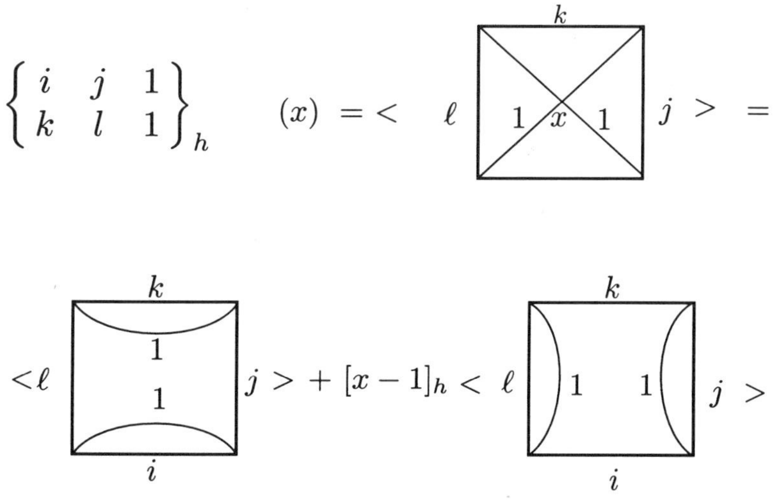

Figure 3.1

The remaining part of the proof goes by induction. Denote the right hand side of (3.2.a) by

(3.2.c) $\qquad \begin{Bmatrix} i & j & n \\ k & l & m \end{Bmatrix}_h^* (x).$

It suffices to establish for this function inductive formulas analogous to
(1.3.c) and (1.3.d). Moreover, by [FT, formula (6.3.g)] the right hand side
of (3.2.a) is invariant under the involution $m \mapsto n, n \mapsto m, l \mapsto j, j \mapsto l$.
Therefore it is enough to establish the following analogue of (1.3.c)
(3.2.d)

$$\left\{\begin{matrix} i & j & n \\ k & l & m \end{matrix}\right\}_h^*(x) = \sum_g \left\{\begin{matrix} i & g & n_1 \\ p & l & m \end{matrix}\right\}_h^*(x + \frac{n_2}{2}) \left\{\begin{matrix} g & j & n_2 \\ k & p & m \end{matrix}\right\}_h^*(x - \frac{n_1}{2})$$

where (i, j, k, l, m, n) is an admissible tuple of integers with $m, n \geq 1$, $n = n_1 + n_2$ with positive $n_1, n_2 \in \mathbb{Z}$, p is a non-negative integer such that the
triples $(l, n_1, p), (p, n_2, k)$ are admissible, and $x \in \mathbb{C}$. Such a formula was
essentially obtained in [FT, Section 8]. Indeed, by [FT, formula (8.3.b)],

(3.2.e) $(\theta(k, l, n))^{-1} \left\{\begin{matrix} n_1 & n_2 & n \\ k & l & p \end{matrix}\right\}_h \left\{\begin{matrix} i & j & n \\ k & l & m \end{matrix}\right\}_h (x) =$

$$\sum_g (-1)^g [g+1] \frac{\left\{\begin{matrix} n_1 & n_2 & n \\ j & i & g \end{matrix}\right\}_h \left\{\begin{matrix} i & g & n_1 \\ p & l & m \end{matrix}\right\}_h (x - \frac{n_2}{2}) \left\{\begin{matrix} g & j & n_2 \\ k & p & m \end{matrix}\right\}_h (x + \frac{n_1}{2})}{\theta(g, p, m)\, \theta(g, n_1, i)\, \theta(g, n_2, j)}.$$

Here

$$\left\{\begin{matrix} n_1 & n_2 & n \\ k & l & p \end{matrix}\right\}_h \quad \text{and} \quad \left\{\begin{matrix} n_1 & n_2 & n \\ j & i & g \end{matrix}\right\}_h$$

are the trigonometric $6j$-symbols without spectral parameter corresponding
to $q = \exp(\pi\sqrt{-1}h)$, cf. [FT, Section 6.2].

The role of the numerical factor appearing in front of the trigonometric
$6j$-symbol in (3.2.a) is to transform formula (3.2.e) into (3.2.d). Since
$n = n_1 + n_2$, we have

$$\left\{\begin{matrix} n_1 & n_2 & n \\ k & l & p \end{matrix}\right\}_h = \frac{\theta(k, l, n)\, [\frac{l+p-n_1}{2}]_h!\, [\frac{k+p-n_2}{2}]_h!}{[p]_h!\, [\frac{k+l-n}{2}]_h!}$$

(cf. [FT, formula 6.2.c]). Similarly,

$$\left\{\begin{matrix} n_1 & n_2 & n \\ j & i & g \end{matrix}\right\}_h = \frac{\theta(j, i, n)\, [\frac{i+g-n_1}{2}]_h!\, [\frac{j+g-n_2}{2}]_h!}{[g]_h!\, [\frac{j+i-n}{2}]_h!}.$$

Substituting these expressions in (3.2.e) we obtain a formula equivalent to
(3.2.d).

3.3. *Proof of Lemma 2.3.* We first briefly discuss relationships between
elliptic and trigonometric monomials (cf. Section 1.2). A trigonometric
monomial is a product

$$\pm \prod_{r=2,3,\ldots} ([r]_h)^{a_r} \prod_{s \in \mathbb{Z}} ([x + s]_h)^{b_s}$$

with integer a_r, b_s having non-zero values only for a finite set of r, s. (The factor $[1]_h$ may also occur but we can ignore it since $[1]_h = 1$.) Clearly, there are no non-trivial identities between trigonometric monomials: two trigonometric monomials are equal as functions of x and q if and only if they correspond to the same sequences $\{a_r\}_r, \{b_s\}_s$. The formula $[...]_{h,\tau} \mapsto [...]_h$ establishes a bijective correspondence between elliptic and trigonometric monomials. For instance, the elliptic monomial $\Theta(i, j, k)$ corresponds to the trigonometric monomial $\theta(i, j, k)$.

By [DJKMO, formulas 2.1.14, 2.1.22a],

$$(3.3.a) \qquad \begin{bmatrix} i & k+m & n \\ k & l & m \end{bmatrix}(x)$$

is an elliptic monomial. (Formula (2.1.22a) in [DJKMO] holds for $m \le n$, the case $n \le m$ is similar.) By (3.1.c), the corresponding trigonometric monomial represents

$$(3.3.b) \qquad \lim_{\mathrm{Im}(\tau) \to +\infty} \begin{bmatrix} i & k+m & n \\ k & l & m \end{bmatrix}(x).$$

We can compute the latter monomial using Theorem 3.2. By [FT, formula (8.8.a)],

$$\begin{Bmatrix} i & k+m & n \\ k & l & m \end{Bmatrix}_h (-x) = (-1)^{\frac{l+m-i}{2}+mn} \begin{Bmatrix} i & k+m & n \\ k & l & m \end{Bmatrix}_h \times$$

$$\times \pi_{m,n-1}\left(x+\frac{1}{2}\right) \prod_{r=1}^{\frac{i+m-l}{2}} \left[x+1+\frac{k+m-l}{2}-r\right]_h \prod_{r=1}^{\frac{l+m-i}{2}} \left[x+2+\frac{k+l+m}{2}-r\right]_h$$

where $\pi_{m,n}$ is defined by formula (1.3.g) with the elliptic square brackets $[...]$ replaced with the trigonometric brackets $[...]_h$. Substituting here

$$\begin{Bmatrix} i & k+m & n \\ k & l & m \end{Bmatrix}_h = \begin{Bmatrix} k & m & k+m \\ i & n & l \end{Bmatrix}_h = \frac{\theta(i, n, k+m)\,[\frac{n+l-k}{2}]_h!\,[\frac{i+l-m}{2}]_h!}{[l]_h!\,[\frac{i+n-k-m}{2}]_h!}$$

and substituting the resulting expression in (3.2.a) we obtain a trigonometric monomial representing (3.3.b). By the argument above, the corresponding elliptic monomial represents (3.3.a). A direct computation shows that this monomial is exactly the one on the right hand side of (2.3.a). The proof of (2.3.b) is similar using [FT, formula (8.7.a)].

3.4. An elliptic deformation of the trigonometric identities. Theorem 3.2 clarifies the relationships between the elliptic and trigonometric $6j$-symbols and allows us to compare the results of Section 1 with similar results of [FT]. It is easy to check that the Yang-Baxter identity for the elliptic $6j$-symbols (Theorem 1.3.2) in the limit $\mathrm{Im}(\tau) \to +\infty$ becomes the Yang-Baxter identity for the trigonometric $6j$-symbols [FT, Theorem 6.5]. The orthonormality relation formulated in Section 1 degenerates to the orthonormality relation for the trigonometric $6j$-symbols [FT, Theorem 6.4]. The symmetries of the elliptic $6j$-symbols degenerate to the symmetries of the trigonometric $6j$-symbols established in [FT]. We expect that any algebraic identity for the trigonometric $6j$-symbols with spectral parameter admits an elliptic deformation. On the other hand, the identities for quantum $6j$-symbols (which are limits of the trigonometric $6j$-symbols when the spectral parameter converges to infinity) break down when one replaces the trigonometric numbers with elliptic ones. This phenomena is deeply related to the modular properties of the elliptic $6j$-symbols studied in Sections 4 and 5.

The trigonometric $6j$-symbols and the identities between them have a geometric interpretation given in [FT] in terms of nets. This interpretation does not admit any straightforward elliptic generalization. It is an outstanding problem to find a geometric interpretation of the elliptic $6j$-symbols and the identities between them.

The trigonometric $6j$-symbol (3.1.a) was explicitly computed in [FT, Theorem 8.10]: in the notation of Theorem 2.2, we have

$$\begin{Bmatrix} i & j & n \\ k & l & m \end{Bmatrix}_h (x) =$$

$$(-1)^{\frac{m-i-l}{2}} \pi_{m,n-1}\left(x - \frac{1}{2}\right) \prod_{r=\frac{i-k}{2}}^{\frac{i+l}{2}} [x+r]_h^{-1} \prod_{r=-\frac{i+k}{2}-1}^{\frac{i-l}{2}-1} [x+r]_h^{-1} \begin{Bmatrix} i & j & n \\ k & l & m \end{Bmatrix}_h' (x)$$

where

$$\pi_{m,n}(x) = \prod_{s=0}^{n-1} \prod_{t=0}^{m-1} \left[x + \frac{m+n}{2} - s - t - 1\right]_h$$

and

(3.4.a)
$$\begin{Bmatrix} i & j & n \\ k & l & m \end{Bmatrix}_h' (x) =$$

$$= \sum_g [g+1]_h \begin{Bmatrix} i & \alpha & g \\ \beta & l & m \end{Bmatrix}_h \begin{Bmatrix} l & \beta & g \\ \gamma & n & k \end{Bmatrix}_h \begin{Bmatrix} n & \gamma & g \\ \alpha & i & j \end{Bmatrix}_h \times$$

$$\times \left(\theta(g,\beta,l)\,\theta(g,\gamma,n)\,\theta(g,\alpha,i)\right)^{-1} \prod_{s=0}^{\sigma} \left[x + \frac{\sigma-g}{2} - s - 1\right]_h \left[x + \frac{\sigma+g}{2} - s\right]_h.$$

Note that three $6j$-symbols without spectral parameter, appearing on the right hand side, are monomials computed as at the end of Section 3.2.

The trigonometric $6j$-symbol (3.4.a) has the same symmetries as the elliptic $6j$-symbol (2.2.a) (cf. formulas (2.5.a) and [FT, (7.4.a)]). These $6j$-symbols are related by the following formula that can be derived from the definitions and (3.1.c):

$$\lim_{\mathrm{Im}(\tau)\to+\infty} \begin{bmatrix} i & j & n \\ k & l & m \end{bmatrix}'(x,h,\tau) =$$

$$= \frac{\theta(i,l,m)\,\theta(k,l,n)\,\theta(i,j,n)}{[\frac{i+l-m}{2}]_h!\,[\frac{l+n-k}{2}]_h!\,[\frac{i+n-j}{2}]_h!} \begin{Bmatrix} i & j & n \\ k & l & m \end{Bmatrix}'_h (-x).$$

We would like to use this opportunity to correct a few errors in the figures in [FR]. On the right hand side of Figure 4.2, the number $x_{i,j}$ assigned to the (i,j)-crossing should stand not on the right of the crossing but exactly below it (as on the left hand side of this figure), here $i = 1, ..., m$ and $j = 1, ..., n$. In Figures 6.8 and 6.9 the symbols x and $-x$ should stand below the corresponding crossings. The same remark concerns the positions of z on the left hand side of Figure 6.10, x_2 in Figures 8.2, 8.3, 8.4, 8.5, and x_1 in Figures 8.4, 8.6. In Figure 8.1 the symbols m, n should be replaced with 1. In Figure 8.8 the symbol Y_z should be replaced with Y_r.

4. Transformation properties of the elliptic $6j$-symbols

4.1. Basic transformations. Denote by $\mathcal{I}$ the set of triples $(z, h, \tau) \in \mathbb{C}^3$ such that $h \notin \mathbb{Q} + \mathbb{Q}\tau$ and $\mathrm{Im}(\tau) > 0$. Fix three complex numbers L, M, N. We define a right action of the group $SL_2(\mathbb{Z})$ on the functions $f : \mathcal{I} \to \mathbb{C}$ by the formula

$$\left(f \begin{bmatrix} a & b \\ c & d \end{bmatrix}\right)(z,h,\tau) =$$

$$\exp\left(-\pi\sqrt{-1}\,c\,\frac{Lz^2 + Mzh + Nh^2}{c\tau + d}\right)\, f(\frac{z}{c\tau+d}, \frac{h}{c\tau+d}, \frac{a\tau+b}{c\tau+d}).$$

Analogous actions of $SL_2(\mathbb{Z})$ on functions of one variable $g(\tau)$ and functions of two variables $g(z, \tau)$ given by

$$\left(g \begin{bmatrix} a & b \\ c & d \end{bmatrix}\right)(\tau) = g(\frac{a\tau+b}{c\tau+d})$$

and

$$\left(g \begin{bmatrix} a & b \\ c & d \end{bmatrix}\right)(z,\tau) = \exp\left(-\pi\sqrt{-1}\frac{cLz^2}{c\tau+d}\right)\, g(\frac{z}{c\tau+d}, \frac{a\tau+b}{c\tau+d})$$

appear in the theory of modular forms and Jacobi forms (see [Mu], [EZ]).

Fix $\varepsilon \in \mathbb{C}\backslash\{0\}$. We consider four actions of the infinite cyclic group on the functions $f : \mathcal{I} \to \mathbb{C}$. These actions are defined by the following formulas, where $a \in \mathbb{Z}$:

$$(fa)(z, h, \tau) = \varepsilon^{-a} f(z + a, h, \tau),$$

$$(fa)(z, h, \tau) = \varepsilon^{-a} \exp(\pi\sqrt{-1}a(La\tau + 2Lz + Mh)) f(z + a\tau, h, \tau),$$

$$(fa)(z, h, \tau) = f(z, h + a, \tau),$$

$$(fa)(z, h, \tau) = \exp(\pi\sqrt{-1}a(Na\tau + 2Nh + Mz)) f(z, h + a\tau, \tau).$$

Observe that a function $f : \mathcal{I} \to \mathbb{C}$ is invariant under the action of $SL_2(\mathbb{Z})$ and the actions of $\mathbb{Z}$ if and only if it satisfies the following six identities:

(4.1.a) $$f(z + 1, h, \tau) = \varepsilon\, f(z, h, \tau),$$

(4.1.b) $$f(z, h + 1, \tau) = f(z, h, \tau),$$

(4.1.c) $$f(z, h, \tau + 1) = f(z, h, \tau),$$

(4.1.d) $$f(z + \tau, h, \tau) = \varepsilon \exp(-\pi\sqrt{-1}(L\tau + 2Lz + Mh)) f(z, h, \tau),$$

(4.1.e) $$f(z, h + \tau, \tau) = \exp(-\pi\sqrt{-1}(N\tau + 2Nh + Mz)) f(z, h, \tau),$$

(4.1.f) $$f(\frac{z}{\tau}, \frac{h}{\tau}, -\frac{1}{\tau}) = \exp(\pi\sqrt{-1}\frac{Lz^2 + Mzh + Nh^2}{\tau}) f(z, h, \tau).$$

4.2. Theorem. *Let (i, j, k, l, m, n) be an admissible tuple of integers. Then the function*

(4.2.a) $$f(z, h, \tau) = \begin{bmatrix} i & j & n \\ k & l & m \end{bmatrix} (h^{-1}z, h, \tau)$$

is invariant under the action of $SL_2(\mathbb{Z})$ and the first three actions of $\mathbb{Z}$ corresponding to

$$L = mn,$$

(4.2.b) $$M = mn + \frac{(j+1)^2 + (l+1)^2 - (i+1)^2 - (k+1)^2}{2},$$

$$(4.2.c) \quad N = \frac{mn(m-n)^2}{12} + \frac{m}{2}\frac{(j+1)^2 + (k+1)^2 - (i+1)^2 - (l+1)^2}{2} +$$

$$+ \frac{n}{2}\frac{(k+1)^2 + (l+1)^2 - (i+1)^2 - (j+1)^2}{2},$$

and $\varepsilon = (-1)^{mn}$. If $m+n$ is even then the function (4.2.a) is invariant also under the fourth action of $\mathbb{Z}$.

Note that $M \in \mathbb{Z}$. Indeed, the admissibility of the triples $(i,j,n), (k,l,n)$ implies that $j + l = i + k \,(\mathrm{mod}\,2)$. In the case $m = n$ the formula for N simplifies to

$$N = m\frac{(k+1)^2 - (i+1)^2}{2}.$$

Proof of Theorem. It is convenient to write down an explicit formula for the function $f = f_{i,j,k,l,m,n}$ defined by (4.2.a). In the case $m = n = 1$, we have

$$f_{l\pm1,l\pm2,l\pm1,l,1,1}(z,h,\tau) = \frac{\theta_{11}(z+h,\tau)}{\theta_{11}(h,\tau)},$$

$$f_{l\mp1,l,l\pm1,l,1,1}(z,h,\tau) = \frac{\theta_{11}((l+1\pm1)h,\tau)\,\theta_{11}(z,\tau)}{\theta_{11}((l+1)h,\tau)\,\theta_{11}(h,\tau)},$$

$$f_{l\pm1,l,l\pm1,l,1,1}(z,h,\tau) = \mp\frac{\theta_{11}(z\mp(l+1)h,\tau)}{\theta_{11}((l+1)h,\tau)}.$$

Formula (1.3.b) yields the following formula for $f = f_{i,j,k,l,m,n}$ in the case $m \geq 2$ or $n \geq 2$:

$$(4.2.d) \quad f(z,h,\tau) = \sum_A \prod_{s=0}^{n-1}\prod_{t=0}^{m-1} f_{\alpha_{s,t},\alpha_{s+1,t},\alpha_{s,t+1},\alpha_{s+1,t+1},1,1}(z+R_{s,t}h,h,\tau)$$

where

$$R_{s,t} = \frac{m+n}{2} - s - t - 1$$

and $A = \{\alpha_{s,t}\}_{s,t}$ runs over the same set of $(n+1) \times (m+1)$-matrices as in (1.3.b).

Now we can check equalities (4.1.a) - (4.1.f). For $m = n = 1$, formula (4.1.a) with $\varepsilon = (-1)^{mn} = -1$ directly follows from (1.1.a) and the formulas for $f_{i,j,k,l,1,1}$ given above. The general case follows from (4.2.d).

For $m = n = 1$, equality (4.1.b) directly follows from (1.1.a) and the formulas for $f_{i,j,k,l,1,1}$ given above. The general case follows from (4.2.d).

In the case $m = n = 1$, equality (4.1.c) follows from (1.1.c). The general case follows from (4.2.d).

For $m = n = 1$, formula (4.1.d) is verified straightforwardly case by case using (1.1.b). The general case follows from (4.2.d) using the equalities

(4.2.e)
$$\sum_{s=0}^{n-1}\sum_{t=0}^{m-1} R_{s,t} = 0$$

and
(4.2.f)
$$M = mn + \frac{(j+1)^2 + (l+1)^2 - (i+1)^2 - (k+1)^2}{2} = \sum_{s=0}^{n-1}\sum_{t=0}^{m-1}(1 + d_{s,t})$$

where

$$d_{s,t} = d_{s,t}(A) = \frac{(\alpha_{s,t+1} + 1)^2 + (\alpha_{s+1,t} + 1)^2 - (\alpha_{s,t} + 1)^2 - (\alpha_{s+1,t+1} + 1)^2}{2}.$$

Let us prove (4.1.f). First consider the case $m = n = 1$. A direct case be case computation based on (1.1.d) shows that
(4.2.g)
$$f(\frac{z}{\tau}, \frac{h}{\tau}, -\frac{1}{\tau}) = \exp\left(\pi\sqrt{-1}\frac{z^2 + Mzh + \frac{(k+1)^2-(i+1)^2}{2}h^2}{\tau}\right) f(z, h, \tau)$$

where M is given by (4.2.b) with $m = n = 1$. This proves (4.1.f) for $m = n = 1$. Consider now the general case. We fix sequences of non-negative integers $\alpha_{0,m} = l, \alpha_{1,m}, ..., \alpha_{n,m} = k$ and $\alpha_{n,m} = k, \alpha_{n,m-1}, ...\alpha_{n,0} = j$ such that the neighboring terms differ by ± 1. Both $f(\frac{z}{\tau}, \frac{h}{\tau}, -\frac{1}{\tau})$ and $f(z, h, \tau)$ are sums over matrices $A = \{\alpha_{s,t}\}$ extending these sequences. It follows from (4.2.g) that the summands corresponding to one and same matrix A are proportional with coefficient $\exp(\frac{\pi\sqrt{-1}}{\tau}P)$ where

$$P =$$

$$\sum_{s=0}^{n-1}\sum_{t=0}^{m-1}\left((z + R_{s,t}h)(z + R_{s,t}h + h + d_{s,t}h) + \frac{(\alpha_{s+1,t+1} + 1)^2 - (\alpha_{s,t} + 1)^2}{2}h^2\right)$$

$$= mnz^2 + Mzh + \sum_{s=0}^{n-1}\sum_{t=0}^{m-1}\left((R_{s,t})^2 + R_{s,t}d_{s,t} + \frac{(\alpha_{s+1,t+1} + 1)^2 - (\alpha_{s,t} + 1)^2}{2}\right)h^2$$

where the last equality follows from (4.2.e) and (4.2.f).

A direct computation shows that

$$\sum_{s=0}^{n-1}\sum_{t=0}^{m-1}(R_{s,t})^2 = \frac{mn(m-n)^2}{12}.$$

It is easy to check that all the entries of the matrix $\{\alpha_{s,t}\}$ contribute 0 to the sum

$$\sum_{s=0}^{n-1}\sum_{t=0}^{m-1}\left(R_{s,t}d_{s,t}+\frac{(\alpha_{s+1,t+1}+1)^2-(\alpha_{s,t}+1)^2}{2}\right)$$

except the four corner entries $\alpha_{0,0}=i, \alpha_{n,0}=j, \alpha_{0,m}=l, \alpha_{n,m}=k$ which contribute respectively

$$-\frac{(\alpha_{0,0}+1)^2}{2}R_{0,0}-\frac{(\alpha_{0,0}+1)^2}{2}=-\frac{(i+1)^2}{2}\frac{m+n}{2},$$

$$\frac{(\alpha_{n,0}+1)^2}{2}R_{n-1,0}=\frac{(j+1)^2}{2}\frac{m-n}{2},$$

$$\frac{(\alpha_{0,m}+1)^2}{2}R_{0,m-1}=\frac{(l+1)^2}{2}\frac{n-m}{2},$$

$$-\frac{(\alpha_{n,m}+1)^2}{2}R_{n-1,m-1}+\frac{(\alpha_{n,m}+1)^2}{2}=-\frac{(k+1)^2}{2}\frac{m+n}{2}.$$

Combining these formulas together we obtain $P=mnz^2+Mzh+Nh^2$. In particular, P does not depend on the choice of the matrix $\{\alpha_{s,t}\}$. This implies (4.1.f).

To prove (4.1.e) we note first that for any $R\in\mathbb{Z}$,

$$(4.2.h)\quad \theta_{11}(u+R\tau,\tau)=(-1)^R\exp(-\pi\sqrt{-1}R^2\tau-2\pi\sqrt{-1}Ru)\,\theta_{11}(u,\tau).$$

This follows by induction from (1.1.b). As above, we first prove (4.1.e) in the case $m=n=1$. Using (4.2.h) we obtain

$$\frac{f_{l\pm1,l\pm2,l\pm1,l,1,1}(z,h+\tau,\tau)}{f_{l\pm1,l\pm2,l\pm1,l,1,1}(z,h,\tau)}=\frac{\theta_{11}(z+h+\tau,\tau)\,\theta_{11}(h,\tau)}{\theta_{11}(h+\tau,\tau)\,\theta_{11}(z+h,\tau)}=$$

$$=\frac{\exp(-\pi\sqrt{-1}(\tau+2z+2h))}{\exp(-\pi\sqrt{-1}(\tau+2h))}=\exp(-2\pi\sqrt{-1}z).$$

Similar computations show that

$$\frac{f_{l\mp1,l,l\pm1,l,1,1}(z,h+\tau,\tau)}{f_{l\mp1,l,l\pm1,l,1,1}(z,h,\tau)}=\exp(\mp2\pi\sqrt{-1}(l+1)(\tau+2h)),$$

$$\frac{f_{l\pm1,l,l\pm1,l,1,1}(z,h+\tau,\tau)}{f_{l\pm1,l,l\pm1,l,1,1}(z,h,\tau)}=\exp(\pm2\pi\sqrt{-1}(l+1)z).$$

In all cases

$$\frac{f_{i,j,k,l,1,1}(z,h+\tau,\tau)}{f_{i,j,k,l,1,1}(z,h,\tau)}=\exp\left(-\pi\sqrt{-1}\frac{(k+1)^2-(i+1)^2}{2}(\tau+2h)-\pi\sqrt{-1}Mz\right)$$

where M is given by (4.2.b) with $m = n = 1$. This proves (4.1.e) for $m = n = 1$. We shall need a more general formula concerning $f_{i,j,k,l,1,1}$. It follows from (4.1.d) that for $R \in \mathbb{Z}$,

$$\frac{f_{i,j,k,l,1,1}(z + R\tau, h, \tau)}{f_{i,j,k,l,1,1}(z, h, \tau)} = (-1)^R \exp(-\pi\sqrt{-1}(R^2\tau + 2Rz + RMh)).$$

This implies

(4.2.i) $f_{i,j,k,l,1,1}(z + R\tau, h + \tau, \tau) =$

$$= (-1)^R \exp(-\pi\sqrt{-1}(R^2\tau + 2Rz + RM(h + \tau)))\, f_{i,j,k,l,1,1}(z, h + \tau, \tau) =$$

$$= (-1)^R \exp(-\pi\sqrt{-1}(R^2\tau + 2Rz + RM(h + \tau)+$$

$$+ \frac{(k + 1)^2 - (i + 1)^2}{2}(\tau + 2h) + Mz))\, f_{i,j,k,l,1,1}(z, h, \tau).$$

Now we prove (4.1.e) assuming that $m + n \in 2\mathbb{Z}$. This assumption implies that $R_{s,t} \in \mathbb{Z}$ (in the notation introduced above) which will allow us to use (4.2.i) with $R = R_{s,t}$. By (4.2.d), both $f(z, h + \tau, \tau)$ and $f(z, h, \tau)$ are sums over one and the same set of matrices $A = \{\alpha_{s,t}\}$. It follows from (4.2.i) and (4.2.e) that the summands corresponding to a matrix A are proportional with coefficient

$$\prod_{s,t}(-1)^{R_{s,t}} \exp(\frac{\pi\sqrt{-1}}{\tau}Q) = \exp(\frac{\pi\sqrt{-1}}{\tau}Q)$$

with

$$Q = \sum_{s=0}^{n-1} \sum_{t=0}^{m-1} ((R_{s,t})^2\tau + 2R_{s,t}(z + R_{s,t}h) + R_{s,t}(1 + d_{s,t})(h + \tau)+$$

$$+ \frac{(\alpha_{s+1,t+1} + 1)^2 - (\alpha_{s,t} + 1)^2}{2}(\tau + 2h) + (1 + d_{s,t})(z + R_{s,t}h)).$$

The computations made above imply that

$$Q = \sum_{s=0}^{n-1} \sum_{t=0}^{m-1} \left((R_{s,t})^2 + R_{s,t}d_{s,t} + \frac{(\alpha_{s+1,t+1} + 1)^2 - (\alpha_{s,t} + 1)^2}{2} \right)(\tau + 2h)+$$

$$+ \sum_{s=0}^{n-1} \sum_{t=0}^{m-1} (1 + d_{s,t})z = N(\tau + 2h) + Mz.$$

This yields (4.1.e).

4.3. Remark. The function f given by (4.2.a) satisfies the identity

(4.3.a) $f(z, h + 2\tau, \tau) = \exp(-\pi\sqrt{-1}(4N\tau + 4Nh + 2Mz)) f(z, h, \tau).$

For even $m + n$ this formula follows from (4.1.e), for odd $m + n$ it is proven similarly to (4.1.e).

4.4. Corollary. it Let (i, j, k, l, m, n) be an admissible tuple of integers. The function $F : \mathcal{I} \to \mathbb{C}$ given by

$$F(z, h, \tau) = \begin{bmatrix} i & j & n \\ k & l & m \end{bmatrix} (z, h, \tau)$$

is invariant under the action of $SL_2(\mathbb{Z})$ defined by

$$\left(F \begin{bmatrix} a & b \\ c & d \end{bmatrix} \right) (z, h, \tau) =$$

$$\exp\left(-\pi\sqrt{-1}\, ch^2\, \frac{Lz^2 + Mz + N}{c\tau + d} \right) F(z, \frac{h}{c\tau + d}, \frac{a\tau + b}{c\tau + d}).$$

This is a reformulation of the first claim of Theorem 4.2.

5. Modular hypergeometric functions

5.1. Classical hypergeometric series. The hypergeometric series with p numerator parameters $\alpha_1, ..., \alpha_p \in \mathbb{C}$ and r denominator parameters $\beta_1, ..., \beta_r \in \mathbb{C}$ is the formal power series in variable z defined by

$$_pF_r(\alpha_1, ..., \alpha_p; \beta_1, ..., \beta_r; z) = \sum_{n=0}^{\infty} \frac{(\alpha_1)_n \cdots (\alpha_p)_n}{n!(\beta_1)_n \cdots (\beta_r)_n} z^n,$$

where $(\alpha)_n$ denotes the shifted factorial:

$$(\alpha)_n = \begin{cases} 1, & n = 0 \\ \alpha(\alpha + 1) \cdots (\alpha + n - 1), & n = 1, 2, ... \end{cases}$$

One assumes that the numbers $\beta_1, ..., \beta_r$ are such that the denominator factors in the terms of the series are not zero. For $|z| < 1$, the series converges absolutely and yields a hypergeometric function. If one of the numerator parameters $\alpha_1, ..., \alpha_p$ is zero or a negative integer, the series terminates and becomes a polynomial in z.

    In 1846/47, E. Heine introduced basic hypergeometric series $_p\phi_r$ essentially by replacing ordinary numbers by trigonometric numbers in the definition of $_pF_r$. Fix a non-zero complex number q. We have

(5.1.a) $_p\phi_r(a_1, ..., a_p; b_1, ..., b_r; q, z) =$

$$= \sum_{n=0}^{\infty} \frac{(a_1; q)_n \cdots (a_p; q)_n}{(q; q)_n (b_1; q)_n \cdots (b_r; q)_n} z^n,$$

where

$$(a; q)_n = \begin{cases} 1, & n = 0 \\ (1 - a)(1 - aq^2) \cdots (1 - aq^{n-1}), & n = 1, 2, \ldots \end{cases}$$

The expression $(a; q)_n$ can be easily rewritten as the q-shifted factorial for the trigonometric numbers. This allows us to view $_p\phi_r$ as the trigonometric analogue of $_pF_r$.

Heine and his followers were able to generalize various results about hypergeometric series to the basic case, see [S] and [GR]. According to the philosophy of three types of numbers outlined in the Introduction and supported by the study of $6j$-symbols, one can replace ordinary numbers (resp. trigonometric numbers) in the definition of hypergeometric series (resp. basic hypergeometric series) by their elliptic analogues. In the full generality, it is hard to expect that the classical properties of hypergeometric functions extend to this new class of elliptic series. Still, as we shall see below, the fundamental results concerning the narrower class of balanced very-well-poised series do admit elliptic generalizations.

Recall the definition of very-well-poised basic hypergeometric series. From now on, we restrict ourselves to the case $p = r + 1$, which plays the most important role in the theory of hypergeometric series. The series $_{r+1}\phi_r(a_1, \ldots, a_{r+1}; b_1, \ldots, b_r; q, z)$ is very-well-poised if

(5.1.b) $qa_1 = a_2b_1 = a_3b_2 = \cdots = a_{r+1}b_r, \quad a_2 = qa_1^{1/2}, \quad a_3 = -qa_1^{1/2}$

where we fixed a square root $a_1^{1/2}$. Following [GR], we use a more compact notation

$$_{r+1}W_r(a_1; a_4, a_5, \ldots, a_{r+1}; q, z)$$

for the very-well-poised basic hypergeometric series (5.1.a). The remaining parameters $a_2, a_3, b_1, \ldots, b_r$ are excluded from the notation because they are determined from (5.1.b) (we assume that $a_i \neq 0$ for $i = 1, \ldots, r + 1$). It was observed in [FT, Lemma 9.2], that for

$$a_1 = q^{2\alpha_1}, \, a_4 = q^{2\alpha_4}, \, a_5 = q^{2\alpha_5}, \ldots, a_{r+1} = q^{2\alpha_{r+1}}$$

with $\alpha_1, \alpha_4, \alpha_5, \ldots, \alpha_{r+1} \in \mathbb{C}$ and

(5.1.c) $z = q^{2\zeta} \quad \text{with} \quad \zeta = \frac{r - 3}{2}(1 + \alpha_1) - \alpha_4 - \alpha_5 - \cdots - \alpha_{r+1},$

we have the following identity

(5.1.d) $_{r+1}W_r(q^{2\alpha_1}; q^{2\alpha_4}, \ldots, q^{2\alpha_{r+1}}; q^2, q^{2\zeta}) =$

$$\sum_{n\geq 0} \frac{[\alpha_1;h]_n}{[n]_h!} \cdot \frac{[\alpha_4;h]_n}{[1+\alpha_1-\alpha_4;h]_n} \cdots \frac{[\alpha_{r+1};h]_n}{[1+\alpha_1-\alpha_{r+1};h]_n} \cdot \frac{[\alpha_1+2n]_h}{[\alpha_1]_h}$$

where $q = \exp(\pi\sqrt{-1}h)$ with $h \in \mathbb{C}$ and

$$[\alpha;h]_n = [\alpha]_h\,[\alpha+1]_h \cdots [\alpha+n-1]_h.$$

Here we use the notation of Section 3, in particular, the square brackets $[...]_h$ are defined by (3.1.b). (We correct here a misprint in the expression for ζ given in [FT], Lemma 9.2.)

If, additionally to the assumption (5.1.c), $\zeta = 1$, i.e., if

(5.1.e) $$\frac{r-5}{2} + \frac{r-3}{2}\alpha_1 - \alpha_4 - \alpha_5 - \cdots - \alpha_{r+1} = 0$$

the very-well-poised basic hypergeometric series $_{r+1}W_r(q^{2\alpha_1};q^{2\alpha_4},...,q^{2\alpha_{r+1}};$ $q^2,q^{2\zeta})$ is then *balanced*, cf. [GR].

5.2. Definition of modular hypergeometric functions.

We use formula (5.1.d) to define an elliptic generalization of balanced very-well-poised basic hypergeometric series. Let $\alpha_1,\alpha_4,\alpha_5,\cdots,\alpha_{r+1}$ be complex numbers satisfying (5.1.e). Let $h,\tau \in \mathbb{C}$ with $\mathrm{Im}(\tau) > 0$ and $h \notin \mathbb{Z} + \mathbb{Z}\tau$. The corresponding modular hypergeometric function is the series

(5.2.a) $$_{r+1}\omega_r(\alpha_1;\alpha_4,...,\alpha_{r+1};h,\tau) =$$

$$\sum_{n\geq 0} \frac{[\alpha_1;h,\tau]_n}{[n]_{h,\tau}!} \cdot \frac{[\alpha_4;h,\tau]_n}{[1+\alpha_1-\alpha_4;h,\tau]_n} \cdots \frac{[\alpha_{r+1};h,\tau]_n}{[1+\alpha_1-\alpha_{r+1};h,\tau]_n} \cdot \frac{[\alpha_1+2n]_{h,\tau}}{[\alpha_1]_{h,\tau}}$$

where the square brackets are defined by (1.1.f) and

(5.2.b) $$[\alpha;h,\tau]_n = [\alpha]_{h,\tau}\,[\alpha+1]_{h,\tau} \cdots [\alpha+n-1]_{h,\tau}.$$

To avoid the problem of convergence, we shall consider only the case when the series (5.2.a) is terminating. To ensure this condition it suffices to assume that at least one of the parameters $\alpha_1,\alpha_4,...,\alpha_{r+1}$ is a non-positive integer.

The term "modular" is justified by the following theorem.

5.3. Theorem.

(i) Modular hypergeometric functions are invariant under the natural action of $SL(2,\mathbb{Z})$ on the variables h,τ, i.e.,

$$_{r+1}\omega_r(\alpha_1;\alpha_4,...,\alpha_{r+1};\frac{h}{c\tau+d},\frac{a\tau+b}{c\tau+d}) = {}_{r+1}\omega_r(\alpha_1;\alpha_4,...,\alpha_{r+1};h,\tau),$$

for all

$$\begin{pmatrix} a & b \\ c & d \end{pmatrix} \in SL(2,\mathbf{Z}).$$

(ii) If $\alpha_1, \alpha_4, ..., \alpha_{r+1} \in \mathbb{Z}$, then function (5.2.a) is invariant under the natural action of $\mathbb{Z}^2$ on the variable h, namely

$$_{r+1}w_r(\alpha_1; \alpha_4, ..., \alpha_{r+1}; h + m + n\tau, \tau) = {}_{r+1}w_r(\alpha_1; \alpha_4, ..., \alpha_{r+1}; h, \tau)$$

for all $(m, n) \in \mathbb{Z}^2$.

Proof. (i) It follows from the transformation properties of the theta function θ_{11} that the elliptic number $[\alpha]_{h,\tau}$ is invariant under the transformation $(h, \tau) \to (h, \tau + 1)$ and acquires the factor $\exp\frac{\pi\sqrt{-1}h^2}{\tau}(\alpha^2 - 1)$ under the transformation $(h, \tau) \to (\frac{h}{\tau}, -\frac{1}{\tau})$. Since the quantity of elliptic brackets in the numerator and the denominator of each term in $_{r+1}w_r$ is the same, we only need to check that the sums of squares of the numbers in the numerator and the denominator are equal. The shifted factorial $[\alpha; h, \tau]_n$ contributes to the sum of squares

$$\alpha^2 + (\alpha + 1)^2 + \cdots + (\alpha + n - 1)^2 = n\alpha^2 + n(n-1)\alpha + \frac{1}{6}n(n-1)(2n-1).$$

This implies that the product

(5.3.a) $$\prod_{k=4}^{r+1} \frac{[\alpha_k; h, \tau]_n}{[1 + \alpha_1 - \alpha_k; h, \tau]_n}$$

contributes

$$\sum_{k=4}^{r+1} \left(n\alpha_k^2 + n(n-1)\alpha_k - n(1 + \alpha_1 - \alpha_k)^2 - n(n-1)(1 + \alpha_1 - \alpha_k)\right) =$$

$$= n(n + \alpha_1) \sum_{k=4}^{r+1} (2\alpha_k - 1 - \alpha_1).$$

The terms

(5.3.b) $$\frac{[\alpha_1; h, \tau]_n}{[n]_{h,\tau}!}, \quad \frac{[a_1 + 2n]_{h,\tau}}{[\alpha_1]_{h,\tau}}$$

contribute

$$n(n + \alpha_1)(\alpha_1 - 1) + (\alpha_1 + 2n)^2 - \alpha_1^2 = n(n + \alpha_1)(\alpha_1 + 3).$$

Thus the difference of the sums of squares in the numerator and the denominator is

$$2n(n + \alpha_1)(\sum_{k=4}^{r+1} \alpha_k - \frac{r-5}{2} - \frac{r-3}{2}\alpha_1) = 0$$

by the balancing condition (5.1.e).

(ii) It follows from the transformation properties of θ_{11} that the elliptic number $[\alpha]_{h,\tau}$ with integer α is multiplied by $(-1)^{\alpha-1}$ under the transformation $h \to h + 1$. Thus we need to check that the sums of numbers in the numerator and the denominator of each term in $_{r+1}\omega_r$ are equal mod 2. The shifted factorial $[\alpha; h, \tau]_n$ contributes

$$\alpha + (\alpha + 1) + \cdots + (\alpha + n - 1) = n\alpha + \frac{n(n-1)}{2}.$$

The product (5.3.a) contributes

$$\sum_{k=4}^{r+1} (n\alpha_k - n(1 + \alpha_1 - \alpha_k)) = \sum_{k=4}^{r+1} n(2\alpha_k - 1 - \alpha_1).$$

The terms (5.3.b) contribute $n(\alpha_1 - 1) + 2n = n(\alpha_1 + 1)$. Now the balancing condition implies that

$$\sum_{k=4}^{r+1} n(2\alpha_k - 1 - \alpha_1) + n(\alpha_1 + 1) = 2n(\sum_{k=4}^{r+1} \alpha_k - \frac{r-3}{2}(1 + \alpha_1)) = -2n = 0 (\text{mod } 2).$$

Invariance of $_{r+1}\omega_r$ under the transformation $h \to h + \tau$ is verified similarly using formula (4.2.h).

5.4. Elliptic 6j-symbols as modular hypergeometric functions. In [FT, Theorem 9.4] we identified the trigonometric 6j-symbols (with spectral parameter) with balanced very-well-poised basic hypergeometric series $_{10}\phi_9$. As it was shown in Section 3, the trigonometric 6j-symbols are limits of elliptic 6j-symbols when $\text{Im}(\tau) \to +\infty$. This suggests to compare the elliptic 6j-symbols with the modular hypergeometric functions $_{10}\omega_9$.

Let a, b, c, A, B, C be non-negative integers such that the triples (a, b, c), (A, B, c), (A, b, C), (a, B, C) are admissible. Set

$$\alpha = \frac{b+c-a}{2}, \quad \beta = \frac{a+c-b}{2}, \quad \gamma = \frac{a+b-c}{2}, \quad \sigma = \frac{a+b+c}{2}.$$

Note that $\alpha, \beta, \gamma, \sigma$ are non-negative integers. Set

$$\mu = \max(A - \alpha, B - \beta, C - \gamma) = \max(|A - \alpha|, |B - \beta|, |C - \gamma|) \geq 0$$

where the second equality follows from the admissibility of the triples (A, B, c), (A, b, C), (a, B, C).

Consider the elliptic $6j$-symbol

$$\begin{bmatrix} A & b & C \\ a & B & c \end{bmatrix}' (x, h, \tau) =$$

$$\sum_g \left(\frac{[g+1]([\frac{g+C-\gamma}{2}]! \, [\frac{g+A-\alpha}{2}]! \, [\frac{g+B-\beta}{2}]!)^2}{([g]!)^3 \Theta(g,C,\gamma)\Theta(g,A,\alpha)\Theta(g,B,\beta)} \prod_{s=0}^{\sigma} [x - \frac{g+\sigma}{2} + s][x + \frac{g-\sigma}{2} + 1 + s] \right)$$

and denote by $\lambda(A, B, C, a, b, c \mid x, h, \tau)$ the first non-zero summand on the right-hand side. This summand corresponds to $g = \mu$.

5.4.1. Theorem. *For any $x \in \mathbb{C}$,*

$$\begin{bmatrix} A & b & C \\ a & B & c \end{bmatrix}' (x) = \lambda(A, B, C, a, b, c \mid x, h, \tau) \times {}_{10}\omega_9(\alpha_1; \alpha_4, ..., \alpha_{10}; h, \tau)$$

where ${}_{10}\omega_9$ is the modular hypergeometric series with the following parameters: $\alpha_1, \alpha_4, \alpha_5$ are the numbers

$$\frac{\mu + A - \alpha}{2} + 1, \quad \frac{\mu + B - \beta}{2} + 1, \quad \frac{\mu + C - \gamma}{2} + 1$$

so that α_1 is the maximal of them (and equals $\mu + 1$) and $\alpha_6, ..., \alpha_{10}$ are the numbers

$$\frac{\mu - A - \alpha}{2}, \quad \frac{\mu - B - \beta}{2}, \quad \frac{\mu - C - \gamma}{2}, \quad \frac{\mu + \sigma}{2} - x + 2, \quad \frac{\mu + \sigma}{2} + x + 1.$$

Note that at least one of the parameters

$$\frac{\mu - A - \alpha}{2}, \quad \frac{\mu - B - \beta}{2}, \quad \frac{\mu - C - \gamma}{2}$$

is a non-positive integer so that the function ${}_{10}\omega_9(\alpha_1; \alpha_4, ..., \alpha_{10}; h, \tau)$ is given by a terminating series. The proof of Theorem 5.4.1 is rather straightforward, it suffices to replace the trigonometric numbers by the elliptic numbers in the proof of Theorem 9.4 in [FT]. To verify balancing condition (5.1.e) it suffices to note that

$$\alpha_1 + \alpha_4 + ... + \alpha_{10} = 4\mu + 6 = 4\alpha_1 + 2.$$

Theorems 5.3 and 5.4.1 imply that the transformation properties of the elliptic $6j$-symbol with respect to the standard action of $SL(2, \mathbb{Z})$

on (h, τ) are completely determined by the transformation properties of the monomial λ. This suggests another proof of Corollary 4.4 and the first claim of Theorem 4.2.

If $x \in \frac{1}{2}\mathbb{Z}$, and $x + \frac{\mu + \sigma}{2} \in \mathbb{Z}$, then all the parameters of $_{10}\omega_9$ are integers so that the transformation properties of the elliptic $6j$-symbol under the standard action of $\mathbb{Z}^2$ on h are also determined by the transformation properties of λ. This suggests another proof of equalities (4.1.b) and (4.1.e) in Theorem 4.2.

5.5. Identities for modular hypergeometric functions.

In [FT], we used the identification of the trigonometric 6j-symbols with $_{10}\phi_9$ to derive from the properties of $6j$-symbols two fundamental identities in the theory of basic hypergeometric functions: Bailey's transformation and the Jackson summation formula. The identification of modular hypergeometric functions with the elliptic $6j$-symbols allows to generalize these identities to the elliptic case.

5.5.1. Theorem.

Modular hypergeometric functions satisfy the following generalization of Bailey's transformation

$$(5.5.\text{a}) \qquad {}_{10}\omega_9(\alpha_1; \alpha_4, ..., \alpha_{10}; h, \tau) = {}_{10}\omega_9(\beta_1; \beta_4, ..., \beta_{10}; h, \tau) \times$$

$$\times \frac{[\alpha_1 + 1; h, \tau]_n [\beta_1 + 1 - \alpha_7; h, \tau]_n [\beta_1 + 1 - \alpha_8; h, \tau]_n [\alpha_1 + 1 - \alpha_7 - \alpha_8; h, \tau]_n}{[\beta_1 + 1; h, \tau]_n [\alpha_1 + 1 - \alpha_7; h, \tau]_n [\alpha_1 + 1 - \alpha_8; h, \tau]_n [\beta_1 + 1 - \alpha_7 - \alpha_8; h, \tau]_n}$$

where $n = -\alpha_{10}$ is a non-negative integer,

$$\beta_1 = 2\alpha_1 + 1 - \alpha_4 - \alpha_5 - \alpha_6 = -\alpha_1 - 1 + \alpha_7 + \alpha_8 + \alpha_9 + \alpha_{10}$$

$$\beta_4 = \beta_1 - \alpha_1 - \alpha_4, \quad \beta_5 = \beta_1 - \alpha_1 + \alpha_5, \quad \beta_6 = \beta_1 - \alpha_1 + \alpha_6$$

and $\{\beta_7, \beta_8, \beta_9, \beta_{10}\}$ is an arbitrary permutation of $\{\alpha_7, \alpha_8, \alpha_9, \alpha_{10}\}$.

The proof of this theorem reproduces the proof of formula (9.5.a) in [FT] with the obvious changes. Note however that this argument, based on Theorem 5.4.1, yields Theorem 5.5.1 only in the case when the parameters $\alpha_1, \alpha_4, ..., \alpha_{10}$ are integers. (In fact, two parameters, say α_8, α_9, may be non-integers but their sum should be an integer.) In [FR] we dealt with trigonometric polynomials so that an equality for all integer values of parameters implies the equality in general. Here we need an additional argument as follows. Multiplying the equality (5.5.a) by the denominators of the left and right hand sides we can reduce it to the form $l(h, \tau) = r(h, \tau)$ where both sides are elliptic polynomials. Present the difference $f(h, \tau) = l(h, \tau) - r(h, \tau)$ as a power series in variable h:

$$f(h, \tau) = \sum_{n \geq 0} f_n(\tau) h^n.$$

Then each coefficient $f_n(\tau)$ is a cusp form of weight n since

$$f(\frac{h}{c\tau + d}, \frac{a\tau + b}{c\tau + d}) = f(h, \tau),$$

for all

$$\begin{pmatrix} a & b \\ c & d \end{pmatrix} \in SL(2, \mathbf{Z})$$

and $\lim_{\mathrm{Im}(\tau) \to +\infty} f_n(\tau) = 0$. The linear space of the cusp forms of a given weight is finite dimensional. Since $f_n(\tau) = 0$ for an infinite number of parameters $\alpha_1, \alpha_4, ..., \alpha_{10}$, the function f_n is identically zero.

5.5.2. Theorem. *Modular hypergeometric functions satisfy the following generalization of the Jackson summation formula*

$$_8\omega_7(\alpha_1; \alpha_4, ..., \alpha_8; h, \tau) = \frac{[\alpha_1 + 1; h, \tau]_n \, [\alpha_1 + 1 - \alpha_4 - \alpha_5; h, \tau]_n}{[\alpha_1 + 1 - \alpha_4; h, \tau]_n \, [\alpha_1 + 1 - \alpha_5; h, \tau]_n} \times$$

$$\times \frac{[\alpha_1 + 1 - \alpha_4 - \alpha_6; h, \tau]_n \, [\alpha_1 + 1 - \alpha_5 - \alpha_6; h, \tau]_n}{[\alpha_1 + 1 - \alpha_6; h, \tau]_n \, [\alpha_1 + 1 - \alpha_4 - \alpha_5 - \alpha_6; h, \tau]_n}$$

where $n = -\alpha_8$ is a non-negative integer.

The proof reproduces the proof of formula (9.5.d) in [FT] with the obvious changes.

5.6. More on elliptic numbers. It is easy to verify that all identities in the ring generated by trigonometric numbers $[x]_h$ (considered as functions of h) follow from the quadratic identity (0.b). We believe that all identities in the ring generated by elliptic numbers $[x] = [x]_{h,\tau}$

(considered as functions of h, τ) follow from Riemann's theta identity of degree 4 mentioned in the Introduction. If true, this would yield a different deduction of Theorems 5.5.1 and 5.5.2 from the case of integer parameters.

In this connection, we briefly discuss the rings of trigonometric and elliptic integers. The ring of trigonometric integers $\mathcal{T}$ is generated by the variables $[n], n \in \mathbf{Z}$ subject to the relations

(5.6.a) $[0] = 0, \ [1] = 1, \ [n] = -[n] \ \text{ for } \ n \in \mathbf{Z}$

and the relation

$$[m + n + r] [r] = [m + r] [n + r] - [m] [n],$$

for any $m, n, r \in \mathbf{Z}$. (This relation is equivalent to (0.b) via the substitution $n = y + z, m = y - z, r = x - y$.) It is easy to observe that $\mathcal{T}$ is the one-variable polynomial ring $\mathbf{Z}[t]$ with $t = [2]$. Similarly, the ring of elliptic integers $\mathcal{E}$ is generated by the variables $[n], n \in \mathbb{Z}$ subject to (5.6.a) and the relation

$$[m + n + r + 2l] [m] [n] [r] =$$

$$= [m + n + r + l] [m + l] [n + l] [r + l] - [m + n + l] [m + r + l] [n + r + l] [l]$$

for any $m, n, r, l \in \mathbb{Z}$. (This relation is equivalent to Riemann's theta identity via $n = y + z, m = y - z, r = x - w, l = w - y$.) The structure of $\mathcal{E}$ seems to be rather involved. Observe that the quotient ring obtained by inverting $[2]$ is generated by $[2], [3], [4]$. Indeed, setting $m = n = r = 1$ we obtain

$$[2l + 3] = [l + 1]^3 [l + 3] - [l] [l + 2]^3$$

which allows to express $[2l + 3], l = 1, 2, \ldots$ as a polynomial of $[n], 1 < n \le 2l + 2$. Similarly, setting $m = n = 1, r = 2$ we obtain

$$[2] [2l + 4] = [l + 1]^2 [l + 2] [l + 4] - [l] [l + 2] [l + 3]^2.$$

This yields an expression of $[2l + 4], l = 1, 2, \ldots$ as a polynomial of $[n], 1 < n \le 2l + 3$ divided by $[2]$. We do not know if $[2], [3], [4]$ are algebraically independent in $\mathcal{E}$.

5.7. Remark. Some of the existing proofs of Bailey's transformation and the Jackson summation formula are perfectly suitable for generalization to the elliptic case. In particular, the proof of the Jackson formula given in [S] relies entirely on a comparison of zeros and poles on both sides of the formula and can be reproduced in the elliptic case. Taking into account the fundamental role of the elliptic functions in the theory of basic hypergeometric series it is surprising that this generalization was not noticed by the experts. We leave it as a challenging problem to generalize various known identities for balanced very-well-poised basic hypergeometric functions to modular hypergeometric functions (for example, Bailey's four-term $_{10}\phi_9$ transformation for non-terminating series, etc.).

References

[Ba] R. J. Baxter, *Exactly Solved Models in Statistical Mechanics.* Academic Press, 1982

[BL] L.C. Biedenharn, J.D. Louck, Angular Momentum in Quantum Physics. *Encyclopedia of mathematics and its applications*, v. 8. Addison-Wesley, 1981

[DJMO] E. Date, M. Jimbo, T. Miwa, M. Okado, Fusion of the eight-vertex
        SOS model, *Lett. Math. Phys.* **12** (1986), 209 - 215. Erratum
        and Addendum: *Lett. Math. Phys.* **14** (1987), 97

[DJKMO] E. Date, M. Jimbo, A. Kuniba, T. Miwa, M. Okado, Exactly solv-
        able SOS models, II: Proof of the star-triangle relation and com-
        binatorial identities. *Advanced Studies in Pure Math.* **16** (1988),
        17 - 122 in Conformal Field Theory and Solvable Lattice Models,
        Tokyo

   [EZ] M. Eichler, D. Zagier, *The theory of Jacobi forms*, Progress in
        Math., 55, Birkhauser, Boston, 1985

   [FT] I. B. Frenkel, V. G. Turaev, Trigonometric solutions of the Yang-
        Baxter equation, nets, and hypergeometric functions, In: *Func-
        tional Analysis on the Eve of the 21st Century*, vol.1. Progress in
        Math., 131, Birkhäuser, Boston, 1995, pp. 65-118

   [FR] I. B. Frenkel, N. Yu. Reshetikhin, Quantum affine algebras and
        holonomic difference equations, *Comm. Math. Phys.* **146** (1992),
        1 - 60

   [GR] G. Gasper, M. Rahman, Basic Hypergeometric Series. *Encyclope-
        dia of mathematics and its applications*, v. 35. Cambridge Univ.
        Press, 1990

   [KL] L. Kauffman, S.L. Lins, *Temperley-Lieb recoupling theory and in-
        variants of 3-manifolds*, Princeton Univ. Press, Princeton, N. J.,
        1994

   [KR] A.N. Kirillov, N.Y. Reshetikhin, Representations of the algebra
        $U_q(sl_2)$, q-orthogonal polynomials and invariants of links. In: In-
        finite dimensional Lie algebras and groups, (ed. by V.G. Kac),
        285-339. Adv. Ser. in Math. Phys. 7, World Scientific, Singapore
        1988

   [Mu] D. Mumford, *Tata lectures on Theta, I*, Progress in Math. 28,
        Birkhauser, Boston, 1983

   [Sl] L. J. Slater, *Generalized hypergeometric functions*, Cambridge Univ.
        Press, Cambridge, 1966

   [Tu] V. G. Turaev, *Quantum Invariants of Knots and 3-Manifolds*, de
        Gruyter Studies in Math. 18, Berlin, 1994

I. B. Frenkel   V. G. Turaev
Dept. of Math.   Dept. of Math.
Yale University   Louis Pasteur University - CNRS
New Haven, CT 06520, USA   Strasbourg 67084, France

Received March 1996

Combinatorics of hypergeometric functions associated with positive roots

Israel M. Gelfand, Mark I. Graev and Alexander Postnikov

ABSTRACT. In this paper we study the hypergeometric system on unipotent matrices. This system gives a holonomic D-module. We find the number of independent solutions of this system at a generic point. This number is equal to the famous Catalan number. An explicit basis of Γ-series in solution space of this system is constructed in the paper. We also consider restriction of this system to certain strata. We introduce several combinatorial constructions with trees, polyhedra, and triangulations related to this subject.

CONTENTS

1. General Hypergeometric Systems

In this paper we use the following notation: $[a, b] := \{a, a+1, \ldots, b\}$ and $[n] := [1, n]$.

Recall several definitions and facts from the theory of general hypergeometric functions (see [GGZ, GZK, GGR2]).

Consider the following action of the complex n-dimensional torus $T = (\mathbb{C}^*)^n$ with coordinates $t = (t_1, t_2, \ldots, t_n)$ on the space $\mathbb{C}^N$

$$(1.1) \qquad x = (x_1, x_2, \ldots, x_N) \longmapsto x \cdot t = (x_1 t^{a_1}, \ldots, x_N t^{a_N}),$$

where $a_j = (a_{1j}, \ldots, a_{nj}) \in \mathbb{Z}^n$, $j = 1, 2, \ldots, N$ and t^{a_j} denotes $t_1^{a_{1j}} \ldots t_n^{a_{nj}}$.

Definition 1.1. The *General Hypergeometric System* associated with the action of torus (1.1) is the following system of differential equations on $\mathbb{C}^N$

$$(1.2) \qquad \sum_{j=1}^{N} a_{ij} x_j \frac{\partial f}{\partial x_j} = \alpha_i f, \qquad i = 1, 2, \ldots, n;$$

$$(1.3) \qquad \prod_{j:\, l_j > 0} \left(\frac{\partial}{\partial x_j} \right)^{l_j} f = \prod_{j:\, l_j < 0} \left(\frac{\partial}{\partial x_j} \right)^{-l_j} f,$$

where $\alpha = (\alpha_1, \alpha_2, \ldots, \alpha_n) \in \mathbb{C}^n$ and $l = (l_1, l_2, \ldots, l_N)$ ranges over the lattice L of integer vectors such that $l_1 a_1 + l_2 a_2 + \cdots + l_N a_N = 0$.

Solutions of the system (1.2), (1.3) are called *hypergeometric functions* on $\mathbb{C}^N$ associated with the action of torus (1.1). The numbers α_i are called *exponents*.

Remark 1.2 Equations (1.2) are equivalent to the following homogeneous conditions

$$(1.4) \qquad\qquad f(x \cdot t) = t^\alpha f(x),$$

where $t = (t_1, t_2, \ldots, t_n) \in T$ and $t^\alpha = t_1^{\alpha_1} t_2^{\alpha_2} \ldots t_n^{\alpha_n}$.

Remark 1.3 For a generic α system (1.2), (1.3) is equivalent to the subsystem, where L ranges over any set of generators for the lattice L.

By A denote the set of integer vectors $a_1, a_2, \ldots, a_N$. Let H_A be the sublattice in $\mathbb{Z}^n$ generated by $a_1, a_2, \ldots, a_N$ and let $m = \dim H_A$ be the dimension of H_A. Let P_A denote the convex hull of the origin 0 and $a_1, a_2, \ldots, a_N$. Then P_A is a polyhedron with vertices in the lattice H_A.

Let Vol_{H_A} be the form of volume on the space $H_A \otimes_{\mathbb{Z}} \mathbb{R}$ such that volume of the identity cube is equal to 1. The volume of a polyhedron with vertices in the lattice H_A times $m!$ is an integer number. In particular, $m! \, \mathrm{Vol}_{H_A} P_A$ is an integer.

Theorem 1.4. *The general hypergeometric system (1.2), (1.3) gives a holonomic D-module. The number of linearly independent solutions of this system in a neighborhood of a generic point is equal to $m! \, \mathrm{Vol}_{H_A} P_A$.*

If there exist an integer covector h such that

$$(1.5) \qquad\qquad h(a_j) = 1 \qquad \text{for all} \qquad j = 1, 2, \ldots, N$$

then we call the corresponding system (1.2), (1.3) *flat* or *nonconfluent*.

Theorem 1.4 in the nonconfluent case was proved in [GZK]. Very close results were found by Adolphson in [Ad], but his technique is quite different from ours. In this paper we study one special case of systems (1.2), (1.3) when condition (1.5) does not hold. We define these systems in the following section.

2. Hypergeometric system on unipotent matrices

Let $R \subset \mathbb{Z}^n$ be a *root system* and $R^+ \subset R$ the set of *positive roots* (see [Bo]). Then we can define the hypergeometric system (1.2), (1.3) associated with the set of integer vectors $A = R^+$.

We consider the case of the root system A_n in more details. Let $\epsilon_0, \epsilon_1, \ldots, \epsilon_n$ be the standard basis in the lattice $\mathbb{Z}^{n+1}$. The root system A_n is the set of all vectors (roots) $e_{ij} = \epsilon_i - \epsilon_j$, $i \neq j$. Let $A = A_n^+$ be the set of all positive roots $A = \{e_{ij} \in A_n : 0 \leq i < j \leq n\}$. It is clear that positive roots generate the n-dimensional lattice $H_A \simeq \mathbb{Z}^n$ of all vectors $v = v_0 \epsilon_0 + v_1 \epsilon_1 + \cdots + v_n \epsilon_n$, $v_i \in \mathbb{Z}$ such that $v_0 + v_1 + \cdots + v_n = 0$.

By Z_n denote the group of unipotent matrices of order $n+1$, i.e. the group of upper triangular matrices $z = (z_{ij})$, $0 \leq i \leq j \leq n$ with 1's on the diagonal $z_{ii} = 1$. The n-dimensional torus T presented as the group of diagonal matrices $t = \operatorname{diag}(t_0, t_1, \ldots, t_n)$, $t_0 \cdot t_1 \ldots t_n = 1$ acts on Z_n by conjugation $z \in Z_n \to tzt^{-1}$, or in coordinates

$$(2.1) \qquad z = \{z_{ij}\} \longmapsto \{z_{ij} t_i t_j^{-1}\}.$$

Clearly, the action of torus (1.1) associated with the set of vectors $A = A_n^+$ is the same as action (2.1). Here $N = \binom{n+1}{2}$ and z_{ij}, $0 \leq i < j \leq n$ are coordinates in $\mathbb{C}^N$.

The main object of this paper is the hypergeometric system associated with action (2.1). Write down this system explicitly.

Definition 2.1. The *Hypergeometric System on the Group of Unipotent Matrices* is the following system of differential equation on the space $Z_n \simeq \mathbb{C}^N$ with coordinates z_{ij}, $0 \leq i < j \leq n$

$$(2.2)$$

$$-\sum_{i=0}^{j-1} z_{ij} \frac{\partial f}{\partial z_{ij}} + \sum_{k=j+1}^{n} z_{jk} \frac{\partial f}{\partial z_{ij}} = \alpha_j f, \qquad j = 0, 1, \ldots, n;$$

$$(2.3) \qquad \frac{\partial f}{\partial z_{ik}} = \frac{\partial^2 f}{\partial z_{ij}\, \partial z_{jk}}, \qquad 0 \leq i < j < k \leq n,$$

where $\alpha = (\alpha_0, \alpha_1, \ldots, \alpha_n) \in \mathbb{C}^{n+1}$ is a vector such that $\sum \alpha_j = 0$.

Solutions of system (2.2), (2.3) are called *hypergeometric functions on the group of unipotent matrices*.

In order to prove that system (2.2), (2.3) is a special case of the system (1.2), (1.3) we need the following simple lemma.

Lemma 2.2. *It follows from equations (2.3) that*

$$(2.4) \qquad \prod_{(i,j):\, l_{ij} > 0} \left(\frac{\partial}{\partial z_{ij}}\right)^{l_{ij}} f = \prod_{(i,j):\, l_{ij} < 0} \left(\frac{\partial}{\partial z_{ij}}\right)^{-l_{ij}} f,$$

for all $l = (l_{ij})$, $0 \leq i < j \leq n$, $l_{ij} \in \mathbb{Z}$ *such that* $\sum_i l_{ij} - \sum_k l_{jk} = 0$, $j = 0, 1, \ldots, n$.

Proof. It follows from (2.3) that

$$\frac{\partial f}{\partial z_{ij}} = \frac{\partial^{j-i} f}{\partial z_{ii+1} \, \partial z_{i+1i+2} \ldots \partial z_{j-1j}}$$

Now change in (2.4) all occurrences of $\dfrac{\partial}{\partial z_{ij}}$ to $\dfrac{\partial^{j-i}}{\partial z_{ii+1} \partial z_{i+1i+2} \ldots \partial z_{j-1j}}$. We get the same expressions in LHS and in RHS.

Let $P_n = P_{A_n^+}$ be the convex hull of the origin 0 and of e_{ij}, $0 \leq i < j \leq n$. The first part of the following theorem is a special case of Theorem 1.4.

Theorem 2.3.

  (1) *The hypergeometric system (2.2), (2.3) gives a holonomic D-module. The number of linearly independent solutions of this system in a neighborhood of a generic point is equal to* $n! \operatorname{Vol} P_n$.

  (2) $n! \operatorname{Vol} P_n$ *is equal to the Catalan number*

$$C_n = \frac{1}{n+1} \binom{2n}{n}.$$

3. Integral expression for hypergeometric functions

In this section we present an integral expression for hypergeometric functions on unipotent matrices (see [GG1]).

Consider the following integral

$$(3.1) \qquad f(z) = \int_C \exp\left(\sum z_{ij} t_i t_j^{-1}\right) t^{-\alpha} \frac{dt}{t},$$

where the sum in exponent is over $0 \leq i < j \leq n$; t is a point of torus $T = \{(t_0, \ldots, t_n) : t_0 \cdot \ldots \cdot t_n = 1\} \simeq (\mathbb{C}^*)^n$; $t^{-\alpha} dt/t = t_1^{-\alpha_1} \ldots t_n^{-\alpha_n} dt_1/t_1 \ldots dt_n/t_n$; and C is a real n-dimensional cycle in $2n$-dimensional space T.

Theorem 3.1. *The function $f(z)$ given by integral (3.1) is a solution of the hypergeometric system (2.2), (2.3).*

4. Γ-series and admissible bases

In this section we construct an explicit basis in the solution space of system (1.2), (1.3). In the case of nonconfluent systems this construction was given in [GZK]. In this section we basically follow [GZK].

Recall that $A = \{a_1, a_2, \ldots, a_N\}$, where $a_j \in \mathbb{Z}^n$. Without loss of generality we can assume that vectors a_j generate the lattice $\mathbb{Z}^n$, i.e. $H_A = \mathbb{Z}^n$. Let $\gamma = (\gamma_1, \gamma_2, \ldots, \gamma_N) \in \mathbb{C}^N$. Consider the following formal series

$$(4.1) \qquad \Phi_\gamma(x) = \sum_{l \in L} \frac{x^{\gamma+l}}{\prod_{j=1}^{N} \Gamma(\gamma_j + l_j + 1)},$$

where $x = (x_1, x_2, \ldots, x_N)$, L is the lattice such as in Definition 1.1, and $x^{\gamma+l} = \prod_{j=1}^{N} x_j^{\gamma_j + l_j}$.

Lemma 4.1. *The series $\Phi_\gamma(x)$ formally satisfies system (1.2), (1.3) with $\alpha = \sum_j \gamma_j a_j$.*

For a fixed vector of exponents $\alpha = (\alpha_1, \ldots, \alpha_n)$ the vector $\gamma = (\gamma_1, \ldots, \gamma_N)$ ranges over the affine $(N - n)$-dimensional plane $\Pi(\alpha) = \{(\gamma_1, \ldots, \gamma_N) : \sum_j \gamma_j a_j = \alpha\}$. In this section we construct several vectors γ such that all series $\Phi_\gamma(x)$ converge in a certain neighborhood and form a basis in the space of solutions of system (1.2), (1.3) in this neighborhood.

A subset $\mathcal{I} \in [N]$ is called a *base* if vectors a_j, $j \in \mathcal{I}$ form a basis of the linear space $H_A \otimes \mathbb{R}$. So we get a *matroid* on the set $[N]$. Let $\Delta_\mathcal{I}$ be the n-dimensional simplex with vertices 0 and a_j, $j \in \mathcal{I}$.

Let $\mathcal{I}$ be a base. By $\Pi(\alpha, \mathcal{I})$ denote the set of $\gamma \in \Pi(\alpha)$ such that $\gamma_j \in \mathbb{Z}$ for $j \notin \mathcal{I}$. It is clear that for every $l \in L$ (see Definition 1.1) $\Phi_\gamma(x) = \Phi_{\gamma+l}$.

The following lemma was proven in [GZK].

Lemma 4.2. *Let $\mathcal{I}$ be a base. Then $|\Pi(\alpha, \mathcal{I})/L| = n! \operatorname{Vol}(\Delta_\mathcal{I})$.*

Definition 4.3. We call a base $\mathcal{I} \in [N]$ *admissible* if the $(n - 1)$-dimensional simplex with vertices a_j, $j \in \mathcal{I}$ belongs to the boundary ∂P_A of the polyhedron P_A. In this case the simplex $\Delta_\mathcal{I}$ is also called *admissible*.

Remark 4.4 If vectors a_j satisfy condition (1.5) then all bases are admissible.

Let $B = \{b_1, b_2, \ldots, b_{N-n}\}$ be a $\mathbb{Z}$-basis in the lattice L. We say that a base $\mathcal{I}$ is *compatible* with a basis B if whenever $l = (l_1, \ldots, l_N) \in L$ such that $l_j \geq 0$ for $j \notin \mathcal{I}$ then l can be expressed as $l = \sum \lambda_k b_k$, where all $\lambda_k \geq 0$. Clearly, the set $\Pi_B(\alpha, \mathcal{I}) = \{\gamma \in \Pi(\alpha, \mathcal{I}) : \gamma = \sum \lambda_k b_k$, where $0 \leq \lambda_k < 1\}$ is a set of representatives in $\Pi(\alpha, \mathcal{I})/L$.

Let $y_k = x^{b_k}$, $k = 1, 2 \ldots, N - n$.

Proposition 4.5. *Let an admissible base $\mathcal{I}$ be compatible with a basis B. Then for all $\gamma \in \Pi_B(\alpha, \mathcal{I})$ the series $\Phi_\gamma(x)$ is of the form $\Phi_\gamma(x) = x^\gamma \sum_m c(m) y^m$, where the sum is over $m = (m_1, \ldots, m_{N-n})$, $m_k \geq 0$. The series $\sum c(m) y^m$ converges for sufficiently small $|y_k|$.*

Proof. Let $b_k = (b_{k1}, \ldots, b_{kN}) \in L$, $k = 1, \ldots, N-n$. By definition, $\Phi_\gamma(x) = x^\gamma \sum_m c(m) y^m$, where $c(m) = \prod_j \Gamma(\gamma_j + \sum_k m_k b_{kj} + 1)^{-1}$, $m =$

$(m_1, \ldots, m_{N-n}) \in \mathbb{Z}^{N-n}$. Let $\gamma \in \Pi_B(\alpha, \mathcal{I})$. Then $\gamma_j + \sum_k m_k b_{kj} + 1 \in \mathbb{Z}$, for $j \notin \mathcal{I}$. Hence, if $c(m) \neq 0$ then $\gamma_j + \sum_k m_k b_{kj} + 1 \geq 0$, $j \notin \mathcal{I}$. Since $\mathcal{I}$ is compatible with B, we can deduce that $c(m) \neq 0$ only if $m_k \geq 0$, $k = 1, \ldots, N-n$ (see details in [GZK]). Convergence of the series $\sum c(m) y^m$ follows from the next lemma.

Lemma 4.6. Let $c(m) = \prod_j \Gamma(\mu_j(m) + \gamma_j + 1)^{-1}$, $m = (m_1, \ldots, m_r)$, $m_k \geq 0$, where μ_j are linear functions of m such that $\sum \mu_j(m) = s_1 m_1 + \cdots + s_r m_r$, $s_k \geq 0$. Then $|c(m)| \leq R\, c_1^{m_1} \ldots c_r^{m_r}$ for some positive constants $R, c_1, \ldots, c_r$.

It is not difficult to prove this lemma using Stiltjes formula. Thus, by Proposition 4.5 for every admissible base I we have $n! \operatorname{Vol}(\Delta_\mathcal{I})$ series $\Phi_\gamma(x)$, $\gamma \in \Pi_B(\alpha, \mathcal{I})$ with nonempty common convergence domain.

Remark 4.7 It can be shown that if $\gamma \in \Pi(\alpha, \mathcal{I})$, where $\mathcal{I}$ is not admissible, then $\Phi_\gamma(x)$ diverges.

Recall that P_A is the convex hull of 0 and a_j, $j = 1, 2, \ldots, N$.

Definition 4.8. The set of bases Θ is called a *local triangulation* of P_A if

(1) $\cup_{\mathcal{I} \in \Theta} \Delta_\mathcal{I} = P_A$;
(2) $\Delta_{\mathcal{I}_1} \cap \Delta_{\mathcal{I}_2}$ is the common face of $\Delta_{\mathcal{I}_1}$ and $\Delta_{\mathcal{I}_2}$ for all $\mathcal{I}_1, \mathcal{I}_2 \in \Theta$.

We call such triangulation Θ local because all simplices $\Delta_\mathcal{I}$, $\mathcal{I} \in \Theta$ contain the origin 0.

Remark 4.9 Note that if Θ is a local triangulation then all bases $\mathcal{I} \in \Theta$ are admissible

Definition 4.10. A local triangulation Θ is called *coherent* if there exist a piecewise linear function ϕ on P_A such that ϕ is linear on simplices $\Delta_\mathcal{I}$, $\mathcal{I} \in \Theta$ and ϕ is strictly convex on P_A.

Lemma 4.11. *There exists a coherent local triangulation of P_A.*

Lemma 4.12. *Let Θ be a coherent local triangulation of P_A. Then there exists a basis B of H_A such that B is compatible with every base $\mathcal{I}$ in Θ.*

Theorem 4.13. *Let Θ be a coherent local triangulation of P_A; and $B = \{b_1, b_2, \ldots \ldots, b_{N-n}\}$ a basis such as in Lemma 4.12. Let $y_k = x^{b_k}$. Then for every $\gamma \in \Pi_B(\alpha, \mathcal{I})$, $\mathcal{I} \in \Theta$ the series $\Phi_\gamma(x)$ is equal x^γ times a series of variables y_k, which converges for sufficiently small $|y_k|$. If exponents $\alpha_1, \alpha_2, \ldots, \alpha_n$ are generic then all these series $\Phi_\gamma(x)$ are linearly independent.*

Hence, for generic $\alpha = (\alpha_1, \alpha_2, \ldots, \alpha_n)$ we constructed $n! \operatorname{Vol}(P_A)$ independent solutions of system (1.2), (1.3), which converge in common domain. Therefore, by Theorem 1.4, these series form a basis in the space of solutions of system (1.2), (1.3).

5. Admissible trees

In this section we describe admissible bases in the case of the hypergeometric system (2.2), (2.3).

It is well known that a subset $\mathcal{I} \subset \{(i,j) : 0 \leq i < j \leq n\}$ is a base in the set of positive roots $A = A_n^+$ if and only if $\mathcal{I}$ is the set of edges of a tree $T_{\mathcal{I}}$ on $[0, n]$.

Definition 5.1. A tree T on the set $[0, n]$ is called *admissible* if there are no $0 \leq i < j < k \leq n$ such that both (i, j) and (j, k) are edges of T.

Proposition 5.2. *A subset* $\mathcal{I} \subset \{(i, j) : 0 \leq i < j \leq n\}$ *is an admissible base in* $A = A_n^+$ *if and only if* $T_{\mathcal{I}}$ *is an admissible tree.*

Lemma 5.3. $n! \operatorname{Vol} \Delta_{\mathcal{I}} = 1$ *for any base* $\mathcal{I}$.

Therefore, by Lemma 4.2 $|\Pi(\alpha, \mathcal{I})/L| = 1$ and Proposition 4.5 for every admissible tree T we have a series $\Phi_T(z) = \Phi_\gamma(z)$, where $\gamma \in \Pi(\alpha, \mathcal{I})$, $T = T_{\mathcal{I}}$. The series $\Phi_T(z)$ converges in some domain and presents a solution of the system (2.2), (2.3). There exists a formula for the number of all admissible trees on the set $[0, n]$.

Theorem 5.4. *The number* F_n *of admissible trees on the set of vertices* $[0, n]$ *is equal to*

$$F_n = \frac{1}{2^n(n+1)} \sum_{k=1}^{n+1} \binom{n+1}{k} k^n.$$

The proof of this formula is given in [Po].
The first few numbers of F_n are given below.

n	0	1	2	3	4	5	6	7
F_n	1	1	2	7	36	246	2104	21652

6. Standard triangulation of P_n

Recall that P_n is the convex hull of 0 and e_{ij}, $0 \leq i < j \leq n$.

In this section we construct a coherent triangulation of the polyhedron P_n. This will give us an explicit basis in the solution space of system (2.2), (2.3).

Let T be a tree on the set $[0, n]$. We say that two edges (i, j) and (k, l) in T form an *intersection* if $i < k < j < l$.

Definition 6.1. A tree T on the set $[0, n]$ is called *standard* if T is admissible and does not have intersections. The corresponding base $\mathcal{I} \subset \{(i, j) : 0 \leq i < j \leq n\}$ is also called *standard*.

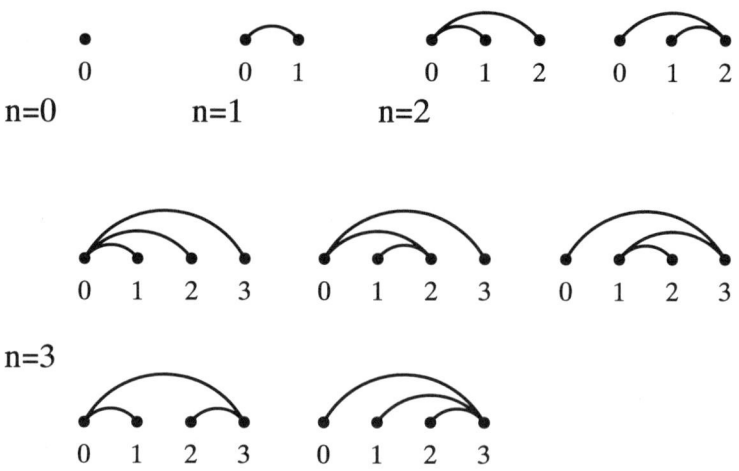

FIGURE 6.1.   Standard trees.

Example 6.2. All standard trees for $n = 0, 1, 2, 3$ are shown on Figure 6.1.

Theorem 6.3. *The set Θ_n of standard bases forms a coherent local triangulation of the polyhedron P_n.*

Theorem 6.4. *The number of standard trees on the set $[0, n]$ is equal to the Catalan number*

$$C_n = \frac{1}{n+1}\binom{2n}{n}.$$

As a consequence of these two theorems we get Theorem 2.3.(2).

Proof of Theorem 6.4.   Construct by induction an explicit 1–1 correspondence ψ_n between the set ST_n of standard trees on $[0, n]$ and the set BT_n of binary trees with n unmarked vertices $\psi_n : \mathrm{ST}_n \to \mathrm{BT}_n$.

If $n = 1$ then ψ_1 maps a unique element of ST_1 to a unique element of BT_1.

Let $n > 1$. Every standard tree $T \in \mathrm{ST}_n$ has the edge $(0, n)$. Delete this edge. Then T splits into two standard trees $T_1 \in \mathrm{BT}_k$ and $T_2 \in \mathrm{BT}_l$, $k + l + 1 = n$ on the sets $[0, k]$ and $[k+1, n]$. Let as define $\psi_n(T)$ as the binary tree whose left and right branches are equal to $\psi_k(T_1)$ and $\psi_l(T_2)$ correspondingly. See the example in Figure 6.2.

It is well known that the number of binary trees is equal to the Catalan number (e.g. see [SW]).

Now prove Theorem 6.3.

Proof of Theorem 6.3.   Recall that $\epsilon_0, \epsilon_1, \ldots, \epsilon_n$ is the standard basis in $\mathbb{Z}^{n+1}$; and $e_{ij} = \epsilon_i - \epsilon_j$.

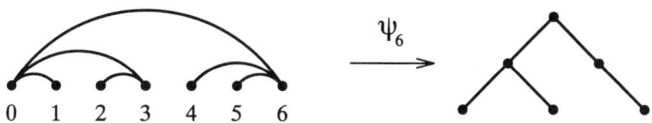

FIGURE 6.2. Bijection between standard and binary trees.

Let $\widetilde{P}_n \subset \mathbb{Z}^{n+1} \otimes \mathbb{R}$ denote the cone with vertex at 0 generated by all positive roots e_{ij}, $i < j$. Let $\widetilde{\Delta}_{\mathcal{I}}$ denote the simplicial cone generated by e_{ij}, $(i,j) \in \mathcal{I}$, where $\mathcal{I}$ is a base (the cone over the simplex $\Delta_{\mathcal{I}}$).

First, prove that the collection of cones $\widetilde{\Delta}_{\mathcal{I}}$, where $\mathcal{I}$ range over all standard bases, is a conic triangulation of $\widetilde{P}_n$. Then it follows that Θ_n is a local triangulation. It is not difficult to show that the cone $\widetilde{P}_n$ is the set of $v = (v_0, v_1, \ldots, v_n) \in \mathbb{R}^{n+1}$ such that

(6.1) $$v_0 + v_1 + \cdots + v_i \geq 0, \qquad i = 1, 2, \ldots, n-1;$$

(6.2) $$v_0 + v_1 + \cdots + v_n = 0.$$

We must show that every generic point v subject to (6.1), (6.2) can be uniquely presented in the form

(6.3) $$v = \sum_{(ij) \in \mathcal{I}} \rho_{ij} e_{ij}, \qquad \rho_{ij} \geq 0,$$

for some standard base $\mathcal{I}$. Prove it by induction on n.

Let $v' = (v_0', v_1', \ldots, v_{n-1}') \in \mathbb{R}^n$ be a vector such that $v_i' = v_i$, $i = 0, 1, \ldots, n-2$, and $v_{n-1}' = v_{n-1} + v_n$. Then $v' \in \widetilde{P}_{n-1}$. By induction we may assume that v' is expressed in the form

$$v' = \sum_{(ij) \in \mathcal{I}'} \rho_{ij}' e_{ij}, \qquad \rho_{ij}' \geq 0,$$

for a standard base $\mathcal{I}' \subset \{(i,j) : 0 \leq i < j \leq n-1\}$.

Let $i_1 < i_2 < \ldots < i_s$ be all vertices of $T' = T_{\mathcal{I}'}$ connected with the vertex $n-1$ in T'.

Consider two cases.

1. $v_{n-1} \geq 0$. Define $\mathcal{I} = \mathcal{I}' \cup \{(n-1, n)\} \cup \{(i_k, n) : k \in [s]\} \setminus \{(i_k, n-1) : k \in [s]\}$. And $\rho_{ij} = \rho_{ij}'$ for $0 \leq i < j \leq n-2$; $\rho_{i_k n} = \rho_{i_k n-1}'$ for $k \in [s]$; $\rho_{n-1\,n} = v_{n-1}$. Then we get expression (6.3) for v.

2. $v_{n-1} < 0$. Then $-v_n \leq \sum_{k=1}^{s} \rho_{i_k n-1}'$. Let t be the minimal integer $0 \leq t \leq s$ such that $\sum_{k=1}^{t} \rho_{i_k n-1}' \geq -v_n$. Then define $\mathcal{I} = \mathcal{I}' \cup \{(i_k, n) : k \in [t]\} \setminus \{(i_k, n-1) : k \in [t-1]\}$. And $\rho_{ij} = \rho_{ij}'$ for $0 \leq i < j \leq n-2$; $\rho_{i_k n} = \rho_{i_k n-1}'$ for $k \in [t-1]$; $\rho_{i_t n} = -\sum_{k=1}^{t-1} \rho_{i_k n-1}' - v_n$; $\rho_{i_k n-1} = \rho_{i_k n-1}'$ for $k \in [t+1, s]$;

$\rho_{i_t n-1} = -\sum_{k=t+1}^{s} \rho'_{i_k n-1} - v_{n-1}$. Then we get expression (6.3) for v. Therefore, Θ_n is a local triangulation.

Prove that Θ_n is coherent triangulation (see Definition 4.10). We must present a piecewise linear function ϕ on P_n such that ϕ is linear on all simplices in Θ_n and ϕ is strictly convex on P_n.

It is sufficient to define ϕ on vertices of P_n. Let $\phi(0) = 0$ and $\phi(\epsilon_{ij}) = (i-j)^2$. It is not difficult to show that such ϕ satisfy the condition of Definition 4.10.

Now we can complete the proof of Theorem 2.3.

Proof of Theorem 2.3.

The first part of Theorem 2.3 is a special case of Theorem 1.4.

The second part follows from Theorems 6.3, 6.4 and Lemma 5.3.

To conclude this section we present a construction of another coherent triangulation of P_n.

Let T be a tree on the set $[0, n]$. We say that two edges (i, j) and (k, l) in T are *enclosed* if $i < k < l < j$.

Definition 6.5. A tree T on the set $[0, n]$ is called *anti-standard* if T is admissible and does not have enclosed edges. The corresponding base $\mathcal{I} \subset \{(i, j) : 0 \le i < j \le n\}$ is also called *anti-standard*.

Theorem 6.6. *The set of anti-standard bases forms a coherent local triangulation of the polyhedron P_n.*

The proof of this theorem is analogous to the proof of Theorem 6.3.

Corollary 6.7. *The number of anti-standard trees on the set $[0, n]$ is equal to the Catalan number C_n.*

7. Coordinate strata

Let Z_n be the group of unipotent matrices z_{ij}, $0 \le i \le j \le n$, $z_{ii} = 1$ (see Section 2).

Consider a subset $S \subset \{(i, j) : 0 \le i < j \le n\}$. By Z_S denote the set of all $z = \{z_{ij}\} \in Z_n$ such that $z_{ij} \ne 0$ if and only if $(i, j) \in S$. We call Z_S *coordinate strata* in the space Z_n. Let $\overline{Z}_S \simeq \mathbb{C}^{|S|}$ be the closure of the stratum Z_S.

We can construct two sheaves of hypergeometric functions on the manifold $\overline{Z}_S$, where $S \subset \{(i, j) : 0 \le i < j \le n\}$. First, the sheaf Res_S of restrictions of hypergeometric functions on Z_n to the manifold $\overline{Z}_S$. Second, the sheaf Sol_S of solutions of the hypergeometric system (1.2), (1.3) associated with $A = A_S = \{e_{ij} : (i, j) \in S\}$ (equivalently, associated with action (2.1) of torus on $\overline{Z}_S$).

The question is: when do these two sheaves coincide?

Definition 7.3. Let $\mathcal{P} = \{b_0, b_1, \ldots, b_n\}$ be a partially ordered set (poset) such that if $b_i <_{\mathcal{P}} b_j$ then $i < j$. Consider the set $S_{\mathcal{P}} = \{(i, j) : b_i <_{\mathcal{P}} b_j\}$. We call this set *associated* with poset $\mathcal{P}$.

Theorem 7.4. *Let $S = S_{\mathcal{P}}$ be the set associated with a poset. Then sheaf* Res$_S$ *coincides with sheaf* Sol$_S$ *for generic exponents* $\alpha_0, \ldots, \alpha_n$, $\sum \alpha_i = 0$.

Remark 7.5 By Theorem 1.4 the dimension of Sol$_S$ in a neighborhood of a generic point is equal to $m! \operatorname{Vol}_{H(S)} P(S)$, where $H(S)$ is the lattice generated by e_{ij}, $(i, j) \in S$, $m = \dim H_S$, and $P(S)$ is the convex hull of the origin and e_{ij}, $(i, j) \in S$.

Proposition 7.6. *A set $S \subset \{(i, j) : 0 \leq i < j \leq n\}$ is associated with a poset $\mathcal{P}$ if and only if there exists a cone C with vertex at 0 such that $S = \{(i < j) : e_{ij} \in C\}$.*

Proof. A set S is associated with a poset if and only if S satisfies the following transitivity: if $(i, j), (j, k) \in S$ then $(i, k) \in S$. The set $S = \{(i < j) : e_{ij} \in C\}$ satisfies transitivity because if $e_{ij}, e_{jk} \in C$ then $e_{ik} = e_{ij} + e_{jk} \in C$. Inversely, let C be the cone generated by all e_{ij}, $(i, j) \in S$. If S satisfies transitivity then $S = \{(i < j) : e_{ij} \in C\}$.

Now we can prove Theorem 7.4

Proof of Theorem 7.4. Clearly, Res$_S$ is a subsheaf of Sol$_S$. Suppose for simplicity that e_{ij}, $(i, j) \in S$ generate $\mathbb{Z}^n$. The dimension of the sheaf Sol$_S$ at a generic point is equal to $n! \operatorname{Vol}(P(S))$ (see Remark 7.5). Hence, it is sufficient to prove that the dimension of Res$_S$ at a generic point is greater than or equal to $n! \operatorname{Vol}(P(S))$.

Let Θ be a coherent local triangulation of $P(A)$. It follows from Proposition 7.6 that Θ extends to a coherent local triangulation Θ' of P_n. Consider $n! \operatorname{Vol}(P(S))$ Γ-series $\Phi_\gamma(z)$ on Z_n, where $\gamma \in \Pi(\alpha, \mathcal{I})$, $\mathcal{I} \in \Theta \subset \Theta'$. By Theorem 4.13 these series are linearly independent and have a common convergence domain. Then restrictions of these series to $\overline{Z}_S$ give $n! \operatorname{Vol}(P(S))$ independent sections of the sheaf Res$_S$ in some neighborhood. Therefore, Res$_S = $ Sol$_S$.

8. Face strata

Describe faces of the polyhedron P_n.

Let $I, J \subset [0, n]$, $I \cap J = \emptyset$. Let S_{IJ} be the set of all (i, j), $0 \leq i < j \leq n$ such that $i \in I$ and $j \in J$.

Proposition 8.1. *Faces f of the polyhedron P_n such that $0 \notin f$ are in 1-1 correspondence with sets S_{IJ}. And $(i, j) \in S_{IJ}$ whenever e_{ij} is a vertex of the corresponding face f.*

Clearly, we may assume that $\min(I \cup J) \in I$ and $\max(I \cup J) \in J$ (if S_{IJ} is nonempty).

Construct a coordinate stratum associated with a face f of P_n $0 \notin f$. Let $S = S_{IJ}$. By Z_{IJ} denote the stratum Z_S (see Section 3). We will call such strata *face strata*.

Note that condition (1.5) holds for vectors e_{ij}, $(i, j) \in S_{IJ}$, because all such e_{ij} belong to a supporting hyperplane of the corresponding face f.

Definition 8.2. The *Hypergeometric System on* $\overline{Z}_{IJ}$ is the hypergeometric system (1.2), (1.3) associated with the set of vectors $A = \{e_{ij} : (i,j) \in S_{IJ}\}$. Solutions of this system are called *Hypergeometric Functions on* $\overline{Z}_{IJ}$.

Remark 8.3 Let $0 \leq p < n$, $I = \{0, 1, \ldots, p\}$, and $J = \{p+1, p+2, \ldots, n\}$. Then $\overline{Z}_{IJ}$ is the space of rectangular matrices $z = \{z_{ij}\}$, $i \in [0, p]$, $j \in [p+1, n]$. The hypergeometric system on $\overline{Z}_{IJ}$ is also called the *Hypergeometric System on the Grassmannian* $G_{n+1\,p+1}$. This system was studied in the works [GGR1, GGR2, GGR3].

It is clear that the set $S = S_{IJ}$ is associated with a poset (see Definition 7.3). Therefore, by Theorem 8.4, the sheaf Res_S coincides with the sheaf Sol_S of hypergeometric functions on $\overline{Z}_{IJ}$ (for generic α).

We will find the dimension of this sheaf in a neighborhood of a generic point. Denote this dimension by D_{IJ}. In other words, D_{IJ} is the number of independent solutions of the hypergeometric system on $\overline{Z}_{IJ}$ in a neighborhood of a generic point.

Let P_{IJ} be the convex hull of 0 and e_{ij}, $(i,j) \in S_{IJ}$. Let H_{IJ} be the sublattice generated by e_{ij}, $(i,j) \in S_{IJ}$, and $m = \dim H_{IJ}$. By Theorem 1.4 the number D_{IJ} is equal to $m! \, \text{Vol}_{H_{IJ}}(P_{IJ})$.

We present an explicit combinatorial interpretation of this number D_{IJ}.

Definition 8.4.

(1) A *word* w of *type* (p, q) is the sequence $w = (w_1, w_2, \ldots, w_{p+q})$, $w_r \in \{1, 0\}$ such that $|\{r : w_r = 0\}| = p$ and $|\{r : w_r = 1\}| = q$.

(2) Let $w = (w_1, w_2, \ldots, w_{p+q})$ and $w' = (w'_1, w'_2, \ldots, w'_{p+q})$ be two words of type (p, q). We say that w' is *exceeds* w if $w'_1 + \cdots + w'_r \geq w_1 + \cdots + w_r$ for all $r = 1, 2, \ldots, p+q$.

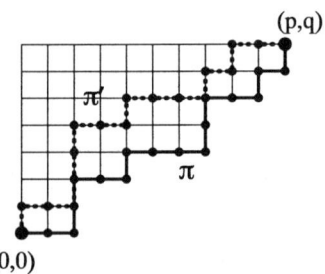

$$w = (0, 0, 1, 1, 0, 0, 1, 0, 0, 0, 1, 1, 0, 0, 1, 0, 1)$$
$$w' = (0, 0, 1, 1, 0, 0, 1, 0, 0, 0, 1, 1, 0, 0, 1, 0, 1)$$

FIGURE 8.1.   The word w' exceeds the word w.

We can present a word w of type (p,q) as the path $\pi = (\pi_0, \pi_1, \ldots, \pi_{p+q})$ in $\mathbb{Z}^2$ such that $\pi_s = (i_s, j_s)$ for all $s = 0, 1, \ldots, p+q$, where i_s (correspondingly, j_s) is the number of 0's (correspondingly, 1's) in $w_1, w_2, \ldots, w_s$. See example for $(p,q) = (10, 7)$ on Fig. 8.1.

Clearly, a word w' exceeds a word w if and only if the path π' corresponding to w' is above the path π corresponding to w. (See Fig. 8.1.)

Let $a = \min I$ and $b = \max J$. Then $D_{IJ} \neq 0$ if ond only if $a < b$. Suppose that $a < b$, $I = \{a\} \cup I'$ and $J = \{b\} \cup J'$, where $I', J' \subset [a+1, b-1]$, $I' \cap J' = \emptyset$. Let $|I'| = p$, $|J'| = q$ and $I' \cup J' = \{t_1 < t_2 < \cdots < t_{p+q}\}$. Associate with the pair (I, J) the word $w_{IJ} = (w_1, \ldots, w_{p+q})$ of type (p, q) such that $w_r = 0$ if $t_r \in I$ and $w_r = 1$ if $t_r \in J$ for all $r = 1, 2, \ldots, p+q$.

Theorem 8.5. *The number D_{IJ} is equal to the number of words w' of type (p, q) which exceed the word $w = w_{IJ}$. In other words, D_{IJ} is the number of paths π' from $(0,0)$ to (p, q) such that π' is above the path $\pi = \pi_{IJ}$ corresponding to w_{IJ}.*

Corollary 8.6. *Let $I = \{0, 2, 4, \ldots, 2k\}$ and $J = \{1, 3, 5, \ldots, 2k+1\}$ then D_{IJ} is equal to the Catalan number C_k.*

Proof. Words $w' = (w'_1, w'_2, \ldots, w'_{2k})$ of type (k, k) which exceed the word $w = (1, 0, 1, 0, \ldots, 1, 0)$ are called *Dyck words*. It is well known (see e.g. [SW]) that the Catalan number C_k is equal to the number of Dyck words.

9. Standard triangulation of P_{IJ}

Let $I, J \subset [0, n]$, $I \cap J = \emptyset$ be two subsets such that $\min(I \cup J) \in I$ and $\max(I \cup J) \in J$ (see Section 8).

Recall that $P_{IJ} = \text{Conv}(0, e_{ij} : (i, j) \in S_{IJ})$. In this section we present a coherent local triangulation of the polyhedron P_{IJ} and prove Theorem 8.5.

Definition 9.1. *Let T be a tree on the set $I \cup J$. We say that T is of type (I, J) if for every edge (i, j) in T $i \in I$ and $j \in J$. The base $\mathcal{I} \subset \{(i, j) : 0 \le i < j \le n\}$ corresponding to T is also called of type (I, J). (Do not confuse $\mathcal{I}$ with I.)*

Clearly, all trees of type (I, J) are admissible (see Definition 5.1).

Theorem 9.2. *The set Θ_{IJ} of all standard (see Definition 6.1) bases of type (I, J) forms a coherent local triangulation of the polyhedron P_{IJ}.*

The proof of this theorem is essentially the same as the proof of Theorem 6.3.

It is clear that $D_{IJ} = m! \, \text{Vol}(P_{IJ})$ is equal to the number of all standard bases (trees) of type (I, J). Prove that this number coincides with the number given by Theorem 8.5.

Theorem 9.3. *Let $|I| = p+1$ and $|J| = q+1$. Then the number of all standard trees T of type (I, J) is equal to the number of words w' of type (p, q) which exceed the word $w = w_{IJ}$.*

Proof. Let D_{IJ} be the number of all standard trees of type (I, J) and $\widetilde{D}_{IJ}$ be the number of words w' of type (p, q) which exceed the word $w = w_{IJ}$ (we use the same notation as in Theorem 8.5).

We prove that $D_{IJ} = \widetilde{D}_{IJ}$ by induction on $p+q$. Obviously, this is true for $p = q = 0$. Let d be the minimal element of J and c be the maximal element of I such that $c \leq d$. Let $\tilde{I} = I \setminus \{c\}$ and $\tilde{J} = J \setminus \{d\}$. Prove that if $p + q > 0$ then

$$(9.1) \qquad\qquad D_{IJ} = D_{\tilde{I}J} + D_{I\tilde{J}}.$$

Every standard tree of type (I, J) has the edge (c, d). In every such tree either c or d is an end-point. The first choice corresponds to the term $D_{\tilde{I}J}$ and the second choice corresponds to the term $D_{I\tilde{J}}$ in (9.1).

The numbers $\widetilde{D}_{IJ}$ also satisfy the relation (9.1). The first term corresponds to the case when the word w' starts with 0 and the second term to the case when w' starts with 1. Therefore, we get by induction $D_{IJ} = \widetilde{D}_{IJ}$.

Theorem 8.5 is a corollary of Theorem 9.3.

10. Examples

In this and the next sections we present several examples which illustrate the notions introduced in the paper and show the direction for following study.

10.1. Case $n = 2$.

In this case the solutions f of the system (2.2), (2.3) are functions of variables z_{01}, z_{02}, z_{12}.

Let $\beta_1 = \frac{1}{3}(\alpha_2 - 2\alpha_0)$ and $\beta_2 = \frac{1}{3}(2\alpha_2 - \alpha_0)$. Because of homogeneous conditions (1.4) we can write $f(z_{01}, z_{02}, z_{12}) = z_{01}^{\beta_1} z_{12}^{\beta_2} F(y)$, where $y = \frac{z_{02}}{z_{01} z_{12}}$. Now system (2.2), (2.3) is equivalent to the following equation on $F(y)$.

$$(10.1) \qquad\qquad \frac{dF}{dy} = \left(y\frac{d}{dy} - \beta_1 \right) \left(y\frac{d}{dy} - \beta_2 \right) F.$$

This is the degenerate hypergeometric equation and its solutions can be written in terms of the degenerate hypergeometric function $_1F_1$ (see [BE]). This system has two dimensional space of solutions, which is compatible with the fact that $C_2 = 2$.

10.1. Upper triangular matrices.

Let $I = \{0, 2, \ldots, 2n\}$ and $J = \{1, 3, \ldots, 2n+1\}$. It is natural to identify the space $\overline{Z}_{IJ}$ with the space of all upper triangular matrices with arbitrary elements on the diagonal. Consider the hypergeometric system on $\overline{Z}_{IJ}$. We call this system *the hypergeometric system on upper triangular matrices.*

This system has the same dimension C_n of solution space as system (2.2), (2.3) (see Corollary 8.6). But it is nonconfluent unlikely system (2.2), (2.3). If fact, system (2.2), (2.3) can be obtained as a limit of the hypergeometric system on upper triangular matrices. For example, if $I = \{0, 2, 4\}$ and $J = \{1, 3, 5\}$ then the corresponding hypergeometric system on $\overline{Z}_{IJ}$ can be reduced to the Gauss hypergeometric equation. And equation (10.1) is a limit of the Gauss hypergeometric equation.

11. Concluding remarks and open problems

11.1. Characteristic manifold.

We do not prove Theorem 1.4 here. There exists a proof of this theorem generalizing the proof from [GZK] for nonconfluent case.

This proof is based on the consideration of *characteristic manifold Ch* for system (1.2), (1.3). The characteristic manifold for system (1.2), (1.3) is the submanifold in the space $\mathbb{C}^N \times \mathbb{C}^N$ with coordinates (x, ξ), $x = (x_1, \ldots, x_N)$, $\xi = (\xi_1, \ldots, \xi_n)$ given by the following algebraic equations.

$$\sum_{j=1}^{N} a_{ij} x_j \xi_j = 0, \qquad i = 1, 2, \ldots, n;$$

$$\prod_{j:l_j>0} \xi_j{}^{l_j} = \prod_{j:l_j<0} \xi_j{}^{-l_j} \qquad \text{if } \sum_j l_j = 0;$$

$$\prod_{j:l_j>0} \xi_j{}^{l_j} = 0 \qquad \text{if } \sum_{j:l_j>0} l_j > \sum_{j:l_j<0} l_j,$$

where $l = (l_1, l_2, \ldots, l_N)$ ranges over the lattice L of integer vectors such that $l_1 a_1 + l_2 a_2 + \cdots + l_N a_N = 0$.

Then system (1.2), (1.3) is holonomic if $\dim Ch = N$. The number of independent solutions at a generic point is equal to degree of Ch along the zero section $\{(0, \xi) : \xi \in \mathbb{C}^N\}$ (see [Ka]).

11.2. Other root systems.

We can define (see Section 2) the hypergeometric system for arbitrary root system R. It is interesting to find analogues of all results in this paper for other root systems.

Let P_{R^+} be the convex hull of 0 and all positive roots $r \in R^+$. Then by Theorem 1.4 the dimension of the system at a generic point is equal to $D(R) = n! \operatorname{Vol}(P_{R^+})$, where n is the dimension of R. These numbers $D(R)$ can be viewed as a generalization of the Catalan numbers for arbitrary root system.

11.3. Discriminant and Triangulations of P_n.

We can associate with system (2.2), (2.3) the discriminant $\mathcal{D}_n(z)$. The discriminant $\mathcal{D}_n(z)$ is a polynomial of $z = (z_{ij})$, $0 \leq i < j \leq n$ such that $\mathcal{D}_n(z) = 0$ if and only if there exists $(z, \xi) \in Ch$ such that $\xi \neq 0$, where Ch is the characteristic manifold for system (2.2), (2.3).

It is an interesting problem to find an explicit expression for $\mathcal{D}_n(x)$ and describe all monomials in $\mathcal{D}_n(x)$. The Newton polytope S_n for $\mathcal{D}_n(x)$ is called *Secondary polytope*. Vertices of S_n correspond to coherent local triangulations of P_n (cf. [GKZ]).

In Section 6 we constructed two coherent local triangulations of P_n. The important problem is to find all such triangulations. Analogously, one can define discriminant $\mathcal{D}_{IJ}(z)$ associated with face strata Z_{IJ} (see Section 8). Vertices of the Newton polyhedron for $\mathcal{D}_{IJ}(z)$ correspond to coherent triangulations of P_{IJ}. (Note that all triangulations of P_{IJ} are local.) How to describe triangulations of P_{IJ}?

The special case of this problem for the pair (I, J) such as in Remark 8.3 (the hypergeometric system on the grassmannian) is connected with triangulations of the product of two simplices $\Delta^p \times \Delta^q$, $p + q = n + 1$. In this case $\mathcal{D}_{IJ}$ is the product of all minors of $(p+1) \times (q+1)$-matrix z (see [GKZ], cf. [SZ, BZ]).

References

[Ad] Adolphson Hypergeometric functions and rings generated by monomials, *Duke Math. Jour.* **73**: 2 (1994), 269–290

[BE] H. Bateman, A Erdélyi *Higher Transcendent Functions, Vol. 1*, Mc Graw-Hill Book Company, Inc., New York, 1953

[BZ] D. Bernstein, A. Zelevinsky, Combinatorics of maximal minors *Jour. of Algebraic Combinatorics* **2**(1993), 111–121

[Bo] N. Bourbaki, *Éléments de Mathématique, Groupes et Algèbres de Lie*, Ch. 6, Hermann, Paris, 1968

[GG1] I. M. Gelfand, M. I. Graev, Hypergeometric functions on flag manifolds, *Doklady Acad. Nauk SSSR*, **338**:3 (1994), 298–301

[GGR1] I. M. Gelfand, M. I. Graev, V. S. Retakh, Γ-series and general hypergeometric functions on the manifold of $k \times m$-matrices, 1990, Preprint IPM no. 64

[GGR2] I. M. Gelfand, M. I. Graev, V. S. Retakh, General hypergeometric systems of equations and series of hypergeometric type *Uspekhi Mat. Nauk SSSR*, **47**:4 (286), 1992, 3–82

[GGR3] I. M. Gelfand, M. I. Graev, V. S. Retakh, Reduction formulas for hypergeometric functions associated with the grassmannian G_{nk} and description of these functions on strata of small codimension in G_{nk}, *Russian Jour. of Math. Physics*, **1**: 1 (1993), 19–56

[GGZ] I. M. Gelfand, M. I. Graev, A. V. Zelevinsky, Holonomic systems of equations and series of hypergeometric type *Doklady Acad. Nauk SSSR*, **295**:1 (1987), 14–19

[GKZ] I. M. Gelfand, M. M. Kapranov, A. V. Zelevinsky *Discriminants, Resultants and Multidimensional Determinants*, Birkhäuser Boston, 1994

[GZK] I. M. Gelfand, A. V. Zelevinsky, M. M. Kapranov, Hypergeometric functions and toric varieties *Funct. Anal. and its Appl.*, **23**: 2 (1989), 12–26

[Ka] M. Kashivara, *System of Microdifferential Equations*, Birkhäuser, Boston, 1983

[Po] A. Postnikov, Intransitive trees, preprint, 1994

[SW] D. Stanton, D. White, *Constructive Combinatorics*, Springer-Verlag, 1986

[SZ] B. Sturmfelds, A. Zelevinsky, Maximal minors and their leading terms, *Advances in Math.*, **98**:1 (1993), 65–112

I. M. Gelfand
Dept. of Math.
Rutgers University
New Brunswick, NJ 08903 USA
igelfand@math.rutgers.edu

M. I. Graev
Dept. of Math.
Research Inst. for System Studies RAS
23 Avtozavodskaya St
Moscow 109280, Russia
Graev@systud.msk.su

and

A. Postnikov
Dept. of Math.
MIT
Cambridge, MA 02139, U.S.A.
apost@math.mit.edu

January 1995

Local invariants of mappings
of surfaces into three-space

Victor V. Goryunov *

Abstract

Following Arnold's and Viro's approach to order 1 invariants of
curves on surfaces [1, 2, 3, 20], we study invariants of mappings
of oriented surfaces into Euclidean 3-space. We show that, besides
the numbers of pinch and triple points, there is exactly one integer
invariant of such mappings that depends only on local bifurcations
of the image. We express this invariant as an integral similar to
the integral in Rokhlin's complex orientation formula for real algebraic curves. As for Arnold's J^+ invariant [1, 2, 3], this invariant
also appears in the linking number of two legendrian lifts of the
image. We discuss a generalization of this linking number to higher
dimensions.

Our study of local invariants provides new restrictions on the
numbers of different bifurcations during sphere eversions.

In [17, 18] Vassiliev introduced the notion of a finite order invariant
of knots. Finite order invariants can be defined for mappings in more
general settings so long as the discriminant has codimension 1 in the
function space. In [1] Arnold defined three invariants of order 1 for plane
curves. These invariants have a local nature: they do not distinguish
different components of the same top strata of the discriminant. Recently
Viro generalized Arnold's invariants to the case of curves on surfaces [19].
Viro's generalization is based on Rokhlin's complex orientation formula
for real algebraic curves [12, 13].

In this paper we consider mappings of oriented surfaces to Euclidean
3-space. We study special order 1 invariants which we call local. The
notion of locality is slightly more rigid than the one used by Arnold: we

---

*Partially supported by a grant from The Danish Natural Science Research Council

require that the values of an invariant get the same increment once the images of the mappings experience the same local bifurcation.

We give a complete list of integer and mod 2 local invariants of generic mappings and of generic immersions (Sections 2 and 3). For example, we show that, besides the numbers of pinch and triple points, there is exactly one more integer local invariant, which we call I_3, of generic mappings of surfaces into $\mathbf{R}^3$. Our invariants provide new restrictions on the numbers of bifurcations during sphere eversions (subsection 3.2).

In Section 4 we relate I_3 to a local invariant I_f which is very similar to the integral form of Rokhlin's complex orientation formula [14, 19, 20].

Arnold found a formula for his direct selftangency plane curve invariant J^+ using linking numbers of corresponding legendrian curves [2, 3]. In Section 5 I_f is expressed in terms of linking numbers in $ST^*\mathbf{R}^3$ of two varieties homeomorphic to the source surface. One of the varieties is a modified legendrian lift of the image of a generic mapping, the second one is the 'negative' of the first.

In Section 6, following [2, 3] and Section 5, we introduce two linking numbers for a generic immersed hypersurface in $\mathbf{R}^n$. These numbers are local invariants dual to the direct and inverse selftangencies of a hypersurface.

Section 7 contains proofs of all theorems of the previous sections.

The author is very grateful to The University of Georgia, Athens, and to Matematisk Institut, Aarhus Universitet, the institutions where the work was done, for their kind hospitality and support.

1   Generic degenerations

Locally, the image of a generic mapping of a fixed closed surface M to $\mathbf{R}^3$ is either a smooth sheet, or transversal intersection of either 2 or 3 smooth sheets, or a Whitney umbrella (the image of $(x, y) \mapsto (x, y^2, xy)$) (see Fig.1). Mappings with more complicated images form a *discriminant* hypersurface Δ in the infinite-dimensional space Ω of all C^∞ maps $M \to \mathbf{R}^3$.

The discriminant subdivides Ω into connected components. A numerical *invariant* is a way to assign numbers to each of these components.

Moving along a generic path in Ω, we watch the jumps, as we pass the discriminant, of the values of an invariant. We say that our *invariant is local* if every jump is completely determined by the diffeomorphism type of the local bifurcation of the image at the instant of crossing of the discriminant.

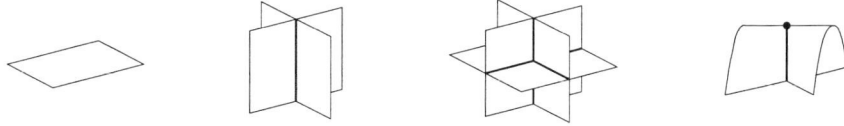

Figure 1: *Local singularities of images of generic maps of surfaces into 3-space*

Remark. Arnold's plane curve invariant St [1, 2, 3] is not local in our sense since the coorientation of the corresponding stratum of the discriminant involves global information about the image.

1.1 The top strata

There are 7 types of local events on the image that take place at generic points of Δ, i.e. occur along generic curves in Ω [11, 10, 8] (Figs. 2 and 3):

(E) elliptic tangency of two smooth sheets;

(H) hyperbolic tangency of two smooth sheets;

(T) (for 'triple'), tangency of the line of intersection of two smooth sheets to a third smooth sheet (all the three sheets are pairwise transversal);

(Q) (for 'quadruple'), four smooth sheets intersecting at the same point;

(C) (for 'cup'), a smooth sheet passes through a pinch point;

(B) birth of a bubble with two pinch points joined by an interval of selfintersection;

(K) (for 'cones' in Russian), the hyperbolic version of B.

A local coordinate form for B and K is $(x, y) \mapsto (x, y^2, y(x^2 \pm y^2 - \lambda))$ (the real parameter λ increases from the left to the right in Fig.3 being zero in the middle).

The seven bifurcations define seven top (i.e. top-dimensional) strata of the discriminant.

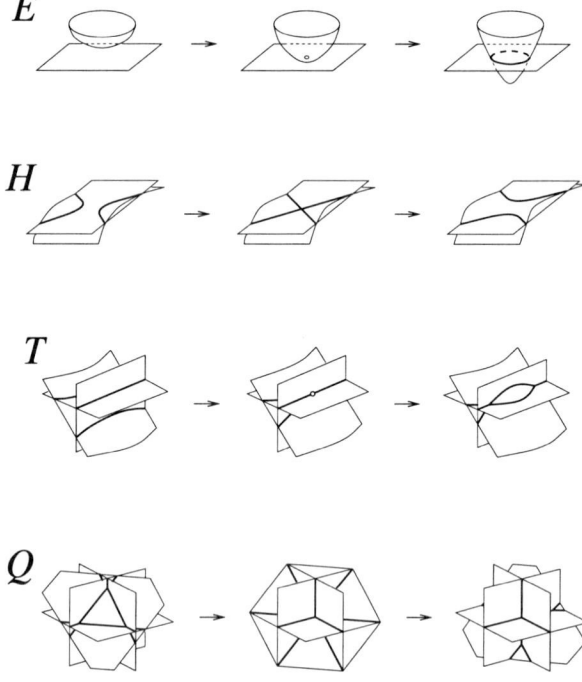

Figure 2: *Local bifurcations of images in generic 1-parameter families of mappings involving only smooth sheets*

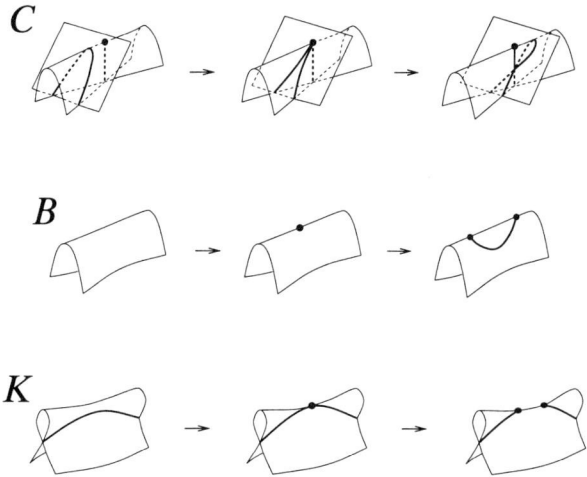

Figure 3: *Local bifurcations of images in generic 1-parameter families of mappings involving pinch points*

1.2 Coorientation of the strata

In order to assign a jump of an invariant to a stratum, we will need to coorient the stratum (unless we work with a $\mathbf{Z}_2$-invariant).

There is a natural way to coorient 5 of our 7 strata. This corresponds to moving from the left to the right in Figs. 2 and 3. Namely, we say that a *passing* through the stratum *is positive* if the passing gives birth to something new on a generic image:

- E, a new circle of selfintersection;

- T, two new triple points;

- C, one new triple point;

- B, K, two new pinch points.

In all these cases, except K, this is equivalent to the appearence of a new 2-cycle on a generic image.

In the two remaining cases, H and Q, the similar procedure fails: the initial and final pictures of Fig.2 in each of the two series are locally diffeomorphic.

1.3 Refinement for oriented surfaces

From now on we assume the source surface M oriented. This allows us to refine the stratification of the discriminant.

Let us fix orientations of M and $\mathbf{R}^3$. We obtain a canonical coorientation of the image at immersive points requiring that the frame ⟨normal of the coorientation, the image of a positive frame of M⟩ is a positive frame in $\mathbf{R}^3$.

We now add the induced coorientations of the sheets in all the 7 bifurcations of the previous subsection. This splits them in 20 subcases listed below (see Fig.4 for 15 of them). Considering a 2-sphere that consists of pieces of several smooth sheets (with their own coorientations), we call a piece *positive* if it has the outer coorientation. The outer coorientation is also called *positive*, the inner one is called *negative*.

We split the strata:

- E, T, Q in $E^j, j = 0, 1, 2$, $T^j, j = 0, 1, 2, 3$, $Q^j, j = 2, 3, 4$. Here j is the number of positive pieces of the appearing sphere (there is the vanishing tetrahedron in the Q-cases which has $4 - j$ positive pieces);

- H in H^+ and H^-: the sheets, at the tangency point, have coinciding, respectively opposite, coorientations;

- C in $C^{\alpha,\beta}, \alpha, \beta = +, -$: for the appearing cup-shaped 2-sphere, α and β are the signs of the coorientations of the lateral (having the pinch point) and bottom pieces respectively;

- B in B^+ and B^-: the superscript is the coorientation of the appearing sphere;

- K in K^+ and K^-: the index is the coorientation of the local 'tubular' part before the bifurcation.

We continue with the coorientation of the discriminant.

Q. We coorient the Q^4 and Q^3 strata in the direction of increase of the number of positive faces of the local tetrahedron (Fig.5). But there is still no local (in the sense introduced above) way to distinguish between the two sides of the Q^2 stratum.

H. To coorient H^- we consider the two points of M at which we get tangency for the degenerate mapping. We say that the crossing of the stratum is positive if the relative motion of the images of the two points occurs opposite to the coorientations of the sheets (Fig.5).

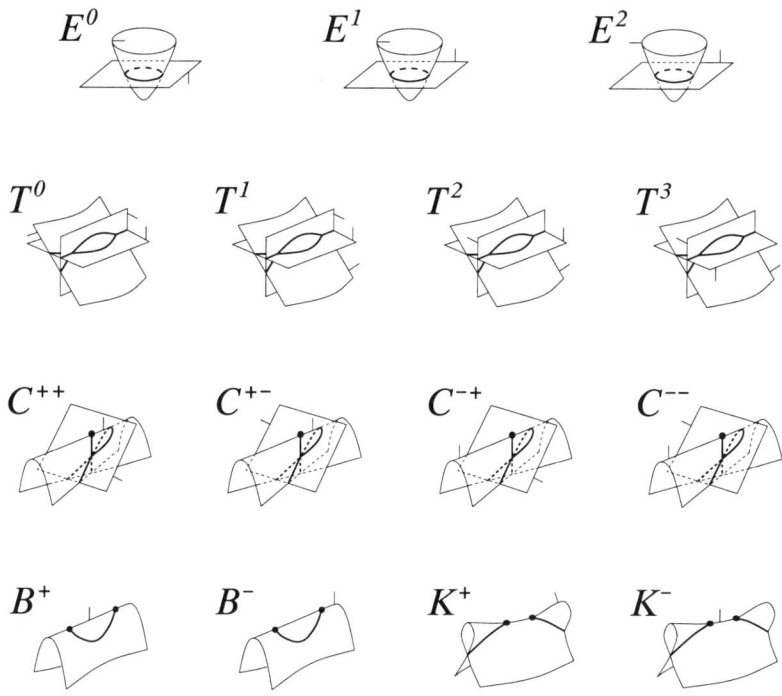

Figure 4: *Local images of oriented surfaces to the positive sides of the top strata of the discriminant*

Once again, there is no way to coorient H^+ by only local means.

In what follows the top strata of the discriminant are always taken with the coorientations introduced in this section. Q^2 and H^+ are not cooriented.

2 Lists of local invariants

2.1 Local invariants in terms of jumps along the top strata

Given an invariant I of generic mappings, we can enlarge its domain $\Omega \backslash \Delta$ to include generic points f_s of a cooriented top stratum $S \subset \Delta \subset \Omega$ [17, 18]. For this we take two points, f_+ and f_-, close to f_s in Ω, to the positive and negative sides of S respectively. We set $I(f_s) =$

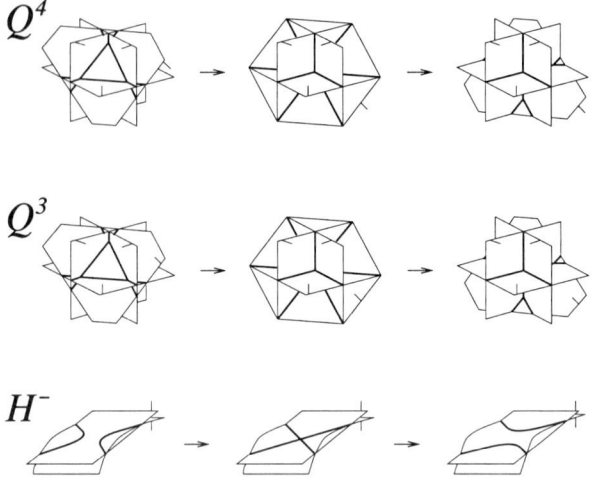

Figure 5: *Positive directions of bifurcations Q^4, Q^3 and H^-*

$I(f_+) - I(f_-)$. If I is a local invariant, the value obtained does not depend on a particular choice of generic $f_s \in S$.

Definition 2.1 $s(I) = I(f_s)$ is the *jump* of the local invariant I along the stratum S.

To assign a jump to a local $\mathbf{Z}_2$-invariant we do not need any coorientation of S.

On the other hand, a local invariant I on $\Omega \setminus \Delta$ is defined, up to an additive constant, by its jumps along the 20 strata of the discriminant. To calculate the value $I(f)$ in this situation, we need to:

(1) choose the value of I on a distinguished generic mapping f_0;

(2) join f_0 and f by a generic path γ in Ω;

(3) calculate the indices of intersection $\gamma \cap S$ of γ with each of the 20 strata S;

(4) set $I(f) = \sum_S (\gamma \cap S)s(I) + I(f_0)$.

Of course, this must be independent of the choice of γ. Since Ω is contractible, the independence is equivalent to vanishing of the increment of I (counted in the above way) along any small loop in Ω around the set of nongeneric points of Δ. This set has codimension 2 in Ω.

To shorten the notation, we will write a local invariant as a linear combination of the strata with the coefficients equal to the jumps. For the moment the invariants are considered up to the additive constants.

Checking of the independence condition (which is a rather routine study of generic 2-parameter families of mappings) provides:

Theorem 2.2 *The space of integer local invariants of smooth mappings of an oriented surface into* $\mathbf{R}^3$ *is 3-dimensional. The following are basic invariants:*

(1) $I_t = 2T + C$;

(2) $I_p = B + K$;

(3) $I_3 = E^2 - E^0 + H^- + T + C^{++} + C^{+-} + B^+ + K^+$.

Here T, C, B, K *are the sums of all the corresponding 4 or 2 substrata.*

Of course, the best choice of the additive constants for the first two invariants is the vanishing on an embedding of the surface. Then I_t is the number of triple points and I_p is the number of pairs of pinch points of the image of a mapping. We fix this normalizations of these two invariants for the rest of the paper. We will make comments on the constant for I_3 in subsection 4.4.

Definition 2.3 Tangency of two smooth sheets of the image that have coinciding (resp. opposite) coorientations is called *direct* (resp. *inverse*) *selftangency*.

Difference of values of I_3 (more precisely, of $2I_3 - I_t$) on two mappings is a modification of the number of opposite selftangencies in a generic homotopy between these mappings.

Theorem 2.4 *The space of* mod 2 *local invariants of smooth mappings of an oriented surface into* $\mathbf{R}^3$ *is 4-dimensional. Basic invariants are the ones of Theorem 2.2 and*

(4) $I_4 = E^1 + H^+ + C^{+-} + C^{-+}$.

In a way similar to I_3, I_4 measures a modified mod 2 number of direct selftangencies in a generic homotopy between two mappings.

The proof of both theorems is given in subsection 7.1.

2.2 mod 2 winding numbers

An oriented circle immersed into a plane has a winding number. By one of many equivalent definitions, this is the number of rotations of the tangent vector when its point of application runs along the circle once. A circle immersed into a 2-sphere has a well-defined mod 2 winding number: to obtain a plane, we puncture the sphere at any point not on the circle and take the parity of the plane winding number. This parity does not depend on the choice of the puncture and, for a generic curve, is opposite to the parity of the number of its double points.

Now take a generic mapping of a surface into $\mathbf{R}^3$. Consider the inverse image $\mathcal{D}$ of the set of singular points of the image. This is a collection of circles immersed into the source surface.

Proposition 2.5 *For generic mappings of a sphere, the mod 2 invariant $I_t + I_p$ is the parity $w_2(\mathcal{D})$ of the total winding number of $\mathcal{D}$.*

Proof. $w_2(\mathcal{D})$ is the mod 2 sum of the number of connected components and the number of double points of $\mathcal{D}$. The parity of the number of double points of $\mathcal{D}$ is the same as the parity of the number of triple points of the mapping. The latter changes only across the stratum C. The number of connected components mod 2 of $\mathcal{D}$ changes only across B and K. Since both $w_2(\mathcal{D})$ and $I_t + I_p$ vanish on embeddings, the claim follows.

3 Immersions

3.1 Local invariants of immersions

Let $\Omega_i \subset \Omega$ be the space of C^∞ immersions of an oriented surface to $\mathbf{R}^3$. Restriction of local invariants from Ω to Ω_i defines local invariants of immersions. Almost always this exhausts all local invariants of immersions (as before, the invariants are considered modulo additive constants):

Theorem 3.1 *The space of integer local invariants of immersions of an oriented surface to $\mathbf{R}^3$ is two-dimensional. Basic invariants are:*

$$I_{t/2} = T \qquad and \qquad I_3 = E^2 - E^0 + H^-.$$

With the normalization $I_{t/2}(embedding) = 0$, $I_{t/2}(f)$ is the number of pairs of triple points of the image of f.

There is something extra over $\mathbf{Z}_2$:

Theorem 3.2 *The space of mod 2 local invariants of immersions of a 2-sphere to 3-space is four-dimensional. In addition to the basic invariants of Theorem 3.1 there are two more:*

$$I_4 = E^1 + H^+ \qquad and \qquad I_q = Q.$$

Here I_4 is the restriction of the invariant of Theorem 2.4 and will appear for any oriented surface, not only for the sphere. The reason why we formulate the theorem for spheres only is as follows.

Considering bifurcations in all possible generic 2-parameter families of immersions of a surface (as in proofs of the first three theorems) it is easy to show that a small loop in Ω_i around the set of nongeneric points of the discriminant intersects the stratum Q an even number of times. But Ω_i has nontrivial fundamental group. A well-defined invariant should have zero increments along all elements of $\pi_1(\Omega_i)$. I do not know generators of this fundamental group in the case of an arbitrary surface. But for the 2-sphere, π_1 of the space of *marked* immersions (i.e. the ones that send a distinguished ordered 2-frame of the tangent space to S^2 at a distinguished point to a distinguished ordered 2-frame at a distinguished point of $\mathbf{R}^3$; for our needs it is enough to consider only marked immersions) is freely generated by a loop [15] described in [9]. This loop intersects Q twice.

It would be very interesting to work out whether I_q survives the $\pi_1(\Omega_i)$-test for surfaces of positive genus.

Of course, there are no such difficulties with π_1 for the restrictions of the local invariants from Ω, which is contractible.

Proofs of Theorems 3.1 and 3.2 are given in subsection 7.2.

3.2 Sphere eversions

Definition 3.3 An *eversion* of a sphere is a turning of a positive sphere inside out in $\mathbf{R}^3$ by a generic regular (i.e. with no pinch points) homotopy.

Eversions of surfaces are possible due to the classical result of Smale [15, 16].

Let $N(S)$ be the number of crossings of the stratum S (the signs of the crossings of a cooriented stratum are respected) during a sphere eversion.

Corollary 3.4 *The numbers of bifurcations during a generic sphere eversion are subject to the following relations:*

$$N(T) = 0, \qquad N(E^2) - N(E^0) + N(H^-) = -1, \qquad N(E^1) + N(H^+) = 0 \bmod 2$$

Proof. Consider a generic path in the space of mappings of S^2 into $\mathbf{R}^3$ along which a positive sphere becomes a negative one via the birth of a negative bubble and death of a positive one (Fig.6). The right-hand sides of the first three relations are the increments of the local invariants of Theorems 2.2 and 2.4 along this path.

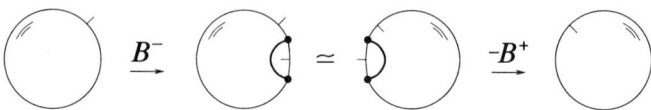

Figure 6: *Nonregular homotopy turning a sphere inside out*

Another way to prove the Corollary, not using any particular path, follows from the description of an integral invariant of the next section.

Remarks. a) There is no suprise in the restriction $N(T) = 0$: having generated triple points, we must kill them to complete an eversion.

b) One more restriction, $N(Q) = 1 \bmod 2$, was proved by Max and Banchoff by explicit presentation of an eversion with one quadruple point [9]. This implies, by the way, that, during an eversion, there are at least two positive crossings of T to create the vertices of the tetrahedron vanishing on Q. The basic loop in Ω_i, mentioned in the comments after Theorem 3.2, is a composition of the eversion of [9] with its 'mirror image'.

4   Integral invariant

We introduce now an invariant of generic mappings from oriented surfaces to 3-space that is very similar to the integral in Rokhlin's complex orientation formula for real algebraic plane curves [12, 13, 14, 19]. This invariant turns out to be local, very closely related to I_3.

4.1   Degrees

Let $\mathrm{Im}f$ be the image of a generic mapping of a surface $f : M \to \mathbf{R}^3$. Take a point u in $\mathbf{R}^3$ not on the image. Consider a small 2-sphere, with the outer coorientation, centered at u. The radial contraction of the image onto the sphere defines a through mapping from M to the sphere. We denote by $deg(u)$ the degree of this mapping.

Imf subdivides the ambient 3-space into a finite number of connected components D. $deg(u)$ is constant on each of them. We denote the corresponding value by $deg(D)$.

We define an integral of function deg against Euler charactestics χ setting [19, 20]:

$$\int_{\mathbf{R}^3\backslash\mathrm{Im}f} deg(u)d\chi(u) = \sum_D deg(D)\chi(D),$$

where D runs through all the connected components of $\mathbf{R}^3 \backslash \mathrm{Im}f$.

There are 8 (resp. 3) local connected components of the complement to the image around a triple point t (resp. a pinch point p). We set the degrees $deg(t)$ and $deg(p)$ to be the arithmetical means of the corresponding 8 or 3 degrees. $deg(t)$ is a semi-integer that also coincides with the arithmetical mean of degrees of any of four pairs of the 'opposite' local components. $deg(p)$ is the degree of the 'largest' of the three components around p (Fig.7).

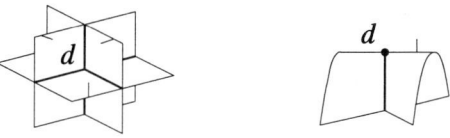

Figure 7: *A triple point of degree $d - \frac{3}{2}$ and a pinch point of degree d. In both cases d is the degree of the distinguished component of $\mathbf{R}^3 \backslash Image$*

4.2   The invariant

We set

$$I_f(f) = \int_{\mathbf{R}^3\backslash\mathrm{Im}f} deg(u)d\chi(u) - \sum_t deg(t) - \frac{1}{2}\sum_p deg(p),$$

where t and p run through all the triple and pinch points of the image.

Example 4.1 The value of I_f on a standard immersion of a surface of genus g (Fig.8), with the outer coorientation, is $1 - g$.

Theorem 4.2 I_f *is a local invariant. Up to an additive constant,*

$$I_f = 2I_3 - I_t - I_p.$$

Figure 8: *Standard immersion of genus 3 surface*

Thus all the three local integer invariants of Theorem 2.2 may be defined only in terms of geometry of the image of the mapping f. There is no need to choose any homotopy in Ω between the distinguished mapping and f.

The assertion of Theorem 4.2 is parallel to the assertion of Viro's Corollary 3.2.B in [20] for the invariant J^- of plane curves.

The proof of Theorem 4.2 consists in the comparison of the jumps of I_f and the above combination of the local invariants. It will be given in subsection 7.3.

Remark. The definition of I_f is similar to the formula for calculating the winding number w of a generic plane curve [20]:

$$w = \sum deg(D) - \sum deg(d),$$

with D and d running through all the connected components of the complement to the curve and all the double points, $deg(D)$ defined in the obvious way, and $deg(d)$ being the arithmetical mean of the four numbers.

4.3   Smoothings of images

The intergal invariant I_f has another description, in terms of a smoothed image.

Lemma 4.3 *There is a canonical way to smooth singularities of the image of a generic mapping of an oriented surface to 3-space. This smoothing is given by local pictures of Fig.9.*

Let $\widetilde{\mathrm{Im}}f$ be the smoothed image of a generic map f. As in subsection 4.1, for each point u of $\mathbf{R}^3 \setminus \widetilde{\mathrm{Im}}f$, we define $\widetilde{deg}(u)$.

Theorem 4.4

$$I_f(f) = \int_{\mathbf{R}^3 \setminus \widetilde{\mathrm{Im}}f} \widetilde{deg}(u)\chi(u) - \sum_p deg(p).$$

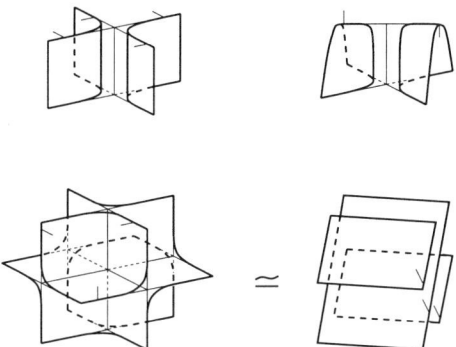

Figure 9: *Local smoothings of the image of a generic map*

Here p runs through all the pinch points of $\mathrm{Im}f$.

The assertion of Theorem 4.4 is parallel to the assertion of Viro's Theorem 3.1.A in [20] for the invariant J^- of plane curves. The proof of Theorem 4.4 is given in subsection 7.4.

4.4 Connected summations

Let us choose a system of constants $a(g)$ for the system of local invariants $I_f + a(g)$ on the spaces of mappings of surfaces of genus g to $\mathbf{R}^3$, in order to get a good invariant on the union of all these spaces. As in [1], for the notion of "goodness" let us take additivity with respect to the connected summation. It turns out that we have to distinguish two types of such summation. The sets of the constants $a(g)$ for the two types are distinct.

A *connected sum* of two surfaces $\mathcal{M}_1$ and $\mathcal{M}_2$, separated by a plane, in $\mathbf{R}^3$ is defined as shown in Fig.10: we cut a small disc out from the 'exterior' part of each of the surfaces and join the surfaces by a thin cylinder circle$\times$interval embedded in $\mathbf{R}^3 \setminus (\mathcal{M}_1 \cup \mathcal{M}_2)$. If we want to save the orientations of the summands, there is an obvious restriction on $\mathcal{M}_1$ and $\mathcal{M}_2$. Coherently with the coorientations of the summands, the connecting cylinder must have either the outer or inner coorientation. We call these two options the *positive* and *negative summations* respectively.

Theorem 4.5 *The invariant $I_f + \alpha g + g - 1$, where α is an arbitrary number, is additive with respect to the positive connected summation. The invariant $I_f + \beta g - g + 1$, where β is an arbitrary number, is additive with respect to the negative connected summation.*

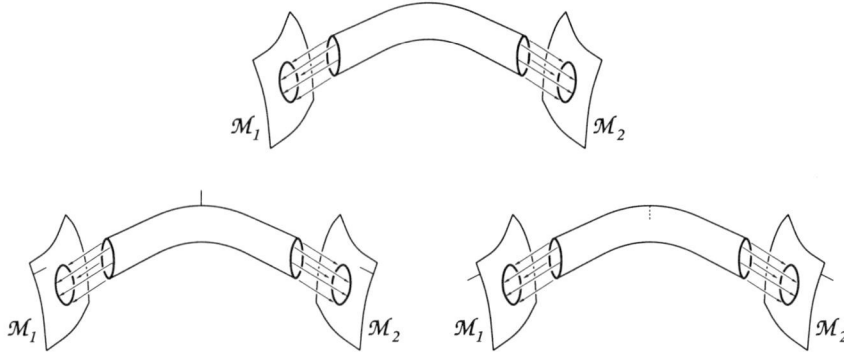

Figure 10: *Connected summation of surfaces and its two types: positive and negative*

The Theorem easily follows from the summations of the standard surfaces (Fig.8). α and β are values of the invariants on the standard tori, with the outer and inner coorientations respectively.

Remark. The invariants I_t and I_p normalized to vanish on embeddings are obviously additive under both connected summations. For I_3 to be positively (resp. negatively) additive we set its value on the standard surface with the outer (resp. inner) coorientation to be $(\alpha g + g - 1)/2$ (resp. $(\beta g - g + 1)/2$).

The best behaviour of I_f is under *twisted* summation, when we use a tube with two pinch points as a connecting bridge (Fig.11). I_f is additive with respect to such a summation. But now we need to shift I_p: to make it twisted-additive we set $I'_p = I_p + 2$.

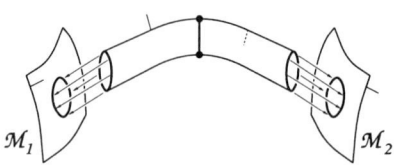

Figure 11: *Twisted connected summation*

5   Linking invariant

In [2, 3] Arnold gave an interpretation of his plane curve invariant J^+ in terms of linking numbers of corresponding legendrian curves in different covers of $ST^*\mathbf{R}^2$. Below we introduce a similar presentation for I_f. We need there some extra work since a generic mapping of a surface into 3-space has critical points, while a generic plane curve does not.

5.1   Legendrian lift of a generic image

Let f be a generic mapping of an oriented surface M into oriented $\mathbf{R}^3$. We are going to associate to f a subvariety L_f^+ of the contact variety $ST^*\mathbf{R}^3$ of cooriented tangent 2-planes of $\mathbf{R}^3$. L_f^+ will be homeomorphic to M, and the image of L_f^+ under the projection $\pi : ST^*\mathbf{R}^3 \to \mathbf{R}^3$ will be the image $\mathrm{Im} f$ of f.

The part of L_f^+ over nonpinch points of $\mathrm{Im} f$, is the usual legendrian lift against π: it associates to a point a the cooriented (as in subsection 1.3) 2-plane tangent to a sheet of $\mathrm{Im} f$ passing through a. This divorces transversal selfintersections of $\mathrm{Im} f$.

Over a pinch point of $\mathrm{Im} f$ the legendrian lift gets a hole. Namely, easy computation shows:

Lemma 5.1 *Consider a Whitney umbrella W cooriented on the complement to its pinch point p. Consider the closure of the legendrian lift of $W \setminus p$ to $ST^*\mathbf{R}^3$. Then the boundary of the closure is an equator of the 2-sphere S_p^2. The equator consists of all covectors vanishing on the image of the differential of a parametrization of the umbrella by plane.*

Here S_p^2 is the fibre of the bundle π over a point p.

To patch the hole in a canonical way, we need to recall the orientation of the fibre induced by an orientation of $\mathbf{R}^3$ [4]. The induced orientation of the fibre S^2 comes from the orientation of the boundary of a small 3-disc in $\mathbf{R}^3$. The set of outer normals of the boundary becomes a fibre when the radius of the ball tends to zero (vectors are identified with covectors by means of a metric on $\mathbf{R}^3$).

Let $\nu : \mathbf{R}^2 \to W$ be a parametrization of the umbrella by an oriented plane. We patch the hole on the legendrian lift of the punctured umbrella with a half of the fibre S_p^2. We chose the half so that the orientation induced from the plane by the lift extends to the patched surface as the canonical orientation of the fibre.

Lemma 5.2 *Choose a vector v in the image of the differential of ν at the pinch point so that v is directed to the half of the umbrella with the outer coorientation (Fig.12). Then the patching hemi-sphere is the set of covectors nonpositive on v.*

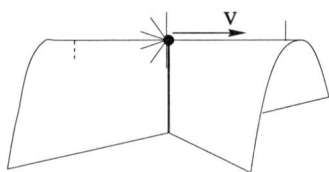

Figure 12: *Covectors added at a pinch point to patch the legendrian lift*

Doing the patching for all pinch points of the image of a generic mapping f of M, we get the subvariety $L_f^+ \subset ST^*\mathbf{R}^3$.

5.2   Intersection with the 3-chain

There is another way to lift $\mathrm{Im} f$ to $ST^*\mathbf{R}^3$: we send a point to the *negative* of the coorientation normal (cf. [2, 3]). Let us lift a punctured neighbourhood of a pinch point is this negative way. The closure of the lift has the equator of Lemma 5.1 as the boundary. The orientation of the boundary equator is easily seen to be the same as for the lift in the previous subsection. Thus, the patches compatible with the orientations are also the same. We denote the new lift of the image (with the old patches used) by L_f^- (it is homeomorphic to M as well).

Now we define the linking number of L_f^+ and L_f^-.

Let us choose a direction in $\mathbf{R}^3$ and shift L_f^- by all possible vectors of this direction. We get in $ST^*\mathbf{R}^3$ a 3-chain V_f^- with the boundary L_f^-. We orient it as the image of $\mathbf{R}_+ \times M$, where $\mathbf{R}_+$ is the nonnegative axis oriented from 0 to ∞.

In order to bring L_f^+ in more or less general position with respect to V_f^-, we slightly shift every point of L_f^+ in the direction of the corresponding normal vector. Namely, a point $(a, n) \in ST^*\mathbf{R}^3$ moves to $(a + \varepsilon n, n)$, for small constant $\varepsilon > 0$ (a is a point in 3-space, n is a (co)vector applied at a). Let $L_{f,\varepsilon}^+$ be the shifted variety.

We take the canonical orientation of $ST^*\mathbf{R}^3$ by the frame $\langle$positive frame of the base, positive frame of the fibre$\rangle$ [4].

Definition 5.3 *The linking invariant* $\ell^-(f)$ *is the intersection number* of $L^+_{f,\varepsilon}$ *and* V^-_f.

Since the cohomology of $\mathbf{R}^3$ is trivial, the definition is independent of the chosen direction.

Example 5.4 The value of ℓ^- on the standard genus g surface (Fig.8) with the outer coorientation is $g - 1$. For the inner coorientation it is $1 - g$.

It is easily seen that in regular homotopies $\ell^-(f)$ can change only at inverse selftangencies. In general the following holds:

Theorem 5.5 $\qquad \ell^- = -I_f + 2I_p.$

A wrong choice of the patch for the lift of an umbrella would imply appearance of an extra term, like the sum of degrees of pinch points, in the right hand side of this formula. Such a sum is not a local invariant.

A sketch of the proof of Theorem 5.5 is given in subsection 7.5.

6 Higher dimensions

The constructions of linking invariants in [2, 3] and the previous section suggest introduction of the following two linking numbers for generic immersed oriented closed hypersurface $\mathcal{M}$ in Euclidean space of arbitrary dimension n. These numbers count direct and inverse selftangencies in generic regular homotopies.

As above, just changing 3 for n, $\mathcal{M}$ defines two legendrian subvarieties, L^+ and L^-, in $ST^*\mathbf{R}^n$. There are no patches now. Choosing a direction in $\mathbf{R}^n$, we get n-chains V^+ and V^-. As in subsection 5.2 we perturb and get L^+_ε.

Definition 6.1 $\ell^+(\mathcal{M})$ and $\ell^-(\mathcal{M})$ are indices of intersection of L^+_ε with V^+ and V^- respectively.

There is a discriminant in the space of C^∞ immersions of a fixed hypersurface into $\mathbf{R}^n$. The top strata correspond to tangency of a smooth sheet to the transversal intersection of $r \leq n$ other smooth sheets (any r of these $r + 1$ sheets are in general position; for $r = n$ 'tangency' means 'passing through') [11].

Lemma 6.2 *The invariant* ℓ^+ *(resp ℓ^-) can change only across the strata of direct (resp. inverse) selftangency of two smooth sheets.*

The claim is obvious from the definition.

We can locally represent the selftangency bifurcation as moving a hyperplane $x = t$, $t \in \mathbf{R}$, through the graph $x = -y_1^2 - \ldots - y_u^2 + z_1^2 + \ldots + z_v^2$, $u + v = n - 1$. Assume the frame $(x, y_1, \ldots, y_u, z_1, \ldots, z_v)$ gives the orientation of the ambient space and x is the direction to produce the n-chains V^+ and V^-. Changing the parameter t from negative to positive, with coorientations of the sheets as in Fig.13, we see:

(1) in inverse selftangency bifurcation ℓ^- has a jump ± 2;

(2) in direct selftangency bifurcation in even-dimensional ambient space ℓ^+ has a jump ± 2;

(3) in direct selftangency bifurcation in odd-dimensional ambient space ℓ^+ does not change.

dim y = u, dim z = v

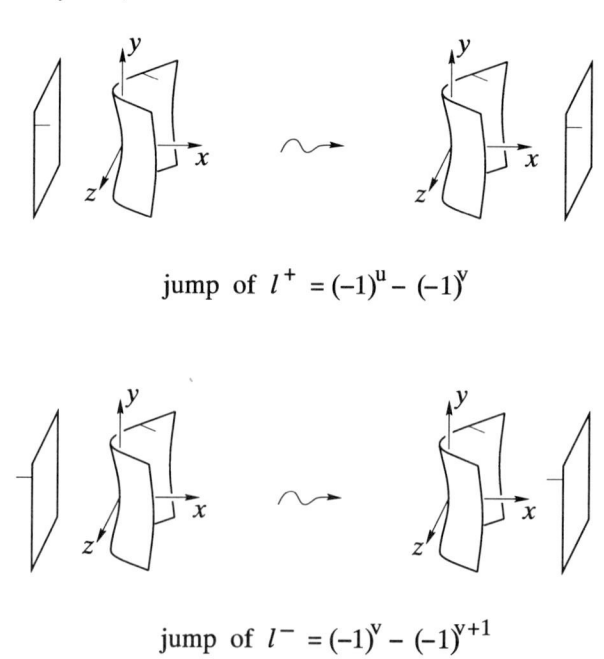

jump of $l^+ = (-1)^u - (-1)^v$

jump of $l^- = (-1)^v - (-1)^{v+1}$

Figure 13: *Jumps of local values of the linking selftangency invariants*

This shows why there are two integer selftangency invariants, Arnold's $J^\pm$ [1, 2, 3], for plane curves, and only one, ℓ^-, for surfaces in 3-space. The fact of (3) reflects existence of locally noncoorientable direct selftangency substratum, like H^+, in higher dimensions.

As in the case of plane curve invariants, we can coorient the inverse selftangency stratum in the direction of decrease of ℓ^- and the direct selftangency stratum, for $\mathbf{R}^{even}$, in the direction of increase of ℓ^+. For plane curves this gives us $J^\pm$ up to additive constants [2, 3]. For ambient $\mathbf{R}^3$, the first choice gives exactly $E^2 + H^- - E^0$.

Example 6.3 On the standard $(n-1)$-sphere in $\mathbf{R}^n$, with the outer coorientation, $\ell^+ = (-1)^{n-1}$ and $\ell^- = -1$. On the sphere with the inner coorientation $\ell^+ = \ell^- = (-1)^{n-1}$.

Since the 6-sphere in $\mathbf{R}^7$ is evertible [16], we get

Corollary 6.4 *During a generic eversion of a 6-sphere, the number of opposite selftangencies is odd.*

Counting the signs of crossings of the discriminant, this number is -1.

7   Proofs

7.1   Proofs of Theorems 2.2 and 2.4 on enumeration
of local invariants of generic maps

Consider a function ϕ on the set of the top strata S of the discriminant. Let us *integrate* ϕ along a generic loop γ in Ω. For this, taking zero as the initial sum, we follow γ and, every time when it intersects a stratum S, add either $\phi(S)$ (if the intersection is done in the direction of the coorientation of the stratum) or $-\phi(S)$ (if the intersection is done in the opposite direction).

$\{\phi(S)\}$ is the set of jumps of a local invariant along the strata if and only if the integral of ϕ along any generic loop is zero (then the invariant is uniquely defined by $\{\phi(S)\}$ up to an additive constant). This gives a system of linear equations on the values $\phi(S)$.

Since Ω is contractible, the equations should only express vanishing of the integrals along small loops in Ω around the set of nongeneric points of the discriminant (the latter has codimension 2 in Ω). Such a loop is realized as a loop around the origin in the parameter space of a germ of a generic 2-parameter family of mappings from our oriented surface to $\mathbf{R}^3$. So we will look through all such deformation and write out the linear system on the jumps.

The jump along a stratum S will be denoted by s. The coorientations of the strata are as in Sect.1. Unknowns h^+ and q^2, the jumps along the noncoorientable strata, may be nonzero only for $\mathbf{Z}_2$-invariants. The complete list of the 2-parameter families, particular normal forms and bifurcation diagrams are provided by general machinery of the Theory of Singularities [10, 5, 6, 7], and in what follows we do not consider any details of the classification procedure.

7.1.1   Uni-germs

The following is the complete list of generic 2-parameter families of map-germs $(\mathbf{R}^2, 0) \to \mathbf{R}^3$ [10]:

$$
\begin{array}{ll}
A_2 : & x, y^2, y(y^2 + x^3 + \lambda x + \mu) \\
B_2^\pm : & x, y^2, y(x^2 \pm y^4 + \lambda y^2 + \mu) \\
H_2 : & x, xy + y^2(y^3 + \lambda y + \mu), y^3 + \lambda y
\end{array}
$$

The bifurcation diagrams for the two different coorientations of the sheet, along with generic members of the families, are shown in Fig.14. Walking counterclockwise around the origins in the parameter planes we read the equations:

$$
\begin{array}{llll}
A_2 : & (1) \quad b^+ - k^+ = 0, & (2) \quad b^- - k^- = 0 \\
B_2^- : & (3) \quad k^+ - h^- - b^- = 0, & (4) \quad k^- + h^- - b^+ = 0 \\
B_2^+ : & (5) \quad e^2 + k^- - b^+ = 0, & (6) \quad e^0 + k^+ - b^- = 0 \\
H_2 : & (7) \quad b^+ + c^{--} - c^{++} - b^- = 0
\end{array}
$$

7.1.2   Bi-germs

Degenerate tangency of two smooth sheets

$$
z = 0 \quad \text{and} \quad z = x^2 + y^3 + \lambda y + \mu
$$

provides (Fig.15):

$$
(8) \quad e^2 = h^-, \qquad (9) \quad e^0 = -h^-, \qquad (10) \quad e^1 = h^+.
$$

Interaction of a smooth sheet with an umbrella. At the pinch point, the smooth sheet may be nontransversal either to the tangent line to the handle or to the image of the differential. The latter case has two subcases. All this is shown in Fig.15. The families are families of parallel

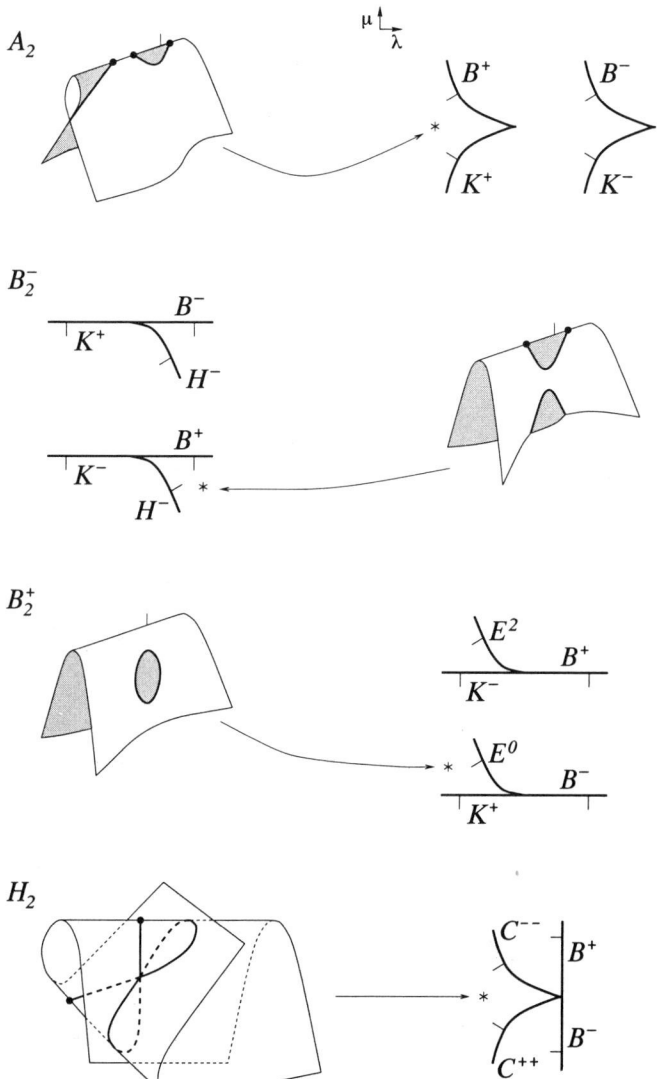

Figure 14: *Bifurcations in generic 2-parameter families of uni-germs*

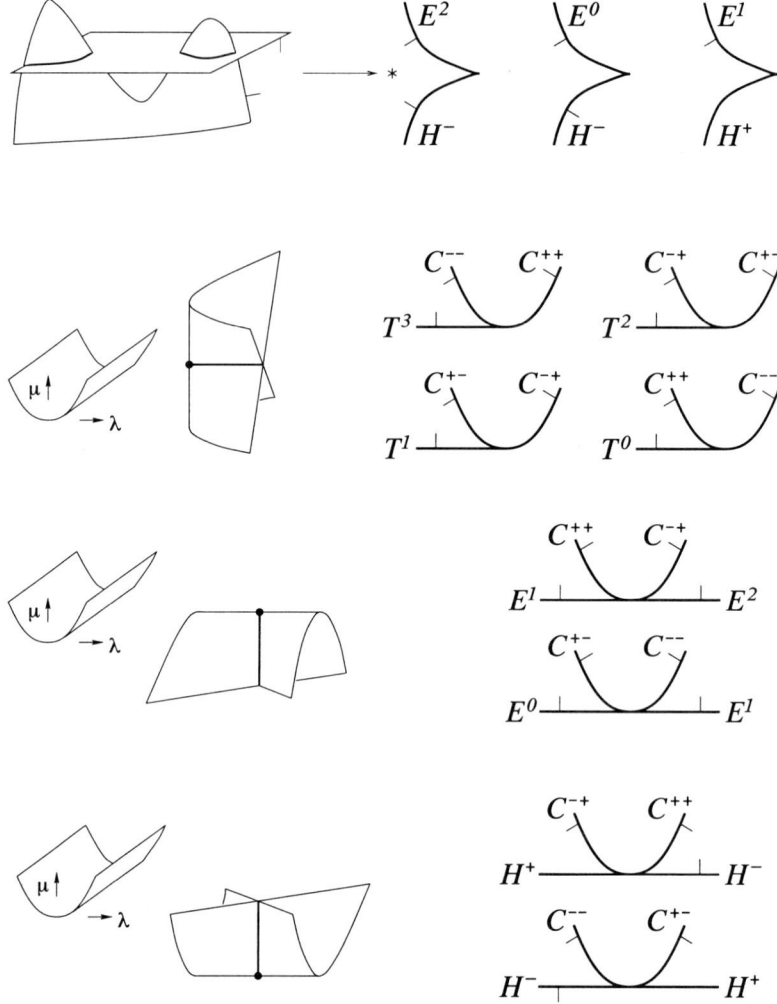

Figure 15: *Bifurcations in generic 2-parameter families of bi-germs*

translations of the umbrella. (λ, μ) are the coordinates of the pinch point. We get 8 equations:

(11) $\quad c^{++} + c^{--} = t^3$

(12) $\quad c^{+-} + c^{-+} = t^2$

(13) $\quad c^{-+} + c^{+-} = t^1$

(14) $\quad c^{--} + c^{++} = t^0$

(15) $\quad c^{++} + e^1 = c^{-+} + e^2$

(16) $\quad c^{+-} + e^0 = c^{--} + e^1$

(17) $\quad c^{++} = c^{-+} + h^+ + h^-$

(18) $\quad c^{+-} = c^{--} + h^+ + h^-$.

7.1.3 Some 3-germs

Lemma 7.1 $\quad$ (20) $\quad t^0 = t^1 = t^2 = t^3$.

Proof. Consider the interaction of three smooth sheets in Fig.16. λ and μ are the coordinates of the vertex of the parabolic sheet. From the bifurcation diagram we get $t^3 = t^2$. Changing the coorientations of the sheets we obtain $t^2 = t^1 = t^0$.

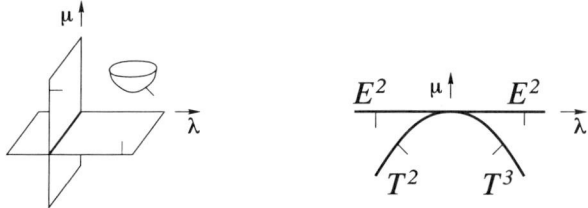

Figure 16: *Bifurcation showing that $t^3 = t^2$*

Lemma 7.2 $\quad$ (19) $\quad q = 0$.

Proof. Consider the interaction of an umbrella and two transversal smooth sheets (Fig.17, parameters λ and μ are the coordinates of the pinch point). Walking counterclockwise around the origin in the parameter plane we read the equation:

$$c^{-+} + t^2 + c^{++} + q^3 - c^{-+} - t^3 - c^{++} = 0.$$

Thus, by the previous Lemma, $q^3 = 0$. Changing the coorientations of the smooth sheets, we get $q^4 = q^2 = 0$.

Remark. We have distinguished two directions at the pinch point: the tangent line to the handle and the image of the differential. In the

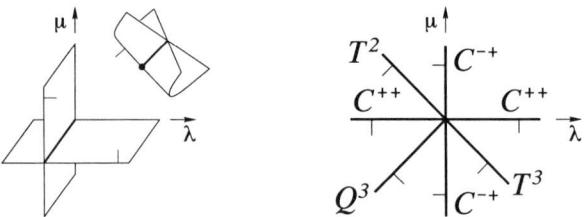

Figure 17: *Interaction of an umbrella and two transversal smooth sheets*

proof of Lemma 7.2 we could take a different umbrella: with these two directions, for $\lambda = \mu = 0$, not separated by the smooth sheets. Similarly, in the proof of Lemma 7.1 we could take a hyperbolic sheet instead of the elliptic one. These changes would have no influence on the final results.

There are some more generic 2-parameter families of mappings involving only smooth sheets: quintuple point, a sheet passing through three coordinate planes with tangency to one of the coordinate lines, second order (degenerate) tangency of a smooth sheet and the line of transversal intersection of two others. But, it follows from the two lemmas above that these families provide no further independent equations.

Solving the system (1–20) over $\mathbf{Z}$ and over $\mathbf{Z}_2$ we get the systems of jumps claimed in Theorems 2.2 and 2.4 respectively.

7.2   Proofs of Theorems 3.1 and 3.2 on enumeration
of local invariants of immersions

Due to the comments made after the formulation of Theorem 3.2 in Sect.3, we have to consider only local events in generic 2-parameter families of immersions. This leaves us with the twelve 'smooth' unknowns and the above equations (8–10) and (20). But now we will obtain extra information from some bifurcations that we did not need to study attentively at the end of the previous subsection.

Lemma 7.3   (21) $q^2 = q^3 = q^4.$

Proof. Fig.18 shows bifurcations of a sheet $u + w = (v + \lambda)^2 + \mu$ with respect to the coordinate planes in the (u, v, w)-space (for $\lambda = \mu = 0$, the sheet passes through the origin being tangent to the v-axis). This implies $q^3 = q^4$. Changing the coorientation of, say, the curved sheet we get the other equality.

Figure 18: *Bifurcation showing that $q^3 = q^4$*

Since Q^2 cannot be cooriented by local means, an integer invariant has no jumps along the whole of the Q-stratum. Jumps of a $\mathbf{Z}_2$-invariant must be the same along all three substrata of Q.

The bifurcation diagram of a quintuple point (five plane passing through one point) consists of 5 Q-lines. This introduces no new equations.

The bifurcation diagram of a smooth sheet $u = v^3 + \lambda v + \mu$ with respect to two sheets $u^2 = w^2$ (for $\lambda = \mu = 0$, the first sheet has second order tangency with the line of intersection of the two others) is a cusp of T. Following the coorientations of the sheets we get either $t^0 = t^3$ or $t^1 = t^2$. Once again, nothing is new.

Now Theorems 3.1 and 3.2 follow as solutions to the system (8–10), (20), (21).

7.3 Proof of Theorem 4.2 about the integral invariant

The equality $I_f = 2I_3 - I_t - I_p$ up to an additive constant is equivalent to the equality of jumps of both sides along all the strata. So, let us compare the jumps. We use the obvious

Lemma 7.4 *Crossing a sheet against its coorientation increases by 1 the degree of a connected component of the complement of the image.*

B^+, Fig.19. Here d is the degree of one of the components of the complement to the image. It determines the degrees of the other components. We calculate the difference of values of I_f on two mappings whose images differ only inside a small ball, where we substitute the right picture of Fig.19 for the left one. Due to the additivity of Euler characteristics, the calculations can be done locally [19, 20]:

$$b^+(I_f) = ((d-1) + d + (d+1) - \frac{1}{2}(d+d)) - ((d-1) + d) = 1.$$

Similarly, $b^-(I_f) = -1$.

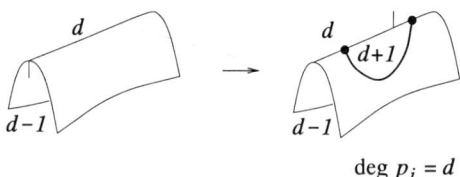

$$\deg p_i = d$$

Figure 19: *Distribution of degrees for B^+ bifurcation*

$C^{\pm,+}$, Fig.20. We have:

$$c^{\pm,+}(I_f) = ((d\pm 1 + 1) - (d + \frac{1}{2}) - \frac{1}{2}((d+1)) - \frac{1}{2}d = \pm 1.$$

Similarly, $c^{\pm,-}(I_f) = \pm 1$.

All the calculated jumps coincide with the jumps of $2I_3 - I_t - I_p$. Note that jumps $c^{\alpha,+}$ and b^+ specify a local invariant on Ω uniquely. Thus it only remains to prove locality of I.

There are no more direct calculations needed.

Indeed, assume we need to calculate a jump of I_f along K^+, with a given distribution of degrees for the components of $\mathbf{R}^3 \setminus \mathrm{Im} f$ (the distribution is completely defined by the degree of any of the components). We can realize such a jump within the 2-parameter A_2-family (Fig.14) with the appropriate distribution of the degrees (Fig.21). The difference of the values of I_f on two particular mappings is determined independently of any path connecting the mappings in Ω. So, the jumps of I_f in the A_2-family along K^+ and B^+ are equal (see the bifurcation diagram of A_2 in Fig.14). This means that the jumps of I_f and $2I_3 - I_t - I_p$ along K^+ coincide.

In the similar way, jumps of I_f known by now successively determine the following jumps of I_f (for any possible distribution of the degrees).

(1) from Fig.14: $k^- = -1,$ $h^- = 2,$ $e^2 = 2,$ $e^0 = -2;$

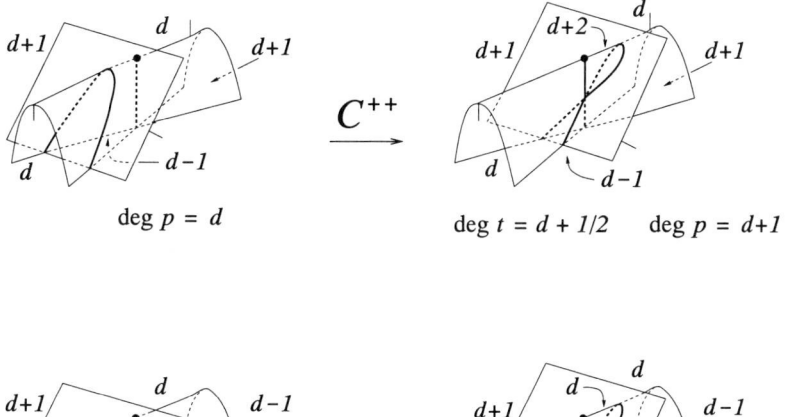

Figure 20: *Distribution of degrees for $C^{\pm,+}$ bifurcations*

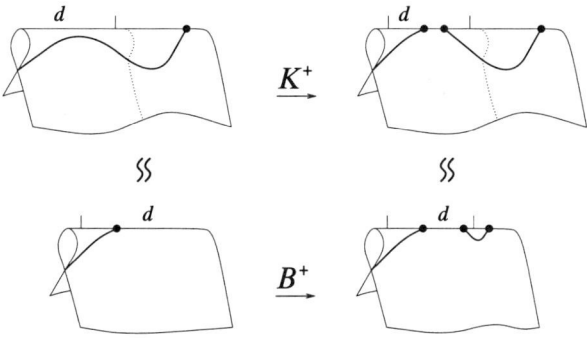

Figure 21: *Embedding of K^+ bifurcation into A_2-family*

(2) from Fig.15: $t^j = 0$, $j = 0, 1, 2, 3$, $e^1 = 0$, $h^+ = 0$;

(3) from Fig.17: $q^4 = q^3 = q^2 = 0$.

Thus, I_f is a local invariant.

7.4   Proof of Theorem 4.4 about the smoothed form of the integral invariant

Values of I_f and $\tilde{I}_f = \sum \widetilde{deg}(\widetilde{D})\chi(\widetilde{D}) - \sum deg(p)$ on embeddings coincide. So, as in the proof of Theorem 4.2, it is sufficient to show that the jumps of I_f and $\tilde{I}_f$ coincide along $B^\pm$ and $C^{\alpha,\beta}$. Fig.22, that shows smoothed B^+ and $C^{\pm,+}$ transformations, gives the following jumps of $\tilde{I}_f$:

$$b^+ = ((d-1) \cdot 1 + d \cdot 2 + (d+1) \cdot 1 - (d+d)) - $$
$$-((d-1) \cdot 1 + d \cdot 1) = 1,$$

$$c^{++} = ((d-1) \cdot 1 + d \cdot 0 + (d+1) \cdot 2 + (d+2) \cdot 1 - (d+1)) - $$
$$-((d-1) \cdot 1 + d \cdot 1 + (d+1) \cdot 2 - d) = 1,$$

$$c^{-+} = ((d-1) \cdot 1 + d \cdot 1 + (d+1) \cdot 1 + (d+2) \cdot 1 - (d+1)) - $$
$$-((d-1) \cdot 1 + d \cdot 1 + (d+1) \cdot 1 + (d+2) \cdot 1 - d) = -1.$$

Changing the coorientations of the sheets to the opposite ones interchanges the signs $+$ and $-$ in the expressions of the degrees. This gives for $\tilde{I}_f$: $b^- = -1$, $c^{--} = -1$, $c^{+-} = 1$.

All the calculated jumps are the same as for I_f. Thus, $\tilde{I}_f = I_f$.

7.5   Sketch of the proof of Theorem 5.5 about the linking invariant

The invariant ℓ^- certainly has zero jumps along the strata T, Q, E^1, H^+. Direct computations show:

Lemma 7.5 $c^{\pm,+}(\ell^-) = \mp 1$, $b^+(\ell^-) = 1$.

As in subsection 7.3, Lemma 7.5 defines all the other jumps of ℓ^- via 2-parameter bifurcations. Thus ℓ^- is local. The values of t, $c^{+,+}$ and b^+, together with the initial conditions checked on standard embeddings of surfaces of genus g (subsections 4.2 and 5.2), identify the invariant ℓ^- as $-I_f + 2I_p$.

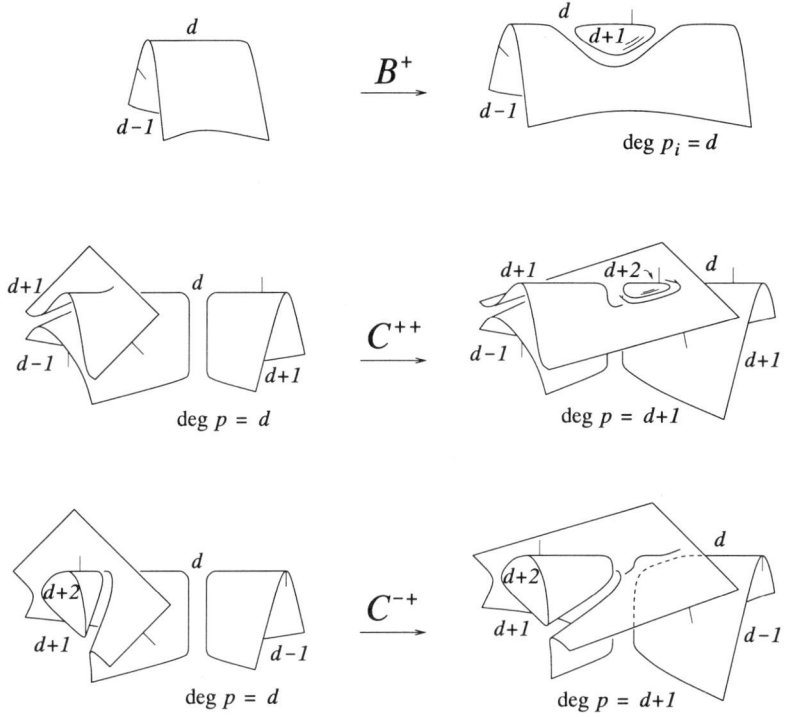

Figure 22: *Bifurcations of smoothed images*

References

[1] V. I. Arnold, *Plane curves, their invariants, perestroikas and clas-sifications*, Adv. Sov. Math. **21** (1994) 33–91

[2] V. I. Arnold, *Invariants and perestroikas of wave fronts on the plane*, in: *Singularities of smooth mappings with additional struc-tures*, Proc. V. A. Steklov Inst. Math. **209** (1995) 14–64 (in Russian)

[3] V. I. Arnold, *Topological invariants of plane curves and caustis*, Uni-versity Lecture Series **5** (1994) AMS, Providence, RI

[4] V. I. Arnold, *Exact lagrangian curves on a sphere — 1 : indices of points and of pairs of points with respect to hypersurfaces*, Letter 4-1994, 8 April 1994

[5] V. I. Arnold, V. V. Goryunov, O. V. Lyashko and V. A. Vassiliev, *Singularities I. Local and global theory*, Encyclopaedia of Mathematical Sciences **6**, Dynamical Systems VI, Springer Verlag, Berlin a.o., 1993

[6] V. I. Arnold, V. V. Goryunov, O. V. Lyashko and V. A. Vassiliev, *Singularities II. Classification and Applications*, Encyclopaedia of Mathematical Sciences **39**, Dynamical Systems VIII, Springer Verlag, Berlin a.o., 1993

[7] V. I. Arnold, S. M. Gusein-Zade and A. N. Varchenko, *Singularities of differentiable maps*, vol.I, Birkhäuser, Boston, 1985

[8] V. V. Goryunov, *Monodromy of the image of a mappping* $\mathbf{C}^2 \to \mathbf{C}^3$, Funct. Anal. Appl. **25** (1991) no.3, 174–180

[9] N. Max and T. Banchoff, *Every sphere eversion has a quadruple point*, in: Contributions to Analysis and Geometry, The Johns Hopkins University Press, Baltimore and London, 1981, 191–209

[10] D. Mond, *On the classification of germs of maps from* $\mathbf{R}^2$ *to* $\mathbf{R}^3$, Proc. London Math. Soc. **50** (1985) no.2, 333–369

[11] F. Pham, *Introduction à l'étude topologique des singularités de Landau*, Gauthier-Villars, Paris, 1967

[12] V. A. Rokhlin, *Complex orientations of real algebraic curves*, Funct. Anal. Appl. **8** (1974) no.4, 331–334

[13] V. A. Rokhlin, *Complex topological characteristics of real algebraic curves*, Russian Math. Surveys **33** (1978) no.5, 85–98

[14] R. W. Sharpe, *On the ovals of evendegree plane curves*, Mich. Math. J. **22** (1976) no.3, 285–288

[15] S. Smale, *A classification of immersions of the two-sphere*, Trans. AMS **90** (1959) 281–290

[16] S. Smale, *The classification of immersions of spheres in Euclidean spaces*, Ann. Math. **69** (1959) no.2, 327–344

[17] V. A. Vassiliev, *Cohomology of knot spaces*, Adv. Sov. Math. **1** (1990) 23–69, AMS, Providence, RI

[18] V. A. Vassiliev, *Complements of discriminants of smooth maps: topology and applications*, AMS, Providence, RI, 1992

[19] O. Y. Viro, *Some integral calculus based on Euler characteristics*, LNM **1346** (1988) 127–138, Springer

[20] O. Y. Viro, *First degree invariants of generic curves on surfaces*, Preprint 1994:21, Dept. of Maths., Uppsala University

Department
  of Mathematical Sciences,
Division of Pure Mathematics,
The University of Liverpool,
Liverpool L69 3BX, UK

E-mail: goryunov@liv.ac.uk

Department
  of Applied Mathematics,
Moscow Aviation Institute,
Volokolamskoe shosse, 4
125871 Moscow, Russia

Theorem on six vertices of a plane curve via Sturm theory

L. Guieu, E. Mourre and V. Yu Ovsienko

Abstract

We discuss the theorem on the existence of six points on a convex closed plane curve in which the curve has a contact of order six with the osculating conic. (This is the "projective version" of the well known four vertices theorem for a curve in the Euclidean plane.) We obtain this classical fact as a corollary of some general Sturm-type theorems.

1. Introduction

The well known classical theorem states that a convex curve on the Euclidean plane has at least four vertices (critical points of its curvature). This theorem has been frequently discussed in mathematical literature (see [1,8]). Beautiful applications of this theorem to symplectic geometry were discovered by V.I. Arnol'd [1,2,3]. The relation to the Sturm theory is given by S. Tabachnikov [8]. His proof of the four vertices theorem is based on the fact that a function on S^1 without n first harmonics of the Fourier decomposition vanishes at least $2n$ times.

A point on a locally convex plane curve c is called *sextactic* if the oscullating conic has a contact of order ≥ 6 with c in this point. (Recall, that in a generic point the contact is of order 5 since a conic is defined by 5 points.) Sextactic points can be defined also as critical points of the affine curvature or by the fact that the projective length element of curve c vanishes in these points.

Sextactic points are invariant under projective transformations. This kind of singular points is an analogue of vertices in projective (or affine) geometry. (Recall that in the Euclidean case the osculating circle has a contact of order ≥ 4 with the curve in any vertex.)

The following classical theorem can be considered as the "projective analogue" of the four vertices theorem.

Theorem. [Six vertices theorem] *A closed convex curve on $\mathbf{R}^2$ has at least six sextactic points.*

Corollary. *The affine curvature of a convex closed curve on $\mathbf{R}^2$ has at least six critical points.*

The proof can be found in [5].

The main result of this paper is a series of general Sturm-type theorems (in the spirit of the Tabachnikov theorem). We estimate the number of zero points of a function on S^1 orthogonal to solutions of a disconjugate linear differential equation. This approach contains at the same time the four vertices theorem of Euclidean geometry and the six vertices theorem.

2. The Sturm theorems

Consider a linear differential equation on S^1:

$$A\phi(x) = \phi^{(n)}(x) + u_{n-1}(x)\phi^{(n-1)}(x) + \cdots + u_0(x)\phi(x) = 0 \qquad (1)$$

Here $u_i(x) \in C^\infty(S^1)$ (this means, all the potentials u_i are periodic: $u_i(x + 1) = u_i(x)$).

Definition. Equation (1) is called *disconjugate on S^1* if:

1. Order $n = 2k + 1$: all the solutions are periodic: $\phi(x + 1) = \phi(x)$ and have at most $2k$ zeros (with multiplicity) on S^1.

2. Order $n = 2k$ all the solutions are anti-periodic: $\phi(x + 1) = -\phi(x)$ and have at most $2k - 1$ zeros (with multiplicity) on S^1.

In these cases, A is called a *disconjugate operator*.

2.1. Theorem 1. *Given a function $f \in C^\infty(S^1)$ orthogonal to all the solutions of a $2n + 1$-order disconjugate equation:*

$$\int_{S^1} f(x)\phi(x)dx = 0$$

then f has at least $2n + 2$ distinct zero points on S^1.

This is a generalization of the Tabachnikov theorem [8] stating the same fact for a function $f(x) \in C^\infty(S^1)$ without n first harmonics. Indeed, such a function is orthogonal to the solutions of the equation $\partial_x(\partial_x^2 + 1)(\partial_x^2 + 4) \cdots (\partial_x^2 + n^2)\phi = 0$.

Corollary. A function f in the image of a $2n + 1$-order disconjugate operator A ($f = Ag$ where $g \in C^\infty(S^1)$ is any function) vanishes at least $2n + 2$ times on S^1.

Indeed, f is orthogonal to the solutions of the equation $A^*\phi = 0$, where A^* is the operator adjoint to A. It is sufficient to remark that the operator A^* is disconjugate if A is disconjugate.

Theorem 2. *Given a function $f \in C^\infty(S^1)$ orthogonal to all the products* *of solutions of a n-order disconjugate equation:*

$$\int_{S^1} f(x)\phi_1(x)\phi_2(x)dx = 0$$

then f has et least 2n distinct zero points on S^1.

Remark. There exist straightforward generalizations of Theorems 1 and 2. It is sufficient to consider a function orthogonal to a product of any 3, 4 etc. solutions of a disconjugate equation.

2.2. Proof of Theorem 1. Consider a function $f \in C^\infty(S^1)$ orthogonal to all the solutions of a disconjugate equation $A\phi = 0$.

First, observe that f has at least one zero. Indeed, there exists a solution ϕ positive almost everywhere on S^1 (take for example a solution vanishing in some point with order $2n$, then the disconjugacy condition implies that it has no more zero points). From $\int_{S^1} f(x)\phi(x)dx = 0$ one concludes that function f changes its sign at least once.

Let us prove that the number of points of S^1 in which function f has odd-order zeros (changes its sign) is superior to $2n$. Suppose that f has $2k$ odd-order zero points $x_1, \ldots, x_{2k}$ on S^1 and $k \le n$. Consider a solution ϕ with two properties:

a) ϕ has a zero of order $2(n-k)+1$ in x_1,

b) ϕ vanishes in all points $x_1, \ldots x_{2k}$.

The existence of such a solution is evident. In fact, there exists a $2k-1$-dimensional space of solutions vanishing with order $2(n-k)+1$ in x_1. The subspace of this space which consists of solutions vanishing in x_2 has the dimension $\ge 2k-2$, etc. Now, the disconjugacy condition implies that

a) Points $x_1, \ldots x_{2k}$ are simple zeros of ϕ,

b) ϕ has no more zeros on S^1.

Finally, (replacing if necessary ϕ by $-\phi$) one obtains that functions $f(x)$ and $\phi(x)$ have the same sign sequence on the segments $]x_1, x_2[,]x_2, x_3[, \ldots]x_{2k}, x_1[$ which implies the contradiction: $\int_{S^1} f(x)\phi(x)dx > 0$. The theorem is proven.

2.3. Proof of Theorem 2 is analogue to those of Theorem 1. Suppose that f has $2k$ odd-order zero points $x_1, \ldots, x_{2k}$ on S^1 and $k \le n-1$. Take any number s which is even if n is odd and odd if n is even, such that $k \le s \le n-1$. Then there exists a solution ϕ_1 having odd order zero points in $x_1, \ldots, x_s$ and such that it has no more zero points on S^1 (see above). In the same way, there exists a solution ϕ_2 having odd order zero points in $x_{s+1}, \ldots, x_{2k}$ and such that it has no more zero points on S^1. Their product $\phi_1\phi_2$ has the same sign sequence as f on the segments

$]x_1, x_2[,]x_2, x_3[, \ldots]x_{2k}, x_1[$ which implies the contradiction: $\int_{S^1} f(x)\phi_1\phi_2(x)dx \neq 0$. The theorem is proven.

3. Affine and projective lengths;
affine and projective curvatures

We recall some classical definitions of affine and projective geometry of curves. It is very interesting to compare the notion of length in the Euclidean, affine and projective cases. If in the Euclidean case it measures in some sense the distance between the curve and a fixed point, then the affine and the projective lengths measure respectively: the distance between the curve and a straight line, and the distance between the curve and a conic.

3.1. Affine length. Consider a parametrised *locally convex* curve $c(x) = (c_1(x), c_2(x))$ in $\mathbf{R}^2$ (a curve without inflection points). For any x, vectors $c'(x)$ and $c''(x)$ are linearly independent. Define the element of affine length by

$$
d\sigma = \begin{vmatrix} c_1'(x) & c_2'(x) \\ c_1''(x) & c_2''(x) \end{vmatrix}^{\frac{1}{3}} dx
$$

Then σ is called the *affine parameter*.

3.2. Affine curvature. Vector $c'''(x)$ is a linear combination of $c'(x)$ and $c''(x)$: $c'''(x) = a(x)c''(x) + b(x)c'(x)$. Moreover, the affine parameter σ is characterized by the fact that $c'''(\sigma)$ is collinear to c':

$$
c'''(\sigma) = k(\sigma)c'(\sigma) \tag{2}
$$

Function $k(\sigma)$ is called the *affine curvature*.

Theorem 3.3. Wilczynski-Cartan construction [6], [9] (see also [7]).
(i) A parametrised locally convex curve $c(x) \in \mathbf{RP}^2$ canonically defines a linear differential equation of the form:

$$
\phi'''(x) = \kappa(x)\phi'(x) + v(x)\phi(x) \tag{3}
$$

(ii) Any equation (3) uniquely defines a locally convex curve $c(x) \subset \mathbf{RP}^2$ (modulo projective transformations of $\mathbf{RP}^2$).

Proof. To associate a locally convex curve with an equation (3), consider space E of solutions of (3). Let $V_x \subset E$ consists of solutions vanishing at the moment x. One has a family of 2-dimensional subspaces in a 3-dimensional linear space, or in other words, a curve in $\mathbf{RP}^2$. It is locally convex (which is easy to verify). In homogeneous coordinates, $c = (\phi_1(x) : \phi_2(x) : \phi_3(x))$ where $\phi_1(x), \phi_2(x), \phi_3(x)$ are any linearly independent solutions of (3). Therefore, the equation (3) is uniquely defined by the corresponding curve.

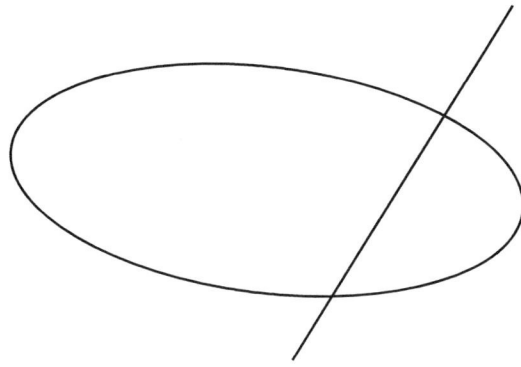

Figure 1.

Lemma 1. *The equation (3) corresponding to a closed convex curve is disconjugate.*

Proof. Consider a closed convex curve $c \subset \mathbf{RP}^2$ (see fig.1). Such a curve has at most two points of intersection with any projective line $\mathbf{RP}^1 \subset \mathbf{RP}^2$. In homogeneous coordinates $c = (\phi_1(x) : \phi_2(x) : \phi_3(x))$ where $\phi_1(x), \phi_2(x), \phi_3(x)$ are solutions of the corresponding equation (3) (see Sec. 2.3). Therefore, any solution of (3) is periodic and has at most 2 zeros on S^1.

3.4. Projective length. Rewrite (3) in more symmetric form:

$$\phi'''(x) = \frac{1}{2}\left[\kappa(x)\phi'(x) + (\kappa(x)\phi(x))'\right] + h(x)\phi(x) \tag{4}$$

where $h(x) = v(x) - \kappa'(x)/2$. Remark here that the operator $A_0 = \partial_x^3 -$

$\frac{1}{2}(\kappa(x)\partial_x + \partial_x\kappa(x))$ is antisymmetric.

Definition [6]. The 1-form on c $d\sigma = h(x)^{\frac{1}{3}}dx$ is called the projective length element.

Remark. The quantity $h(x)$ transforms as a cubic differential $h(x)(dx)^3$ by coordinate transformations. Therefore, the 1-form $d\sigma$ is well defined (see [7]).
     The projective length shows how much the curve differs from a conic.

Lemma 2. *c is a conic if and only if $h \equiv 0$.*
     Proof. *Consider a second order equation*

$$\psi''(x) = \frac{\kappa(x)}{4}\psi(x)$$

Verify that the solutions of the equation $A_0\phi = 0$ are given by quadratic polynomials in its solutions. In particular, $\phi_1 = \psi_1^2, \phi_2 = \psi_1\psi_2, \phi_3 = \psi_2^2$ (where ψ_1, ψ_2 are linearly independent) is a basis of solutions. Thus, $\phi_2^2 = \phi_1\phi_3$ and the curve $c = (\phi_1(x) : \phi_2(x) : \phi_3(x))$ is a conic.

3.5. Projective curvature. Let us suppose that $d\sigma \neq 0$ and so σ defines a local parameter on c. Then, the function $\kappa(\sigma)/4$ is called the *projective curvature* of the curve $c(x)$.

3.6. An affine curve as a projective curve. Consider a standard embedding $\mathbf{R}^2 \hookrightarrow \mathbf{RP}^2$ preserving the projective structure on $\mathbf{R}^2$ (see fig.2).

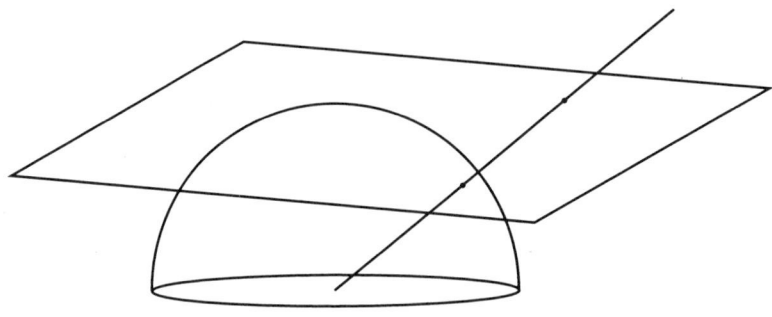

Figure 2.

An affine locally convex curve $c \subset \mathbf{R}^2$ is embedded to $\mathbf{RP}^2$ as a projective locally convex curve. To define its projective length and projective curvature, represent the equation (2) in the form (4):

$$c'''(\sigma) = \frac{1}{2}[k(\sigma)c'(\sigma) + (k(\sigma)c(\sigma))'] - \frac{1}{2}k'(\sigma)c(\sigma)$$

Therefore, the projective length of c can be defined by the relation:

$$h(\sigma) = -\frac{1}{2}k'(\sigma).$$

On the other hand, any projective curve can be considered (locally) as an affine curve. The equation (3) reduces to the form (2) by a changing of the parameter.

4. Sextactic points

Definition. A point of a locally convex curve $c \subset \mathbf{RP}^2$ is called *sextactic* if there exists a conic in $\mathbf{RP}^2$ which has a contact of order ≥ 6 with c in this point.

4.1. Critical points of the projective length. The notion of a sextactic point can be expressed in terms of the curvature (in affine case) and in terms of the length element (in projective case).

Proposition 1. *A point of a locally convex affine curve $c \subset \mathbf{R}^2$ is sextactic if and only if it is a critical point of the affine curvature.*

Corollary. *A point of a locally convex curve $c \subset \mathbf{RP}^2$ is sextactic if and only if the projective length element $d\sigma$ vanishes at this point.*

Remark here that this statement is just an infinitesimal version of Lemma 2.

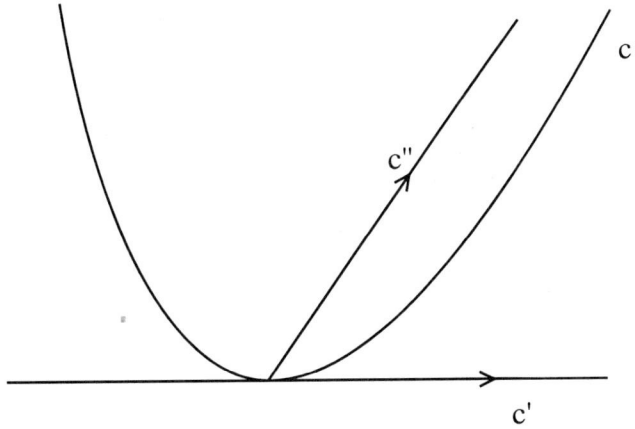

Figure 3.

Proof of the proposition. Consider a locally convex curve $c \subset \mathbf{RP}^2$. Take the affine parameter on c, then the coordinates of c satisfies the equation (2). In the neighborhood of point $c_0 = c(0)$ curve c is given by the Taylor series:

$$c(\sigma) = \sigma c_0' + \frac{\sigma^2}{2}c_0'' + \frac{\sigma^3}{6}c_0''' + \frac{\sigma^4}{24}c_0^{IV} + \frac{\sigma^5}{120}c_0^{V} + \dots$$

From (2) one has:

$$
\begin{aligned}
c''' &= & kc' \\
c^{IV} &= & k'c' + kc'' \\
c^{V} &= & (k'' + k^2)c' + 2k'c''
\end{aligned}
$$

and finally

$$c(\sigma) = (\sigma + k_0\frac{\sigma^3}{6} + k_0'\frac{\sigma^4}{24} + \dots)c_0' + (\frac{\sigma^2}{2} + k_0\frac{\sigma^4}{24} + k_0'\frac{\sigma^5}{120} + \dots)c_0''$$

Fix coordinates (x, y) on $\mathbf{RP}^2$ generated respectively by vectors $c'(0)$ and $c''(0)$ (see fig.3).

Consider the following conic:

$$x^2 - 2y + k_0 y^2$$

Satisfy the coordinates of $c(\sigma)$ to function $F(x, y) = x^2 - 2y + k_0 y^2$. One obtains:

$$F(x(\sigma), y(\sigma)) = k_0' \frac{\sigma^5}{20} + \dots$$

Thus, the conic has a contact of order 5 with c and so this is the *osculating conic* to c. If $k_0' = 0$, then the order of contact is 6. The proposition is proven.

4.2. Proof of the six vertices theorem. Let us show how the six vertices theorem follows from Theorem 1.

Lemma 3. *The parameter $h(x)$ in equation (3) satisfies the following condition:*

$$\int_{S^1} \phi_1(x)\phi_2(x)h(x)dx = 0$$

where $\phi_1(x), \phi_2(x)$ are any two solutions of (3).

Proof. Let $\phi(x)$ be a solution of (3), then $\phi h = A_0 \phi$. Lemma 3 follows now from the fact that A_0 is antisymmetric. Indeed,

$$\int_{S^1} \phi_1(x)\phi_2(x)h(x)dx = \int_{S^1} \phi_1(x)A_0\phi_2(x)dx =$$
$$-\int_{S^1} A_0(\phi_1(x))\phi_2(x)dx = -\int_{S^1} \phi_1(x)\phi_2(x)h(x)dx = 0$$

The six vertices theorem follows now from Theorem 2 and Lemma 1. In fact, the function $h(x)$ is orthogonal to all the products of solutions of a disconjugate equation of order 3. Thus, it has at least 6 distinct zero points on S^1 (Theorem 2). Sextactic points of a locally convex curve $c \subset \mathbf{RP}^2$ coincide with zero points of h (Proposition 1). One obtains, that a closed convex curve on $\mathbf{RP}^2$ has at least 6 distinct sextactic points. The theorem is proven.

4.3. Geometrical properties of sextactic points. Let us give here two geometrical descriptions of sextactic points.

A. Any curve c in general position has almost everywhere a contact of order 5 with its osculating conic. Nondegenerate sextactic points can be characterized by the fact that c does not cross its osculating conic in such noints (see fig.4).

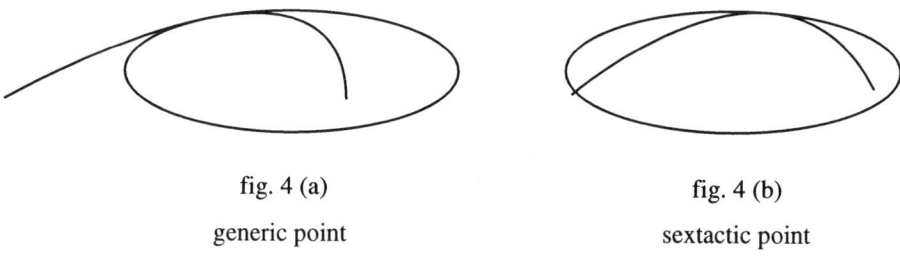

fig. 4 (a)   fig. 4 (b)

generic point   sextactic point

Figure 4.

B. Dual curves. Let c_1 and c_2 be locally convex curves, take any two points $p_1 \in c_1$ and $p_2 \in c_2$. Then, there exists a projective transformation $Q \in PGL(3, \mathbf{R})$ such that $Qp_2 = p_1$ and the curve Qc_2 has a contact of order ≥ 5 with c_1 in p_1. Let $\bar{c}$ be a *projectively dual curve* to the curve c. We show that c has a contact of order ≥ 5 with $\bar{c}$ in sextactic points.

Lemma 4. *A point p of a locally convex curve $c \subset \mathbf{RP}^2$ is sextactic if and only if c there exists a projective isomorphism $I : \mathbf{RP}^{2*} \overset{\cong}{\to} \mathbf{RP}^2$ such that c has a contact of order ≥ 6 with $I(\bar{c})$ in p.*

    Proof. *Let C be the osculating conic of a locally convex curve c in a point p. Then the dual conic $\bar{C} \in \mathbf{RP}^{2*}$ is the osculating conic of $\bar{c}$. Take* *an isomorphism* $I : \mathbf{RP}^{2*} \to \mathbf{RP}^2$ *which maps* $\bar{C}$ *to* C *and the point of contact of* $\bar{C}$ *with* $\bar{c}$ *to the point of contact of* C *with* c.

After completion of this paper we received the preprint [4] containing the proof of Theorem 1 and its applications to the theory of space curves. We also discovered unpublished results of A. Viro who gave another proof of the six vertices theorem using the Sturm-Tabachnikov approach.

Acknowledgments. It is a pleasure to acknowledge fruitful discussions with V.I. Arnol'd and S. Tabachnikov.

References

[1]   V.I. Arnol'd, "Ramified covering $CP^2 \to S^4$, hyperbolicity and projective topology." Sib. Math. Journal **29** (1988) n.5, 36–47.

[2]   V.I. Arnol'd, "On topological properties of Legendre projections in Contact geometry of wave fronts." St. Petersburg Math. Journal (1994), n.3.

[3]   V.I. Arnol'd, "Sur les propriétés topologiques des projections Lagrangiennes en géométrie symplectique des caustiques." Cahiers de Math. de la Decision, CEREMADE, n.9320, 1–9.

[4]   V.I. Arnol'd, "On the number of flattening points on space curves." Preprint.

[5]   G. Bol, "Projective Differentialgeometrie." Studia Mathematica / Mathematische Lehrbücher - Vandenhoeck & Ruprecht -Göttingen, 1955.

[6]   E. Cartan, "Leçons sur la théorie des espaces à connexion projective." Gauthier-Villars, Paris, 1937.

[7]   L. Guieu, V.Yu. Ovsienko, "Structures symplectiques sur les espaces de courbes projectives et affines." J. Geom. Phys., to appear.

[8]   S. Tabachnikov, "Around four vertices." Russian Math. Surveys **45** (1990), n.1, 229–230.

[9]   E.J. Wilczynski, "Projective differential geometry of curves and ruled surfaces". Leipzig-Teubner 1906.

L. Guieu
Université d'Aix-Marseille III
guieu@math.univ-montp2.fr

E. Mourre and V. Yu Ovsienko
Centre de Physique Théorique
Unité Propre de Recherche 7061
CNRS - Luminy, Case 907
F–13288 Marseille Cedex 9 - France

Received June 1995

The Arf–invariant and
the Arnold invariants of plane curves

S.M.Gusein-Zade, S.M.Natanzon

In [A] V.I.Arnold considered closed generic plane curves, i.e. immersions $S^1 \to \mathbf{R}^2$, the images of which have no singularities except simple (double) self-intersections. In a generic one-parameter family of immersions three types of modifications ("perestroykas") of generic curves can be met. They correspond to three natural strata in the set of non-generic immersions (the discriminant). These strata consist of immersions with a direct self-tangency (J^+), with an inverse self-tangency (J^-), and with a triple crossing (St) respectively. All three of them are coorientable. An invariant of generic plane curves is an invariant of the first order (in the sense of Vassiliev) if its change under a modification of crossing a stratum (in the positive direction) depends only on the stratum, but not on a (nonsingular) point of its crossing. For closed plane curves V.I.Arnold defined basic invariants of the first order J^+, J^-, and St corresponding to the described strata. The invariants J^+, J^-, and St can be defined for so-called "long" curves (that is for curves going from the infinity to the infinity) as well.

The method of real morsifications describes the intersection form (and a Dynkin diagram) of an isolated critical point of a function germ $f : (\mathbf{C}^2, 0) \to (\mathbf{C}, 0)$ in terms of a real plane curve with only simple (double) self-intersections — the zero-level set of an appropriate deformation of a real singularity from the same stratum $\{\mu = const\}$. (Generally speaking this curve is determined by the germ f not in the unique way.) If the germ f is irreducible then its intersection form is non-degenerate, the corresponding plane curve consists of one component (i.e. is the image of a generic immersion $\mathbf{R} \to \mathbf{R}^2$). In this case the monodromy group over the field Z_2 is determined by the Milnor number $\mu = \mu(f)$ and the *Arf*-invariant, which is an element of Z_2 ([J]). The *Arf*-invariant (as well as the intersection form) can be defined for any long curve (in such a way that for a curve associated to a singularity of two variables in accordance with the method of real morsifications it coincides with the *Arf*-invariant of the singularity).

The object of the paper is to describe a relation between the *Arf*-invariant and the Arnold invariants of plane curves.

Theorem . *The Arf-invariant of (long) curves is an invariant of the first order,*

$$\mathrm{Arf} = J^-/2 \bmod 2.$$

This statement seems to be the first known relation between a "conventional" invariant of a plane curve singularity (the monodromy group over the field Z_2 here) and the basic Arnold invariants of a corresponding (according to the method of real morsifications) generic real plane curve.

Remark. The invariants J^+ and J^- can be defined for real irreducible germ functions $f : (\mathbf{R}^2, 0) \to (\mathbf{R}, 0)$ (the invariant St can not be defined for them). As a corollary of the Theorem we have the fact that for real irreducible singularities of functions of two variables the *Arf*-invariant has an integer lifting (that is $J^-/2$).

1. Basic invariants of long curves.

Definition. A long curve L (or for short simply a curve) is an immersion $L : \mathbf{R}^1 \to \mathbf{R}^2$ of the line $\mathbf{R}^1$ into the plane $\mathbf{R}^2$, which is generic (i.e. its image has no singularities except simple double self-intersections) and which coincides with the map $x \to (x, 0)$ outside a bounded segment.

Two curves are considered as identical if they are topologically equivalent, i.e. if one can be transformed into the other by a diffeomorphism of the plane, which coincides with the identy outside a bounded region. On the set of curves there are two natural involutions. The involution Σ_1 is induced by the reflection $(x, y) \to (x, -y)$ of the plane $\mathbf{R}^2$ with respect to the x–axis. The involution Σ_2 is induced by the reflection $(x, y) \to (-x, y)$ of the plane $\mathbf{R}^2$ with respect to the y–axis (with the simultaneous reflection $x \to (-x)$ of the line $\mathbf{R}^1$). On the set of curves there can be defined a sum operation. The sum of two curves is constructed as it is shown in Fig.1. This operation is well-defined, associative, but non-commutative.

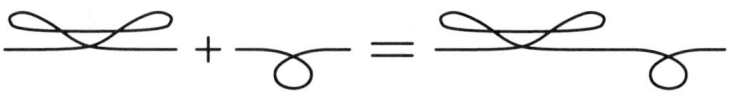

Figure 1.

The basic invariants J^+, J^-, and St of (long) curves can be defined in the same way as for closed curves in [A]. The only difference is the following. The space of all immersions $\mathbf{R}^1 \to \mathbf{R}^2$ is the union of infinitely many contractable components. Each component consists of immersions with the fixed index (rotation number). An invariant of the first order is determined by its jumps for the modifications of crossing the (defined) strata and its values for representatives from all the components. As these representatives it is natural to use curves L_i from Fig.2 (the index of the curve L_i is equal to i). The table of jumps for the invariants J^+, J^-, and St can be found in [A]. Let $I(L) = (J^+(L), J^-(L), St(L))$. Then the jump $\Delta I(L)$ is equal to $(2,0,0)$ for the stratum J^+ of direct tangencies, $\Delta I(L) = (0,-2,0)$ for the stratum J^- of inverse tangencies, $\Delta I(L) = (0,0,1)$ for the stratum St of triple points. This table defines the invariants up to additive constants depending on the index. The values of these invariants for the curves L_i are determined by the following statement.

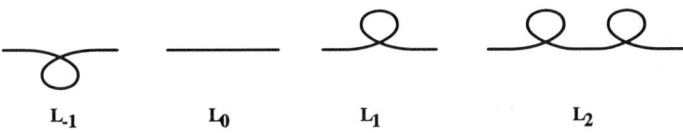

L_{-1} L_0 L_1 L_2

Figure 2.

Lemma 1. *There is a unique way to define the invariants J^+, J^-, and St (i.e. their values for the curves L_i) so that they are invariant with respect to the involutions Σ_1 and Σ_2 and additive with respect to the sum operation. In this case $I(L_1) = (-1, -2, 1/2)$.*

Remark. It is sufficient to require only the additiveness and the invariance with respect to the involution Σ_1. The invariance with respect to the involution Σ_2 takes place automatically.

Proof. It is clear that $I(L_i) = |i| \cdot I(L_1)$. We have $I(L_1 + L_{-1}) = 2I(L_1)$. The index of the curve $L_1 + L_{-1}$ is equal to zero and hence this curve can be deformed (in the class of immersions) into the trivial curve L_0 (Fig.3). In Fig.3 on each stage of the deformation except the last one we have exactly one generic modification (of types J^+, St, J^- and J^- respectively). Therefore $I(L_1 + L_{-1})$ can be found as sum of jumps for crossings of the discriminant. It is only necessary to keep track of signs of the modifications which are met in the course of the deformation).

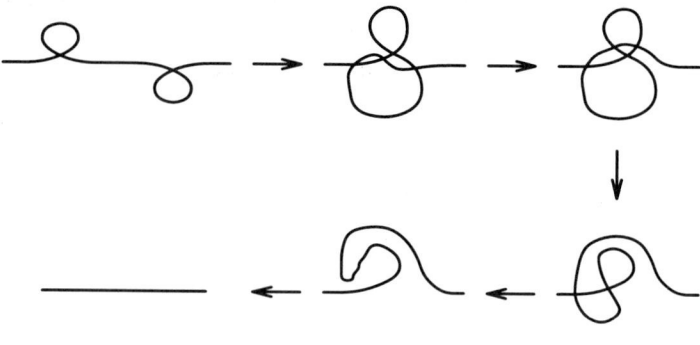

Fig.3.

Remarks. 1. The values of J^+ and of J^- are alwais integers (moreover J^- is always even). However the value of St can be rational. It can have (and if the index of the curve is odd it actually has) 2 as the denominator. Therefore it would be somewhat more convenient to define the jump $\Delta I(L)$ to be equal to $(0,0,2)$ for the stratum of triple points. However it is reasonable to follow the initial definition of V.Arnold.

2. To a (long) curve L there correspond two closed curves L_+ and L_-. They can be obtained from L by connecting its "ends" in the upper and in the lower half-plane respectively (Fig.4). It is not difficult to see that $I(L) = (I(L_+) + I(L_-))/2$.

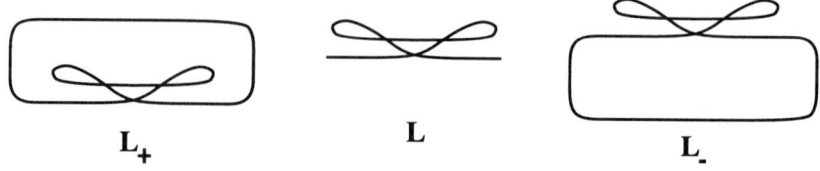

Fig.4.

3. The invariance with respect to the involutions Σ_1 and Σ_2 implies that the basic invariants J^+, J^-, and St are well-defined not only for generic immersions $\mathbf{R}^1 \to \mathbf{R}^2$ standard at the infinity but for any proper generic immersion $\mathbf{R}^1 \to \mathbf{R}^2$ with finite number of self-intersection points. Thus they are defined for a real plane curve corresponding to an irreducible germ $f : (\mathbf{C}^2, 0) \to (\mathbf{C}, 0)$ in accordance with the method of real morsifications.

2. The *Arf*–invariant of a long curve.

The method of real morsifications describes two bilinear forms on an (integer) lattice associated to a plane curve (a symmetric and a skew-symmetric ones). The general definition can be found e.g. in [AGV] (pp.115-116). These two forms coincide over the field Z_2. Only this case will be used here.

Let L be a plane curve (i.e. a generic immersion $\mathbf{R}^1 \to \mathbf{R}^2$ standard at the infinity) with self-intersection points p_i ($i = 1, 2, \ldots, n$). Let U_j be the bounded components of the complement to the curve L (actually there are just n of them too).

Definition. The Milnor lattice (over the field Z_2) associated with the curve L is the vector space H over the field Z_2 generated by (formal) elements $\delta_i^{(1)}$ and $\delta_j^{(0)}$ corresponding to the self-intersection points p_i of the curve L and to the bounded components U_j of its complement. The elements $\delta_i^{(1)}$ and $\delta_j^{(0)}$ are called the basic vanishing cycles.

Definition. The intersection form associated with the curve L is the bilinear form $(\cdot, \cdot)$ on the Milnor lattice H (with the values in Z_2) defined by the following list of the scalar products (the intersection numbers) of the basic vanishing cycles:
1) the scalar square of each element is equal to zero, i.e. the form is even;
2) the scalar product $(\delta_i^{(1)}, \delta_{i'}^{(1)})$ of two basic vanishing cycles with the upper index (1) is equal to zero;
3) $(\delta_i^{(1)}, \delta_j^{(0)})$ is equal to the number of verticies of the component U_j which coincide with the self-intersection point p_i (this number can be equal to 0, 1, or 2; in the last case the intersection number is equal to $0 \in Z_2$ again);
4) the scalar product $(\delta_j^{(0)}, \delta_{j'}^{(0)})$ of two basic vanishing cycles with the upper index (0) is equal to the number of common sides of the corresponding components U_j and $U_{j'}$ of the complement to the curve L.

The point 2) means that the subspace, generated by the cycles $\{\delta_i^{(1)}\}$, is an isotropic subspace of the Milnor lattice H, i.e. the restriction of the intersection form to it is equal to zero. In fact it is a maximal isotropic subspace: see Lemma 2 below.

Definition. The Picard-Lefschetz transformation corresponding to an element $\delta \in H$ is the linear map $h : H \to H$ defined by the formula

$$h(a) = a + (a, \delta) \cdot \delta \ (a \in H).$$

Definition. The monodromy group of the curve L is the subgroup of the group $\mathrm{Aut}(H)$ of the automorphisms of the Milnor lattice H generated by the Picard-Lefschetz transformations $h_i^{(1)}$ and $h_j^{(0)}$ corresponding to the basic vanishing cycles $\delta_i^{(1)}$ and $\delta_j^{(0)}$.

Definition. A vanishing cycle is an element of the Milnor lattice H which can be obtained from a basic vanishing cycle by the action of a transformation from the monodromy group.

Remark. We consider the basis $\{\delta_i^{(1)}, \delta_j^{(0)}\}$ of the Milnor lattice H as a weakly distinguished one (see [AGV]). Actually it can be considered as a distinguished one. However it requires a suitable ordering of its elements. The description can be found in [AGV] again.

Lemma 2. *The intersection form of a (long) curve is non-degenerate.*

Proof. Let $\mathbf{C}^1$ be the complexification of $\mathbf{R}^1$ and let M be a real surface (or rather a complex curve), obtained from $\mathbf{C}^1$ by identification of the preimages of the self-intersection points (which lie in $\mathbf{R}^1 \subset \mathbf{C}^1$). Let $\tilde{M}$ be a smoothing of the curve M (as a real surface it can be obtained from M by cutting out neighbourhoods of the double points and replacing them by cylinders). It is possible to show (in fact it has been done in [GZ]) that the intersection form corresponding to the curve L coincides with the (usual) intersection form in the homology group $H_1(\tilde{M}, Z_2)$ in a certain basis. The last intersection form is non-degenerate (because the surface $\tilde{M}$ has "one component at the infinity").

In order to define an *Arf*-invariant one needs to have a so-called *arf*-function (or a quadratic function), i.e. a function $\alpha : H \to Z_2$ with the property

$$\alpha(x + y) = \alpha(x) + \alpha(y) + (x, y).$$

Such a function is determined by its values for the basic elements.

Definition. The *arf*-function of the curve L is defined to be equal to 1 for all the basic vanishing cycles.

It isn't difficult to see that the value $\alpha(\delta)$ of the *arf*-function of the curve L is equal to 1 for any vanishing cycle δ. Since the intersection form is non-degenerate there exists a standard basis $\{u_i, v_i\}$ $(i = 1, 2, \ldots, n)$ of the lattice H, i.e. a basis for which $(u_i, u_j) = (v_i, v_j) = 0$, $(u_i, v_j) = \delta_{ij}$ (the Kronecker symbol).

Definition. The *Arf*-invariant $\mathrm{Arf}(L)$ of the curve L is equal to $\sum_i \alpha(u_i)\alpha(v_i)$.

The *Arf*-invariant is well-defined, i.e. it doesn't depend on the choice of a standard basis of the lattice H. The monodromy group (over the field Z_2) of an isolated critical point of an irreducible function germ $f : (\mathbf{C^2}, \mathbf{0}) \to (\mathbf{C}, \mathbf{0})$ is determined by its Milnor number and the *Arf*-invariant ([J]). This *Arf*-invariant coincides with the *Arf*-invariant of a curve L corresponding to the germ f in accordance with the method of real morsifications.

3. The relation between the *Arf*–invariant and the basic invariants.

In this section we prove the Theorem formulated in the Introduction. It isn't difficult to see that the equality from it takes place for the representatives L_i (Fig.2) of curves with a fixed index (the Dynkin diagram of the curve L_i is the disjoint union of $|i|$ copies of the diagram A_2). Therefore in order to prove it for the general case it is necessary to show that the *Arf*-invariant has the jumps equal to 0 for the stratum St of triple points and for the stratum J^+ of direct tangecies and the jump equal to 1 for the stratum J^- of inverse tangecies. For the stratum St it follows from the fact that the action of a triple point modification on the Dynkin diagram of a plane curve is equivalent to a change of the distinguished basis in the Milnor lattice ([AGV], pp.120-121). In this sense the *arf*-functions (and hence the *Arf*-invariants) of the curves before and after the modification coincide with each other. This fact is local. It means that its proof consists of an analysis of the behaviour only of the vanishing cycles corresponding to the self-intersection points of the curve and to the bounded components of its complement adjacent to the point of the modification. For the strata J^+ and J^- it can not be so. The change of the Dynkin diagram of a curve for a tangency modification (locally) is one and the same for the direct tangency and for the inverse one (see Fig.7 below). It is possible to distinguish between these cases only globally. However the changes of the *Arf*-invariant for these two cases are different.

In order to prove the Theorem we shall use two lemmas.

The intersection form associated with a curve L is non-degenerate. Therefore to each basis of the Milnor lattice H of the curve L there corresponds the dual basis. The dual to the basis $\{\delta_i^{(1)}, \delta_j^{(0)}\}$ is the basis $\{\delta_i^{(1)*}, \delta_j^{(0)*}\}$ such that $(\delta_i^{(1)}, \delta_j^{(0)*}) = (\delta_j^{(0)}, \delta_i^{(1)*}) = 0$, $(\delta_k^{(0)}, \delta_{k'}^{(0)*}) = (\delta_k^{(1)}, \delta_{k'}^{(1)*}) = \delta_{kk'}$ $(k, k' = 1, 2, \ldots, n)$. Let U_j be the bounded component of the complement to the curve L corresponding to the basic vanishing cycle $\delta_j^{(0)}$ and let p_i be a self-intersection point of L. The point p_i cuts the curve L into three parts: two curves going from the infinity to the point p_i and the loop Ω_i which goes from p_i to itself. This loop is smooth everywhere except at the point p_i itself, where it has a corner. Let $\ell_j^i = \ell(\Omega_i, U_j)$ be the index of the loop Ω_i with respect to the region U_j (it is equal to the linking number of the loop Ω_i with a point from U_j).

Lemma 3. *The element $\delta_j^{(0)*}$ corresponding in the dual basis to the basic vanishing cycle $\delta_j^{(0)}$ is equal to the sum $s_j = \sum\limits_{i=1}^{n} \ell_j^i \cdot \delta_i^{(1)}$ of the cycles corresponding to the self-intersection points of the curve L.*

Proof. It is clear that $(s_j, \delta_{j'}^{(1)}) = 0$ for each basic vanishing cycle $\delta_{j'}^{(1)}$ corresponding to a self-intersection point. Hence it is sufficient to check that $(s_j, \delta_k^{(0)}) = \delta_{jk}$. Let as consider the loops Ω_i as elements of the homology group $H_1(\mathrm{Im}\,L; Z_2)$. We have $(s_j, \delta_k^{(0)}) =$

$(\sum\limits_{i=1}^{n} \ell(\Omega_i, U_j)\delta_i^{(1)}, \delta_k^{(0)}) = \sum\limits_{i=1}^{n}(\delta_i^{(1)}, \delta_k^{(0)}) \cdot \ell(\Omega_i, U_j) = \ell(\sum\limits_{i=1}^{n}(\delta_i^{(1)}, \delta_k^{(0)})\Omega_i, U_j) =$

$\ell(\sum\limits_{i:p_i \in \partial U_k} (\delta_i^{(1)}, \delta_k^{(0)})\Omega_i, U_j)$. In the last sum the coefficient $(\delta_i^{(1)}, \delta_k^{(0)})$ can be equal to 0 if the self-intersection point p_i of the curve L is on the boudary ∂U_k of the component U_k twice. The desired fact follows from the formula

$$\sum\limits_{i:p_i \in \partial U_k} (\delta_i^{(1)}, \delta_k^{(0)})\Omega_i = \partial U_k$$

in the homology group $H_1(\mathrm{Im}\,L; Z_2)$; ∂U_k is the element of the homology group represented by the boundary of the component U_k. This formula gives the equality $(s_j, \delta_k^{(0)}) = \delta_{jk}$ because $\ell(\partial U_k, U_j) = \delta_{kj}$, i.e. it is equal to 1 if $k = j$ and is equal to 0 othewise. The cycle $S = \sum\limits_{i:p_i \in \partial U_k} (\delta_i^{(1)}, \delta_k^{(0)})\Omega_i$ is equal to the sum of smooth segments of the curve L with some coefficients. Let us pass along the curve L from $-\infty$ to $+\infty$ and keep track of the coefficients. The coefficient is equal to zero for the first (non-bounded) segment of the curve L. Suppose that we pass a self-intersection point p_{i_0} of the curve L (say from the segment a to the segment b). If this point is not on the boundary of the region U_k, then all the loops Ω_i with $(\delta_i^{(1)}, \delta_k^{(0)}) \neq 0$ which include the segment a include the segment b as well. Hence the coefficients at a and b in S are equal to each other. If the point p_{i_0} is on the boundary of the region U_k twice (and hence the loop Ω_{i_0} does not participate in the sum S), then we have just the same situation and the corresponding coefficients are equal to each other again. At last if the point p_{i_0} is on the boundary of the region U_k once, then $(\delta_{i_0}^{(1)}, \delta_k^{(0)}) = 1$ and in the sum S only the loop Ω_{i_0} goes along one of the segments a or b and does not go along the other one. Hence the coefficients at a and b in S differ by 1. This proves the lemma.

The self-intersection points cut the curve L (or rather Im L) into smooth segments (two of which are unbounded). The preimages of the self-intersection points cut the line $\mathbf{R}^1$ into corresponding segments. Let us define the sign $sgn(a) \in Z_2$ of a segment a of the curve L as the number of the preimages of the self-intersection points between the corresponding segment of the line $\mathbf{R}^1$ and the minus infinity (or equivalently between the segment and the plus infinity). For a component U of the complement to the curve L (bounded or not) let us take a (smooth) path in the plane $\mathbf{R}^2$ which goes from a point of the component to the infinity in the (say) upper half-plane (e.g. which coincides with the y-axis outside a bounded region) and which is transversal to the curve L (in particular it does not go through the self-intersection points of L). Let us define the sign $sgn(U) \in Z_2$ of the component U to be equal to the sum of signs of all the segments of the curve L intersecting the path. It is not difficult to see that the sign $sgn(U)$ is well-defined, i.e. it does not depend on the choice of the path from the component to the infinity. We shall call a segment of the curve L or a component of its complement even or odd if its sign is even or odd (i.e. if it is equal to $0 \in Z_2$ or to $1 \in Z_2$ respectively). It is possible to say that the sign of a component of the complement to the curve L is equal to the number of odd segments between the component and the infinity. The both unbounded segments of the curve L and the both unbounded components of its complement are even.

Suppose that U_{j_1} and U_{j_2} are components of the complement to the curve L (possibly unbounded and possibly coinciding with each other) such that there exists a (smooth) path from a point of U_{j_1} to a point of U_{j_2} which intersects (transversally) the curve L in two points (say on the segments a and b, which also can coincide with each other). If at the intersection points the velocity vectors of the curve L point to one and the same side of the path, then the components U_{j_1} and U_{j_2} have different signs (Fig.5, left). If they point to different sides of the path, then the components U_{j_1} and U_{j_2} have equal signs (Fig.5, right). This follows from the fact that the number of (the preimages of) the self-intersection points between the segments a and b is odd in the first case and is even in the second one. Hence in the first case one of the segments is even and the other one is odd. So the signs of the components are different. In the second case either both of the segments are even or both of them are odd. So the signs of the components are equal to each other.

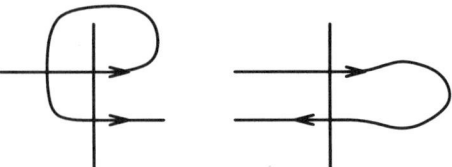

Fig.5.

Lemma 4. *The sum $\sum\limits_{i=1}^{n} \ell_j^i$ of all the coefficients in the expression for the dual vanishing cycle $\delta_j^{(0)*}$ corresponding to the bounded component U_j of the complement to the curve L is equal to the sign $sgn(U_j)$ of the component.*

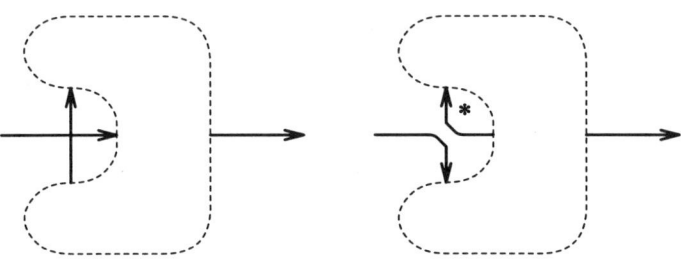

Fig.6.

Proof. We shall use the method of (mathematical) induction. The fact is trivial for the curve without self-intersection points (that is for the curve L_0 from Fig.2). Let us suppose that the statement has been proved for all curves with less than n self-intersection points and let L be a curve with n self-intersection points. Let us take the first self-intersection point of the curve L, that is the end point of the left unbounded segment of it. Without loss of generality we can suppose that this is the self-intersection point p_n. One of the two natural smoothings of this self-intersection point gives a long curve (say $\tilde{L}$) with $(n-1)$ self-intersection points (the other one gives a union of a long curve and a closed curve). See Fig.6; inside the regions bounded by the dotted lines the curve is arbitrary (and one and the same). Let the component of the complement to the curve L marked by $*$ in Fig.6 be U_{j_0}. This is the component which does not become connected with an unbounded component of the complement after the smoothing. However it is possible that it is unbounded initially; in this case we suppose the corresponding dual vanishing cycle $\delta_{j_0}^{(0)*}$ to be equal to 0. To each segment of the curve L or to each component of its complement there corresponds a segment or a component of the complement of the curve $\tilde{L}$. The sign of a segment a of the curve L coincides with (respectively is different from) the sign of the corresponding segment of the curve $\tilde{L}$ if and only if the segment a does not lie (respectively lies) on the loop Ω_n corresponding to the self-intersection point p_n. It implies that the sign of a component U_j of the complement to the curve L coincides with (respectively is different from) the sign of the corresponding component for the curve $\tilde{L}$ if and only if $\ell(\Omega_n, U_j) = 0$ (respectively $\ell(\Omega_n, U_j) = 1$). Let $\tilde{\Omega}_i$ be the loop corresponding to the self-intersection point p_i of the curve $\tilde{L}$, $\tilde{\ell}_i^j = \ell(\tilde{\Omega}_i, U_j)$ $(i = 2, \ldots, n)$. If the loop Ω_i corresponding to the self-intersection point p_i of the curve L passes the self-intersection point p_n (the point of the smoothing) then (in the natural sense) $\tilde{\Omega}_i = \Omega_i + \Omega_n$. Othewise $\tilde{\Omega}_i = \Omega_i$. Let $\tilde{J} = \{i : 1 \leq i \leq (n-1)$, the loop Ω_i passes the self-intersection point p_n of the curve $L\}$, $J = \tilde{J} \cup \{n\}$. The number $\#\tilde{J}$ of the elements of the set $\tilde{J}$ is equal to the sum $\sum_{i=1}^{n-1} \ell(\tilde{\Omega}_i, U_{j_0}) = \sum_{i=1}^{n-1} \tilde{\ell}_i^{j_0}$. By the induction hypothesis it is equal (mod 2) to the sign of the region U_{j_0} which is equal to 0 (see the paragraph just before Lemma 4). Therefore the number $\#J$ of the elements of J is odd.

We have $\sum_{i=1}^{n-1} \tilde{\ell}_i^j = \sum_{i=1}^{n-1} \ell(\tilde{\Omega}_i, U_j) = \sum_{i=1}^{n} \ell(\Omega_i, U_j) + \sum_{i \in J} \ell(\Omega_n, U_j) = \sum_{i=1}^{n} \ell_i^j +$ $\ell(\Omega_n, U_j)$. Therefore the sum $\sum_{i=1}^{n} \ell_i^j$ coincides with (respectively is different from) the sum $\sum_{i=1}^{n-1} \tilde{\ell}_i^j$ if and only if $\ell(\Omega_n, U_j) = 0$ (respectively $\ell(\Omega_n, U_j) = 1$). It means that this sum changes just in the same way as the sign $sgn(U_j)$ of the component U_j. This proves the lemma.

Corollary 1. *The value $\alpha(\delta_j^{(0)*})$ of the arf-function α for the dual vanishing cycle $\delta_j^{(0)*}$ is equal to the sign $sgn(U_j)$ of the corresponding component.*

It follows from the fact that the restriction of the intersection form to the subspace, generated by the cycles $\{\delta_j^{(1)}\}$, is equal to zero and hence the arf-function α is linear on it.

Corollary 2. *Suppose we have the situation described in the paragraph preceding Lemma 4. Let $\delta_{j_1}^{(0)*}$ and $\delta_{j_2}^{(0)*}$ be the dual vanishing cycles corresponding to the components U_{j_1} and U_{j_2}. If one (or both) of the components is unbounded we shall suppose the corresponding dual vanishing cycle (cycles) to be equal to 0. Then $\alpha(\delta_{j_1}^{(0)*} + \delta_{j_2}^{(0)*})$ is equal to 1 in the first case (if the velocity vectors point to one and the same side of the path) and is equal to 0 in the second one (if the velocity vectors point to different sides of the path).*

Proof of the Theorem. We have to consider a (direct or inverse) tangency modification and to determine the corresponding jump of the *Arf*-invariant. In a general case such a modification leads to the following change of the Dynkin diagram of the curve (see Fig.7; the thin lines show the curves and the thick ones show the edges of the corresponding Dynkin diagrams; only the indicies of the corresponding basic vanishing cycles are indicated). Before the modification (Fig.7, left) we have cycles δ_*, δ_+, and δ_- (and some other cycles which do not really participate in the modification). The words "In a general case" above mean that in particular cases some of these cycles can be absent (the corresponding components can be unbounded) or can coincide with each other (for the cycles δ_+ and δ_-). The necessary changes of the proof are clear (we shall indicate them in the course of the proof). After the modification (Fig.7, right) we have cycles δ_1, δ_2, δ_3, δ_4, δ_5, δ_+, and δ_-. For a basic vanishing cycle δ outside the picture (that is which does not really participate in the modification) we have: 1) the intersection numbers (δ_+, δ) and (δ_-, δ) are the same as those before the modification (it is the reason to denote the cycles δ_+ and δ_- before and after the modification in the same way); 2) $(\delta_i, \delta) = 0$ for $i = 2, 3, 4$; 3) $(\delta_*, \delta) = (\delta_1, \delta) + (\delta_5, \delta)$ (if the cycle δ_*, existed before the modification, i.e. if the corresponding component of the complement was bounded). As earlier let us suppose the corresponding dual vanishing cycle δ_+^* or/and δ_-^* to be equal to zero if one (or both) of the corresponding components is unbounded.

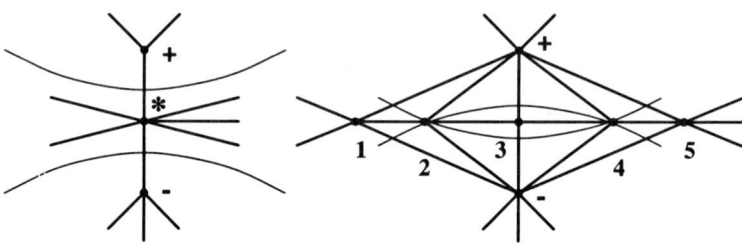

Fig.7.

Let us make the following change of the basis: $\delta_i \mapsto \delta_i' = \delta_i + \delta_+^* + \delta_-^*$, $1 \le i \le 4$ (all the other basic cycles are unchanged; generally speaking the new basis is not a weakly distinguished one). This is a change of the basis indeed because of the following. It is not difficult to see that the coefficients at δ_2 and δ_4 in $\delta_j^{(0)*}$ (in the Milnor lattice H of the curve after the modification) may be (and in fact are) different only for the component of the complement corresponding to the cycle δ_3. In particular they are equal to each other in $\delta_+^* + \delta_-^*$. In the Dynkin diagram this change cuts all the connections (edges) between the cycles δ_i with $1 \le i \le 4$ and the cycles δ_+ and δ_-. From Corollary 2 (after Lemma 4) it follows that for $1 \le i \le 4$ $\alpha(\delta_i') = 1$ if the modification is of the type J^- (i.e. an inverse tangency modification) and $\alpha(\delta_i') = 0$ if the modification is of the type J^+ (i.e. a direct tangency modification). This is the crucial point of the proof.

The change of the basis of the form $\delta_5 \mapsto \delta_5' = \delta_5 + \delta_1' + \delta_3'$ cuts the connection between the cycle δ_5 and the cycle δ_4'. It changes the connections between the cycle δ_5 and the cycles outside the picture. The intersection number (δ_5', δ) is equal to $(\delta_5, \delta) + (\delta_1, \delta)$ and hence it just coincides with the intersection number (δ_*, δ) (before the modification) for each basic vanishing cycle δ outside the picture. This takes place in the case if the cycle δ_* existed before the modification. If it is not the case, the cycle δ_5 does not exist after the modification as well and we simply omit this change of the basis. This change does not influence the corresponding values of the *arf*-function.

At last if we add the cycle $\tilde{\delta}_2' + \tilde{\delta}_4'$ to each basic cycle δ outside the picture with $(\delta_1', \delta) = 1$, we do not change the values of the *arf*-function and cut all the conections between the part of the diagram, corresponding to the cicles δ_i' with $1 \le i \le 4$, and the remaining part of the diagram. This splits the Dynkin diagram into two parts (and the *Arf*-invariant into two summands). The first part (which includes the cycles δ_+, δ_-, δ_5', and all the basic vanishing cycles outside the picture) coincides with the Dynkin diagram of the curve before the modifications. The values of the *arf*-function for all the basic cycles in it are equal to 1 and therefore the *Arf*-invariant of it is equal to the *Arf*-invariant of the curve before the modification. The second part of the diagram coincides with the Dynkin diagram A_4. The values of the *arf*-function for all the basic cycles in it are equal to 1 if the considered modification is of the type J^- and are equal to 0 if the considered modification is of the type J^+. The *Arf*-invariant of this part of the diagram is equal to 1 in the first case and is equal to 0 in the second. This proves the Theorem.

The first author expresses his gratitute to the University Complutense of Madrid and to the University of Kaiserslautern, the hospitality of which gave a possibility to work on the problem. Research partially supported by the ISF Grants M91000 and MD8000.

References.

[A] V.I.Arnold. Topological invariants of plane curves and caustics. University Lecture Series, Vol.5, Providence R.I., 1994.

[AGV] V.I.Arnold, S.M.Gusein-Zade, A.N.Varchenko. Singularities of

Differentiable Maps, Vol.II, Birkhauser: Boston, Basel, Berlin, 1988

[GZ] S.M.Gusein-Zade. Intersection matrices for certain singularities of functions of two variables. Functional Anal. and its Appl., 8, 1974, pp. 10-13.

[J] W.A.M.Janssen. Skew-symmetric vanishing lattices and their monodromy groups. Mathematische Annalen, 266 (1983), pp.115-133.

Independent University of Moscow
B. Vlas'evskiy per., 11,
Moscow 121002, Russia

Produit cyclique d'espaces
et opérations de Steenrod

Max Karoubi

Dans cet article nous poursuivons notre étude des opérations de Steenrod commencée dans [3], en utilisant le produit cyclique des espaces et, indirectement, leur produit symétrique infini par Dold et Thom [1]. Afin de faciliter sa lecture, nous commencerons par rappeler quelques définitions fondamentales.

Soit X un ensemble simplicial pointé, de point base $*$, et soit R un anneau commutatif. Nous désignerons par $L(X; R)$ (ou simplement $L(X)$ s'il n'y a pas de risque de confusion) le R-module simplicial libre de base X avec la relation $* = 0$. Il est bien connu que les groupes d'homotopie de $L(X)$ sont (presque par définition) les groupes d'homologie réduits de X à coefficients dans R. En particulier, si X est un modèle simplicial de la sphère S^n (par exemple l'ensemble simplicial des applications continues des Δ_r dans S^n, $r = 0, 1, 2, \ldots$), $L(X)$ est un modèle de l'espace d'Eilenberg–Mac Lane $K(R; n)$: tous ses groupes d'homotopie sont nuls sauf le $n^{\text{ième}}$ qui est isomorphe à R.

Si Y est un deuxième ensemble simplicial connexe pointé, un accouplement R-bilinéaire

$$L(X) \wedge L(Y) \longrightarrow L(X \wedge Y)$$

est défini en associant à $x = \sum_i \lambda_i x_i$ et $y = \sum_j \mu_j y_j$ l'expression suivante[1] :

$$x \cup y = \sum_{i,j} \lambda_i \mu_j x_i \wedge y_j.$$

Pour $X = S^q$ et $Y = S^r$ par exemple, il permet de définir une application

$$K(R, q) \wedge K(R, r) \longrightarrow K(R, q + r).$$

Elle induit le **cup-produit** usuel en cohomologie

$$H^q(X; R) \times H^r(X; R) \longrightarrow H^{q+r}(X; R)$$

car $H^m(X; R) \approx [X, K(R, m)]$ en général.

Un cas particulier important de cup-produit est la **puissance p^{ième}** d'un élément $x = \sum_i \lambda_i x_i$ de $L(X)$: elle s'écrit

$$x^p = \sum_{i_1, i_2, \ldots, i_p} \lambda_{i_1} \lambda_{i_2} \ldots \lambda_{i_p} \, x_{i_1} \wedge x_{i_2} \wedge \cdots \wedge x_{i_p}$$

---

[1] où λ_i et $\mu_j \in R$ et $x_i \in X$, $y_j \in Y$.

dans $L(X^{\wedge p}) = L(X \wedge X \wedge \cdots \wedge X)$ (p facteurs X).

Supposons maintenant que p soit un nombre premier. Le **p-produit cyclique normalisé** de X est le quotient de $X^{\wedge p}$ par l'action du groupe cyclique C_p permutant circulairement les facteurs, la diagonale de $X^{\wedge p}$ étant identifiée au point base. Il est noté $CP_p^+(X)$. Si $\varphi : X^{\wedge p} \longrightarrow CP_p^+(X)$ désigne l'application canonique, il est clair que $\varphi(x^p) = \varphi((\sum_i \lambda_i x_i)^p)$ est la somme de p copies de $\wp(x)$, où l'application simpliciale

$$\wp : L(X) \longrightarrow L(CP_p^+(X))$$

est définie par la formule suivante

$$\wp(x) = \wp(\sum_i \lambda_i x_i) = \sum_{\langle i_1, i_2, \ldots, i_p \rangle} \lambda_{i_1} \lambda_{i_2} \cdots \lambda_{i_p}\, \varphi(x_{i_1} \wedge x_{i_2} \wedge \cdots \wedge x_{i_p}),$$

$\langle i_1, i_2, \ldots, i_p \rangle$ désignant la classe de la suite $(i_1, i_2, \ldots, i_p)$ modulo l'action du groupe cyclique C_p. En suivant la terminologie de Steenrod [5], qui est tout à fait adaptée à notre contexte, nous appellerons $\wp(x)$ la **p-puissance réduite** de $x = \sum_i \lambda_i x_i$. Cette décomposition de $\varphi(x^p)$ est l'analogue simplicial de la propriété algébrique bien connue:

$$\left(\sum_i a_i\right)^p - \sum_i (a_i)^p \quad \text{divisible par } p,$$

où p est un nombre premier, les a_i étant des variables commutant circulairement (dans un anneau non nécessairement commutatif).

Si $p = 2$ par exemple, on a

$$\wp(\sum_i \lambda_i x_i) = \sum_{i,j} \lambda_i \lambda_j \varphi(x_i \wedge x_j).$$

Le **p-produit cyclique** (non normalisé) $CP_p(X)$ est le quotient de $X^{\wedge p}$ par l'action du groupe cyclique C_p (la diagonale X n'étant pas identifiée au point base). La p-puissance réduite $\wp$ ne se factorise *pas* en général par $L(CP_p(X))$ à homotopie près, sauf pour $R = \mathbf{Z}$ et X connexe. En effet, dans ce cas, on peut remplacer $L(X)$ par le produit symétrique infini de la réalisation géométrique $|X|$ de X [1]. La p-puissance réduite peut être alors définie par la formule suivante[2]

$$\wp'(\sum_i y_i) = \sum_{\langle i_1, i_2, \ldots, i_p \rangle} \varphi'(y_{i_p} \wedge y_{i_2} \wedge \cdots \wedge y_{i_p}).$$

---

[2] $\sum_i y_i = (y_1, y_2, \ldots y_n, \cdots)$ appartenant au produit symétrique infini de $|X|$.

Dans cette formule, $\varphi' : |X|^{\wedge p} \longrightarrow CP_p(|X|)$ est l'application quotient, les indices i_α ne sont pas tous égaux et $\langle i_1, i_2, \ldots, i_p \rangle$ désigne la classe de la suite $(i_1, i_2, \ldots, i_p)$ modulo l'action du groupe cyclique C_p. En réalisant géométriquement les ensembles simpliciaux (ce qui nous permet d'identifier X à $|X|$ par abus d'écriture), on peut alors démontrer que le diagramme suivant est commutatif à homotopie près (pour $R = \mathbf{Z}$):

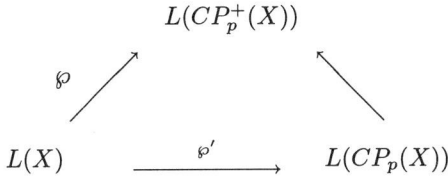

Considérons le cas particulier où X est la sphère S^q. La détermination des homologies de $CP_p(X)$ et $CP_p^+(X)$ est alors aisée, car $CP_p(X)$ est la suspension itérée d'un espace lenticulaire. Par exemple, $CP_2(S^q)$ (resp. $CP_2^+(S^q)$) a le type d'homotopie de $S^{q+1}(RP_{q-1})$ (resp. $S^{q+1}(RP_{q-1}) \vee S^{q+1}$). Si $R = \mathbf{Z}/p$ et si une base des espaces vectoriels d'homologie de $CP_p(X)$ ou $CP_p^+(X)$ à coefficients dans $\mathbf{Z}/p$ est choisie, les espaces $L(CP_p(X))$ et $L(CP_p^+(X))$ s'écrivent comme des produits d'espaces d'Eilenberg–Mac Lane explicites (cf. l'annexe C).

Pour p impair, l'application φ ci-dessus détermine ainsi une opération cohomologique

$$K(\mathbf{Z}/p, q) \longrightarrow K(\mathbf{Z}/p, q+1) \times K(\mathbf{Z}/p, q+2) \times \cdots \times K(\mathbf{Z}/p, pq)$$

dont les composantes $K(\mathbf{Z}/p, q) \longrightarrow K(\mathbf{Z}/p, q+2j)$ sont en fait nulles pour $j \neq i(p-1)$ et égales[3] aux opérations de Steenrod P^i pour $j = i(p-1)$. Pour $p = 2$, l'application

$$\varphi : L(S^q) \longrightarrow L(CP_2^+(S^q))$$

induit une opération cohomologique

$$K(\mathbf{Z}/2, q) \longrightarrow K(\mathbf{Z}/2, q+1) \times K(\mathbf{Z}/2, q+2) \times \cdots \times K(\mathbf{Z}/2, 2q)$$

dont les composantes $K(\mathbf{Z}/2, q) \longrightarrow K(\mathbf{Z}/2, q+j)$ sont les carrés de Steenrod Sq^j. Par ailleurs, considérons la cofibration homotopique

$$BC_p^+ \wedge X \longrightarrow EC_p^+ \wedge_{C_p} X^{\wedge p} \longrightarrow CP_p^+(X)$$

où BC_P^+ (resp. EC_p^+) est l'espace classifiant du groupe C_p (resp. son fibré universel) auquel on a ajouté un point en dehors. L'application du produit

---

[3] à une normalisation près ; cf. 2.6 et [3].

cyclique normalisé $CP_p^+(X)$ dans la suspension de $BC_p^+ \wedge X$ (suite de Puppe) définit un homomorphisme $L(CP_p^+(X)) \longrightarrow L(S^1 \wedge BC_p^+ \wedge X)$. L'application composée $L(X) \xrightarrow{\varphi'} L(CP_p^+(X)) \longrightarrow L(S^1 \wedge BC_p^+ \wedge X)$, induit (en lui appliquant le foncteur π_n avec $n > 1$) des *opérations de Steenrod en homologie* (cf. 2.8 et 2.9)

$$H_n(X) \longrightarrow H_n(S^1 \wedge BC_p^+ \wedge X)$$

pour tout anneau commutatif de coefficients. En homologie mod. p, ces opérations sont duales des opérations de Steenrod usuelles (à une normalisation près).

Ce qui précède est détaillé dans les §1 et 2 de cet article. Dans le §3 (largement conjectural), nous généralisons les considérations précédentes en remplaçant le groupe cyclique par un sous-groupe quelconque du groupe symétrique $\mathfrak{S}_n$, en particulier un sous-groupe de Sylow. Enfin, dans le §4, une autre généralisation est présentée dans le cadre de la cohomotopie: l'analogue de l'espace $L(X)$ (pour $R = \mathbf{Z}$) et X *connexe* est $Q(X) = \Omega^\infty S^\infty(X)$. On utilise ici une description de cet espace détaillée dans [6] par exemple, ainsi que quelques résultats de J. Lannes [4]. Ce §4 n'est pas entièrement original; il recoupe des travaux antérieurs sur les invariants de James–Hopf. Citons notamment un article de N. Kuhn en relation avec le sujet (Transactions American Math. Society, 283, 1984, pp. 303–313).

Enfin, trois annexes sont consacrées à la démonstration de résultats techniques (cohomologie bivariante, transfert et espaces lenticulaires, etc.…), dont l'auteur n'a pas trouvé trace sous cette forme dans la littérature. Ils permettent notamment de mieux conceptualiser les théorèmes démontrés dans les deux premiers paragraphes.

Remerciements. Je remercie W. Dwyer, N. Kuhn, J. Lannes, H. Miller et C. Mouët pour des commentaires fort utiles après une première rédaction de cet article, ainsi que R. Jardine qui a inspiré l'annexe C.

Table des matières

Les notations suivantes sont fréquemment utilisées dans cet article: si G est un groupe discret et X un G-espace, X_{hG} est l'espace de Borel associé, soit $X_{hG} = EG \times_G X, EG$ désignant le G-fibré principal universel sur BG; X_{hG} est le "quotient homotopique" de X par G. De même, X^{hG} désigne l'ensemble des "points fixes homotopiques", c'est-à-dire l'espace $\mathcal{S}(\xi)$ des sections du fibré $\xi = EG \times_G X$ sur BG. Les espaces X_G et X^G sont respectivement le quotient X/G et l'espace des points fixes de X par l'action de G. On a évidemment des applications $X_{hG} \longrightarrow X_G$ et $X^G \longrightarrow X^{hG}$. Elles ne sont pas des équivalences d'homotopie en général.

1. Produit cyclique d'espaces et puissance réduite

1.1. Commençons ce paragraphe par quelques considérations classiques sur le transfert en homologie et en cohomologie usuelles. Soit G un groupe fini opérant à droite sur un ensemble simplicial T et soit T/G l'ensemble quotient (l'action de G n'est pas nécessairement libre). Un homomorphisme de "transfert"

$$\psi : L(T/G) \longrightarrow L(T)$$

est alors défini sur les éléments de base en posant $\psi(t) = \sum_{g \in G} \bar{t}^g$, pour $t \in T/G$, $\bar{t}$ relevé quelconque de t, $\bar{t}^g$ désignant le transformé de $\bar{t}$ par g. En fait, cette formule (qui est indépendante du choix de $\bar{t}$) montre que l'image de ψ est contenue dans $L(T)^G$, ensemble des points fixes de $L(T)$ par l'action de G. Dans le cadre topologique, X étant connexe, $L(X)$ peut être remplacé par le produit symétrique infini (pour $R = \mathbf{Z}$): il convient de remarquer que ψ est alors continue, induite en fait par une application continue de T/G dans $L(T)^G$.

1.2. Considérons la somme directe

$$\Delta(T) = L(T^G) \oplus L(T/G) \ .$$

Une application (notée encore ψ) de $\Delta(T)$ dans $L(T)^G$ est définie en posant

$$\psi(t_0, t) = t_0 + \sum_{g \in G} \bar{t}^g \ .$$

Si T et U sont deux G-ensembles *pointés*, le "cup-produit"

$$\Delta(T) \wedge \Delta(U) \longrightarrow \Delta(T \wedge U)$$

associe aux deux couples (t_0, t) et (u_0, u) le couple (w_0, w) suivant:
1) $w_0 = t_0 \wedge u_0$
2) w est la classe dans $L((T_\wedge U)/G)$ de $\tilde{w}$ dans $L(T_\wedge U)$ défini par

$$\tilde{w} = \sum_{g \in G} \left[(\bar{t}^g \wedge u_0) + (t_0 \wedge \tilde{u}^g) + (\bar{t}^g \wedge \tilde{u}) \right] \ .$$

Un accouplement

$$L(T)^G \wedge L(U)^G \longrightarrow L(T \wedge U)^G$$

est défini aussi de manière évidente. Enfin, si $s : BG \longrightarrow EG \times_G T$ (resp. $s' : BG \longrightarrow EG \times_G U$) est une section de la fibration $EG \times_G T \longrightarrow BG$ (resp. $EG \times_G U \longrightarrow BG$), le "cup-produit" de s et s' est la section $u \mapsto (\tilde{u}, s \wedge s')$, où $u \mapsto (\tilde{u}, s)$ (resp. $u \mapsto (\tilde{u}, s')$) est une section de la première fibration (resp. la seconde). Il est clair que les applications $\Delta(T) \longrightarrow L(T)^G$ et $L(T)^G \longrightarrow L(T)^{hG}$ sont compatibles avec ces structures "multiplicatives."

1.3. Voici une variante de ce qui précède lorsque $p = 0$ dans R, p étant l'ordre du groupe G. Pour un ensemble T, posons de manière générale $T_0 = T/T^G$. Le transfert précédent induit alors un homomorphisme

$$L(T_0/G) \longrightarrow L(T)^G .$$

En effet, si $x \in L(T^G)$, où T^G est considéré comme sous-ensemble de T, son transfert est un multiple de p, donc est nul dans $L(T)$. Si on pose de même

$$\Delta_0(T) = L(T^G) \oplus L(T_0/G)$$

l'homomorphisme de $\Delta(T)$ dans $L(T)^G$ induit un homomorphisme

$$\Delta_0(T) \longrightarrow L(T)^G .$$

Si U est un deuxième G-ensemble, le "cup-produit" $\Delta(T) \wedge \Delta(U) \longrightarrow \Delta(T \wedge U)$ défini plus haut induit un accouplement analogue sur les ensembles Δ_0 compatible avec l'homomorphisme de transfert $\Delta_0(T) \longrightarrow L(T)^G$ comme il a été explicité en 1.2.

1.4. Si X est un ensemble pointé, la "puissance $p^{\text{ième}}$" est une application définie par la formule suivante, déjà utilisée en [3]:

$$\varphi_p \left(\sum \lambda_i x_i \right) = \sum_{i_1, i_2, \ldots, i_p} \lambda_{i_1} \lambda_{i_2} \cdots \lambda_{i_p} x_{i_1} \wedge x_{i_2} \cdots \wedge x_{i_p} . \tag{1}$$

Soient maintenant X et Y deux ensembles pointés. On définit un accouplement (ou produit) R-bilinéaire:

$$L(X^{\wedge p}) \wedge L(Y^{\wedge p}) \longrightarrow L((X \wedge Y)^{\wedge p}) .$$

C'est celui associé à l'accouplement évident d'ensembles $X^{\wedge p} \wedge Y^{\wedge p} \longrightarrow (X \wedge Y)^{\wedge p}$. Ces définitions étant posées, nous avons l'identité

$$\varphi_p(x, y) = \varphi_p(x) \varphi_p(y)$$

avec des notations évidentes. Si X est une sphère S^q et Y une sphère $S^{q'}$, celle-ci doit être légèrement modifiée par un automorphisme de $S^{p(q+q')}$, car le "produit"

$$L(S^{pq}) \wedge L(S^{pq'}) \longrightarrow L(S^{p(q+q')})$$

peut être défini indépendamment de la décomposition de chacune des sphères en p facteurs. Cet automorphisme est de degré $(-1)^{qq'p(p-1)/2}$.

1.5. Considérons maintenant le quotient de $X^{\wedge p}$ par l'action du groupe cyclique $C_p = \mathbf{Z}/p$ opérant par permutation circulaire des facteurs, la diagonale X étant identifiée au point base. L'ensemble obtenu est le **p-produit cyclique normalisé** de X; il est noté $CP_p^+(X)$. Le groupe cyclique C_p opère de même sur l'ensemble des suites $(i_1, i_2, \ldots, i_p)$. On note $\langle i_1, i_2, \ldots, i_p \rangle$ la classe de $(i_1, i_2, \ldots, i_p)$ dans l'ensemble quotient. Si p est un *nombre premier*[4], l'application composée

$$\psi_p : L(X) \longrightarrow L(X^{\wedge p}) \longrightarrow L(CP_p^+(X))$$

s'écrit alors[5]

$$\psi_p\left(\sum \lambda_i x_i\right) = p \sum_{\langle i_1, i_2, \ldots, i_p \rangle} \lambda_{i_1} \lambda_{i_2} \cdots \lambda_{i_p} x_{i_1} \wedge x_{i_2} \cdots \wedge x_{i_p} .$$

Cette identité permet de décrire la *p-puissance réduite*

$$\wp : L(X) \longrightarrow L(CP_p^+(X))$$

par la formule suivante:

$$\wp\left(\sum \lambda_i x_i\right) = \sum_{\langle i_1, i_2, \ldots, i_p \rangle} \lambda_{i_1} \lambda_{i_2} \cdots \lambda_{i_p} x_{i_1} \wedge x_{i_2} \cdots \wedge x_{i_p}$$

avec les conventions précisées dans la Note 5 (le point base étant toujours identifié à 0). Notons $\Delta_p^+(X)$ l'ensemble $\Delta_0(X^{\wedge p}) = L(X) \oplus L(CP_p^+(X))$ (avec les notations de 1.3) et considérons la composition suivante:

$$\gamma : \Delta_p^+(X) \longrightarrow L(X^{\wedge p})^{C_p} \longrightarrow L(X^{\wedge p})^{hC_p}$$

où la première application est le transfert (cf. 1.3).

1.6 Théorème. *L'application γ est multiplicative: on a l'identité $\gamma(z \cdot t) = \gamma(z) \cdot \gamma(t)$ pour $z \in \Delta_p^+(X)$ et $t \in \Delta_p^+(Y)$.*

---

[4] Le cas où p est non premier sera envisagé au §3.

[5] Si cette convention ne prête pas à confusion, on écrira de la même manière un élément de $L(X^{\wedge p})$ et sa classe dans $L(CP_p^+(X))$.

Démonstration. C'est une conséquence immédiate des considérations générales de 1.2-4.

1.7. Supposons maintenant que $p = 0$ dans R. Pour $x = \sum \lambda_i x_i \in L(X)$, posons $x_{(p)} = \sum (\lambda_i)^p x_i$. L'application $\overline{\wp} : L(X) \longrightarrow L(X) \oplus L(CP_p^+(X)) = \Delta_p^+(X)$ définie par $\overline{\wp}(x) = (x_{(p)}, \wp(x))$ est alors compatible avec la structure multiplicative sur le deuxième groupe définie en 1.2 et 1.3 (il convient de poser $T = X^{\wedge p}, U = Y^{\wedge p}, G = C_p$ avec les notations de 1.2). De manière précise, calculons $\wp [(\sum \lambda_i x_i) \cdot (\sum \mu_j y_j)] = \wp (\sum \lambda_i \mu_j x_i \wedge y_j)$. Il vient le développement suivant sur chaque simplexe:

$$\wp \left(\sum \lambda_i \mu_j x_i \wedge y_j \right) = \sum_{\langle (i_1, j_1) \cdots (i_p, j_p) \rangle} \lambda_{i_1} \mu_{j_1} \cdots \lambda_{i_p} \mu_{j_p} (x_{i_1} \wedge y_{j_1}) \wedge \cdots \wedge (x_{i_p} \wedge y_{j_p}) ,$$

où la somme est étendue à toutes les suites de couples (i_α, j_α) non tous égaux (le groupe C_p opérant ainsi avec des orbites de cardinal p). D'après les considérations générales de 1.2., les éléments de cette somme se répartissent en 3 types (cf. plus généralement le §3):

1. Les i_α et les j_β ne sont pas tous égaux; on trouve alors

$$\wp \left(\sum \lambda_i x_i \right) \cdot \wp \left(\sum \mu_j y_j \right)$$

pour le produit défini en 1.2.

2. Les i_α sont égaux et les j_β ne sont pas tous égaux; il vient alors

$$\sum_{(i, \langle j_1, j_2, \ldots, j_p \rangle)} (\lambda_i)^p \mu_{j_1} \cdots \mu_{j_p} (x_i \wedge y_{j_1}) \wedge \cdots \wedge (x_i \wedge y_{j_p}) .$$

Cette dernière expression sera notée $x_{(p)} \cdot \wp(y)$.

3. Les j_β sont égaux et les i_α ne sont pas tous égaux; on trouve alors

$$\sum_{(\langle i_1, i_2, \ldots, i_p \rangle, j)} \lambda_{i_1} \cdots \lambda_{i_p} (\mu_j)^p (x_{i_1} \wedge y_j) \wedge \cdots \wedge (x_{i_p} \wedge y_j)$$

expression qui sera notée de même $\wp(x) \cdot y_{(p)}$.

En résumé, nous obtenons ainsi le théorème suivant:

1.8 Théorème. *Supposons que $p = 0$ dans R. Alors l'application*

$$\overline{\wp} : L(X) \longrightarrow L(X) \oplus L(CP_p^+(X)) = \Delta_p^+(X)$$

définie par $\overline{\wp}(x) = (x_{(p)}, \wp(x))$ vérifie la propriété suivante

$$\overline{\wp}(xy) = \overline{\wp}(x) \overline{\wp}(y)$$

pour $x \in L(X)$ *et* $y \in L(Y)$, *les deux membres de la formule appartiennent à* $\Delta_p^+(X \wedge Y)$, *pour la structure multiplicative définie en 1.2 et 1.3.*

Considérons le cas particulier où X est la sphère S^q. Alors $X^{\wedge p} \cong S^{pq}$, le groupe cyclique C_p opérant de manière naturelle sur S^{pq}. Le lemme suivant est évident (remarquer que S^{pq} est le compactifié d'Alexandrof de $\mathbf{R}^{pq}$):

1.9 Lemme. *L'espace quotient S^{pq}/C_p s'identifie à la $(q+1)^{\text{ième}}$ suspension (non réduite) de l'espace lenticulaire S^{pq-q-1}/C_p, où C_p opère sur S^{pq-q-1} par l'involution antipodique si $p = 2$ et grâce à la représentation induite sur $\mathbf{C}^{q(p-1)/2}$ par $\rho \oplus \rho^2 \oplus \cdots \oplus \rho^{(p-1)/2}$, où ρ est la représentation usuelle de $\mathbf{Z}/p = \mu_p$ dans $\mathbf{C}^*$, si p est impair.*

1.10. *Remarque.* Si X est une sphère, il en résulte que l'application canonique de X dans $CP_p^+(X)$ est homotopiquement triviale. Donc $CP_p^+(X)$ a le type d'homotopie du "wedge" $SX \vee CP_p^+(X)$, où SX désigne la suspension de X. En particulier, $CP_2^+(S^q)$ a le type d'homotopie de $S^{q+1} \vee S^{q+1}(RP_{q-1})$. Par conséquent, si $R = \mathbf{Z}/2$, l'espace vectoriel simplicial $L(CP_p^+(S^q))$ est le produit suivant d'espaces d'Eilenberg–Mac Lane[6]:

$$K(\mathbf{Z}/2, q+1) \times K(\mathbf{Z}/2, q+2) \times \cdots \times K(\mathbf{Z}/2, 2q) \ .$$

1.11 Proposition. *Soit $R = \mathbf{Z}/2$ et soient P^n et P^m deux quotients de S^n et S^m, compactifiés d'Alexandrof de $\mathbf{R}^n$ et $\mathbf{R}^m$, par une action linéaire du groupe $G = C_2$. Alors l'application*

$$L(P^n) \wedge L(P^m) \longrightarrow L(P^{n+m})$$

définie en 1.2 induit le cup-produit usuel sur les espaces d'Eilenberg–Mac Lane $K(\mathbf{Z}/2, \alpha)$, composantes des espaces $L(P^s)$.[7]

Démonstration. Les espaces P^n et P^m sont des suspensions itérées d'espaces projectifs réels et l'application est déduite d'un accouplement

$$S^{r+1}(RP_{q-1}) \wedge S^{t+1}(RP_{s-1}) \longrightarrow L(S^{r+t+1}(RP_{q+s-1}))$$

avec $r + q = n$ et $t + s = m$. Si q est pair, $H_n(S^{r+1}(RP_{q-1})) \cong \mathbf{Z}/2$ et l'application "dernière cellule" de $S^{r+1}(RP_{q-1})$ dans S^n est de degré un. On peut aussi remarquer qu'une autre interprétation de l'application "dernière cellule" $L(RP_q) \longrightarrow L(S^q)$ est le transfert mod. 2 qui est de degré 1 mod. 2. Ceci est clair si q est impair (car l'action de $\mathbf{Z}/2$ est compatible avec l'orientation).

---

[6] La base de l'homologie de $CP_2^+(S^q)$ est évidente (cf. l'annexe C).

[7] Chaque espace d'homologie étant de dimension un, il n'y a pas d'ambiguïté sur le choix de la base (cf. encore l'annexe C).

Dans le cas général, il suffit d'écrire le diagramme commutatif

$$
\begin{array}{ccccc}
RP_{q-1} & \longrightarrow & RP_q & \longrightarrow & S^q \\
\uparrow & & \uparrow & & \uparrow \\
S^{q-1} & \longrightarrow & S^q & \longrightarrow & S^q \vee S^q
\end{array}
$$

en remarquant que, dans tous les cas, l'application $RP_q \longrightarrow S^q$ induit un iso-morphisme sur les groupes d'homologie à coefficients dans $\mathbf{Z}/2$. Le diagramme commutatif suivant (où les flèches verticales sont canoniquement scindées) permet alors de conclure, en raisonnant par récurrence sur la dimension des espaces projectifs:

$$
\begin{array}{ccc}
L\left(S^{r+1}(RP_{q-1})\right) \wedge L\left(S^{t+1}(RP_{s-1})\right) & \longrightarrow & L\left(S^{r+t+1}(RP_{q+s-1})\right) \\
\downarrow & & \downarrow \\
L(S^{r+q}) \wedge L(S^{t+s}) & \longrightarrow & L(S^{r+t+q+s})
\end{array}
$$

Revenons maintenant au cup-produit qui nous concerne plus particulièrement ici[8] (avec les notations de 1.3):

$$
\Delta_0(S^n) \wedge \Delta_0(S^m) \longrightarrow \Delta_0(S^{n+m}) \ .
$$

1.12 Théorème. *L'accouplement précédent induit le cup-produit usuel au niveau des espaces d'Eilenberg–Mac Lane, composantes des espaces $\Delta_0(S^n)$.*

Démonstration. D'après la proposition 1.11, il induit déjà les accouplements usuels

$$
K(\mathbf{Z}/2, q+\alpha) \wedge K(\mathbf{Z}/2, q'+\alpha') \longrightarrow K(\mathbf{Z}/2, q+q'+\alpha+\alpha') \ ,
$$

sauf peut-être pour α ou $\alpha' = 0$ ou 1. Dans ce cas, utilisons le transfert $\Delta_0(S^n) \longrightarrow L(S^n)^{hC_2}$ décrit en 1.1 et 1.2. Puisqu'il est compatible avec les structures multiplicatives d'après 1.3, il suffit de montrer qu'il induit un iso-morphisme sur les groupes d'homotopie π_q et $\pi_{q+1} \cong \mathbf{Z}/2$. Pour le premier groupe, ceci résulte immédiatement de [3] 1.15. Pour le second, considérons la première opération cohomologique $K(\mathbf{Z}/2, q) \longrightarrow K(\mathbf{Z}/2, q+1)$ induite par la puissance $2^{\text{ième}}$ et qui résulte de la décomposition de $L(S^{2q})^{hC_2}$ en un produit d'espaces d'Eilenberg–Mac Lane (cf. l'annexe C). Cette opération se factorise évidemment par $\Delta_2^+(S^q)$ grâce à la $2^{\text{ième}}$ puissance réduite. Il suffit donc de montrer que l'application $K(\mathbf{Z}/2, q) \longrightarrow K(\mathbf{Z}/2, q+1)$ n'est pas nulle. Puisque c'est l'homomorphisme de Bockstein Sq^1 d'après [3] 1.15, la démonstration du théorème est achevée (cf. aussi l'annexe B).

---

[8] De manière générale, rappelons que pour tout G-espace X, on pose $\Delta_0(X) = L(X^G) \oplus L(X_0/G)$ avec $X_0 = X/X^G$.

1.13 Théorème. *Soit p un nombre premier impair, et soient P^n et P^m deux quotients de S^n et S^m, compactifiés d'Alexandrof de $\mathbf{R}^n$ et $\mathbf{R}^m$, par une action linéaire du groupe C_p, le sous-espace des points fixes étant de dimension strictement positive dans les deux cas. L'application*

$$\Delta(S^n) \wedge \Delta(S^m) \longrightarrow \Delta(S^{n+m})$$

définie en 1.3 (pour $R = \mathbf{Z}$) induit alors le cup-produit usuel sur les espaces d'Eilenberg–Mac Lane facteurs des espaces Δ (compte tenu de diverses normalisations précisées plus loin.)

Démonstration. D'après l'annexe B, avec les notations de 1.3, on sait que le transfert

$$L(S^n/C_p) \longrightarrow L(S^n)^{hC_p}$$

induit un isomorphisme sur les groupes d'homotopie π_i si $i > q+1$, q désignant la dimension de l'espace des points fixes. Par ailleurs, d'après 1.3, il est compatible avec les structures multiplicatives et nous avons déterminé en [3] la structure multiplicative des $L(S^n)^{hC_p}$. Plus précisément, la composante neutre de $L(S^n)^{hC_p}$ a le type d'homotopie d'un produit d'espaces d'Eilenberg–Mac Lane $K(\mathbf{Z}, n) \times \prod_r K(\mathbf{Z}/p, n-2r)$ avec $r > 0$. Pour $n + j$ pair, le choix des générateurs des groupes d'homotopie $\pi_j(L(S^n/C_p)) \approx \pi_j(S^{(n)hC_p})$ équivaut à celui des groupes de cohomologie $H^{n+j}(BC_p; \mathbf{Z})$ du groupe cyclique C_p, choix explicité dans la Note 1 de [3].

Pour être complet, il convient de préciser le choix du générateur de $\pi_q(L(S^q))$. Pour que celui-ci soit compatible avec les structures multiplicatives, il doit correspondre au générateur naturel de $\pi_q(L((S^n)^{hC_p})) \approx H^{n-q}(C_p; \mathbf{Z})$ par l'homomorphisme de transfert

$$\Delta(S^n) = L(S^q) \oplus L(S^n/G) \longrightarrow L(S^n)^{hG} .$$

Un cas particulièrement important est celui où $S^n = S^{pq}$ (produit contracté de p facteurs S^q). Dans ce cas, on a vu au §1 de [3] que l'application $L(S^q) \longrightarrow L(S^{pq})^{hC_p}$ induit sur les générateurs "naturels" des groupes d'homotopie π_q la multiplication par $(-1)^s (m!)^q$ avec $m = (p-1)/2$ et $s = m(q^2 - q)/2$.

1.14 Remarque. Le cup-produit

$$L(CP_p^+(S^q)) \wedge L(CP_p^+(S^{q'})) \longrightarrow L(CP_p^+(S^{q+q'}))$$

introduit un signe au niveau des générateurs canoniques, car l'application

$$S^{pq} \wedge S^{pq'} \longrightarrow S^{p(q+q')}$$

est de degré $(-1)^{qq'p(p-1)/2}$ (comparer avec [3] §1.7).

2. Définition des opérations de Steenrod
par la puissance réduite mod. p

2.1. Soit S l'ensemble des suites non constantes $(\varepsilon_1, \varepsilon_2, \ldots, \varepsilon_p)$ telles que $\varepsilon_i = 0$ ou 1. Le groupe cyclique C_p opère transitivement sur S avec des orbites de cardinal p. Soit $\overline{S}$ l'ensemble quotient et soit $s : \overline{S} \longrightarrow S$ une section arbitraire. On définit une application

$$\Theta_p : L(X) \times L(X) \longrightarrow L(X^{\wedge p})$$

par la formule suivante (où on pose $s(\overline{u}) = (\varepsilon_1, \varepsilon_2, \ldots, \varepsilon_p)$ pour $\overline{u} \in \overline{S}$):

$$\Theta_p \left(\sum \lambda_i^0 x_i^0, \sum \lambda_i^1 x_i^1 \right) = \sum_{\overline{u}(i_1, \ldots, i_p)} \lambda_{i_1}^{\varepsilon_1} \lambda_{i_2}^{\varepsilon_2} \cdots \lambda_{i_p}^{\varepsilon_p} x_{i_1}^{\varepsilon_p} \wedge x_{i_2}^{\varepsilon_2} \wedge \cdots \wedge x_{i_p}^{\varepsilon_p} \ .$$

Si $p = 2$ par exemple, pour un choix évident de s, la formule s'écrit simplement

$$\Theta_2 \left(\sum \lambda_i^0 x_i^0, \sum \lambda_i^1 x_i^1 \right) = \sum_{(i,j)} \lambda_i^0 \lambda_j^1 x_i^0 \wedge x_j^1$$

c'est-à-dire comme le cup-produit usuel. Si $p = 3$, on peut choisir la formule suivante:

$$\Theta_3 \left(\sum \lambda_i^0 x_i^0, \sum \lambda_i^1 x_i^1 \right) = \sum \lambda_{i_1}^0 \lambda_{i_2}^1 \lambda_{i_3}^1 x_{i_1}^0 \wedge x_{i_2}^1 \wedge x_{i_3}^1 + \sum \lambda_{i_1}^0 \lambda_{i_2}^0 \lambda_{i_3}^1 x_{i_1}^0 \wedge x_{i_2}^0 \wedge x_{i_3}^1$$

(car $(0,1,1)$ et $(0,0,1)$ sont deux représentants de $\overline{S}$). Désignons par $\nu :$ $L(X^{\wedge p}) \longrightarrow L(CP_p^+(X))$ l'homomorphisme induit par l'application quotient $X^{\wedge p} \longrightarrow CP_p^+(X)$.

2.2 Théorème. *Soit $\wp : L(X) \longrightarrow L(CP_p^+(X))$ l'application définie par la p-puissance réduite (cf. 1.5) et soient $x^0 = \sum \lambda_i^0 x_i^0$ et $x^1 = \sum \lambda_i^1 x_i^1$ deux éléments de $L(X)$. On a alors la formule suivante:*

$$\wp(x^0 + x^1) = \wp(x^0) + \wp(x^1) + \nu(\Theta_p(x^0, x^1)) \ .$$

Démonstration. Le développement de $\wp(x^0 + x^1)$ donne la somme suivante:

$$\wp(x^0) + \wp(x^1) + \sum_{\substack{(i_1, \ldots, i_p) \\ (\varepsilon_1, \ldots, \varepsilon_p)}} \lambda_{i_1}^{\varepsilon_1} \lambda_{i_2}^{\varepsilon_2} \cdots \lambda_{i_p}^{\varepsilon_p} x_{i_1}^{\varepsilon_1} \wedge x_{i_1}^{\varepsilon_2} \wedge x_{i_2}^{\varepsilon_2} \wedge \cdots \wedge x_{i_p}^{\varepsilon_p} \ ,$$

où (ε_i) est une suite non constante et où le $\sum$ est évidemment égal à $\nu(\Theta_p(x^0, x^1))$. Le théorème en résulte aussitôt.

2.3. Ce dernier résultat a une conséquence importante que nous allons décrire grâce au diagramme cocartésien homotopique

$$
\begin{array}{ccc}
BC_p \times X/BC_p & \longrightarrow & (EC_p \times_{C_p} X^{\wedge p})/BC_p \\
\downarrow & & \downarrow \\
X & \longrightarrow & CP_p(X)
\end{array}
$$

De celui-ci se déduit une application canonique bien définie à homotopie près

$$
CP_p(X) \longrightarrow S^1 \wedge (BC_p^+ \wedge X) = S^1 \wedge (BC_p \times X/BC_p) \,,
$$

où BC_p^+ désigne l'espace classifiant BC_p auquel on a ajouté un point en dehors. Notons que cette application se factorise par $CP_p^+(X) = CP_p(X)/X$ (cf. aussi 2.8). En désignant par $\wp^{\#}$ l'application composée de la puissance réduite $\wp : L(X) \longrightarrow L(CP_p^+(X))$ et de l'homomorphisme $L(CP_p^+(X)) \longrightarrow L(S^1 \wedge BC_p^+ \wedge X)$ induit par l'application précédente, nous pouvons écrire la formule d'additivité suivante (à homotopie près):

$$
\wp^{\#}(x^0 + x^1) = \wp^{\#}(x^0) + \wp^{\#}(x^1)
$$

(car l'application $(x^0, x^1) \mapsto \wp^{\#}(x^0 + x^1) - \wp^{\#}(x^0) + \wp^{\#}(x^1)$ se factorise par $L(X^{\wedge p})$ à homotopie près). En particulier, si X est une sphère S^q et si $L(CP_p^+(S^q))$ est décomposé en un produit d'espaces d'Eilenberg–Mac Lane (cf. l'annexe C), les applications $L(S^q) \longrightarrow L(CP_p^+(S^q)) \longrightarrow K(\mathbf{Z}/p, q + r)$ qu'on en déduit sont homotopiquement additives pour $q + r < pq$. L'application puissance $p^{\text{ième}}$ de $L(S^q) = K(\mathbf{Z}/p, q)$ dans $L(S^{pq}) = K(\mathbf{Z}/p, pq)$ induit aussi une opération additive au niveau de la cohomologie pour des raisons évidentes. Nous allons déterminer ces opérations en commençant par le cas $p = 2$.

2.4 Théorème. *Soit $R = \mathbf{Z}/2$. L'application*

$$
\psi : K(\mathbf{Z}/2, q) = L(S^q) \longrightarrow \Delta_2^+(S^q)
$$

où $\Delta_2^+(S^q) = L(S^q) \oplus [L(S^{q+1}(RP_{q-1}) \vee S^{q+1})]$

$$
\cong K(\mathbf{Z}/2, q) \times K(\mathbf{Z}/2, q+1) \times K(\mathbf{Z}/2, q+2) \times \cdots \times K(\mathbf{Z}/2, 2q)
$$

induit l'opération cohomologique somme des carrés de Steenrod

$$
\varphi(x) = \sum_{i=0}^{\dim(x)} Sq^i(x) \,.
$$

Démonstration. Les opérations cohomologiques *additives* Sq^i sont caractérisées par les axiomes suivants:

1. $Sq^0 = 1$.
2. Si $\dim(x) = n$, $Sq^n(x) = x^2$.
3. $Sq^i(x) = 0$ si $i > \dim(x)$.
4. $Sq^j(xy) = \sum_{k=0}^{j} Sq^k(x) Sq^{j-k}(x)$.

Puisque l'opération cohomologique $\psi(x)$ vérifie les mêmes axiomes d'après ce qui précède, on a bien $\psi(x) = \varphi(x)$, ce qui démontre le théorème. Une autre démonstration peut être aisément construite à partir de [3] §1 et de l'annexe B.4.

2.5. Soit p un nombre premier impair tel que $p = 0$ dans R et soit τ l'endomorphisme idempotent de $L(CP_p^+(X))$ défini par la formule suivante:

$$\tau(x_1 \wedge \cdots \wedge x_p) = \frac{1}{(p-1)!} \sum_{\alpha \in \mathfrak{G}_p/C_p} x_{\sigma(1)} \wedge \cdots \wedge x_{\sigma(p)} \, ,$$

σ parcourant l'ensemble des classes à gauche de l'ensemble quotient $\mathfrak{G}_p/C_p$. Il est clair que l'image de l'application composée

$$L(X) \longrightarrow L(X) \oplus L(CP_p^+(X)) \overset{\tau}{\longrightarrow} L(X) \oplus L(CP_p^+(X))$$

est contenue dans le sous-groupe invariant par τ. Cette remarque implique le résultat suivant:

2.6 Théorème. *Soit p un nombre premier impair et soit $R = \mathbf{Z}/p$. L'application*

$$K(\mathbf{Z}/p, q) = L(S^q) \longrightarrow \Delta_p^+(S^q) \, ,$$

où $\Delta_p^+(S^q) = L(S^q) \oplus L(S^{q+1} \vee (S^{pq}/C_p))$

$$\approx K(\mathbf{Z}/p, q) \times K(\mathbf{Z}/p, q+1) \times K(\mathbf{Z}/p, q+2) \times \cdots \times K(\mathbf{Z}/p, pq)$$

induit des opérations cohomologiques

$$H^q(X; \mathbf{Z}/p) \longrightarrow H^{q+2r}(X; \mathbf{Z}/p) \, .$$

Celles-ci sont triviales si $r \not\equiv 0$ mod. $p - 1$. Si $r \equiv 0$ mod. $p - 1$, l'opération cohomologique obtenue

$$P^i : H^q(X; \mathbf{Z}/p) \longrightarrow H^{q+2i(p-1)}(X; \mathbf{Z}/p)$$

est la puissance réduite de Steenrod (cf. [5] et [3], 1.14), multipliée par le facteur de normalisation $(-1)^r (m!)^q$ explicité en [3] 1.14, m étant égal $(p-1)/2$ et $r = i \cdot m + m(q^2 - q)/2$.

Démonstration. Ce théorème résulte immédiatement de l'annexe B.1-2 et des considérations développées dans [3] §1 (cf. 1.13 plus particulièrement).

2.7 Remarque. Ce qui précède montre que la puissance réduite $L(X) \longrightarrow L(CP_p^+(X))$ *ne se factorise pas* par $L(CP_p^h(X))$ en général, $CP_p^h(X)$ désignant l'espace de Borel $EC_p \times_{C_p} X^{\wedge p}$, l'application $L(EC_p \times_{C_p} X^{\wedge p}) \longrightarrow L(CP_p^+(X))$ étant induite par la deuxième projection. Par contre, nous verrons au §4 que les opérations de Steenrod en *cohomotopie* (définies comme applications de $Q(X)$ dans $Q(CP_p(X))$ se factorisent par $Q(CP_p^h(X))$ (cf. aussi 2.11).

2.8. Réécrivons la cofibration homotopique établie en 2.3:

$$BC_p^+ \wedge X \longrightarrow EC_p^+ \wedge_{C_p} X^{\wedge p} \longrightarrow CP_p^+(X) \,,$$

BC_p^+ (resp. EC_p^+) désignant l'espace classifiant (resp. le fibré principal universel) du groupe cyclique C_p auquel on a ajouté un point en dehors. De cette cofibration, on déduit une application $CP_p^+(X) \longrightarrow S^1 \wedge BC_p^+ \wedge X$ (suite de Puppe). En lui appliquant le foncteur L et en composant par la puissance réduite $L(X) \longrightarrow L(CP_p^+(X))$ comme en 2.3, on obtient une application canonique

$$L(X) \longrightarrow L(S^1 \wedge BC_p^+ \wedge X) \,.$$

Celle-ci permet de retrouver les opérations de Steenrod d'une manière différente pour un anneau commutatif de coefficients R *quelconque*. De manière plus précise, définissons une **théorie de la cohomologie bivariante** sur les espaces pointés X et Y en posant $HH(X,Y) = [Y, L(X)]$ (ensemble des classes d'homotopie d'applications pointées; plus de détails sont proposés dans l'annexe A). Si X (resp. Y) est une sphère S^n, on trouve la cohomologie réduite $\tilde{H}^n(Y)$ (resp. l'homologie réduite $\tilde{H}_n(X)$). Les considérations précédentes permettent de définir une application

$$HH(X,Y) \longrightarrow HH(S^1 \wedge BC_p^+ \wedge X, Y) \,.$$

En particulier, pour $Y = S^n$, on en déduit des *opérations de Steenrod en homologie*

$$H_n(X) \longrightarrow H_n(S^1 \wedge BC_p^+ \wedge X)$$

l'homologie étant prise à coefficients dans un anneau commutatif *arbitraire R*. Celles-ci sont à comparer aux opérations *cohomologiques*

$$H^n(X; R) \longrightarrow H^{np}(X \times B\mathfrak{G}_p; R_\varepsilon)$$

où R_ε est le système local induit par la signature sur $\mathfrak{G}_p$ à valeurs dans $\pm 1 \subset R$, définies dans [3]. Si $R = \mathbf{Z}/p$, la formule de Künneth permet d'écrire les opérations de Steenrod homologiques comme morphismes

$H_n(X) \longrightarrow \bigoplus_r H_{n-r}(X)$ (les générateurs de l'homologie du groupe discret C_p étant choisis de manière explicite: cf. la Note 1 de [3]).

2.9 Théorème. *Soit $n - r > 0$. Si $R = \mathbf{Z}/p$, le morphisme $H_n(X) \longrightarrow H_{n-r}(X)$ défini ci-dessus est dual de l'opération de Steenrod non normalisée $D_s : H^{n-r}(X) \longrightarrow H^n(X)$ définie en [3] §1.3 et 1.5 avec $s = p(n - r) - r$.*

Démonstration. Ecrivons $L(S^1 \wedge BC_p^+)$ comme somme des groupes abéliens simpliciaux $\bigoplus_{\alpha=1}^{\infty} L(S^\alpha)$ (à homotopie près; cf. l'annexe C). On a alors $L(S^1 \wedge BC_p^+ \wedge X) \approx \bigoplus_{\alpha=1}^{\infty} L(S^\alpha) \otimes L(X)$. Soit maintenant $u \in H^q(X)$, représentée par une application $f_u : X \longrightarrow L(S^q)$. L'opération de Steenrod $L(S^q) \longrightarrow L(S^1 \wedge BC_p^+ \wedge S^q) \approx \bigoplus_{\alpha=1}^{\infty} L(S^\alpha) \otimes L(S^q)$ induit une application $L(S^q) \longrightarrow L(S^q \wedge S^\alpha)$, donc, par composition avec f_u, une application $X \longrightarrow L(S^q \wedge S^\alpha)$, c'est-à-dire un élément $P_\alpha(u) \in H^{q+\alpha}(X)$. La correspondance $u \mapsto P_\alpha(u)$ définit l'opération de Steenrod *cohomologique* D_s avec $s = (p-1)q - \alpha$. Par ailleurs, soit $v \in H_{q+\alpha}(X)$, représentée par une application $S^{q+\alpha} \longrightarrow L(X)$. Alors, l'opération *homologique* associée à α, soit Q_α, s'obtient par la composition des morphismes suivants:

$$S^{q+\alpha} \longrightarrow L(X) \longrightarrow L(X \wedge S^\alpha) = L(X) \otimes L(S^\alpha) .$$

On a ainsi $Q_\alpha(v) \in H_{q+\alpha}(X \wedge S^\alpha) \cong H_q(X)$.

Le produit scalaire $\langle v, P_\alpha(u) \rangle$ est le degré de la composition

$$S^{q+\alpha} \longrightarrow L(X) \longrightarrow L(S^q) \longrightarrow L(S^q \wedge S^\alpha)$$

tandis que le produit scalaire $\langle Q_\alpha(v), u \rangle$ est le degré de la composition

$$S^{q+\alpha} \longrightarrow L(X \wedge S^\alpha) \longrightarrow L(S^q \wedge S^\alpha)$$

ou encore

$$S^{q+\alpha} \longrightarrow L(X) \longrightarrow L(X \wedge S^\alpha) \longrightarrow L(S^q \wedge S^\alpha) .$$

Il suffit donc de vérifier la commutativité du diagramme suivant:

$$
\begin{array}{ccc}
L(X) & \longrightarrow & L(S^q) \\
\downarrow & & \downarrow \\
L(X \wedge S^\alpha) & \longrightarrow & L(S^q \wedge S^\alpha)
\end{array}
$$

Elle résulte de la naturalité de l'application $L(X) \longrightarrow L(X \wedge S^\alpha)$.

2.10 Remarque. Soit x un élément du noyau de l'application de Steenrod en homologie $H_n(X) \longrightarrow H_n(S^1 \wedge BC_p^+ \wedge X)$. Grâce à la cofibration homotopique

$$BC_p^+ \wedge X \longrightarrow EC_p^+ \wedge_{C_p} X^{\wedge p} \longrightarrow CP_p^+(X)$$

on peut appliquer à x une *opération de Steenrod secondaire*. Le résultat appartient au conoyau de l'application $H_n(BC_p^+ \wedge X) \longrightarrow H_n(EC_p^+ \wedge_{C_p} X^{\wedge p})$, soit $H_n(BC_p \times X/BC_p) \longrightarrow H_n(EC_p \times_{C_p} X^{\wedge p})/BC_p)$. Ce cas se présente par exemple si x appartient à l'image de l'homomorphisme de Hurewicz $\pi_n^s(X) \longrightarrow H_n(X)$ (cf. le §4).

2.11 Remarque. Dans ce paragraphe, à l'exception des dernières considérations, nous nous sommes restreints à $R = \mathbf{Z}/p$. Cependant, la 2-puissance réduite par exemple (pour $R = \mathbf{Z}$) définit une application

$$K(\mathbf{Z}, q) \cong L(S^q) \longrightarrow \Delta_2(S^q) \cong K(\mathbf{Z}, q) \times K(\mathbf{Z}/2, q+2) \times \cdots$$
$$\cdots \times K(\mathbf{Z}/2, q+4) \times K(\mathbf{Z}/2, 2q)$$

d'après l'annexe C. Cette extension des définitions est néanmoins illusoire. En effet, d'après la fin de l'annexe **C**, l'application canonique $\Delta_2^+(S^q) \longrightarrow S_2^+(S^q)/2$ induit sur les espaces d'Eilenberg–Mac Lane facteurs des Δ le produit de l'application identique par l'homomorphisme de Bockstein. Puisque $Sq^{2i+1} = Sq^1 Sq^{2i}$, les opérations cohomologiques ainsi définies sur $K(\mathbf{Z}, q)$ se déduisent de celles définies sur $K(\mathbf{Z}/2, q)$ par l'application composée $K(\mathbf{Z}, q) \longrightarrow K(\mathbf{Z}/2, q) \xrightarrow{Sq^{2i}} K(\mathbf{Z}/2, q+2i)$.

3. Généralisation

3.1. Nous nous proposons de généraliser partiellement les considérations des deux paragraphes précédents en remplaçant le groupe cyclique C_p, p premier, par un sous-groupe G du groupe symétrique $\mathfrak{G}_n$, n étant un entier quelconque. Pour simplifier les considérations du début, nous supposerons que $R = \mathbf{Z}$, ce qui permet de remplacer $L(X)$, X connexe pointé, par le produit symétrique infini de X, d'un maniement plus simple et plus géométrique [1].

Si $I = (i_1, i_2, \ldots, i_n)$ est une suite d'entiers et si $g \in G$, nous faisons opérer g à *droite* sur cette suite en posant $I^g = (i_{g(1)}, i_{g(2)}, \ldots, i_{g(n)})$. De manière parallèle, g opère à droite sur $X^{\wedge n}$ par $(x_i, x_2, \ldots, x_n)^g = (x_{g(1)}, x_{g(2)}, \ldots, x_{g(n)})$. De manière générale, si U est un G-espace et si H est un sous-groupe de G, on désigne par $U^{(H)}$ la réunion des sous-espaces de U formé des éléments invariants par H ou par un sous-groupe conjugué gHg^{-1}. En particulier, considérons la réunion

$$(X^{\wedge n})^{(H)} = \bigcup_g (X^{\wedge n})^{gHg^{-1}},$$

évidemment invariante par l'action de G. L'espace quotient $(X^{\wedge n})^{(H)}/G$, noté simplement $X_{(H)}$, va jouer le rôle du produit cyclique de X étudié dans le paragraphe précédent.

De manière précise, soit $x = \sum_{i=1}^{m} x_i$ un élément de $L(X)$ (puissance symétrique infinie de X). La puissance $n^{\text{ième}}$, soit x^n, appartient à $L(X^{\wedge n})$ et s'écrit comme la somme de m^n termes $x_{i_1} \wedge x_{i_2} \wedge \cdots \wedge x_{i_n}$, somme qui peut se décomposer de la manière suivante suivant le "type" de la suite $I = (i_1, i_2, \ldots, i_n)$. Le *type* $\langle H \rangle$ est formé de suites I dont le stabilisateur est un sous-groupe conjugué de H. Leur orbite $\langle i_1, i_2, \ldots, i_n \rangle$ est donc de cardinal $|G|/|H|$. Bien entendu, le type $\langle H \rangle$ ne dépend que de la classe de conjugaison (notée aussi $\langle H \rangle$) de H dans G. Si $G = C_p$, p premier par exemple, il n'y a que deux types: le groupe G lui-même (il s'agit alors des suites constantes) et le groupe trivial (les indices i_α ne sont pas tous égaux).

A chaque type $\langle H \rangle$, on associe la "H-puissance réduite" de x comme l'expression suivante appartenant à $L(X_{\langle H \rangle}) := L((X^{\wedge n})^{(H)}/G)$:

$$\sum_{\langle i_1, i_2, \ldots, i_n \rangle} x_{i_1} \wedge x_{i_2} \wedge \cdots \wedge x_{i_n} \ .$$

Cette somme est étendue aux suites de type $\langle H \rangle$. La "puissance réduite totale" est l'application

$$\wp : L(X) \longrightarrow \bigoplus_{\langle H \rangle} L(X_{\langle H \rangle})$$

où chaque composante de $\wp$, soit

$$\wp_H : L(X) \longrightarrow L(X_{\langle H \rangle})$$

correspond à la somme explicitée ci-dessus. En particulier, pour H réduit à l'élément neutre 0, on en déduit une application

$$\wp_0 : L(X) \longrightarrow L(X^{\wedge n}/G) = L(X_{\langle 0 \rangle}) \ .$$

Un homomorphisme de "transfert"

$$\phi_H : L(X_{\langle H \rangle}) \longrightarrow L(X^{\wedge n})^G$$

est défini en associant à la classe de l'élément $x = x_1 \wedge x_2 \wedge \cdots \wedge x_n$ dans $(X^{\wedge n})^{gHg^{-1}}$ la somme suivante

$$y = \sum_{t \in G/H'} x^{t^{-1}}$$

où t parcourt l'ensemble des classes à gauche de G par $H' = gHg^{-1}$. Cette somme est bien invariante par G. Elle est indépendante du choix de g: si r est le cardinal de H, il suffit de remarquer que $r \cdot y$ est la somme des x^t, t parcourant le groupe G. Puisque $L(X^{\wedge n})$ est un $\mathbf{Z}$-module libre, l'élément y est bien défini par la formule ci-dessus.

3.2 Théorème. *L'application composée*

$$L(X) \longrightarrow \bigoplus_{\langle H \rangle} L(X_{\langle H \rangle}) \xrightarrow{\sum \phi_H} L(X^{\wedge n})^G \longrightarrow L(X^{\wedge n})^{hG}$$

coïncide avec la puissance $n^{\text{ième}}$ *équivariante définie dans* [3].

Démonstration. Il suffit de décomposer $\left(\sum x_i\right)^n$ suivant les différentes classes $\langle H \rangle$ de sous-groupes de G.

3.3. Soit K un deuxième sous-groupe de G tel que $H \subset K$. On suppose que G opère à droite sur deux espaces pointés U et V. On peut alors définir de manière générale un "cup-produit" qui généralise celui introduit en 1.2, soit:

$$\varphi : L(U^{(H)}/G) \wedge L(V^{(K)}/G) \longrightarrow L((U \wedge V)^{(H)}/G) ,$$

par la formule suivante

$$\varphi(u, v) = \sum_{g \in G/K'} u^g \wedge v$$

où $K' = \alpha K \alpha^{-1}$ si $u \in \alpha H \alpha^{-1}$. Il convient de vérifier que cette somme ne dépend que des classes de u et v dans $U^{(H)}/G$ et $V^{(K)}/G$ respectivement. Pour u ceci est clair, car l'application $g \mapsto \beta g$ induit une translation de G/K' (défini par la relation d'équivalence $x \sim y$ ssi $x^{-1}y \in K'$). Par ailleurs, $\varphi(u, v^\beta) = \sum_{g \in G/K'} u^{g\beta^{-1}} \wedge v$ et la translation $g \mapsto g\beta^{-1}$ induit un isomorphisme de G/K' sur G/K'' avec $K'' = \beta \alpha K \alpha^{-1} \beta^{-1}$. Enfin, la somme définissant $\varphi(u, v)$ est indépendante du choix de α. En effet, elle peut s'écrire aussi $\sum_{g \in G/K'} u \wedge v^{g^{-1}}$, ou encore $\frac{1}{r} \sum_{g \in G} u \wedge v^{g^{-1}}$, r étant le cardinal de K.

3.4. Un exemple intéressant est celui où G est un p-sous-groupe de Sylow[9] du groupe symétrique $\mathfrak{S}_n$ avec $n = p^r$, p premier. On désigne par $S = S_{p^r}$ un tel sous-groupe. Considérons le quotient (noté $X^{\overline{\wedge} n}$) de $X^{\wedge n}$ par l'action de S. La puissance réduite (correspondant au sous-groupe trivial $H = 0$) est alors définie comme une application

$$\wp_0 : L(X) \longrightarrow L(X^{\overline{\wedge} n})$$

qui associe à $x = \sum x_i$ l'élément suivant de $L(X^{\overline{\wedge} n})$:

$$\wp_0(x) = \sum_{\langle i_1, i_2, \ldots, i_n \rangle} x_{i_1} \wedge x_{i_2} \wedge \cdots \wedge x_{i_n} .$$

---

[9] Rappelons que deux sous-groupes de Sylow sont conjugués.

Dans cette expression $I = (i_1, i_2, \ldots, i_n)$ est une suite d'indices sur lesquels le groupe S opère librement (i.e. $(i_{g(1)}, i_{g(2)}, \ldots, i_{g(n)}) \neq (i_1, i_2, \ldots i_n)$ si $g \neq e$ et $\langle i_1, i_2, \ldots, i_n \rangle$ désigne une orbite ($\wp_0$ a été défini dans un contexte plus général en 3.1). Si S n'opère pas librement sur la suite I, il existe un élément g d'ordre une puissance de p tel que $(i_{g(1)}, i_{g(2)}, \ldots, i_{g(n)}) = (i_1, i_2, \ldots, i_n)$; en décomposant la permutation g en cycles, on voit que p indices i_α au moins sont égaux.

3.5. Il convient de noter la non-trivialité de l'opération cohomologique $\wp_0 : L(X) \longrightarrow L(X^{\wedge n}/S)$. Celle-ci est sans doute reliée aux compositions d'opérations de Steenrod classiques. Vérifions ce fait pour $r = 2$, soit $n = p^2$. Dans ce cas, le groupe $S = S_{p^2}$ est de cardinal p^{p+1}; il est engendré par p cycles d'ordre p, soit $(1, 2, \ldots, p)$, $(p + 1, p + 2, \ldots, 2p), \ldots, (p^2 - p + 1, \ldots, p^2)$ ainsi que par le cycle consistant à permuter circulairement les paquets précédents. Le fait que S opère librement sur une suite $(i_1, i_2, \ldots, i_{p^2})$ revient donc à écrire que chaque paquet $(i_1, \ldots, i_p), (i_{p+1}, \ldots, i_{2_p})$, etc.... n'est pas formé de suites constantes et que les p paquets ne sont pas tous égaux. Considérons maintenant la composition

$$L(X) \longrightarrow L(X^{\wedge n}/S) \longrightarrow L(X^{\wedge n})^S \longrightarrow L(X^{\wedge n})^{C_p \times C_p} \longrightarrow L(X^{\wedge n})^{h(C_p \times C_p)} \tag{1}$$

où la deuxième application est le transfert (cf. 1.1 et 3.1) et la troisième la restriction, le groupe $C_p \times C_p$ étant inclus dans S par la permutation simultanée des paquets précédents entre eux et dans leur intérieur. Nous pouvons de même composer la puissance de Steenrod $L(X) \longrightarrow L(X^{\wedge p})^{hC_p}$ avec elle-même (cf. [3] 2.11). La différence est la somme $\sum [t_0 t_s D_{0,s}(x) + t_r t_0 D_{r,0}(x) - t_0 t_0 D_{0,0}(x)]$ avec les notations de [3] 2.11. On en déduit que la composition (1) ci-dessus est la somme $\sum_{r \neq 0, s \neq 0} t_r t_s D_{r,s}(x)$ qui n'est pas triviale en général.

3.6. Plaçons nous maintenant dans le cadre simplicial, ce qui nous permettra de considérer un anneau commutatif de base quelconque R. Un élément x de $L(X) = L(X; R)$ s'écrit alors comme une somme $\sum \lambda_i x_i$, où les λ_i sont des scalaires appartenant à R (*les x_i étant distincts*). On exprime l'élément $\wp_0(x)$ de manière légèrement différente comme la somme

$$\sum_{\langle i_1, i_2, \ldots, i_n \rangle} \lambda_{i_1} \lambda_{i_2} \cdots \lambda_{i_n} x_{i_1} \wedge x_{i_2} \wedge \cdots \wedge x_{i_n}$$

où la suite $(i_1, i_2, \ldots, i_n)$ est de stabilisateur trivial. Cependant, *l'application ainsi définie n'est pas simpliciale*, car le cardinal des orbites par l'action de $S = S_{p^r}$ peut se modifier par une opération face. Pour remédier à cet inconvénient, notons $X_0^{\wedge n}$ le sous-ensemble de $X^{\wedge n}$ formé des suites $x = (x_1, x_2, \ldots, x_n)$ dont le stabilisateur (pour l'action de S) n'est pas trivial. Ce sous-ensemble est invariant par l'action de S. On pose $X_0^{\overline{\wedge} n} = X_0^{\wedge n}/S$ et $X^{\overline{\wedge} n} = X^{\wedge n}/S$. Si on interprète la puissance réduite comme une application

$$L(X) \longrightarrow L(X^{\overline{\wedge} n}/X_0^{\overline{\wedge} n}) \, ,$$

elle devient alors simpliciale. Cette définition est bien une généralisation de la situation développée dans le paragraphe précédent. En effet, si on choisit $r = 1, S = C_p$, on a $X_0^{\wedge n} = X$ et $X^{\overline{\wedge}n}/X_0^{\overline{\wedge}n} = CP_p^+(X)$.

3.7 Théorème. *La puissance réduite définit un morphisme simplicial*

$$\wp : L(X) \longrightarrow L(X^{\overline{\wedge}n}/X_0^{\overline{\wedge}n}) \, .$$

Avec des notations évidentes, elle peut être définie simplement par la formule suivante

$$\wp(x) = \frac{1}{|S_{p^r}|} \cdot (x)^{p^r} \, .$$

Démonstration. Elle résulte immédiatement des considérations précédentes grâce à la méthode décrite en 1.5.

3.8 Proposition. *L'espace quotient $X^{\overline{\wedge}n}/X_0^{\overline{\wedge}n}$ a le type d'homotopie du quotient des espaces de Borel correspondant*

$$ES \times_S X^{\wedge n}/ES \times_S X_0^{\wedge n} \, .$$

Démonstration. En effet, l'action de S est libre sur le complémentaire $X^{\wedge n} - X_0^{\wedge n}$. L'espace quotient écrit dans l'énoncé est donc égal à $X^{\overline{\wedge}n} - X_0^{\overline{\wedge}n}$, car $X^{\overline{\wedge}n} = X^{\wedge n}/S$ et $X_0^{\overline{\wedge}n} = X_0^{\wedge n}/S$.

3.9. Soit $R = \mathbf{Z}/p^s$, s étant un entier arbitraire. Les espaces $X^{\overline{\wedge}n}$ et $X_0^{\overline{\wedge}n}$ étant difficiles à décrire directement, nous allons introduire une variante en remplaçant S par le groupe symétrique $G = \mathfrak{G}_n$ lui-même. De manière précise, définissons un endomorphisme τ de $L(X^{\overline{\wedge}n})$ en posant

$$\tau(x_1 \wedge \cdots \wedge x_n) = \frac{|S|}{|G|} \sum_{\sigma \in G/S} x_{\sigma(1)} \wedge \cdots \wedge x_{\sigma(n)} \, .$$

Alors τ est idempotent sur $L(X^{\wedge n}/S)$, car c'est la composée $\tau_1 \circ \tau_2$ de deux applications telles que $\tau_2 \cdot \tau_1 = \mathrm{Id}$. D'une part, $\tau_1 : L(X^{\wedge n}/G) \longrightarrow L(X^{\wedge n}/S)$ est définie par la même formule que τ; d'autre part $\tau_2 : L(X^{\wedge n}/S) \longrightarrow L(X^{\wedge n}/G)$ est l'application quotient. L'image de la puissance réduite $L(X) \longrightarrow L(X^{\wedge n}/S)$ est invariante par τ: elle est donc contenue dans $L(X^{\wedge n}/G)$ considéré comme sous-groupe de $L(X^{\wedge n}/S)$ par l'application τ_1. Soit $X_1^{\wedge n}$ la réunion des sous-espaces conjugués de $X_0^{\wedge n}$ par l'action des éléments g de G, c'est-à-dire $X_1^{\wedge n} = \cup g X_0^{\wedge n} g^{-1}$. Alors $X_1^{\wedge n}$ est la réunion des sous-espaces de $X^{\wedge n}$ formé des produits $x_1 \wedge x_2 \wedge \cdots \wedge x_n$, où p éléments x_α au moins sont égaux comme on le voit en décomposant les permutations en cycles (cf. 3.4).

3.10 Théorème. *Soit $R = \mathbf{Z}/p^s$, s étant un entier arbitraire avec $n = p^r$. Alors la n-puissance réduite mod. p^s définit une application simpliciale*

$$L(X) \longrightarrow L(EG \times_G X^{\wedge n}/EG \times_G X_1^{\wedge n})$$

où $G = \mathfrak{S}_n$ et où $X_1^{\wedge n}$ est la réunion de sous-espaces de $X^{\wedge n}$ formé des produits $(x_1 \wedge x_2 \wedge \cdots \wedge x_n)$ dans lesquels p éléments x_α au moins sont égaux.

Considérons le cas particulier où X est un sphère de dimension q. Alors $X^{\wedge n}$ est le compactifié d'Alexandrof de $\mathbf{R}^{qn}$. Le sous-espace $X_1^{\wedge n}$ est réunion de compactifiés d'Alexandrof de sous-espaces de $\mathbf{R}^{qn}$ de dimension $m = q(n-p+1)$ qui sont obtenus en choisissant p coordonnées parmi n (il y a donc $\binom{n}{p}$ tels sous-espaces). Nous allons déterminer une partie importante de la topologie de l'espace $X_1^{\wedge n}$ grâce au lemme suivant:

3.11 Lemme. *L'espace $X_1^{\wedge n}$ contient en facteur direct (à homotopie équivariante près) le bouquet de sphères $\vee_t S^m$, où $m = q(n - p + 1)$, et où t parcourt l'ensemble des parties à p éléments d'un ensemble à n éléments ($n = p^r$ comme plus haut).*

Démonstration. L'espace $X_1^{\wedge n}$ est le compactifié d'Alexandrof de la réunion de sous-espaces de $\mathbf{R}^n$ qui sont de dimension m: chacun de ces sous-espaces est l'image par l'action de G du sous-espace V défini par l'ensemble des vecteurs s'écrivant $(x_1, x_2, \ldots, x_n)$, où x_i appartient à $\mathbf{R}^q$ et où $x_1 = x_2 = \cdots = x_p$. On définit maintenant une application

$$\varphi : \bigvee_t S^m \longrightarrow X_1^{\wedge n}$$

où chaque flèche $S^m \longrightarrow X_1^{\wedge n}$ est induite par l'inclusion d'un sous-espace. Par ailleurs, $W^g = \left(\bigcup_{h \neq g} V^h\right) \cap V^g$ est un sous-ensemble de codimension ≥ 1 de V^g. Il existe donc une application de degré un, soit $\theta_g : (V^+)^g \longrightarrow (V^+)^g$ (où V^+ désigne la sphère S^m, compactifiée d'Alexandrof de $\mathbf{R}^m$) telle que $\theta_g(W^g) = \{\infty\}$. On définit alors une application équivariante

$$\psi : X_1^{\wedge n} \longrightarrow \bigvee_t S^m$$

comme étant égale à θ_g sur chaque cellule de dimension m, $\{\infty\}$ ailleurs. Il est clair que la composée $\psi \cdot \varphi$ est homotope à l'identité de manière équivariante.

3.12 Théorème. *Pour $i < qn$, l'homologie relative $H_i(X^{\wedge n}, X_1^{\wedge n})$ (à coefficients dans R) contient en facteur direct $H_{i-m}(BK; R)$, où $K = \mathfrak{S}_p \times \mathfrak{S}_{n-p}$ est le sous-groupe évident de $G = \mathfrak{S}_n$ ($n = p^r; m = q(n - p + 1)$).*

Démonstration. Cette homologie relative est nulle en degrés $> qn$. Pour les degrés inférieurs à qn, c'est celle de la paire $(EG \times_G X^{\wedge n}, EG \times_G X_1^{\wedge n})$, donc de $EG \times_G X_1^{\wedge n}$, d'après l'isomorphisme de Thom appliqué au fibré $EG \times_G X^{\wedge n}$

sur BG (ce fibré est orienté si p est impair; si $p = 2$, il convient d'utiliser des coefficients locaux). Par ailleurs, le lemme précédent implique que l'espace $EG \times_G X_1^{\wedge n}$ contient en facteur direct $EG \times_G Y$, où $Y = \bigvee_t S^m$, t parcourant l'espace homogène G/K. D'après le lemme de Shapiro et la suite spectrale de Serre, on a par ailleurs $H_i(EG \times_G Y) \cong H_{i-m}(BK)$, ce qui achève la démonstration du théorème.

3.13. *Remarque.* Le théorème précédent permet de définir pour tout espace T des opérations cohomologiques:

$$H^q(T; Z/p^s) \longrightarrow H^i(T; G_i)$$

avec $n = p^r$; $m = q(n - p + 1)$; $i < qn$ et où $G_i = H^{i-m}(BK; \mathbf{Z}/p^s)$, avec $K = \mathfrak{G}_p \times \mathfrak{G}_{n-p}$. Un problème ouvert est d'étudier leur relation avec les opérations de Steenrod usuelles.

4. Opérations de Steenrod en cohomotopie[10]

4.1. Si X est un espace pointé *connexe*, rappelons d'abord que $Q(X) = \Omega^\infty S^\infty X$ peut être décrit à partir du groupe symétrique infini de la manière suivante (cf. [6] par exemple). Dans la réunion disjointe des espaces suivants

$$X_n = \mathbf{R}^{\infty(n)} \times_{\mathfrak{G}_n} X^n$$

(où $\mathbf{R}^\infty$ désigne la réunion de tous les espaces $\mathbf{R}^n$), identifions[11] l'élément de X_{n+m} défini par

$$(t_1, t_2, \ldots, t_n, t_{n+1}, \ldots, t_{n+m}; x_1, x_2, \ldots, x_n, *, *, \ldots, *) \, ,$$

où $*$ désigne le point base, à l'élément de X_n défini par

$$(t_1, t_2, \ldots, t_n; x_1, x_2, \ldots, x_n) \, .$$

Il est alors démontré dans [6] que l'espace ainsi obtenu a le type d'homotopie de $Q(X)$. Cet espace peut être muni d'une loi $+$ de $X_n \times X_m$ dans X_{n+m}. Elle associe à $(t, x) = (t_1, \ldots, t_n; x_1, \ldots, x_n)$ et $(u, y) = (u_1,, \ldots, u_m; y_1,, \ldots, y_m)$ la suite

$$(t, x) + (u, y) = (t_1,, \ldots, t_n, \tilde{u}_1, \ldots, \tilde{u}_m; x_1,, \ldots, x_n, y_1, \ldots, y_m) \, ,$$

---

[10] Comme il a été signalé dans l'introduction, ce paragraphe ne prétend pas être totalement original. Il est relié à de multiples travaux antérieurs sur les invariants de James–Hopf et les scindages de Snaith. Cependant, la comparaison avec les opérations de Steenrod semble être nouvelle.

[11] De manière générale on désigne par $Y^{(n)}$ le sous-espace de Y^n formé des suites $(y_1, y_2, \ldots, y_n)$ telles que les éléments y_i soient *distincts*.

où les $\tilde{u}$ sont les images des u dans un *autre* facteur $\mathbf{R}^\infty$. Cette loi $+$ est indépendante du choix de cet autre facteur à homotopie près.

L'élément $(t_1, t_2, \ldots, t_n; x_1, x_2, \ldots, x_n)$ considéré plus haut sera noté de manière plus suggestive sous la forme

$$t_1 \cdot x_1 + t_2 \cdot x_2 + \cdots + t_n \cdot x_n$$

où les t_i sont *distincts*. En particulier, la projection sur le facteur x du couple (t, x) avec $t = (t_1, t_2, \ldots, t_n)$ et $x = (x_1, x_2, \ldots, x_n)$ (ce qui revient à projeter $\mathbf{R}^\infty$ sur le point base) induit une application

$$Q(X) \longrightarrow L(X) \,,$$

où $L(X)$ désigne le produit symétrique infini (ou son groupe symétrisé). Si X est la sphère S^q par exemple, nous retrouvons l'application canonique $Q(S^q) \longrightarrow K(\mathbf{Z}, q)$: elle induit l'homomorphisme de Hurewicz *cohomologique* $\pi_s^q(X) \longrightarrow H^q(X)$ de la cohomotopie stable vers la cohomologie pour tout espace connexe X. De manière duale, en appliquant le foncteur π_n au morphisme $Q(X) \longrightarrow L(X)$, on trouve l'homomorphisme de Hurewicz *homologique* $\pi_n^s(X) \longrightarrow H_n(X)$.

4.2. Le "cup-produit" classique

$$Q(X) \wedge Q(Y) \longrightarrow Q(X \wedge Y)$$

est induit par l'accouplement $X_m \wedge Y_n \longrightarrow (X \wedge Y)_{mn}$ défini par

$$[(t_i, x_i), (t_j, x_j)] \mapsto (t_i t_j, x_i \wedge x_j)$$

où $t_i t_j$ désigne le couple $(t_i, t_j) \in \mathbf{R}^\infty \oplus \mathbf{R}^\infty \cong \mathbf{R}^\infty$. Le résultat final sera noté de manière abrégée

$$\sum_{(i,j)} t_i t_j x_i \wedge x_j \,.$$

On définit de même une opération "puissance $p^{\text{ième}}$":

$$\phi : Q(X) \longrightarrow Q(X^{\wedge p})$$

par la formule

$$(t, x) \mapsto \sum_{(i_1, i_2, \ldots, i_p)} t_{i_1} \cdot t_{i_2} \cdots t_{i_p} x_{i_1} \wedge x_{i_2} \wedge \cdots \wedge x_{i_p}$$

(somme de n^p produits).

Grâce à l'identification de $\mathbf{R}^\infty$ à $\mathbf{R}^\infty \oplus \mathbf{R}^\infty \cdots \oplus \mathbf{R}^\infty$, le groupe symétrique $\mathfrak{S}_p$ opère *à droite* sur $Q(X^{\wedge p})$ par la formule suivante (σ étant un élément de $\mathfrak{S}_p$):

$$\sum_{(i_1, i_2, \ldots, i_p)} t_{i_1} \cdot t_{i_2} \cdots t_{i_p} x_{i_1} \wedge x_{i_2} \wedge \cdots \wedge x_{i_p}$$

$$\mapsto \sum_{(i_1, i_2, \ldots, i_p)} t_{i_{\sigma(2)}} \cdot t_{i_{\sigma(2)}} \cdots t_{\sigma(p)} x_{\sigma(1)} \wedge x_{\sigma(2)} \wedge \cdots \wedge x_{\sigma(p)} .$$

4.3. Il est clair que l'image de ϕ est invariante par cette action. La puissance $2^{\text{ième}}$ par exemple est définie par

$$(t_i, x_j) \mapsto \sum t_i \cdot t_j \cdot x_i \wedge x_j .$$

L'action du groupe $\mathbf{Z}/2$ sur $\mathbf{R}^\infty \oplus \mathbf{R}^\infty$ et $X \wedge X$ équivaut à permuter (i, j) et (j, i), ce qui définit un élément d'ordre 2 dans $\mathfrak{S}_{n^2}$.

Etudions plus en détail l'action de $\mathfrak{S}_p$ sur $Q(X^{\wedge p})$ décrite plus haut. Elle est en fait le produit de deux actions qui commutent entre elles: celle (notée $\mathcal{X}_1$) induite par l'action de $\mathfrak{S}_p$ sur $\mathbf{R}^\infty \cong \mathbf{R}^\infty \oplus \cdots \oplus \mathbf{R}^\infty$ (p facteurs $\mathbf{R}^\infty$) et une deuxième (notée $\mathcal{X}_2$), induite par l'action de $\mathfrak{S}_p$ sur $X^{\wedge p}$, donc sur $Q(X^{\wedge p})$.

4.4 Lemme. *L'action $\mathcal{X}_1$ est homotopiquement triviale.*

Démonstration. Il est clair que le modèle homotopique de $Q(Y)$ ne change pas si on remplace $\mathbf{R}^\infty$ par $\mathbf{R}^\infty \oplus (\mathbf{R}^\infty \oplus \cdots \oplus \mathbf{R}^\infty)$. Sur le premier facteur $\mathbf{R}^\infty$, $\mathfrak{S}_p$ opère trivialement (action induite sur $Q(Y)$ notée $\mathcal{X}_0$). Sur le second facteur $(\mathbf{R}^\infty \oplus \mathbf{R}^\infty \cdots \oplus \mathbf{R}^\infty)$, $\mathfrak{S}_p$ opère par permutation des facteurs $\mathbf{R}^\infty$. On obtient ainsi des équivalences d'homotopie équivariantes

$$(Q(Y), \mathcal{X}_0) \longrightarrow (Q(Y), \mathcal{X}_0 \oplus \mathcal{X}_1) \longleftarrow (Q(Y), \mathcal{X}_1) .$$

Le lemme en résulte.

4.5. Ainsi, à condition d'inverser formellement dans la catégorie des fractions les équivalences d'homotopie équivariantes, on voit que la puissance $p^{\text{ième}}$ induit l'application suivante (le groupe $\mathfrak{S}_p$ opérant sur $X^{\wedge p}$, donc sur $Q(X^{\wedge p})$, par permutation des facteurs):

$$Q(X) \longrightarrow Q(X^{\wedge p})^{h\mathfrak{S}_p} .$$

Par ailleurs, dans le §2 de [3], nous avons défini une application semblable pour les produits symétriques infinis

$$L(X) \longrightarrow L(X^{\wedge p})^{h\mathfrak{S}_p} .$$

Il est clair que le diagramme suivant commute

$$
\begin{array}{ccc}
Q(X) & \longrightarrow & Q(X^{\wedge p})^{h\mathfrak{G}_p} \\
\downarrow & & \downarrow \\
L(X) & \longrightarrow & L(X^{\wedge p})^{h\mathfrak{G}_p}
\end{array}
$$

4.6. Restreignons-nous maintenant au sous-groupe $C_p = \mathbf{Z}/p$ de $\mathfrak{G}_p$, p étant un nombre premier. Pour un CW complexe Y avec C_p action, J. Lannes [4] a démontré l'équivalence d'homotopie suivante (à condition de compléter les espaces considérés au nombre premier p):

$$
Q(Y)^{hC_p} \approx Q(Y^{C_p}) \times Q(\tilde{Y}_{hC_p})^{p-1}
$$

où $\tilde{Y}_{hC_p} = Y_{hC_p}/BC_p$ (cf. les notations dans l'introduction). Appliquons ce théorème à $Y = X^{\wedge p}$, le groupe C_p opérant sur $X^{\wedge p}$ par permutation circulaire des facteurs, le premier facteur $Q(Y^{C_p})$ s'identifiant ainsi à $Q(X)$. Déterminons le second facteur du produit précédent si X est la sphère S^q, donc $Y = S^{pq}$, le groupe C_p opérant sur S^{pq}, considéré comme compactifié d'Alexandrof de $\mathbf{R}^{pq}$. Si p est impair, $\tilde{Y}_{hC_p}$ s'identifie à l'espace de Thom du fibré qV sur BC_p, où V est le fibré suivant:

$$
V = \mathbf{1} \oplus L \oplus L^2 \oplus \cdots \oplus L^{(p-1)/2}
$$

(cf. [3], §1.13: $\mathbf{1}$ désigne le fibré réel trivial de rang un et L le fibré canonoique sur BC_p qui est de rang complexe 1, donc de rang réel 2). Cet espace de Thom est la limite inductive (suivant N) de la $q^{\text{ième}}$ suspension du compactifié d'Alexandrof de $(\mathbf{C}^N - \mathbf{C}^{q(p-1)/2})/\mathcal{R}$. Ici $\mathcal{R}$ est la relation d'équivalence sur $\mathbf{C}^N$ qui identifie deux suites (x_i) et (y_i) si $y_i = \lambda_i x_i$ avec $\lambda_i = \lambda$ si $i > q(p-1)/2$ et $\lambda_i =$ une puissance de λ comprise entre 1 et $(p-1)/2$ pour chaque espace $\mathbf{C}^q$ contenu dans $\mathbf{C}^{q(p-1)/2}$. La limite inductive s'identifie ainsi à $S^q(BC_p/L_{q,p})$, $q^{\text{ième}}$ suspension du quotient de l'espace classifiant BC_p par un espace lenticulaire $L_{q,p}$. Cet espace lenticulaire $L_{q,p}$ est obtenu comme quotient de l'ensemble des vecteurs non nuls de l'espace vectoriel $\mathbf{C}^q \oplus \cdots \oplus \mathbf{C}^q (m = (p-1)/2$ facteurs $\mathbf{C}^q)$ par la relation d'équivalence qui identifie les suites

$$
(x_1, x_2, \ldots, x_m) \text{ et } (\lambda x_1, \lambda^2 x_2, \ldots, \lambda^m x_m)
$$

pour λ racine $p^{\text{ième}}$ de l'unité et $x_i \in \mathbf{C}^q$. Le résultat final est donc une application "puissance $p^{\text{ième}}$"

$$
Q(S^q) \longrightarrow Q(S^q) \times Q(S^q(BC_p/L_{q,p}))
$$

(du moins après complétion des espaces considérés en p). En particulier, l'application canonique de $Q(S^q(BC_p/L_{q,p}))$ dans $Q(S^{q+1}(L_{q,p}))$ permet de définir le morphisme suivant après complétion en p:

$$Q(S^q) \longrightarrow Q(S^{q+1}(L_{q,p})) \ .$$

4.7. On peut procéder de même pour $p = 2$ en remplaçant $\mathbf{C}^n$ par $\mathbf{R}^n$. Après complétion des espaces au nombre 2, on obtient une application

$$Q(S^q) \longrightarrow Q(S^q) \times Q(S^q(RP_\infty/RP_{q-1})) \ ,$$

où RP_{n-1} (resp. RP_∞) désigne l'espace projectif réel de dimension $n-1$ (resp. infinie). Comme en 4.6, cette application se déduit d'un théorème de J. Lannes [4]. Il en résulte une application de $Q(S^q)$ dans $Q(S^{q+1}(RP_{q-1}))$, du moins après complétion au nombre 2.

4.8. Le lecteur attentif aura remarqué des similitudes frappantes entre les considérations de 4.6-7 et celles développées dans les paragraphes précédents. En particulier, pour tout espace X, se pose le problème de l'existence d'une "puissance réduite:"

$$Q(X) \longrightarrow Q(CP_p(X)) \ .$$

Pour cela, reconsidérons la puissance $p^{\text{ième}}$ ordinaire

$$R^{\infty(n)} \times_{\mathfrak{G}_n} X^n \longrightarrow R^{\infty(n^p)} \times_{\mathfrak{G}_{n^p}} (X^{\wedge p})^{n^p}$$

qui a été définie formellement par la formule suivante

$$\sum t_i x_i \mapsto \sum t_{i_1} t_{i_2} \cdots t_{i_p} x_{i_1} \wedge x_{i_2} \wedge \cdots \wedge x_{i_p} \ .$$

Le quotient de $\mathbf{R}^\infty \approx (\mathbf{R}^\infty)^p$ par l'action de C_p est encore un espace contractile. Cependant, les classes des éléments $t_{i_1} t_{i_2} \cdots t_{i_p}$ dans ce quotient ne sont pas distinctes en général. On a $t_{i_1} t_{i_2} \cdots t_{i_p} = t_{j_1} t_{j_2} \cdots t_{j_p}$ si et seulement si les suites $(i_1, i_2, \ldots, i_p)$ et $(j_1, j_2, \ldots, j_p)$ se déduisent l'une de l'autre par permutation circulaire. Quotientons alors *simultanément* $(\mathbf{R}^\infty)^p$ et $X^{\wedge p}$ par l'action du groupe cyclique C_p et restreignons-nous aux suites $(i_1, i_2, \ldots, i_p)$ telles que les i_α ne soient pas tous égaux. L'expression

$$\sum_{\langle i_1, i_2, \ldots, i_p \rangle} t_{i_1} t_{i_2} \cdots t_{i_p} x_{i_1} \wedge x_{i_2} \wedge \cdots \wedge x_{i_p}$$

a alors un sens dans l'espace $((\mathbf{R}^\infty)^p/C_p)^{(m)} \times_{\mathfrak{G}_m} (X^{\wedge p}/C_p)^m$, avec $m = (n^p - n)/p$. En passant à la limite inductive suivant n, on en déduit bien une application

$$\wp : Q(X) \longrightarrow Q(CP_p(X)) \ ,$$

désignée aussi par "puissance réduite mod. p" (la relation entre cette application et celle déduite du théorème de J. Lannes sera faite en 4.15, à la fin de ce paragraphe, grâce à un argument de transfert).

4.9. Contrairement à ce que nous avons vu en homologie et cohomologie ordinaires (cf. 2.3 et 2.7), l'application $\wp : Q(X) \longrightarrow Q(CP_p(X))$ ainsi construite se factorise par $Q(EC_p \times_{C_p} X^{\wedge p}/BC_p) := Q(EC_p^+ \wedge_{C_p} X^{\wedge p})$, où EC_p désigne l'espace $EC_p \cup \{\infty\}$ (L'espace $EC_p^+ \wedge_{C_p} X^{\wedge p}$ sera noté simplement $CP_p^h(X)$: c'est le **produit cyclique homotopique** de X.) En effet, si $\sum t_i x_i \in Q(X)$, on peut lui associer l'élément suivant de $Q(EC_p^+ \wedge_{C_p} X^{\wedge p})$

$$\sum_{\langle i_1, i_2, \ldots, i_p \rangle} c\left(t_{i_1} \cdot t_{i_2} \cdots t_{i_p}\right) \left[t_{i_1} t_{i_2} \cdots t_{i_p}, x_{i_1} \wedge x_{i_2} \wedge \cdots \wedge x_{i_p}\right] .$$

Dans cette expression, $c(t_{i_1} \cdot t_{i_2} \cdots t_{i_p})$ désigne la classe de $t_{i_1} \cdot t_{i_2} \cdots t_{i_p}$ dans $(\mathbf{R}^\infty)^p/C_p$ et le couple $\left[t_{i_1} \cdot t_{i_2} \cdots t_{i_p}, x_{i_1} \wedge x_{i_2} \wedge \cdots \wedge x_{i_p}\right]$ représente un élément de $EC_p^+ \wedge_{C_p} X^{\wedge p}$ qui ne dépend que de la classe $\langle i_1, i_2, \ldots, i_p \rangle$. Puisque les $c(t_{i_1} \cdot t_{i_2} \cdots t_{i_p})$ sont distincts, on a bien défini une application $\wp^h : Q(X) \longrightarrow Q(CP_p^h(X))$ telle que le diagramme suivant commute

$$
\begin{array}{ccc}
Q(X) & \longrightarrow & Q(CP_p^h(X)) \\
\| & & \downarrow \\
Q(X) & \longrightarrow & Q(CP_p(X))
\end{array}
$$

Comme nous l'avons vu en 2.3 et 2.7, l'espace $CP_p^h(X)$ est lié au produit cyclique de X par un diagramme cocartésien

$$
\begin{array}{ccc}
BC_p \times X/BC_p & \longrightarrow & CP_p^h(X) \\
\downarrow & & \downarrow \\
X & \longrightarrow & CP_p(X)
\end{array}
$$

Puisque le cône de l'application $BC_p \times X/BC_p \longrightarrow X$ est la suspension de $BC_p^+ \wedge X$, on a une cofibration homotopique

$$CP_p^h(X) \longrightarrow CP_p(X) \longrightarrow S^1 \wedge BC_p^+ \wedge X .$$

La factorisation montrée plus haut montre donc que l'application composée

$$Q(X) \longrightarrow Q(CP_p(X)) \longrightarrow Q(S^1 \wedge BC_p^+ \wedge X)$$

est homotope à 0, en contraste avec la situation homologique (cf. 2.7). En particulier, l'application composée

$$Q(S^n) \longrightarrow L(S^n) \longrightarrow L(CP_p(S^n)) \longrightarrow L(S^1 \wedge BC_p^+ \wedge S^n)$$

est homotope à 0: ceci implique que les opérations de Steenrod $H^q(X) \longrightarrow$ $H^{q+r}(X; \mathbf{Z}/p)$ sont nulles sur l'image de l'homomorphisme de Hurewicz $\pi_s^q(X) \longrightarrow H^q(X)$ si $r > 0$. Bien entendu, cette nullité des opérations de Steenrod sur l'image de l'homomorphisme de Hurewicz peut se démontrer de manière élémentaire. Cependant, le caractère plus précis que nous avons donné pour cette annulation laisse espérer des "opérations de Steenrod secondaires" (cf. 2.10).

4.10. Le choix d'un point base dans EC_p (par exemple, l'image de l'élément neutre e de EC_p qui est un groupe abélien) définit une application $X^{\wedge p} \longrightarrow CP_p^h(X)$, soit $x_1 \wedge x_2 \cdots \wedge x_p \mapsto (e, x_1 \wedge x_2 \cdots \wedge x_p)$. L'application composée

$$X^{\wedge p} \longrightarrow CP_p^h(X) \longrightarrow CP_p(X)$$

est alors l'application quotient canonique. En particulier, cherchons l'image de la puissance $p^{\text{ième}}$ de $x = \sum t_i x_i$ dans $Q(CP_p^h(X))$. On a

$$x^p = \sum_{(i_1, i_2, \ldots, i_p)} t_{i_1} t_{i_2} \cdots t_{i_p} x_{i_1} \wedge x_{i_2} \wedge \cdots \wedge x_{i_p} \, .$$

Son image dans $Q(CP_p^h(X))$ est

$$\sum (t_i)^p (x_i)^p + \sum_{(i_1, i_2, \ldots, i_p)} t_{i_1} t_{i_2} \cdots t_{i_p} (e, x_{i_1} \wedge x_{i_2} \wedge \cdots \wedge x_{i_p})$$

où la dernière somme est indexée par des suites $(i_1, i_2, \ldots, i_p)$, les i_α n'étant pas tous égaux. Cet élément est homotope (en tant qu'application de $Q(X)$ dans $Q(CP_p^h(X))$ à la somme suivante:

$$\sum_{(i_1, i_2, \ldots, i_p)} t_{i_1} t_{i_2} \cdots t_{i_p} \left[t_{i_1} t_{i_2} \cdots t_{i_p}, x_{i_1} \wedge x_{i_2} \wedge \cdots \wedge x_{i_p} \right]$$

(utiliser la contractibilité de $\mathbf{R}^{\infty(p)}$). En choisissant un domaine fondamental de l'action de C_p sur $\mathbf{R}^{\infty(p)}$, on voit que cette expression s'écrit aussi comme p fois la somme suivante dans $Q(CP_p^h(X))$:

$$\sum_{\langle i_1, i_2, \ldots, i_p \rangle} c\left(t_{i_1} t_{i_2} \cdots t_{i_p} \right) \left[t_{i_1} t_{i_2} \cdots t_{i_p}, x_{i_1} \wedge x_{i_2} \wedge \cdots \wedge x_{i_p} \right]$$

($c(t_{i_1} t_{i_2} \cdots t_{i_p})$ désignant la classe de $t_{i_1} t_{i_2} \cdots t_{i_p}$ dans le domaine fondamental), c'est-à-dire précisément la "puissance réduite" $\wp^h(x)$ de x dans $Q(CP_p^h(X))$ qui a été définie en 4.9.

4.11. Comme en 4.6-7, considérons le cas particulier où X est une sphère S^q. Pour p impair, d'après ce qu'on a vu plus haut, la p-puissance réduite définit donc une application

$$Q(S^q) \longrightarrow Q(CP_p^h(S^q)) = Q(S^q(BC_p/L_{q,p})) \ .$$

Si $p = 2$, les calculs faits en 4.7 montrent que la puissance réduite définit aussi une application

$$Q(S^q) \longrightarrow Q(CP_2^h(S^q)) = Q(S^q(RP_\infty/RP_{q-1})) \ .$$

La relation entre cette puissance réduite et la puissance $p^{\mathrm{ième}}$ ordinaire sera faite en 4.13.

4.12. Nous nous proposons de faire le lien entre les deux points de vue[12] que nous venons de développer sur les opérations de Steenrod en cohomotopie, grâce à une application de transfert analogue à celle définie en homologie et en cohomologie (cf. 1.1). Pour cela, considérons un groupe fini G opérant librement sur un espace pointé V[13] (sauf au point base) et posons $U = V/G$. On peut alors définir une application "transfert"

$$\tau : U \longrightarrow Q(V)$$

de la manière suivante (communication de J. Lannes). Il convient d'abord de plonger G dans un groupe symétrique $\mathfrak{S}_r$ et V dans $\mathbf{R}^{\infty(r)} \cup \{0\} := \mathbf{R}^{\infty(r)+}$ de manière équivariante (c'est possible car V est un CW-complexe de type fini). On a donc le diagramme commutatif

$$
\begin{array}{ccc}
V & \xrightarrow{\ \lambda\ } & \mathbf{R}^{\infty(r)+} \\
\pi \downarrow & & \downarrow \\
U & \xrightarrow{\quad\quad} & \mathbf{R}^{\infty(r)+}/G
\end{array}
$$

($G \subset \mathfrak{S}_r$ opère sur $\mathbf{R}^{\infty(r)+}$ par permutation des coordonnées). L'application τ cherchée associe alors au point u de U l'élément

$$\tau(u) = \sum_{\pi(v)=u} \lambda(v)v \ ,$$

où $\lambda(v) \cdot v \in \mathbf{R}^{\infty(r)} \times_{\mathfrak{S}_r} V^r$. L'application τ s'étend de manière usuelle à $Q(U)$: il suffit de poser

$$\tau\left(\sum_\alpha t_\alpha u_\alpha\right) = \sum_{\pi(y_\alpha)=x_\alpha} t_\alpha \cdot \lambda(v_\alpha) \cdot v_\alpha \ .$$

---

[12] espace des points fixes homotopiques et quotient homotopique.

[13] On suppose ici que V est un CW-complexe de type fini.

Dans cette formule, l'expression $t_\alpha \cdot \lambda(v)$ doit être comprise comme un élément de $\mathbf{R}^\infty \times \mathbf{R}^{\infty(r)}$. L'image du transfert appartient au sous-espace $Q(V)^G$ de $Q(V)$ formé des éléments invariants par G (G opérant *simultanément* sur V et $\mathbf{R}^{\infty(r)}$). Le transfert peut donc être interprété aussi comme l'application composée

$$QU \longrightarrow Q(V)^G \longrightarrow Q(V)^{hG} \,.$$

Sous cette dernière forme, compte tenu du lemme 4.4, il suffit de faire opérer G sur V.

Une première application du transfert est le théorème suivant.

4.13 Théorème. *L'application puissance $p^{\text{ième}}$ "normalisée" en cohomotopie, soit la composition*

$$Q(X) \longrightarrow Q(X^{\wedge p}) \longrightarrow Q(X^{\wedge p}/X)$$

se factorise à travers $Q(CP_p^h(X))$. De manière précise, on a le diagramme commutatif

$$
\begin{array}{ccccc}
Q(X) & \longrightarrow & Q(X^{\wedge p}) & \longrightarrow & Q(X^{\wedge p}/X) \\
\wp^h \downarrow & & & & \downarrow \approx \\
Q(CP_p^h(X)) & & \overset{\tau'}{\longrightarrow} & & Q(EC_p \times X^{\wedge p}/EC_p \times X)
\end{array}
$$

où τ' désigne la projection $Q(CP_p^h(X)) \longrightarrow Q(CP_p^h(X)/BC_p \times X)$, composée par le transfert

$$\tau : Q(CP_p^h(X)/BC_p \times X) = Q(EC_p \times_{C_p} X^{\wedge p}/EC_p \times_{C_p} X)$$
$$\longrightarrow Q(EC_p \times X^{\wedge p}/EC_p \times X) \,.$$

Démonstration. Soit μ un plongement de X dans $\mathbf{R}^\infty$. D'après 4.9, l'application composée $\tau' \cdot \wp^h$ associe à $x = \sum t_i x_i \in Q(X)$ l'expression suivante

$$\sum_{(i_1, i_2, \dots, i_p)} c\left(t_{i_1}, t_{i_2}, \dots, t_{i_p}\right) \cdot t_{i_1} \cdot t_{i_2}, \dots, t_{i_p} \mu(x_{i_1}) \mu(x_{i_2}) \cdots \mu(x_{i_p})$$
$$\left[t_{i_1} \cdot t_{i_2}, \dots, t_{i_p}, x_{i_1} \wedge x_{i_2} \wedge \cdots \wedge x_{i_p}\right]$$

où les i_α ne sont pas tous égaux. Ce calcul est licite car l'action de C_p sur $X^{\wedge p} - X$ est libre. L'élément $(\tau' \cdot \wp^h)(x)$ ainsi trouvé est canoniquement homotope (par une homotopie affine) à une somme analogue où on remplace l'expression $c\left(t_{i_1}, t_{i_2}, \dots, t_{i_p}\right) t_{i_1} \cdot t_{i_2}, \dots, t_i \mu(x_{i_1}) \mu(x_{i_2}) \cdots \mu(x_{i_p})$ simplement par $t_{i_1} t_{i_2} \cdots t_{i_p}$. Modulo X, nous pouvons aussi supposer que les indices i_α sont quelconques: on retrouve bien la puissance $p^{\text{ième}}$.

4.14 Corollaire. *Si p est un nombre premier impair, on a un diagramme commutatif*

$$Q(S^q(BC_p/L_{q,p})) = Q(CP_p^h(S^q))$$

$$Q(S^q) \nearrow \quad \downarrow$$

$$\searrow$$

$$Q(S^{pq})$$

où la fléche $\searrow$ représente la puissance $p^{\text{ième}}$. Si $p = 2$, on a la factorisation suivante:

$$Q(S^q(RP_\infty/RP_{q-1})) = Q(CP_2^h(S^q))$$

$$Q(S^q) \nearrow \quad \downarrow$$

$$\searrow$$

$$Q(S^{2q})$$

Dans les deux diagrammes, la flèche verticale $Q(CP_p^h(S^q)) \longrightarrow Q(S^{pq})$ est la composée

$$Q(CP_p^h(S^q)) \longrightarrow Q(CP_p(S^q)) \longrightarrow Q(S^{pq})$$

où la deuxième application est induite par le "transfert" $\tau : Q(CP_p(S^q)) \longrightarrow Q(S^{pq})$.

Démonstration. Sans restreindre la généralité, on peut quotienter $CP_p(S^q)$ et S^{pq} par S^q, ce qui revient à ajouter un facteur $Q(S^{q+1})$ à la formule (cf. 1.10). Le transfert n'est défini en toute rigueur que pour des actions libres, sauf au point base, ce qui explique les quillemets dans l'énoncé ci-dessus. En ajoutant le facteur $Q(S^{q+1})$, on se ramène au théorème précédent avec $X = S^q$.

Faisons maintenant le lien avec le travail de J. Lannes mentionné en 4.6 et 4.7.

4.15 Théorème. *On a le diagramme commutatif suivant:*

$$
\begin{array}{ccc}
Q(X) & \xrightarrow{\;\;g\;\;} & Q(X^{\wedge p}/X)^{hC_p} \\
\| & & \uparrow \approx \\
Q(X) & \xrightarrow{\wp^{h+}} & Q(EC_p \times X^{\wedge p}/EC_p \times X)^{hC_p}
\end{array}
$$

où g désigne la puissance $p^{\text{ième}}$ équivariante définie en 4.5 et où $\wp^{h+}$ désigne l'application composée

$$Q(X) \xrightarrow{\wp^h} Q(EC_p \times_{C_p} X^{\wedge p}/EC_p \times_{C_p} *) \longrightarrow Q(EC_p \times_{C_p} X^{\wedge p}/EC_p \times_{C_p} X) \xrightarrow{\tau}$$
$$Q(EC_p \times X^{\wedge p}/EC_p \times X)^{C_p} \longrightarrow Q(EC_p \times X^{\wedge p}/EC_p \times X)^{hC_p},$$

τ désignant le transfert.

Démonstration. Il suffit de répéter la démonstration du théorème 4.13 en tenant compte du caractère équivariant de la puissance $p^{\text{ième}}$.

4.16. Analysons enfin $\wp^h(x^0 + x^1)$, où $x^0 = \sum t_i^0 x_i^0$ et $x^1 = \sum t_i^1 x_i^1$. D'après 4.9, nous pouvons écrire

$$\wp^h(x^0+x^1) = \sum_{\substack{\langle i_1,i_2,\ldots,i_p\rangle \\ (e_1,e_2,\ldots,e_p)}} c\left(t_{i_1}^{\varepsilon_1}, t_{i_2}^{\varepsilon_2}, \ldots, t_{i_p}^{\varepsilon_p}\right) \left[t_{i_1}^{\varepsilon_1} t_{i_2}^{\varepsilon_2} \cdots t_{i_p}^{\varepsilon_p}, x_{i_1}^{\varepsilon_1} \wedge x_{i_2}^{\varepsilon_2} \wedge \cdots \wedge x_{i_p}^{\varepsilon_p}\right]$$

avec $\varepsilon_\alpha = 0$ ou 1. Avec les notations de 2.1, nous pouvons définir de même une application $\varphi_p : Q(X) \times Q(X) \longrightarrow Q(EC_p \times X^{\wedge p}/EC_p)$ par la formule suivante (où les ε_α ne sont pas tous égaux):

$$\varphi_p(x^0, x^1) = \sum_{\overline{u},(i_1,\ldots,i_p)} c\left(t_{i_1}^{\varepsilon_1}, t_{i_2}^{\varepsilon_2}, \ldots, t_{i_p}^{\varepsilon_p}\right) \left[t_{i_1}^{\varepsilon_1} t_{i_2}^{\varepsilon_2} \cdots t_{i_p}^{\varepsilon_p}, x_{i_1}^{\varepsilon_1} \wedge x_{i_2}^{\varepsilon_2} \wedge \cdots \wedge x_{i_p}^{\varepsilon_p}\right] .$$

Comme en 2.1, il est clair que $\wp^h(x^0 + x^1) = \wp^h(x^0) + \wp^h(x^1) + \tau(\varphi_p(x^0, x^1))$, où

$$\tau : Q(EC_p \times X^{\wedge p}/EC_p) \longrightarrow Q(EC_p \times_{C_p} X^{\wedge p}/BC_p)$$

est l'application canonique. Puisque EC_p est contractile, on peut remplacer le premier espace par $Q(X^{\wedge p})$. On en déduit le théorème suivant, qui mesure le "degré d'additivité" de l'opération de Steenrod $\wp^h$.

4.17 Théorème. *Soient x^0 et x^1 deux éléments de $Q(X)$. On a alors la formule suivante*

$$\wp^h(x^0 + x^1) = \wp^h(x^0) + \wp^h(x^1) + \tau(\wp_p(x^0, x^1))$$

où $\tau : Q(X^{\wedge p}) \longrightarrow Q(EC_p \times_{C_p} X^{\wedge p}/BC_p)$ est l'application canonique et où, avec les notations de 2.1, $\varphi_p : Q(X) \times Q(X) \longrightarrow Q(X^{\wedge p})$ est définie par l'expression

$$\varphi_p(x^0, x^1) = \sum_{\overline{u},(i_1,\ldots,i_p)} c\left(t_{i_1}^{\varepsilon_1}, t_{i_2}^{\varepsilon_2}, \ldots, t_{i_p}^{\varepsilon_p}\right) x_{i_1}^{\varepsilon_1} \wedge x_{i_2}^{\varepsilon_2} \wedge \cdots \wedge x_{i_p}^{\varepsilon_p} .$$

En particulier, pour $p = 2$, φ_2 est le cup-produit.

Annexe A. Cohomologie bivariante et isomorphisme de Thom [14]

A.1. Dans cette annexe nous rassemblons quelques résultats sur la cohomologie bivariante et l'isomorphisme de Thom que nous n'avons pas trouvés dans la littérature sous cette forme. Rappelons (cf. 2.8) que si X et Y sont deux CW-complexes pointés, le groupe $HH(X, Y)$ est défini comme l'ensemble des classes d'homotopie d'applications pointées de Y dans $L(X)$. Ce groupe HH vérifie des propriétés analogues à celles de la KK-théorie de Kasparov. En particulier, un accouplement bilinéaire

$$HH(X, Y \wedge A) \times HH(A \wedge X', Y') \longrightarrow HH(X \wedge X', Y \wedge Y')$$

est défini de la manière suivante. A partir de $f' : Y' \longrightarrow L(A \wedge X')$, on déduit une application $Y \wedge Y' \longrightarrow Y \wedge L(A \wedge X') \longrightarrow L(Y \wedge A \wedge X')$. L'application $Y \wedge A \longrightarrow L(X)$ et la précédente se composent en

$$Y \wedge Y' \longrightarrow L(L(X) \wedge X') \longrightarrow L(L(X \wedge X')) \longrightarrow L(X \wedge X') \,.$$

Elles définissent ainsi l'accouplement cherché.

A.2. *Cas particuliers* (ici H_n et H^n désignent respectivement l'homologie et la cohomologie réduites).

1. $A = S^0$, $Y = S^n$, $Y' = S^p$. On en déduit l'accouplement

$$H_n(X) \times H_p(X') \longrightarrow H_{n+p}(X \wedge X') \,.$$

2. $A = S^0, X = S^n$, $X' = S^p$. On obtient alors l'accouplement

$$H^n(Y) \times H^p(Y') \longrightarrow H^{n+p}(Y \wedge Y') \,.$$

Les flèches décrites en 1. et 2. représentent évidemment les cup-produits usuels en homologie et en cohomologie.

3. $X' = S^0$, $X = S^n, Y' = S^p$. On obtient alors le "slant-produit"

$$H^n(Y \wedge A) \times H_p(A) \longrightarrow H^{n-p}(Y) \,.$$

4. Enfin, un autre type de produit

$$HH(X, Y) \times HH(Y, Z) \longrightarrow HH(X \wedge Y, Z)$$

---

[14] La rédaction de cette annexe s'inspire en partie de l'article: "Formes topologiques non commutatives," paru aux *Annales Scientifiques de l'Ecole Normale Supérieure* [2].

est défini comme la composition des flèches suivantes (la première étant induite par la diagonale de Y dans $Y \wedge Y$):

$$HH(X,Y) \times HH(Y,Z) \longrightarrow HH(X,Y) \times HH(Y \wedge Y, Z)$$
$$\longrightarrow HH(X \wedge Y, Z) \, .$$

Si $X = S^n$ et $Z = S^p$, on en déduit le "cap-produit:"

$$H^n(Y) \times H_p(Y) \longrightarrow H_{p-n}(Y) \, .$$

Il est parfois commode de poser

$$HH^j(X,Y) = HH(X \wedge S^j, Y) \ \text{ pour } \ j \geq 0$$
$$HH^j(X,Y) = HH(X, S^{-j} \wedge Y) \ \text{ pour } \ j < 0 \, .$$

Les foncteurs $(X,Y) \mapsto HH^*(X,Y)$ sont alors des foncteurs homologiques en X et cohomologiques en Y respectivement.

A.3. On peut interpréter de manière légèrement différente ces définitions de la manière suivante. Fixons un entier $n \geq 0$ et considérons le bicomplexe $(p,q) \mapsto \Omega^{p,q}(X,Y) = \mathrm{Mor}(B^p \wedge Y, L(X \wedge B^{q+1}))$, où p est restreint aux entiers $\leq n$. Les différentielles de ce bicomplexe sont induites par les applications canoniques $B^n \longrightarrow B^{n+1}$ obtenues par la composition évidente $B^n \longrightarrow S^n \longrightarrow B^{n+1}$. En particulier, le sous-complexe $q \mapsto \Omega^{0,q}(S^0, Y) = \mathrm{Mor}(Y, L(B^{q+1}))$ est le complexe de de Rham des "formes topologiques non commutatives" considérées dans [2]. Le bicomplexe général s'écrit schématiquement ainsi

$$
\begin{array}{ccccccc}
\Omega^{n,2} & \longrightarrow & \Omega^{n-1,2} & \longrightarrow & \Omega^{n-2,2} & \longrightarrow & \cdots \\
\uparrow & & \uparrow & & \uparrow & & \\
\Omega^{n,1} & \longrightarrow & \Omega^{n-1,1} & \longrightarrow & \Omega^{n-2,1} & \longrightarrow & \cdots \\
\uparrow & & \uparrow & & \uparrow & & \\
\Omega^{n,0} & \longrightarrow & \Omega^{n-1,0} & \longrightarrow & \Omega^{n-2,0} & \longrightarrow & \cdots
\end{array}
$$

En fait, les lignes sont exactes car la suite

$$0 \longrightarrow \mathrm{Mor}(S^n \wedge Y, Z) \longrightarrow \mathrm{Mor}(B^n \wedge Y, Z) \longrightarrow \mathrm{Mor}(S^{n-1} \wedge Y, Z) \longrightarrow 0$$

est exacte si Z est un groupe topologique contractile. La cohomologie de ce bicomplexe est donc celle du complexe obtenu en considérant le noyau des premières flèches horizontales, soit le complexe

$$\mathrm{Mor}(S^n \wedge Y, L(X \wedge B^1)) \longrightarrow \mathrm{Mor}(S^n \wedge Y, L(X \wedge B^2))$$
$$\longrightarrow \mathrm{Mor}(S^n \wedge Y, L(X \wedge B^3)) \longrightarrow \cdots$$

dont la cohomologie en degré r est isomorphe à $HH(X \wedge S^r, Y \wedge S^n) \cong HH^{r-n}(X,Y)$.

A.4. Venons en maintenant à l'isomorphisme de Thom que nous allons interpréter dans notre langage. Ici V désigne un fibré vectoriel *orienté* de rang n et de base X compacte (cette hypothèse n'est là que pour simplifier le raisonnement; il est facile de généraliser). Soit $B(V)$ (resp. $S(V)$) le fibré en boules (resp. en sphères) de V pour une métrique quelconque sur V. En particulier, V est homéomorphe à $V' = B(V) - S(V)$.

La classe de Thom de V peut être représentée par une application continue $B(V) \longrightarrow L(S^n)$ égale à 0 sur $S(V)$, grâce à un argument standard du type Mayer–Vietoris. Par ailleurs, l'application diagonale de V' dans $V' \times V'$ induit une application *propre* de V' dans $B(V) \times V'$. En considérant les compactifiés d'Alexandrof, on en déduit les applications continues suivantes (où $\tau(V) = B(V)/S(V)$ désigne l'espace de Thom de V, interprété comme compactifié d'Alexandrof de V'):

$$\tau(V) \longrightarrow B(V)^+ \wedge \tau(V) \xrightarrow{\cong} X^+ \wedge \tau(V) \longrightarrow X^+ \wedge L(S^n) \, .$$

A.5 Theoreme (Thom). *L'application composée précédente*

$$\tau(V) \longrightarrow X^+ \wedge L(S^n)$$

induit les isomorphismes suivants pour tout espace pointé Y:

$$HH^{-n}(Y, X^+) \longrightarrow HH(Y, \tau(V)) \quad et \quad HH(\tau(V), Y) \longrightarrow HH^n(X^+, Y) \, .$$

En particulier, pour $Y = S^{p+n}$, on en déduit des isomorphismes

$$H^p(X^+) \longrightarrow H^{p+n}(\tau(V)) \quad et \quad H_p(\tau(V)) \longrightarrow H_{p-n}(X^+) \, .$$

Démonstration. Elle résulte d'un argument standard de Mayer–Vietoris (on se ramène au cas où V est trivial).

Annexe B. Transfert et espaces lenticulaires.

Le but de cette annexe est de rassembler quelques résultats (sans doute bien connus) sur les espaces lenticulaires qui facilitent la compréhension des §1 et 2 de cet article. Nous en donnons ici des démonstrations complètes.

B.1 Théorème. *Soit $\mathbf{C}^{n+1}$ un espace vectoriel complexe euclidien muni d'une action linéaire du groupe $G = C_p$, p premier impair. On suppose que cette représentation est somme directe de représentations irréductibles non triviales. Soit S^{2n+1} la sphère de $\mathbf{C}^{n+1}$. Alors, si $R = \mathbf{Z}$, l'application de transfert détaillée dans le §1, soit*

$$\alpha : L(S^{2n+1}/G) \longrightarrow L(S^{2n+1})^{hG}$$

est une équivalence d'homotopie.

Démonstration. Nous allons montrer que l'application α induit un isomorphisme sur tous les groupes d'homotopie π_i. Ceux-ci sont égaux à 0 pour i pair, à $\mathbf{Z}/p$ pour i impair $\neq 2n+1$ et à $\mathbf{Z}$ pour $i = 2n+1$. Pour le facteur $\mathbf{Z}$, remarquons d'abord qu'il est déterminé pour le premier groupe par le transfert $L(S^{2n+1}/G) \longrightarrow L(S^{2n+1})$ qui induit un isomorphisme sur les groupes π_{2n+1} (cf. 1.11 et 1.13). Par ailleurs, soit

$$\gamma : L(S^{2n+1}) \longrightarrow L(S^{2n+1}) \times \cdots \times L(S^{2n+1}) = M \ ,$$

où M comprend p facteurs $L(S^{2n+1})$, l'application définie par

$$\gamma(u) = (u, u^\sigma, \ldots, u^{\sigma^{p-1}}) \ ,$$

σ étant le générateur du groupe cyclique C_p. Alors γ est équivariante pour l'action de C_p sur M consistant à permuter circulairement les facteurs. Soit $N = \pi_{2n+1}(M) \approx \mathbf{Z}^p$.

Puisque $H^i(G; N) = 0$ pour $i > 0$ et que $H^0(G; N) = \mathbf{Z}$, γ induit une application $L(S^{2n+1})^{hG} \longrightarrow (M)^{hG} = L(S^{2n+1})$ telle que le diagramme suivant commute (t étant le transfert):

$$
\begin{array}{ccc}
L(S^{2n+1}/G) & \longrightarrow & L(S^{2n+1})^{hG} \\
& \searrow\ t \quad \swarrow & \\
& L(S^{2n+1}) &
\end{array}
$$

L'application α induit donc un isomorphisme $\pi_{2n+1}(L(S^{2n+1}/G)) \approx \pi_{2n+1}(L(S^{2n+1})^{hG}) \approx \mathbf{Z}$. En particulier, si $n = 0$, on a $S^1 \approx (L(S^1/G)) \approx L(S^1)^{hG}$.

Nous allons démontrer le théorème général par récurrence sur n en supposant donc $L(S^{2n-1}/G) \approx L(S^{2n-1})^{hG}$. Pour cela, écrivons le diagramme commutatif

$$
\begin{array}{ccccc}
L(S^{2n-1}/G) & \longrightarrow & L(S^{2n+1}/G) & \xrightarrow{\ t\ } & L(S^{2n+1}) \\
\downarrow & & \downarrow & & \| \\
L(S^{2n-1})^{hG} & \longrightarrow & L(S^{2n+1})^{hG} & \xrightarrow{\ t\ } & L(S^{2n+1})
\end{array}
$$

En appliquant le foncteur π_i, $0 < i \leq 2n - 1$, on obtient le diagramme suivant

$$
\begin{array}{ccc}
H_i(S^{2n-1}/G) & \longrightarrow & H_i(S^{2n+1}/G) \\
\downarrow & & \downarrow \\
\pi_i(L(S^{2n-1})^{hG}) & \longrightarrow & \pi_i(L(S^{2n+1})^{hG})
\end{array}
$$

Pour i pair, tous les groupes sont nuls. Pour i impair, il est bien connu que la première flèche horizontale est un isomorphisme si $i \neq 2n-1$ et que c'est aussi un isomorphisme pour $i = 2n-1$ si on quotiente le premier groupe par p. Pour démontrer le théorème, il suffit donc de montrer la même propriété pour la deuxième flèche horizontale.

Pour accomplir ce programme, considérons la fibration vectorielle

$$V = S^{2n+1} - S^1 \longrightarrow S^{2n-1}\ .$$

Plus précisément, écrivons $C^{n+1} = L_0 \oplus L_1 \oplus \cdots \oplus L_n$ (somme de représentations irréductibles); S^{2n+1} est la sphère de $\mathbf{C}^{n+1}$; S^1 est vue comme la sphère de L_n. La structure vectorielle est déduite du diagramme cartésien classique suivant:

$$
\begin{array}{ccc}
S^{2n+1} - S^1 & \longrightarrow & S^{2n-1} \\
\downarrow & & \downarrow \\
CP_n - \{\infty\} & \overset{\theta}{\longrightarrow} & CP_{n-1}
\end{array}
$$

avec $\theta([x_0, x_1, \ldots, x_n]) = [x_0, x_1, \ldots, x_{n-1}]$, où la fibre au-dessus du "point" L s'identifie à $\mathrm{Hom}(L, L_n)$. Si on munit ce fibré vectoriel d'une métrique invariante, l'espace de Thom V^+ s'identifie au quotient $B(V)/S(V)$ du fibré en boules $B(V)$ par le fibré en sphères $S(V)$, soit aussi S^{2n+1}/S^1. En quotientant par l'action de G, on a encore une fibration vectorielle

$$W = (S^{2n+1} - S^1)/G \longrightarrow S^{2n-1}/G$$

dont l'espace de Thom est $(S^{2n+1}/S^1)/G$. Soit τ la "classe de Thom" de cette fibration, vue comme application $W^+ \longrightarrow L(S^2)$. Elle induit la classe de Thom de V grâce à la composition $V^+ \longrightarrow W^+ \longrightarrow L(S^2)$ (car G opère librement sur la base). L'isomorphisme de Thom de V peut être ainsi réalisé au niveau du produit symétrique infini (cf. la fin de l'annexe précédente) par une application *équivariante* $\tau(V) \longrightarrow (S^{2n-1})^+ \wedge L(S^2)$ qui induit une flèche équivariante

$$\Omega^2(L(S^{2n+1}/S^1) \longrightarrow L(S^{2n-1})\ .$$

On en déduit un isomorphisme

$$\pi_i(L(S^{2n-1})^{hG}) \longrightarrow \pi_{i-2}(L(S^{2n+1})^{hG})$$

et un diagramme commutatif

$$
\begin{array}{ccccc}
H_i(S^{2n-1}/G) & \longrightarrow & H_i(S^{2n+1}/G) & \longrightarrow & H_{i-2}(S^{2n-1}/G) \\
\approx\downarrow & & \downarrow & & \approx\downarrow \\
\pi_i(L(S^{2n-1})^{hG}) & \longrightarrow & \pi_i(L(S^{2n+1})^{hG}) & \longrightarrow & \pi_{i-2}(L(S^{2n-1})^{hG})
\end{array}
$$

L'application composée $H_i(S^{2n-1}/G) \longrightarrow H_{i-2}(S^{2n-1}/G)$ est un isomorphisme si $2 < i < 2n - 1$ et est aussi un isomorphisme pour $i = 2n - 1$ si on quotiente le premier groupe par p. Il en résulte que la même propriété vaut pour l'application composée $\pi_i(L(S^{2n-1})^{hG}) \longrightarrow \pi_{i-2}(L(S^{2n-1})^{hG})$. Donc $\alpha : H_i(S^{2n-1}/G) \longrightarrow \pi_i(L(S^{2n-1})^{hG})$ est aussi un isomorphisme.

B.2. Théorème. *Soit* $\mathbf{R}^m = \mathbf{C}^n \oplus \mathbf{R}^q$, *où* $\mathbf{C}^n$ *(resp.* $\mathbf{R}^q$*) est somme de représentations irréductibles du groupe cyclique* C_p *(resp. est un* C_p*-module trivial). Alors l'application canonique*

$$\alpha : L(S^{m-1}/G) \longrightarrow L(S^{m-1})^{hG}$$

induit un isomorphisme sur les groupes d'homotopie π_i *pour* $i \geq q$. *Si on remplace le foncteur* L *par* $L \otimes \mathbf{Z}/p$, *la même conclusion vaut pour* $i > q$.

Démonstration. Il est plus commode de considérer S^m comme le compactifié d'Alexandrof de $\mathbf{R}^m$ qui s'identifie à la suspension (non réduite) de S^{m-1} en tant que G-espace. Nous pouvons ainsi écrire $S^{2n+q} = S^{2n} \wedge S^q$ et $S^{2n+q}/G = (S^{2n}/G) \wedge S^q$. De manière générale, notons $S'Y$ la suspension non réduite d'un espace Y. On a alors les équivalences d'homotopie suivantes $\Omega(L(S^{2n+q}/G)) \approx \Omega(L(S'(S^{2n+q-1}/G)) \approx L(S^{2n+q-1}/G)$ et $\Omega(L(S^{2n+q})^{hG}) \approx \Omega(L(S'(S^{2m+q-1})^{hG}) \approx L(S^{2n+q-1})^{hG}$. De même, nous avons $\Omega^{q+1}(L(S^{2n+q}/G)) \approx \Omega^{q+1}(L(S'^{q+1}(S^{2n-1}/G))) \approx L(S^{2n-1}/G)$ et $\Omega^{q+1}(L(S^{2n+q})^{hG}) \approx \Omega^{q+1}(L(S'^{q+1}(S^{2n-1})^{hG})) \approx L(S^{2n-1})^{hG}$. La première partie du théorème se déduit ainsi du théorème précédent, en appliquant le foncteur π_i aux isomorphismes ci-dessus pour $i \geq q$. La deuxième partie s'en déduit aussitôt en écrivant les suites exactes d'homotopie reliant $\pi_i(L(X))$ et $\pi_i(L(X)) \otimes \mathbf{Z}/p$.

B.3. Nous nous proposons d'étendre les résultats précédents en homologie mod. 2, c'est-à-dire en supposant $R = \mathbf{Z}/2$. Pour éviter toute confusion, nous noterons ici $\overline{L}$ le foncteur L pour $R = \mathbf{Z}/2$, en conservant la notation première L dans le cas où $R = \mathbf{Z}$. Si X est un G-espace (avec $G = C_2$), nous noterons X_0 le quotient de X par le sous-espace des points fixes X^G. Il est clair que le transfert usuel induit en fait une application (notée encore t)

$$\overline{L}(X_0/G) \longrightarrow \overline{L}(X)^{hG}$$

(car le transfert usuel s'annule sur $\overline{L}(X^G)$; cf. 1.3).

B.4 Théorème. *Soit* $\mathbf{R}^n = \mathbf{R}^m \oplus \mathbf{R}^q$ *un espace euclidien où le groupe* $G = C_2$ *opère par involution antipodique sur* $\mathbf{R}^m$ *et par l'identité sur* $\mathbf{R}^q$ *(avec* $q \geq 1$*). Le groupe* G *opère donc sur la sphère correspondante* S^n *(vue comme compactifiée d'Alexandrof de* $\mathbf{R}^n$*), l'espace des points fixes* $(S^n)^G$ *s'identifiant à* S^q. *On pose* $S_0^n = S^n/(S^n)^G$. *Alors l'application de transfert sur la sphère, soit*

$$\alpha : \overline{L}(S_0^n/C_2) \longrightarrow \overline{L}(S^n)^{hC_2}$$

(où $\overline{L} = L \otimes \mathbf{Z}/2$) induit un isomorphisme sur les groupes d'homotopie π_i pour $i \geq q$.

Démonstration. Elle se divise en deux parties:

1. Supposons d'abord $i \geq q + 2$. Nous avons alors le diagramme commutatif suivant de fibrations (avec $G = C_2$)

$$
\begin{array}{ccccc}
\Omega(\overline{L}(S^{m+q}/G)) & \longrightarrow & L(S^{m+q}/G) & \xrightarrow{\ .2\ } & L(S^{m+q}/G) \\
\downarrow & & \downarrow & & \downarrow \\
\Omega(\overline{L}(S^{m+q})^{hG}) & \longrightarrow & L(S^{m+1})^{hG} & \xrightarrow{\ .2\ } & L(S^{m+q})^{hG}
\end{array}
$$

En écrivant les deux suites d'homotopie correspondantes, ainsi qu'en appliquant les théorèmes B.1 et B.2 (adaptés au cas réel et appliqués en homologie mod. 2), on en déduit que $\pi_i(\overline{L}(S^{m+q}/G)) \approx \pi_i(\overline{L}(S^{m+q})^{hG})$ pour $i \geq q + 2$. Par ailleurs, l'espace des points fixes de S^{m+q} par l'action de G s'identifie à S^q et l'application $S^q \longrightarrow S^{m+q}/G$ est homotopiquement triviale. Par conséquent, $\overline{L}(S_0^n/G) \approx \overline{L}(S^{m+q}/G) \oplus \overline{L}(S^{q+1})$ et le théorème est bien démontré pour $i \geq q + 2$.

2. Supposons maintenant que i soit égal à $q + 1$. Dans ce cas, l'homomorphisme

$$
\alpha_{q+1} : \mathbf{Z}/2 \approx \pi_{q+1}(\overline{L}(S_0^n/G)) \longrightarrow \pi_{q+1}(\overline{L}(S^{m+q})^{hG}) \approx \mathbf{Z}/2
$$

est égal à 0 ou est un isomorphisme pour tous q et n. En effet, il suffit de répéter l'argument exposé dans la démonstration du théorème B.1 (en remplaçant l'homologie entière par l'homologie mod. 2). La non-trivialité de α_{q+1} résultera donc d'un exemple approprié. De manière plus précise, considérons le transfert introduit en 1.3 (avec $p = 2$):

$$
\gamma : \Delta_2^+(S^q) \longrightarrow L(S^{2q})^{hG} .
$$

On doit montrer que γ induit un isomorphisme sur les groupes d'homotopie $\pi_{q+1} \cong \mathbf{Z}/2$. Pour cela, considérons la première opération cohomologique $K(\mathbf{Z}/2, q) \longrightarrow K(\mathbf{Z}/2, q+1)$ induite par la puissance $2^{\text{ième}}$ et qui résulte de la décomposition de $\overline{L}(S^{2q})^{hG}$ en produit d'espaces d'Eilenberg–Mac Lane (cf. annexe C). Cette opération se factorise à travers $\Delta_2^+(S^q)$ par la puissance réduite (cf. 1.10). Il suffit donc de montrer que l'application $K(\mathbf{Z}/2, q) \longrightarrow K(\mathbf{Z}/2, q+1)$ n'est pas nulle: en fait c'est l'homomorphisme de Bockstein d'après [3]. Mais cet homomorphisme n'est pas trivial (considérer des produits d'espaces projectifs par exemple). Ceci achève la démonstration du théorème B.4.

Annexe C. Type d'homotopie des groupes abéliens simpliciaux.

C.1. Il est bien connu qu'un groupe abélien topologique X a le type d'homotopie d'un produit d'espaces d'Eilenberg–Mac Lane (cf. [8]). Comme me l'a signalé R. Jardine (communication privée), cette décomposition en produit n'est pas canonique. Cependant, nous allons voir que cette décomposition peut être rendue "canonique" si $\pi_n(X)$ est un groupe abélien de type fini, nul si n est assez grand, avec les choix supplémentaires suivants (ceci est un cas particulier de résultats plus généraux de R. Jardine à paraître):

1) Choix des générateurs des facteurs cycliques libres de $\pi_n(X)$.

2) Pour chaque facteur cyclique $\mathbf{Z}/r$ de $\pi_n(X)$, choix d'un élément de $\pi_{n-1}(X;\mathbf{Z}/r)$ dont l'image dans $\pi_n(X)$ par l'homomorphisme canonique

$$\pi_{n-1}(X;\mathbf{Z}/r) \longrightarrow \pi_n(X)$$

est un générateur de ce facteur cyclique.

Plus précisément, soient G_n^i les facteurs cycliques de $\pi_n(X)$ et soient x_n^i les générateurs correspondants. Nous définissons une application de l'espace d'Eilenberg–Mac Lane $K(G_n^i, n)$ dans X de la manière suivante.

Premier cas. $G_n^i \cong \mathbf{Z}$. Choisissons comme modèle de $K(G_n^i, n)$ le produit symétrique[15] infini $SP^\infty(S^n)$ de la sphère S^n (cf. 2.2). Le générateur x_n^i définit une application essentielle $\alpha_n^i : S^n \longrightarrow X$ qui s'étend en $\beta_n^i : SP^\infty(S^n) \longrightarrow X$ par la formule

$$\beta_n^i(x_1, x_2, \ldots, x_r, \ldots) = \sum_{r=1}^\infty \alpha_n^i(x_r) \, .$$

Bien entendu, la structure de groupe *abélien* de X intervient ici de manière essentiele.

Deuxième cas. $G_n^i \cong \mathbf{Z}/r\mathbf{Z}$. On procède de manière analogue en choisissant comme modèle de $K(G_n^i, n)$ le groupe abélien $SP^\infty(S^n)/rSP^\infty(S^n)$. Cependant, une difficulté technique apparaît ici: les β_n^i, multipliés par r, ne sont pas égaux à 0, mais *homotopes* à 0. Choisissons alors un relèvement $\tilde{\alpha}_n^i$ de α_n^i dans le groupe d'homotopie $\pi_{n-1}(X;\mathbf{Z}/r)$ par l'homomorhpisme canonique $\pi_{n-1}(X;\mathbf{Z}/r) \longrightarrow \pi_n(X)$. Alors $\tilde{\alpha}_n^i$ définit une classe d'homotopie d'application de l'espace de Moore S_r^n dans X (S_r^n est le cône de l'application de degré r de S^n dans S^n). Le produit symétrique infini de S_r^n ayant le type d'homotopie de $SP^\infty(S^n)/rSP^\infty(S^n) = K(G_n^i, n)$, on peut définir β_n^i comme dans le premier cas, grâce au choix de ce relevé $\tilde{\alpha}_n^i$.

---

[15] On peut aussi choisir le groupe abélien libre de base un modèle simplicial de S^n.

En faisant la somme de tous ces β_n^i, on en déduit une application du produit des $K(G_n^i, n)$ dans X qui est une équivalence d'homotopie d'après le théorème de Whitehead.

C.2. Supposons que X soit un R-module simplicial, les groupes d'homotopie étant des R-modules *libres*. Dans ce cas, le choix d'une base des groupes d'homotopie définit des applications[16] $L(S^n) \longrightarrow X$. On en déduit directement une application d'un produit des $L(S^n)$ dans X qui est une équivalence d'homotopie, sans le détour des espaces de Moore.

C.3. Supposons maintenant que $R = \mathbf{Z}$ et considérons un groupe abélien simplicial *libre* homotopiquement équivalent à un espace d'Eilenberg–Mac Lane $K(\mathbf{Z}/r, n)$. Alors $X/r = X \otimes \mathbf{Z}/r$ est homotopiquement équivalent à $K(\mathbf{Z}/r, n) \times K(\mathbf{Z}/r, n+1)$, comme on le voit en écrivant la suite exacte d'homotopie associée à la suite exacte de groupes simpliciaux

$$0 \longrightarrow X \xrightarrow{\cdot r} X \longrightarrow X/r \longrightarrow 0 \, .$$

Le résultat suivant est utilisé en 2.11.

C.4 Proposition. *Avec les hypothèses précédentes, l'application canonique*

$$X = K(\mathbf{Z}/r, n) \longrightarrow X/r \approx K(\mathbf{Z}/r, n) \times K(\mathbf{Z}/r, n+1)$$

s'identifie au produit de l'application identique par l'homomorphisme de Bockstein.

Démonstration. D'après ce qui précède, il existe une application $L(S_r^n) \longrightarrow$ X qui est une équivalence d'homotopie, S_r^n étant un espace de Moore. Sans restreindre la généralité, on peut donc supposer que X est un espace de Moore. Par ailleurs, on peut écrire le diagramme commutatif à homotopie près suivant (f étant une application de degré r):

$$
\begin{array}{ccccc}
L(S^n) & \xrightarrow{\ L(f)\ } & L(S^n) & \longrightarrow & L(S_r^n) \\
\downarrow & & \downarrow & & \downarrow \\
0 \longrightarrow \ L(S^n) & \xrightarrow{\ \cdot r\ } & L(S^n) & \longrightarrow & L(S^n)/r \ \longrightarrow 0
\end{array}
$$

Dans ce diagramme, les flèches horizontales sont des fibrations homotopiques et l'application "bord" $L(S_r^n) \longrightarrow \Omega^{-1}(L(S^n)) \cong L(S^{n+1})$ (suite de Puppe) se factorise donc en $L(S_r^n) \xrightarrow{\approx} L(S^n)/r \longrightarrow \Omega^{-1}(L(S^n)) \cong L(S^{n+1})$, c'est-à-dire essentiellement par l'homomorphisme de Bockstein. La proposition s'en déduit en réduisant $\Omega^{-1}(L(S^n))$ mod. r.

---

[16] $L(S^n)$ désignant le R-module simplicial libre de base S^n.

Références

[1] A. Dold et R. Thom. Quasifaserungen und unendliche symmetrische produkte, *Annals of Math.* vol. 67, no. 2 (1958), 239–281. Voir aussi: une généralisation de la notion d'espace fibré, Application aux produits symétriques infinis, *C.R. Acad. Sci. Paris* **242** (1956), 1680–1682.

[2] M. Karoubi, Formes différentielles non commutatives et cohomologie à coefficients arbitraires, *Transactions of the American Mathematical Society* **347** (1995), 4277-4299. Voir également *Comptes Rendus Acad. Sci. Paris*, t. 316, p. 833–836 (1993), ainsi que l'article "Formes topologiques non commutatives," *Annales Scientifiques de l'Ecole Normale Supérieure* **28** (1995), 477–492

[3] M. Karoubi, Formes différentielles non commutatives et opérations de Steenrod, *Topology* **34** (1995), 699–715. Voir aussi *Comptes Rendus Acad. Sci. Paris*, t. 316, p. 917–920 (1993).

[4] J. Lannes (communication privée).

[5] N. Steenrod et D. Epstein, Cohomology operations, *Annals of Math. Studies*, No. 50, Princeton University Press (1962).

[6] P. Vogel, Cobordisme d'immersions, *Annales Scientifiques de l'Ecole Normale Supérieure* (1974), p. 317–358.

Max Karoubi
Université Paris 7–UMR 9994 du CNRS
Mathématiques
2, place Jussieu
75251 Paris Cedex 05, France
email: karoubi@mathp7.jussieu.fr
Received May 1995; corrected February 1996

Characteristic Classes of Singularity Theory

1. Introduction

Theorems of the global singularity theory express topological invariants in terms of singularities of some geometrical objects: vector bundle mappings, the Thom-Boardman singularities of smooth mappings, Lagrangian and Legendrian singularities, $\mathbb{RC}$-singularities ect. (cf. [P], [R], [L], [HL2], [AGLV], [V]). A classical example is Hopf's theorem relating the singularities of vector fields to the Euler characteristic.

A lot of problems in this theory can be formulated as follows. Consider a smooth locally trivial bundle $\pi \colon E \to M$ with fiber V and structure group G and a generic section $s \colon M \to E$. A *singularity class* is any G-invariant subset in V. For a given singularity class Ξ, the condition that the point $s(b) \in \pi^{-1}(b) \cong V$, $b \in M$, lies in Ξ defines invariantly the set $\Xi(M) \subset M$ of points at which the section s has singularity Ξ. The problem is to determine under what condition the intersection index with $\Xi(M)$ defines a cohomology class on M and to find an expression for this class.

A general approach was suggested by V. Vassiliev. In his book [V] a cochain complex is constructed generators which correspond to singularity classes and the coboundary operator is given by adjacencies of singularities of neighbor codimensions. For any cohomology class Ξ of this universal complex the intersection with $\Xi(M)$ is well defined characteristic class of the given G-bundle over M.

The homomorphism of the Vassiliev complex to the ring of characteristic classes of G-bundles is neither injective nor surjective. This means that some classes of the Vassiliev complex can give trivial cohomology classes on M and some characteristic classes of G-bundles cannot be expressed as classes dual to singularity classes. In addition, Vassiliev's complex does not take into account that the restriction of the bundle to $\Xi(M)$ defines characteristic classes on $\Xi(M)$ itself and this can give new 'derived' classes on M. In this paper we show that using these 'derived' classes one can express all the characteristic classes of G-bundles.

Formally, we construct a cohomological spectral sequence converging to the cohomology group of characteristic classes of G-bundles. This spectral

[1]Research of the author supported by the grants of International Science Foundation (grant Nr.MSD000) and of the Russian Foundation for Basic Research (Project Nr. 94–01–01203)

sequence is given by the classification of orbits of the action of G on V and described in terms of this classification. Corresponding to the singularity class Ξ subgroup of $E_1^{*,*}$ is isomorphic to the cohomology group of some group G_Ξ associated to this class Ξ, namely, G_Ξ is the stationary group of any representative f of Ξ. The differential δ_1 is given by adjacencies of singularity classes of neighbor codimentions and higher differentials δ_r are described by adjacencies of singularity classes whose codimension differs by r. Though the spectral sequence depends on the classification chosen, its terms beginning with the second one already do not depend on this classification.

There is a natural isomorphism between the universal Vassiliev complex and the row $E_1^{*,0}$ in the first term of the characteristic spectral sequence. Correspondingly, the cohomology of the complex is isomorphic to the row $E_2^{*,0}$ of the second term. Thus, the Vassiliev complex which was originally defined with the help of an abstract algebraic-geometrical construction here appears as a result of calculations in the spectral sequence.

In the previous paper [K2] we applied this approach to the classification of germs of function up to stable right equivalence and V-equivalence which give the characteristic classes of Lagrangian and Legendrian immersions respectively. In this paper we consider an application of the characteristic spectral sequence to the study of the classification of germs of functions up to (non-stable) right equivalence.

Consider a structure of smooth fiber bundle on a manifold M (or, more general, a smooth foliation). Let $f: M \to \mathbb{R}$ be any generic function. Then the singularities of the restriction of f to the fibers can form cycles on M. In this paper we prove that *if the cohomology class dual to a cycle of singularities of some non-Morse type is correctly defined it is always trivial*. In particular, this unexpected result show that the cohomology classes of the Vassiliev complex for R-classification of functions calculated in [AGLV] give always trivial cohomology classes on M.

2. Example: characteristic Gysin exact sequence

Let $\pi: E \to M$ be an n-dimensional vector bundle over a compact manifold M and $s: M \to E$ be a generic section. Zeros of this section form a smooth submanifold $F \subset M$ of codimension n. It is well known that *the mod2-cohomology class given by the intersection index with F coincides with $\omega_n(\pi)$, the n-th Stiefel-Whitney class of the vector bundle π*.

This assertion has the following interpretation. In what follows by the cohomology we mean the cohomology with coefficients in $\mathbb{Z}_2$. The inclusion $i: F \to M$ induces the *Gysin homomorphism*

$$i_*: H^{*-n}(F) \longrightarrow H^*(M)$$

defined as a composition of the following two homomorphisms

$$H^{*-n}(F) \xrightarrow{\sim} H^*(M, M \setminus F) \longrightarrow H^*(M),$$

where the first arrow is the composition of the Thom isomorphism of the normal bundle of F with the excision isomorphism and the second one is induced by the inclusion of topological pairs $(M, \emptyset) \subset (M, M \setminus F)$. An equivalent definition of i_* is the composition of the Poincaré duality in F, the usual homomorphism i_* of the homology groups and the Poincaré duality in M.

The assertion above stays that the Gysin homomorphism takes the fundamental class of F to $\omega_n(\pi)$.

Now, let η be any characteristic class of the restriction of the bundle π to F, say, some Stiefel-Whitney class. Then $i_*(\eta)$ is also a characteristic class i.e. it does not depend on the section S but on the bundle π only. In other words, a homomorphism γ_n exists which makes the following diagram commutative

$$\begin{array}{ccc} H^{*-n}(BO(n)) & \xrightarrow{\gamma_n} & H^*(BO(n)) \\ \downarrow & & \downarrow \\ H^{*-n}(F) & \xrightarrow{i_*} & H^*(M) \end{array}$$

2.1. Lemma. *The homomorphism γ_n is given by the multiplication by $\omega_n \in H^n(BO(n))$.*

Proof. The lemma follows from the so called projection formula

$$i_*(i^*(a)b) = a\, i_*(b)$$

applied to $b = 1 \in H^0(F)$ and $a = \eta \in H^*(M)$ any characteristic class.

The lemma implies that γ_n is injective. The cokernel of γ_n is isomorphic to the ring of polynomials of the $n-1$ first Stiefel-Whitney classes that is to the ring of characteristic classes of $(n-1)$-dimensional vector bundles. This gives the following splitting exact sequence which we call the *characteristic Gysin exact sequence*

$$\ldots \xrightarrow{0} H^{k-n}(BO(n)) \xrightarrow{\gamma_n} H^k(BO(n)) \rightarrow H^k(BO(n-1)) \xrightarrow{0} H^{k-n+1}(BO(n)) \xrightarrow{\gamma_n} \ldots$$

The Gysin homomorphism i_* takes part in the Gysin exact long sequence

$$\ldots \rightarrow H^{k-n}(F) \xrightarrow{i_*} H^k(M) \rightarrow H^k(M \setminus F) \rightarrow H^{k-n+1}(F) \xrightarrow{i_*} \ldots$$

which is nothing but the exact sequence for the pair $(M, M \setminus F)$ with the cohomology groups of the pair $(M, M \setminus F)$ exchanged by the isomorphic to them cohomology groups of F with shift of dimensions by n.

The discussion above leads to the following result which generalizes the assertion stated at the beginning of this section.

2.2. Theorem. *There is a natural functorial homomorphism between the characteristic Gysin exact sequence and the Gysin sequence for the submanifold* $F \subset M$.

A smooth mapping of manifolds $\varphi \colon N \to M$ induces an n-dimensional vector bundle $\varphi^* \pi$ over N, its section $\varphi^* s$ and the homomorphism between the Gysin sequences for the pairs $F \subset M$ and $\varphi^{-1}(F) \subset N$. The functoriality mentioned in the theorem means the functoriality with respect to this homomorphism of the Gysin sequences.

This theorem is a particular case of more general Theorem 3.7. One can give also its direct proof which uses the standard arguments of the theory of characteristic classes.

According to this theorem the calculation of the characteristic classes for the bundle π can be reduced in some sense to that of the restriction of π to F and to that of the $(n-1)$-dimensional factor bundle over $M \setminus F$ of the bundle π over the line bundle generated by the section s.

The same result holds for the Euler-Pontrjagin characteristic classes in the orientable case and for the Chern classes in the complex case. It would be interesting to interpret this results in terms of geometric residue theorems for the Chern-Weil theory [HL1,HL2].

3. Characteristic spectral sequence of G-bundles

In this section by the cohomology we mean the cohomology group with the coefficients in a fixed abelian group A, say $A = \mathbb{Z}$, $\mathbb{Z}_2$ or $\mathbb{Q}$.

Let a Lie group G acts algebraically on a contractible algebraic manifold V. The linear representations theory of Lie groups give rise to a plenty of examples of such actions.

Consider a (G, V)-bundle $\pi \colon F \to M$ i.e. a locally trivial bundle with a fiber diffeomorphic to V and structure group G.

Let $\Xi \subset V$ be one of the orbits of the action of G on V. For $x \in \pi^{-1}(b) \cong V$, $b \in M$, the condition that x belongs to the orbit Ξ is invariantly defined. This gives the classification of points of the bundle space E in accordance with the classification of the orbits of the action G on V. Denote $\Xi(E) \subset E$ the set of points corresponding to the orbit Ξ.

For V is contractible the bundle π has a global smooth section $s \colon M \to E$ and all its sections are homotopic to each other. Any section defines also the classification of points on the base M. Denote $\Xi(M) = s^{-1}\Xi(E)$.

3.1. Lemma. *Suppose s is generic. Then $\Xi(M)$ is a smooth submanifold the codimension of which is equal to the codimension of Ξ in V. The* *structure group of the restriction* $\pi|_{\Xi(M)}$ *of the bundle* π *to* $\Xi(M)$ *can be reduced to* G_Ξ, *where* G_Ξ *is the stabilizer of any point* $v \in \Xi(M)$.

Proof. The first assertion of the lemma follows from the transversality theorem. To prove the second one it is enough to observe that the structure group of $\pi|_{\Xi(M)}$ is reduced to G_Ξ if we choose only such trivializations of $\pi|_{\Xi(M)}$ over domains in $\Xi(M)$ in which the section s takes the form $s(b) \equiv v$, $b \in \Xi(M)$, where $v \in \Xi$ is some fixed point.

A smooth mapping $\varphi: N \to M$ induces a (G, V)-bundle on the manifold N and the homomorphism

$$\varphi^*: H^*(M) \longrightarrow H^*(N).$$

The *characteristic class* associates with every G-bundle over a manifold M a uniquely defined cohomology class in $H^*(M)$ which is invariant under homomorphisms φ^* above.

One could hope that the characteristic classes of the given bundle π can be expressed in terms of characteristic classes of the smaller groups G_Ξ on the smaller manifolds $\Xi(M)$. This idea leads to the characteristic spectral sequence which we describe now.

To avoid difficulties in the case when V contains infinitely many orbits we introduce the following notion.

3.2. Definition [V]. A *cellular G-classification* on a manifold V is any locally finite partition of it into disjoint sets (called *classes*) satisfying the following properties.

1. All the classes are semialgebraic sets and also smooth manifolds (without boundary) and all adjacencies of classes satisfy the Whitney regularity condition.

2. G-equivalent points lie in one class.

3. For any two connected components L_1, L_2 of one class there are G-equivalent points $f_1 \in L$, $f_2 \in L_2$.

4. For each class the codimension of the orbits containing in it is constant.

5. The subset of $\Xi \times G$ consisting of all possible pairs of the form (point of Ξ, element of its stationary subgroup) together with the restriction of the projection $\Xi \times G \to \Xi$ to this space is smooth fiber bundle over Ξ and the quotient space of the class Ξ with respect to G-equivalence is smooth and contractible.

The existence of a cellular classifications is proved in [V]. For any two such classifications there is the third one which can be obtained as a subpartition of the two given ones.

3.3. For a given classification consider the filtration on V given by *open* subsets

$$\mathcal{F}_0 \subset \mathcal{F}_1 \subset \cdots \subset V,$$

where $\mathcal{F}_i$ is the union of all classes the codimension of which is not greater than i. This induces filtrations on E and M

$$\mathcal{F}_0(E) \subset \mathcal{F}_1(E) \subset \cdots \subset E, \qquad \mathcal{F}_0(M) \subset \mathcal{F}_1(M) \subset \cdots \subset M,$$

where $\mathcal{F}_i(E)$ and $\mathcal{F}_i(M)$ are the union of all $\Xi(E)$ (resp. $\Xi(M)$) with $\Xi \subset \mathcal{F}_i$. With this filtration we associate cohomological spectral sequences $E_r^{p,q}(E)$ and $E_r^{p,q}(M)$ converging to the cohomology group $H^*(E) \simeq H^*(M)$.

To justify the spectral sequence $E_r^{p,q}(M)$ we need to impose the generality condition to the section s. The initial term of this spectral sequence $E_1^{p,*}(M)$ is isomorphic to the cohomology of the smooth manifold $\mathcal{F}_p(M) \setminus \mathcal{F}_{p-1}(M)$ of codimension p with the coefficient system orienting the normal bundle of this submanifold.

3.4. Remark. It is more natural to consider this filtration by open sets rather than, say, the filtration formed by the complements of $\mathcal{F}_i(M)$, cf. [K2]. Moreover, while from the point of view of, say, the theory of CW-complexes more complicated spaces are obtained from simpler ones by gluing cells of increasing dimensions from the point of view of the singularity theory more complicated spaces are obtained from simpler ones by cutting off smooth submanifolds of increasing codimensions. It would be interesting to build the general theory realizing such approach. As an example, we give the following construction of the classifying space for a Lie group $G \subset GL(n, \mathbb{R})$. Consider a natural action of G on $\mathbb{R}^n$ and the product action on $\mathbb{R}^n \otimes \mathbb{R}^N = \mathbb{R}^n \oplus \ldots \oplus \mathbb{R}^n$ (N times). Let $W_N \subset \mathbb{R}^n \otimes \mathbb{R}^N$ be the union of all free orbits. Then the codimension of the complement $\mathbb{R}^n \otimes \mathbb{R}^N \setminus W_N$ increases infinitely with N and W_N has zero homotopy groups for small relative to N dimensions. Thus, for any d there is a number $N(d)$ such that the cohomology groups of the orbit space $B_N = W_N/G$ are isomorphic to the cohomology group of the classifying space BG in all dimensions smaller than d.

The filtration on M is induced by the section s from the filtration on E, $\mathcal{F}_i(M) = s^{-1}(\mathcal{F}_i(E))$. Therefore s induces a homomorphism between the spectral sequences

$$s^* : E_*^{*,*}(E) \longrightarrow E_*^{*,*}(M)$$

In the same way the mapping $\varphi : N \to M$ inducing a (G, V)-bundle on N induces also the homomorphism

$$\varphi^* : E_*^{*,*}(E) \longrightarrow E_*^{*,*}(\varphi^* E)$$

where we denote $\varphi^* E$ the total space of the (G, V)-bundle $\varphi^* \pi$ over N.

3.5. Definition. The *characteristic spectral sequence* $E_*^{*,*}$ is the inductive limit of the spectral sequences $E_*^{*,*}(E)$ over all $(G.V)$-bundles and over homomorphisms φ^* above between them.

3.6. Proposition. *The characteristic spectral sequence converges to the group $H^*(BG)$ of the cohomology characteristic classes of G-bundles.*

This follows from the fact that the group of characteristic classes can be defined as the inductive limit of the groups $H^*(M)$ over all G-bundles and homomorphisms φ^* between this groups, and from the properties of inductive limit.

The following proposition is the direct corollary of the definition.

3.7. Theorem. *For a given (G,V)-bundle on M and its section s there is a natural homomorphism*

$$\kappa^*: E_*^{*,*} \longrightarrow E_*^{*,*}(M)$$

such that the limit homomorphism of the cohomology

$$\kappa^*: H^*(BG) \longrightarrow H^*(M)$$

is the characteristic homomorphism which evaluates a characteristic class on the given G-bundle over M.

3.8. Example. The case of arbitrary n-dimensional vector bundles corresponds to the natural action of the group $G = GL(n, \mathbb{R})$ on $V = \mathbb{R}^n$. The space V consists of two orbits, the origin and its complement. Hence, the considered spectral sequence have two nontrivial columns only. A spectral sequence with two nontrivial columns may be reduced to an exact long sequence which is in our case nothing but the Gysin exact sequence discussed in Sect.2. Therefore Theorem 3.7 takes form of Theorem 2.2.

The following explicit construction can be taken as another definition of the characteristic spectral sequence. Let the principal G-bundle $W_N \to B_N$ be as in 3.4. Consider the associated with it (G,V)-bundle $\pi: E_N \to B_N$, $E_N = W_N \times_G V$. Then the spectral sequence $E_*^{*,*}(E_N)$ converges to $H^*(E_N) \simeq H^*(B_N)$.

3.9. Theorem. *If $N > N(d)$ then the homomorphism*

$$\kappa^*: E_r^{p,q} \longrightarrow E_r^{p,q}(E_N)$$

is an isomorphism for $p + q < d$ and all r.

Proof. A given (G,V)-bundle $E \to M$ over a smooth manifold M can be induced from the bundle $\pi: E_N \to B_N$ by a smooth mapping $\varphi: M \to B_N$ for N large enough. This mapping induces a homomorphism

$$\varphi^*: E_r^{p,q}(E_N) \longrightarrow E_r^{p,q}(E)$$

which shows that κ^* in Theorem 3.9 is surjective. As φ is uniquely defined up to a homotopy the homomorphism φ^* does not depend on the choice of the mapping φ which shows that the epimorphism κ^* is, in fact, an isomorphism.

The following theorem describes explicitly the term $E_1^{*,*}$ of the characteristic spectral sequence. Recall that if a class $\Xi \subset V$ consists of the only orbit the structure group of the restriction of a (G,V)-bundle over M to $\Xi(M)$ can be reduced to G_Ξ. The same holds in the general case because the property 5) of Definition 3.2 implies that the stationary subgroups of all the points of the same class are isomorphic to each other (we denote it again by G_Ξ). Hence, the given G_Ξ-bundle over $\Xi(M)$ is induced by some continuos mapping $\kappa_\Xi \colon \Xi(M) \to BG_\Xi$ defined uniquely up to homotopy.

If we realize the group G_Ξ as the stationary subgroup of an element $v \in \Xi \subset V$ then G_Ξ acts on the linear space $T_v V/T_v \Xi$. Some of the elements of G_Ξ do not preserve the orientation of this space. Denote O_Ξ the coefficient system (a locally isomorphic to the abelian group A sheaf) on BG_Ξ which orients the vector bundle over BG_Ξ associated with the given action of G_Ξ on $T_v V/T_v \Xi$. Then the coefficient system $O_{\Xi(M)}$ on $\Xi(M)$ coorienting the normal bundle to $\Xi(M) \subset M$ is isomorphic to the coefficient system induced from O_Ξ by the mapping κ_Ξ and we get an unambiguously defined homomorphism

$$\kappa_\Xi^* \colon H^*(BG_\Xi, O_\Xi) \longrightarrow H^*(\Xi(M), O_{\Xi(M)}).$$

3.10. Theorem. *The term $E_1^{p,*}$ of the characteristic spectral sequence is isomorphic to the direct sum of summands corresponding to the classes of codimension p of the chosen classification of points on V. The direct summand corresponding to the class Ξ is isomorphic to the group of the twisted characteristic classes $H^*(BG_\Xi, O_\Xi)$ of G_Ξ-bundles. The characteristic homomorphism*

$$\kappa^* \colon E_1^{p,*} = \bigoplus_{\mathrm{codim}\,\Xi=p} H^*(BG_\Xi, O_\Xi) \longrightarrow E_1^{p,*}(M) = \bigoplus_{\mathrm{codim}\,\Xi=p} H^*(\Xi(M), O_{\Xi(M)})$$

is the direct sum of the homomorphisms κ_Ξ^*.

3.11. Lemma. *Let E_N and (G,V)-bundle over E_N be as in Theorem 3.8. Then the space $\Xi(E_N)$ is isomorphic to the factorspace of the space with trivial N first homotopy groups over a free action of the group G_Ξ.*

Theorem 3.10 is followed by this lemma and the description of the term $E_1^{*,*}(M)$ given in 3.3.

Proof of Lemma 3.11. If Ξ consists of the only orbit the lemma is evident, under the isomorphism $E_N = V \times_G W_N$ we have

$$\Xi(E_N) = Gv \times_G W_N = \{v\} \times_{G_\Xi} W_N = \{v\} \times (W_N/G_\Xi).$$

In the general case it follows from the property 5) of Definition 3.2 that $\Xi(E_N)$ is the total space of the bundle with the topologically trivial base and the fiber as in the conclusion of the lemma.

The differential $\delta_1 \colon E_1^{p,q} \to E_1^{p+1,q}$ is given by adjacencies of classes of neighbor codimensions (see [K2] for its explicit description). The higher differentials δ_r are given by adjacencies of classes the codimension of which differs by r.

Formally, our spectral sequence depends on the chosen classification on the space V. In fact, for $r \geq 2$ its terms already do not depend on the classification.

3.12. Theorem. *For $r \geq 2$ terms $E_r^{p,q}$ of the characteristic spectral sequence are naturally isomorphic to each other for different cellular classifications on V (with fixed p, q, r).*

The proof word by word repeats the proof of the corresponding theorem in [K2].

4. Vassiliev universal complex and the characteristic spectral sequence

Let $\pi \colon E \to M$ be a (G, V)-bundle and $s \colon M \to E$ its generic section. One possible way to define a characteristic class is to try to define it as an intersection index with the submanifold $\Xi(M)$ corresponding to some class $\Xi \subset V$ of orbits of the action of G on V. The closure $\overline{\Xi(M)}$ may not be smooth and to see if the intersection index with $\Xi(M)$ is well defined we need to study the singularities of $\overline{\Xi(M)}$ of codimension 1. Moreover, if we need to define an integral cohomology class rather then the class mod 2 we should define a coorientation of $\Xi(M)$. To do this, V. Vassiliev introduced in [V] two chain complexes, an integral complex ω of coorientable classes and a complex ν of $\mathbb{Z}_2$-modules of all classes. Generators of these universal complexes are formed by the orbit classes of some cellular classification and the differential is given by adjacencies of classes of neighbor codimensions (see [V] for the details).

4.1. Theorem. *The row*

$$0 \longrightarrow E_1^{0,0} \xrightarrow{\delta_1} E_1^{1,0} \xrightarrow{\delta_1} E_1^{2,0} \xrightarrow{\delta_1} \cdots$$

of the characteristic spectral sequence coincides for the coefficient groups $A = \mathbb{Z}$ and $A = \mathbb{Z}_2$ with the Vassiliev universal complex ω of coorientable classes and the complex ν of all classes respectively.

4.2. Corollary. *Terms $E_2^{p,0}$ of the characteristic spectral sequence coincide with the p-th cohomology group of the Vassiliev complex ω (for $A = \mathbb{Z}$) and that of the complex ν (for $A = \mathbb{Z}_2$).*

The proof follows from Theorem 3.10, the description of the groups $H^0(BG_\Xi, O_\Xi)$ which are isomorphic to A if the sheaf O_Ξ is trivial and $H^0(BG_\Xi, O_\Xi) = 0$ if not, and from the definition of the universal complexes given in [V].

The dimensional arguments show that the higher differentials δ_r, $r \geq 2$, are trivial on terms $E_r^{p,0}$ and the following homomorphisms are well defined

$$H^p(\omega) \simeq E_2^{p,0} \longrightarrow E_\infty^{p,0} \longrightarrow H^p(BG) \longrightarrow H^p(M)$$

(for the integral coefficients; the same holds for the complex ν and $\mathbb{Z}_2$-coefficients). Thus, the cohomology classes of the Vassiliev complex define characteristic classes. The homomorphism obtained $H^p(\omega) \to H^p(BG)$ may have a nontrivial kernel which corresponds to the nontrivial differentials $\delta_r : E_r^{p-r,r-1} \to E_r^{p,0}$, $r \geq 2$. It follows that the cohomology class defined by some nontrivial class of the Vassiliev complex may be always trivial for any (G, V)-bundle and its section on M, and this fact can be established only by considering adjacencies of classes the codimension of which differ by more then 1. Examples of such phenomena are given in [K2] and in Sect.5 below.

5. Characteristic spectral sequence
of right equivalence of functions

It would be interesting to apply the theory described in the previous sections to the concrete examples of Lie group actions. Nevertheless, the main point of our interest belongs to the infinite dimensional cases of classifications which appear in the singularity theory. In this section we consider the case when G is the group of diffeomorphism germs of $(\mathbb{R}^n, 0)$ and V is the space of smooth function germs $(\mathbb{R}^n, 0) \to (\mathbb{R}, 0)$.

More precisely, denote $G(k, n)$ the group of k-jets of diffeomorphisms of the space $(\mathbb{R}^n, 0)$ and $V(k, n)$ the space of k-jets of functions $(\mathbb{R}^n, 0) \to (\mathbb{R}, 0)$. Denote $E_*^{*,*}(k, n)$ the characteristic spectral sequence for the natural action of $G(k, n)$ on $V(k, n)$.

Let $k' > k$ be two integer. Then if k'-jets of two functions are $G(k', n)$-equivalent then their k-jets are $G(k, n)$-equivalent as well. Hence, the classification on the space of k'-jets of functions may be obtained as a subpartition of some classification on the space of k-jets of functions and we get a homomorphism of spectral sequences

$$\tau_k^{k'} : E_r^{p,q}(k, n) \longrightarrow E_r^{p,q}(k', n)$$

5.1. Proposition. *For any integer d there is a number $k(d)$ such that the homomorphisms $\tau_k^{k'}$ above are isomorphisms for any fixed p, q, r, n, and $k' > k > k(p + q)$.*

The proof is similar to that of the corresponding proposition in [K2] for the case of the *stable* right equivalence.

5.2. Definition. The limit spectral sequence

$$E_r^{p,q}(n) = \lim_{k \to \infty} E_r^{p,q}(k,n)$$

is called the *characteristic spectral sequence of right equivalence of functions* of n variables.

The group $G(k,n)$ is contractible to the group $G(1,n) = GL(n,\mathbb{R})$ of linear diffeomorphisms. Hence we have

5.3. Theorem. *The characteristic spectral sequence of right equivalence of functions converges (for the case of $\mathbb{Z}_2$-coefficients) to the ring of polynomials of Stiefel-Whitney classes $\omega_1, \ldots, \omega_n$ of degrees $1, \ldots, n$.*

The following theorem give an explicit description of the limit term $E_\infty^{p,q}(n)$.

5.4. Theorem. *For $k = 1$ we have*

$$E_1^{p,q}(1,n) \simeq E_\infty^{p,q}(1,n) \simeq \begin{cases} 0 & , \quad p \neq 0 \text{ and } p \neq n \\ H^q(BO(n-1)), & p = 0 \\ H^q(BO(n)) & , \quad p = n \end{cases}$$

The natural homomorphism

$$\tau \colon E_r^{p,q}(1,n) \longrightarrow E_r^{p,q}(n)$$

for $r = \infty$ induces an isomorphism of the limit terms $E_\infty^{p,q}$.

Proof. The action of the group $G(1,n)$ on $V(1,n)$ is the standard action of $GL(n,\mathbb{R})$ on $\mathbb{R}^n$. Therefore, the first part of the theorem is a reformulation of Theorem 2.2 as it is explained in Example 3.8.

5.5. Lemma. *For $r = 1$ the homomorphism*

$$\tau \colon E_1^{p,q}(1,n) \longrightarrow E_1^{p,q}(n)$$

is an isomorphism for $p < n$ and a monomorphism for $p \geq n$.

Proof. For $p = 0$ the groups $E_1^{0,*}(k,n)$ correspond to the orbit of k-jets of nondegenerate functions. A nondegenerate function is right equivalent (up to a constant) to a linear function and the stationary group of its k-jet contains the subgroup $O(n-1)$ of orthogonal transformations of the hyperplane given by zeros of the function (with respect to some fixed Euclidean structure). The inclusion of this subgroup is a homotopy equivalence, as one can prove, and we have

$$E_1^{0,*}(1,n) \simeq E_1^{0,*}(k,n) \simeq E_1^{0,*}(n) \simeq H^*(BO(n-1)).$$

Now, for $1 \leq p < n$ there is no orbits of codimension p of the action of $G(k,n)$ on $V(k,n)$ and we have

$$E_1^{p,*}(k,n) = E_1^{p,*}(n) = 0 \qquad 1 \leq p < n.$$

For $p = n$ the groups $E_1^{n,*}(k,n)$ correspond to the orbits of k-jets of Morse singularities. A germ of function with a Morse singularity is right equivalent (up to a constant) to a nondegenerate quadratic form. Consider the orbit for which this quadratic form is positive definite and, hence, defines a Euclidean structure. For any $k \geq 2$ the stationary group of its k-jet contains the subgroup $O(n)$ of linear orthogonal transformations and the inclusion of this subgroup is a homotopy equivalence. Hence, in accordance with Theorem 3.10 one of the direct summands of the group $E_1^{n,1}$ is isomorphic to the group $H^*(BO(n))$ which shows that

$$\tau_1^k \colon E_1^{n,*}(1,n) = H^*(BO(n)) \to E_1^{n,*}(k,n) \qquad k \geq 2,$$

is monomorphic. This completes the proof of Lemma 5.5.

Using Lemma 5.5 we prove the second part of 5.4. Denote $I \subset E_1^{*,*}(n)$ the image of $\tau \colon E_1^{*,*}(1,n) \to E_1^{*,*}(n)$. According to Lemma 5.5 and the first part of Theorem 5.4 all the differentials δ_r vanish on I. The dimensional arguments show that I does not intersect the image of any differential δ_r. Hence, I maps monomorphically to the limit term $E_\infty^{*,*}(n)$. But, according to Theorem 5.3 the ranks of the groups $E_\infty^{*,*}(1,n)$ and those of $E_\infty^{*,*}(n)$ coincide. Therefore the monomorphism $\tau \colon E_\infty^{*,*}(1,n) \to E_\infty^{*,*}(n)$ is, in fact, an isomorphism.

This proof looks rather artificial. In 5.6 we give its geometrical meaning.

Consider an n-dimensional vector bundle $\pi \colon E \to M$ and a generic function $f \colon E \to \mathbb{R}$ in the total space defined in some neighborhood of the zero section. With every point $x \in M$ we associate a function germ of the restriction of f to the fiber $\pi^{-1}(x) \cong \mathbb{R}^n$. The classification of critical function points give a classification of points of the manifold M. This give a spectral sequence $E_*^{*,*}(M)$ converging to the cohomology group of M and a homomorphism

$$\kappa^* \colon E_*^{*,*}(n) \to E_*^{*,*}(M)$$

such that the limit homomorphism of the cohomology groups

$$\kappa^* \colon H^*(BO(n)) \to H^*(M)$$

evaluates the characteristic classes of the vector bundle π.

Theorem 5.4 have the following unexpected corollary which can be considered as its geometrical meaning.

5.6. Proposition. *Any characteristic class defined in terms of non-Morse singularities is trivial.*

Here we give the direct simple proof. The proposition follows from the observation that *for every vector bundle* $\pi: E \to M$ *there is a function* $f_0: E \to \mathbb{R}$ *such that all singularities of its restriction to the fibers* $\pi^{-1}(x)$, $x \in M$ *are of Morse type.* To build such a function one should take any generic section h of the conjugate bundle $E^* \to M$ considered as a function on E with linear restrictions to the fibers of π and any Riemannian structure g on E considered as a function with quadratic restrictions to the fibers of π. Then he function

$$f_0 = h + g$$

have the properties as required.

5.7. Example. An n-dimensional *foliation* on a smooth manifold M is an integrable n-dimensional distribution ν that is an integrable subbundle of the tangent bundle. n-dimensional integral submanifolds of this distribution are called *fibers*.

Let $f: M \to \mathbb{R}$ be a generic function. Then its restriction to a generic fiber is a Morse function, but non-Morse singularities of the restriction of f to the fibers can appear irremovably by small perturbations of the function. In [AGLV] it is suggested to build characteristic classes of foliations as classes dual to the cycles formed by points of some singularity type of the restriction of f to the fibers. Theorem 5.3 stays that the Stiefel-Whitney classes of the bundle ν only can appear in such a way. Moreover, in accordance with Proposition 5.6 any characteristic class defined in terms of singularities which are more complicated than Morse ones is trivial.

Let E be the total space of the bundle ν. Then there is a smooth mapping $\mathrm{Exp}: E \to M$ defined in a neighborhood of the zero section which is identical on the zero section and which maps diffeomorphically the fibers of the bundle ν to the fibers of the foliation. Then the function $\mathrm{Exp}^* f: E \to \mathbb{R}$ produces the same classification of points of M in accordance with singularity type as the function f itself. This explain why the integrability of the distribution ν does play any role.

5.8. Although the limit term of the characteristic spectral sequence has a simple description in Theorem 5.4 the spectral sequence itself has a rather complicated behavior and seems to have no good description in large dimensions. For example, the cohomology groups of the Vassiliev complex which correspond to he row $E_2^{*,0}$ of our spectral sequence are calculated in [V,AGLV]. For $l \leq 6$ the groups $E_2^{n+l,0}(n)$ for $A = \mathbb{Z}$ and $A = \mathbb{Z}_2$, respectively, are given in the following table

	l	0	1	2	3	4	5	6
$A = \mathbb{Z}$	$n = 1$	$\mathbb{Z}$	0	0	0	$\mathbb{Z}_2$	0	0
	$n = 2$	$\mathbb{Z}$	0	0	0	$\mathbb{Z}_2^2$	$\mathbb{Z}$	0
	$n \geq 3$	$\mathbb{Z}$	0	0	0	$\mathbb{Z}_2^n$	$\mathbb{Z}$	$\mathbb{Z}_3^{n-2}$
$A = \mathbb{Z}_2$	$n = 1$	$\mathbb{Z}_2$	0	$\mathbb{Z}_2$	0	$\mathbb{Z}_2$	0	$\mathbb{Z}_2$
	$n \geq 2$	$\mathbb{Z}_2$	0	0	0	$\mathbb{Z}_2^n$	$\mathbb{Z}_2$	$\mathbb{Z}_2$

Generators of these groups are described in the following theorem.

5.9. Theorem ([AGLV]). *For any n-dimensional foliation on M and a generic function on it the following chains equipped with proper signs form integral cycles:* ${}^{r}A_5^+ \cup {}^{r}A_5^-$, ${}^{0}A_6^+ \cup \ldots \cup {}^{n-1}A_6^-$, ${}^{r}P_8^1 \cup {}^{r}P_8^2$. *In addition, there are the following cycles* mod 2: ${}^{0}A_3^+ \cup \ldots \cup {}^{n-1}A_3^-$, ${}^{0}A_7^+ \cup \ldots \cup {}^{n-1}A_7^-$.

The left superscript in this theorem denotes the number of positive squares of the second differential of a function germ, see [V], [AGLV] for the notations. As a corollary of Proposition 5.6 we have

5.10. Theorem. *All the cycles listed in Theorem 5.9 are homological to zero.*

To give an explicit description of this phenomenon one should consider adjacencies of singularities the codimension of which differs by more than 1.

As an example we prove the relation $A_5^+ \cup A_5^- \sim 0$ for the case $n = 1$. The same arguments explain the relation $A_{2k+1}^+ \cup A_{2k+1}^- \sim 0$. In fact, our proof contains the calculation of the (nontrivial!) differential $\delta_2 \colon E_2^{2k-2,1}(1) \to E_2^{2k,0}(1)$. One could give similar description for the other cycles from Theorem 5.9.

Consider a smooth linear bundle $\pi \colon E \to M$ and a function $f \colon E \to \mathbb{R}$ given in a neighborhood of the zero section. Without loss of generality we can assume that M is 4-dimensional and compact and f is generic. In particular, the set of singularities $A_5^{\pm}$ of the restriction of f to the fibers of π consists of finite number of points. We show that *this number is always even.*

The closure $\overline{A_3(M)} \subset M$ of the set of points of singularity type $A_3^{\pm}$ is smooth and two-dimensional. The closure of A_4 divides it into two domains $A_3^+(M)$ and $A_3^-(M)$. For any closed curve $\gamma \subset A_3^+(M)$ we define its index $\widehat{A}_3^+(\gamma) \in \mathbb{Z}_2$ which is equal to 0 or 1 depending on weather orientable or not the restriction of the bundle π to γ. This index does not change when we exchange the curve γ by the homological one and defines the cohomology class $[\widehat{A}_3^+] \in H^1(A_3^+(M))$ which is nothing but the first Stiefel-Whitney class of the restriction of the bundle π to $A_3^+(M) \subset M$ (hereinafter by the cohomology we mean the cohomology mod 2).

Near the points of the type A_4 the function with singularity A_3^+ at the origin has nearby one more critical point. The direction to this additional point defines a particular orientation of the restriction of π to $A_3^+(M)$ in a neighborhood of the curve $A_4(M)$. This allows to extend the index $\widehat{A}_3^+$ to the set of curves in $A_3^+(M)$ with ends in $A_4(M)$: it is equal to 0 or 1 depending on weather this particular orientation passes to itself or not when going along the curve from one its and to another. Define now the index $[\widehat{A}_3^+] \in H^1(\overline{A_3(M)} \setminus \{A_5^{\pm}(M)\})$ which on a closed curve $\gamma \subset \overline{A_3(M)} \setminus \{A_5^{\pm}(M)\}$ is equal to the $\widehat{A}_3^+$-index of the intersection of γ with the closure $\overline{A_3^+(M)}$.

Let $C \subset \overline{A_3^+(M)}$ be the union of small circles around all $A_5^{\pm}$-points. Then C is a boundary of a 2-chain in $\overline{A_3(M)} \setminus \{A_5^{\pm}(M)\}$, and hence, *the sum of $\widehat{A}_3^+$-indices of small circles around all $A_5^{\pm}$-points is equal to zero* (mod 2). It remains to observe that the $\widehat{A}_3^+$-index of a small circle going around any point of the type $A_5^{\pm}$ is equal to 1. The proof of this is seen from Fig.1.

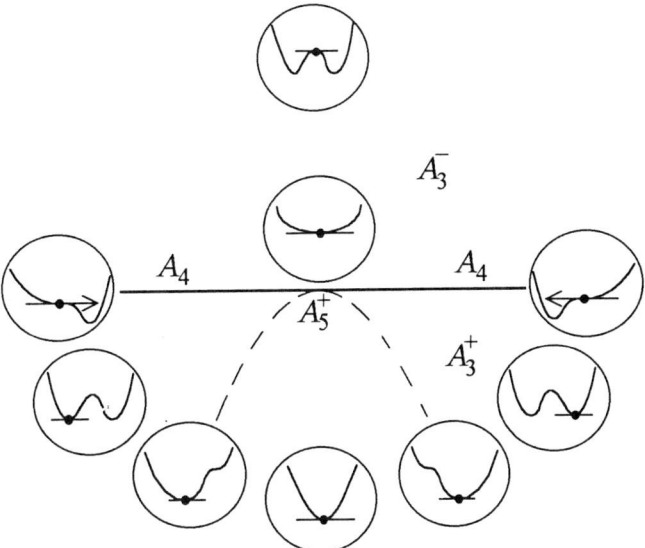

Figure 1. Functions with singularity $A_3^{\pm}$
at the origin near a function with singularity $A_5^{\pm}$

References

[AGLV] Arnold V.I., Goryunov V.V., Lyashko O.V., Vassiliev V.A., *Singularities.* **I.**, in: Dynamical systems, **VI**, Springer Verlag (1991).

[HL1]   Harvey R., Lawson B., *A theory of characteristic currents associated with a singular connection*, Astérisque, **213** (1993), 1–268.

[HL2]   Harvey R., Lawson B., *Geometric Residue Theorems*, preprint (1994).

[K1]   Kazarian, M.È., *Hidden Singularities and Homological Vassiliev Complex of Singularity Classes,* to appear.

[K2]   Kazarian, M.È., *Characteristic Classes of Lagrangian and Legendrean Singularities,* to appear.

[L]   Lai H., *Characteristic classes of real manifolds immersed in complex manifolds*, Trans. A.M., **172** (1972), 1–33.

[P]   Porteous I.R., *Simple singularities of maps*, Lect. Notes Math., **192** (1971), 286–307.

[R]   Ronga F., *Le calcul des classes duales aux singularites de Boardman d'odre deux*, Comm. math. helv., **47** (1972), 15–35.

[V]   Vassiliev, V.A., *Lagrange and Legendre Characteristic Classes.*, 2-d edition, Gordon and Breach (1993).

Maxim Kazarian
Independent Moscow University
Moscow, Russia
kazarian@ium.ac.msk.su

Received March 1995

Value of generalized hypergeometric function at unity

A. Kazarnovski-Krol

Abstract

Value of generalized hypergeometric function at a special point is calculated. More precisely, the value of certain multiple integral over vanishing cycle (all arguments collapse to unity) is calculated. The answer is expressed in terms of Γ-functions. The constant is relevant to the part of ρ in the Gindikin-Karpelevich formula for the c-function of Harish-Chandra. Calculation is an adaptation of classical calculations of Gelfand and Naimark (1950) to the Heckman-Opdam hypergeometric functions in the case of root system of type A_{n-1}.

0. Introduction

In this paper the value of certain multiple integral over a vanishing cycle (all arguments collapse to unity) is calculated. This is a normalization constant. It is related to the part of ρ in the Gindikin-Karpelevich formula for the c-function of Harish-Chandra. The cycle is a distinguished one in the theory of zonal spherical functions. It appeared for the first time in [1] and its role is clarified in [21].

Technically our calculations are an adaptation of formulas of [1] (elliptic coordinates) for the case of generic k, i.e., for the Heckman-Opdam hypergeometric functions of type A_{n-1}. The integral may be regarded as a variation on a theme of the Selberg integral [19], and recently integrals of this type have drawn much attention because of applications to conformal field theory.

The paper has continuation in [20,21].

1. Theorem

Let z_l, $l = 1, ..., n$; $t_{i,j}$, $i = 1, ..., j$, $j = 1, ..., n - 1$ be the set of real variables. It is convenient to organize these variables in the form of a pattern cf. fig.1.

Remark 1.1 This type of variables reflects the flag structure (ith row corresponds to i-dimensional plane in a flag), cf.[1,2]. One should notice that this type of variables is used in [10] for the isomorphism of Heckman-Opdam hypergeometric system with a particular case of trigonometric version of Knizhnik-Zamolodchikov equation.

$$z_1 \qquad\qquad z_2 \qquad\qquad \cdots \qquad\qquad \cdots \qquad\qquad z_n$$

$$t_{1,n-1} \qquad\quad t_{2,n-1} \qquad\quad \cdots \qquad\quad t_{n-1,n-1}$$

$$\cdots \qquad\qquad \cdots \qquad\qquad \cdots$$

$$t_{1,2} \qquad\qquad t_{2,2}$$

$$t_{1,1}$$

Figure 1. Variables organized in a pattern

Definition 1.2 Consider the following multivalued form ω_Δ :

$$
\begin{aligned}
\omega_\Delta := &\prod_{i=1}^{n} z_i^{a_i} \prod_{i_1 > i_2} (z_{i_1} - z_{i_2})^{1-2k} \\
&\times \prod_{i<l} (z_l - t_{i,n-1})^{k-1} \prod_{i \geq l} (t_{i,n-1} - z_l)^{k-1} \\
&\times \prod_{j=1}^{n-2} \prod_{i_1 \geq i_2} (t_{i_1,j} - t_{i_2,j+1})^{k-1} \prod_{j=1}^{n-2} \prod_{i_1 > i_2} (t_{i_1,j+1} - t_{i_2,j})^{k-1} \\
&\times \prod_{j=2}^{n-1} \prod_{i_1 > i_2} (t_{i_1,j} - t_{i_2,j})^{2-2k} \\
&\times \prod_{j=1}^{n-1} \prod_{i=1}^{j} t_{ij}^{a_{ij}} \quad dt_{11} dt_{12} dt_{22} \dots dt_{n-1,n-1}
\end{aligned}
$$

Remark 1.3 For the applications one should let $a_{ij} = \lambda_{n-j+1} - \lambda_{n-j} - k$ and $a_i = \lambda_1 + \frac{k(n-1)}{2}$, but we formulate the theorem in this more general form (with a_{ij} and a_i unspecified).

Definition 1.4 Set

$$\Phi(z_1, \dots, z_n) = \int_\Delta \omega_\Delta,$$

where ω_Δ is defined above and , assuming that $z_1, z_2, \dots, z_n$ are real and satisfy $0 < z_1 < z_2 < \dots < z_n$, define cycle Δ by the following inequalities: $t_{i,j+1} \leq t_{ij} \leq t_{i+1,j+1}$ and $z_i \leq t_{i,n-1} \leq z_{i+1}$.

We assume that phases of the factors of the form ω_Δ are equal to zero provided k , a_{ij}, and a_i are real.

Theorem 1.5 *The limit of* $\Phi(z_1, \dots, z_n)$ *as all* z_i *approach* 1, *while* $z_1 < z_2 < \dots < z_n$, *is equal to:*

$$\Phi(1, \dots, 1) := \lim_{\substack{\text{all } z_i \to 1}} \Phi(z_1, \dots, z_n) = \frac{\Gamma(k)\Gamma(k)^2 \dots \Gamma(k)^n}{\Gamma(k)\Gamma(2k) \dots \Gamma(nk)} \quad.$$

Proof. Following the classical work of I.M. Gelfand and M.A. Naimark cf. [1] , let

$$\tau_{ij} = \frac{\prod\limits_{i_1=1}^{j-1}(t_{i_1,j-1} - t_{ij})}{\prod\limits_{i_1 \neq i}(t_{i_1,j} - t_{ij})} \quad,$$

$i = 1, \ldots, j$ $j = 1, \ldots, n-1$. Note that $\sum\limits_{i=1}^{j} \tau_{ij} = 1$, and

$$\frac{D(\tau_{1,j}, \ldots, \tau_{j-1,j})}{D(t_{1,j-1}, \ldots, t_{j-1,j-1})} = \frac{\prod\limits_{1 \leq i < k \leq j-1}(t_{i,j-1} - t_{k,j-1})}{\prod\limits_{1 \leq i < p \leq j}(t_{ij} - t_{pj})}$$

[see [1] for the details]. Let also

$$\tau_{in} = \frac{\prod\limits_{i=1}^{n-1}(t_{i_1,n-1} - z_i)}{\prod\limits_{i_1 \neq i}(z_{i_1} - z_i)} \qquad i = 1, \ldots, n.$$

One has $\sum\limits_{i=1}^{n} \tau_{in} = 1$ and

$$\frac{D(\tau_{1,n}, \ldots, \tau_{n-1,n})}{D(t_{1,n-1}, \ldots, t_{n-1,n-1})} = \frac{\prod\limits_{1 \leq i < k \leq n-1}(t_{i,n-1} - t_{k,n-1})}{\prod\limits_{1 \leq i < p \leq n}(z_i - z_p)}$$

In the variables τ_{ij} $i = 1, \ldots, j-1$, $j = 1, \ldots, n$ integral for $\Phi(z_1, \ldots, z_n)$ is written as:

$$\Phi(z_1, \ldots, z_n) = \prod_{i=1}^{n} z_i^{a_i} \int \prod_{j=1}^{n-1} \prod_{i=1}^{j} (t_{ij})^{a_{ij}}$$

$$\times \prod_{j=1}^{n} ((\tau_{1j}\tau_{2j} \ldots \tau_{j-1,j})(1 - \tau_{1j} - \ldots - \tau_{j-1,j}))^{k-1} d\tau_{12} d\tau_{13} d\tau_{23} \ldots d\tau_{1n} \ldots d\tau_{n-1,n} \quad .$$

Integration is taken over

$$\tau_{ij} > 0, \qquad \sum_{i=1}^{j-1} \tau_{ij} < 1.$$

As all z_i, $i = 1, \ldots, n$ approach 1, all t_{ij} also approach 1, so

$$\Phi(1, \ldots, 1) = \int 1 \times \prod_{j=1}^{n} ((\tau_{1j}\tau_{2j} \ldots \tau_{j-1,j})(1 - \tau_{1j} - \ldots - \tau_{j-1,j}))^{k-1}$$
$$\times \, d\tau_{12} d\tau_{13} d\tau_{23} \ldots d\tau_{1n} \ldots d\tau_{n-1,n}.$$

So using Dirichlet's formula one gets

$$\Phi(1, \ldots, 1) = \frac{\Gamma(k)\Gamma(k)^2 \ldots \Gamma(k)^n}{\Gamma(k)\Gamma(2k) \ldots \Gamma(nk)}$$

Remark 1.6 The same calculation as in theorem 1.5. shows that $\Phi(z_1, z_2, \ldots, z_n)$ is equal to the same constant as in theorem provided that all $a_{ij} = 0$ and $a_i = 0$. Then this constant is interpreted as the volume of a maximal compact subgroup (under certain normalizations)cf. [1,13,14] in the case of generic k, i.e. when there is no group and no subgroup at all. The fact that this constant does not depend on z_i also implies the monodromy properties of cycle Δ, cf. [21].

Acknowledgments. I would like to express my gratitude to I. M. Gelfand for many stimulating discussions on the representation theory and the theory of hypergeometric functions. I would like to thank S. Shatashvili for helpful discussions.

References

[1] Gelfand I.M., Naimark M.A, Unitary representations of classical groups, *Tr. Mat. Inst. Steklova* **36**(1950), 1–288; [in Russian, see also extract in English in Collected Papers of I.Gelfand, vol. 2, pp.182-211]

Gelfand I.M., Naimark M.A., *Unitare Darstellungen der Klassischen Gruppen*, Akademie Verlag (German translation), 1957

[2] Gelfand I.M., Tsetlin M.L, Finite-dimensional representations of the group of unimodular matrices. *Dokl. Akad. Nauk SSSR* **71** (1950), 825-828.

[3] Gelfand I.M., Spherical functions on symmetric Rimannian spaces. *Dokl. Akad. Nauk SSSR* **70** (1950), 5-8

[4] Gelfand I.M., Berezin F.A., Some remarks on the theory of spherical functions on symmetric Rimannian manifolds, *Tr. Mosk. Mat. O.-va*, **5** (1956), 311-351; [Transl., II.Ser., Am.Math.Soc. 21(1962) 193-238]

[5] Heckman G.J., Opdam E.M., Root systems and Hypergeometric functions I, *Comp. Math.* **64** (1987), 329-352

[6] Cherednik I.V., A unification of Knizhnik-Zamolodchikov and Dunkl operators via affine Hecke Algebras, *Invent. Math.*, **106**, 411-431

[7] Varchenko, A.N., Multidimensional Hypergeometric functions in Conformal Field Theory, Algebraic K-theory, Algebraic Geometry

[8] Gindikin S.G., Karpelevich F.I., Plancherel measure for Rimannian symmetric spaces of nonpositive curvature, *Dokl.Akad. Nauk SSSR*, **145**:2 (1962), 252-255

[9] Schechtman,V.V., Varchenko, A.N., Hypergeometric solutions of Knizhnik- Zamolodchikov equations., *Lett. Math. Phys.* **20**: 4 (1990), 279-283

[10] A.Matsuo, Integrable connections related to zonal spherical functions, *Invent. Math.*, **110** (1992), 95-121

[11] Cherednik I., Integral solutions of trigonometric Knizhnik-Zamolodchikov equations and Kac-Moody algebras, *Publ. RIMS Kyoto Univ.*, **27** (1991), 727-744

[12] V.Knizhnik, A.Zamolodchikov, Current algebra and Wess-Zumino models in two dimensions, *Nucl. Phys. B*, **247** (1984), 83-103

[13] Aomoto K., Sur les transformations d'horisphère et les equations intégrales qui s'y rattachent, *J.Fac.Sci.Univ.Tokyo*, **14**, 1-23

[14] I.G. Macdonald, The volume of a compact Lie group, *Inv. Math.*, **56** (1980), 93-95

[15] Aomoto K., Les équations aux différences linéaires et les intégrales des fonctions multiformes, *J.Fac.Sci.Univ. Tokyo* , **22** (1975)

[16] Etingof P.I., Kirillov A.A. Jr., Macdonald's polynomials and representations of quantum groups, to appear in Math. Res. Let., hep-th/9312103, 1994

[17] Etingof P.I., Kirillov A.A. Jr., A unified representation theoretic approach to special functions, Hep-th/9312101

[18] Alexeev A., Faddeev L., Shatashvili S., Quantisation of symplectic orbits of compact Lie groups by means of functional integral, *Jour. of Geom. and Phys.*, **5** (1989), 391-406

[19] Selberg A., Bemerkninger om et multiplet integral, *Norsk. Mat. Tids.*, **26** (1944), 71-78

[20] Kazarnovski-Krol A., Cycles for asymptotic solutions and Weyl group, 1994

[21] Kazarnovski-Krol A., Decomposition of a cycle, 1994

[22] Opdam E., An analogue of the Gauss summation formula for hypergeometric functions related to root systems, preprint, July 1991

Department of Mathematics
Rutgers University
New Brunswick, NJ 08854, USA

Received March 1995

Harish-Chandra decomposition for zonal spherical function of type A_n

A. Kazarnovski-Krol

0. Introduction

Heckman-Opdam system of differential equations is holonomic , with regular singularities and has locally $|W|$-dimensional space of solutions (cf. corollary 3.9 of [12]), where $|W|$ is the cardinality of the Weyl group W. The system is a generalization of radial parts of Laplace-Casimir operators on symmetric Riemannian spaces of nonpositive curvature and is isomorphic to Calogero-Sutherland model in the integrable systems.

Harish-Chandra asymptotic solution is a unique solution of the system with the prescribed asymptotic behavior:

$$F_w(z) = z^{w\lambda + \rho}(1 + \dots)$$

$(0 < |z_1| < |z_2| < \dots < |z_{n+1}|)$. Here $w \in W$ are elements of the Weyl group. These solutions provide a basis in the space of all the solutions in the chamber $0 < |z_1| < \dots < |z_{n+1}|$. Among all the solutions there is a distinguished one up to the constant multiplier, which admits continuation to analytic function at $z_1 = z_2 = z_3 = \dots = z_{n+1} \neq 0$. This solution is referred to as zonal spherical function. Zonal spherical function is normalized s.t. it is equal to 1 at $z_1 = z_2 = \dots = z_{n+1} = 1$.

Representation of the zonal spherical function as linear combination of elements of the basis (Harish-Chandra asymptotic solutions) is called Harish-Chandra decomposition.

In ref. [11] we provided an integral representation for the solutions of Heckman-Opdam system of differential equations in the case of A_n. We also described contours for integration Δ_w, integrals over them provide Harish-Chandra asymptotic solution $F_w(z)$. In ref. [34] we studied the cycle Δ for integration for zonal spherical function . This paper is devoted to homological treatment of Harish-Chandra decomposition for zonal spherical functions of type A_n. Namely, we explicitly decompose the distinguished cycle Δ cf. [10 ,16] into linear combination of cycles Δ_w described in [11] and check that after normalization this turns out to be the Harish-Chandra decomposition for zonal spherical function of type A_n (theorem 2.2 and 3.1 below). The point of view that linear relations between the solutions reflect the linear relations in homology group is due to B. Riemann. He also

---

[0] This paper is a revised version of the preprint of the author "Decomposition of a cycle"

emphasized the importance of the monodromy. In this case the corresponding homology theory is described in [2 , 37]. Harish-Chandra asymptotic
solutions correspond to conformal blocks in conformal field theory (WA_n
algebras) and provide a basis in the space of conformal blocks, zonal spherical function is a particular conformal block, in the case of A_2 see figs. 3a,3b,
3c,3d,3e,3f and 2 below.

0.1 Notations.

$\alpha_1, \alpha_2, \ldots, \alpha_n$ - simple roots of root system of type A_n

$\Lambda_1, \Lambda_2, \ldots, \Lambda_{n+1}$ - weights of fundamental (vector) representation,
$\Lambda_2 = \Lambda_1 - \alpha_1$, $\Lambda_3 = \Lambda_1 - \alpha_1 - \alpha_2, \ldots, \Lambda_1 + \Lambda_2 + \ldots + \Lambda_{n+1} = 0$.

R_+ - set of positive roots

$\delta = \frac{1}{2} \sum_{\alpha \in R_+} \alpha$ -halfsum of positive roots

k- complex parameter ('halfmultiplicity' of a root)

$$\rho = \frac{k}{2} \sum_{\alpha \in R_+} \alpha$$

$c(\lambda, k)$- c-function of Harish-Chandra

1. MULTIVALUED FORM

Consider the following set of variables:

z_l, $l = 1, \ldots, n+1$, t_{ij}, $i = 1, \ldots, j$, $j = 1, \ldots, n$.

Variables z_l have meaning of arguments, while variables t_{ij} are variables
of integration.

It is convenient to organize variables z_l, t_{ij} in the form of a pattern,
cf. fig 1.

$$
\begin{array}{ccccc}
z_1 & z_2 & \cdots & \cdots & z_{n+1} \\
& t_{1,n} & t_{2,n} & \cdots & t_{n,n} \\
& \cdots & \cdots & \cdots & \\
& t_{1,2} & t_{2,2} & & \\
& & t_{1,1} & &
\end{array}
$$

FIGURE 1. Variables organized in a pattern

The idea of such an organization is borrowed from Gelfand-Zetlin pat-
terns [1].

Definition 1.1. Consider the following multivalued form $\omega(z, t)$:

$$\omega(z,t) := \prod_{i=1}^{n+1} z_i^{\lambda_1 + \frac{kn}{2}} \prod_{i_1 > i_2} (z_{i_1} - z_{i_2})^{1-2k}$$

$$\times \prod_{l=1}^{n+1} \prod_{i=1}^{n} (z_l - t_{i,n})^{k-1}$$

$$\times \prod_{j=1}^{n-1} \prod_{i_1=1}^{j+1} \prod_{i=1}^{j} (t_{ij} - t_{i_1,j+1})^{k-1}$$

$$\times \prod_{j=2}^{n} \prod_{i_1 > i_2} (t_{i_1,j} - t_{i_2,j})^{2-2k}$$

$$\times \prod_{j=1}^{n} \prod_{i=1}^{j} t_{ij}^{\lambda_{n-j+2} - \lambda_{n-j+1} - k} \quad dt_{11} dt_{12} dt_{22} \ldots dt_{nn}$$

Remark 1.2. k is a complex parameter - 'halfmultiplicity' of a restricted root, cf. Heckman, Opdam [12].

In [11] we proved that integrals over the form $\omega(z,t)$ over appropriate cycles provide all the solutions to Heckman-Opdam system of differential equtions and described cycles Δ_w for Harish-Chandra asymptotic solutions (definition 4.3 and theorem 6.3 of [11]).

Definition 1.3. A complex number z can be represented as $z = re^{i\alpha}$, where r, α are real numbers, $r \geq 0$. r is called absolute value of z, while α is called the phase of z. When we say that the phase of a complex number z is equal to 0, we mean that $\alpha = 0$, or the number itself is real and nonnegative.

2. THE DISTINGUISHED CYCLE Δ

Assume that $z_1, z_2, \ldots, z_{n+1}$ are real and

$$0 < z_1 < z_2 < \ldots < z_{n+1}.$$

Definition 2.1. Define cycle $\Delta = \Delta(z)$ by the following inequalities:
$t_{i,j+1} \leq t_{ij} \leq t_{i+1,j+1}$ and
$z_i \leq t_{in} \leq z_{i+1}$.
Define form $\omega_\Delta(z,t)$ as:

$$
\omega_\Delta(z,t) := \prod_{i=1}^{n+1} z_i^{\lambda_1 + \frac{kn}{2}} \prod_{i_1 > i_2} (z_{i_1} - z_{i_2})^{1-2k}
$$

$$
\times \prod_{i \leq l} (z_l - t_{i,n})^{k-1} \prod_{i > l} (t_{i,n} - z_l)^{k-1}
$$

$$
\times \prod_{j=1}^{n-1} \prod_{i_1 > i_2} (t_{i_1,j} - t_{i_2,j+1})^{k-1} \prod_{i_2 \geq i_1} (t_{i_2,j+1} - t_{i_1,j})^{k-1}
$$

$$
\times \prod_{j=2}^{n} \prod_{i_1 > i_2} (t_{i_1,j} - t_{i_2,j})^{2-2k}
$$

$$
\times \prod_{j=1}^{n} \prod_{i=1}^{j} t_{ij}^{\lambda_{n-j+2} - \lambda_{n-j+1} - k} \quad dt_{11} dt_{12} dt_{22} \ldots dt_{nn}
$$

It is assumed that phases of factors in the formula for $\omega_\Delta(z,t)$ are equal to zero if k and $\lambda_1, \lambda_2, \ldots, \lambda_{n+1}$ are real. For the homological meaning of the cycle Δ see fig. 2 below and theorem 5.7 of [34].

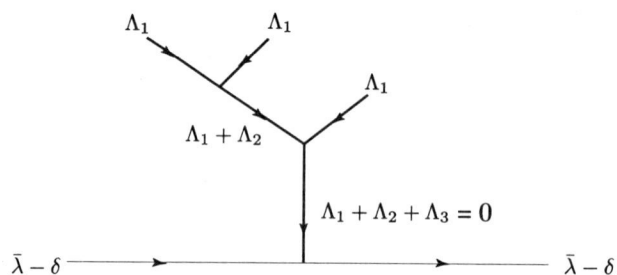

FIGURE 2. Zonal spherical function

In [11] cycles $\Delta_w(z)$ and forms $\omega_w(z,t)$ were described, in the case of A_2 see figs. 3a,3b,3c,3d,3e,3f below. In the case A_n the figures are similar.

The following theorem explains the relation between $\Delta(z)$ and $\Delta_w(z)$.

Theorem 2.2. *(Harish-Chandra decomposition)*

$$
\int_{\Delta(z)} \omega_{\Delta(z,t)} = \sum_{w \in S_{n+1}} b(w, \lambda, k) \int_{\Delta_w(z)} \omega_w(z),
$$

where

$$b(w, \lambda, k) = \frac{e^{2\pi i(\lambda, \delta)} e^{\pi i \; l(w)(k-1)}}{(2i)^{\frac{n(n+1)}{2}} \prod\limits_{\alpha \in R_+} \sin(-\pi(w\lambda, \alpha^\vee))}$$

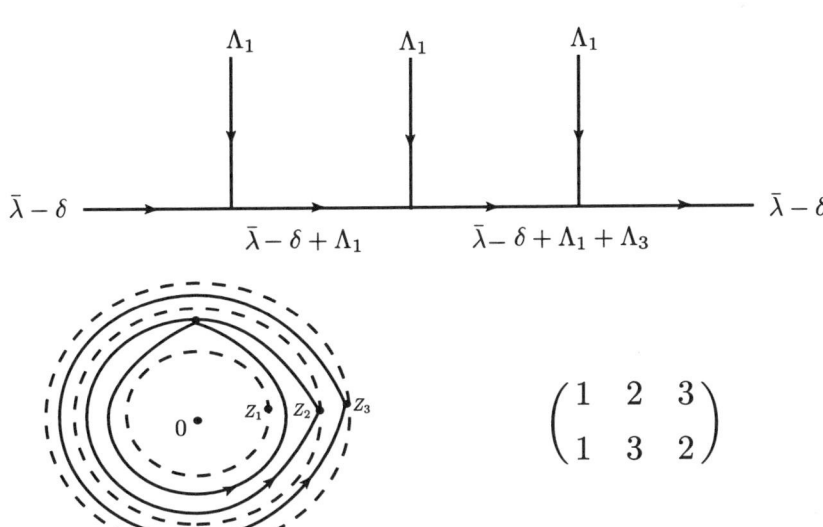

FIGURE 3A.

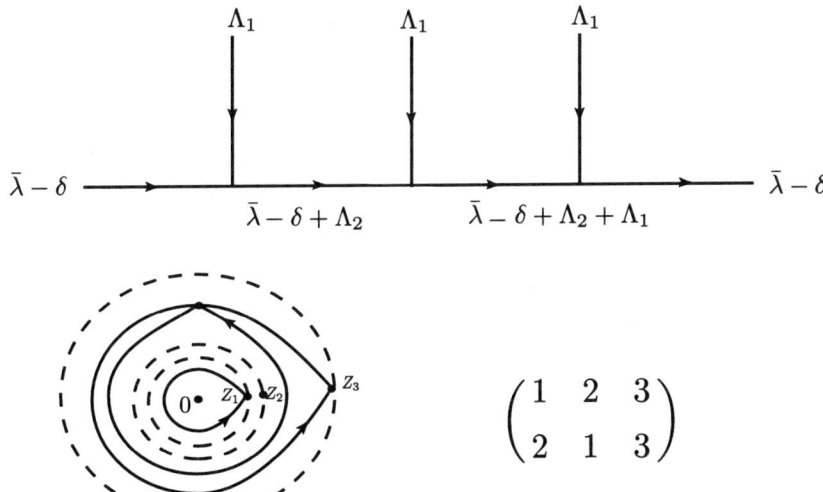

FIGURE 3B.

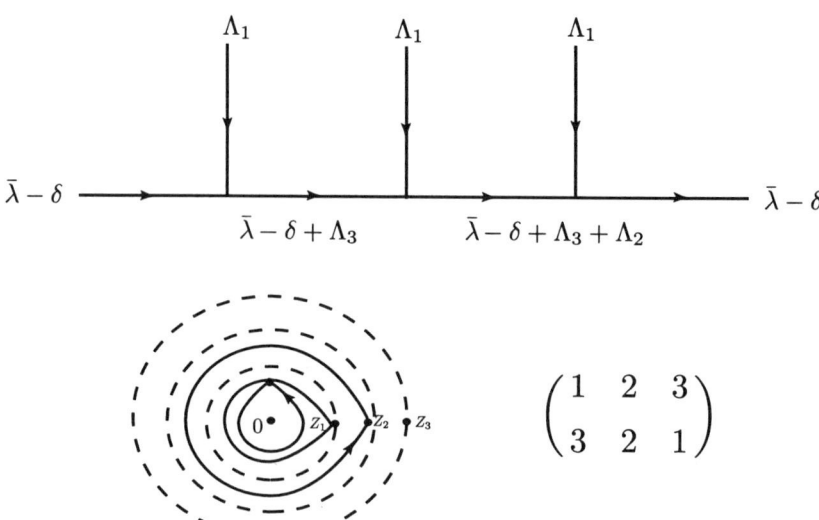

FIGURE 3C.

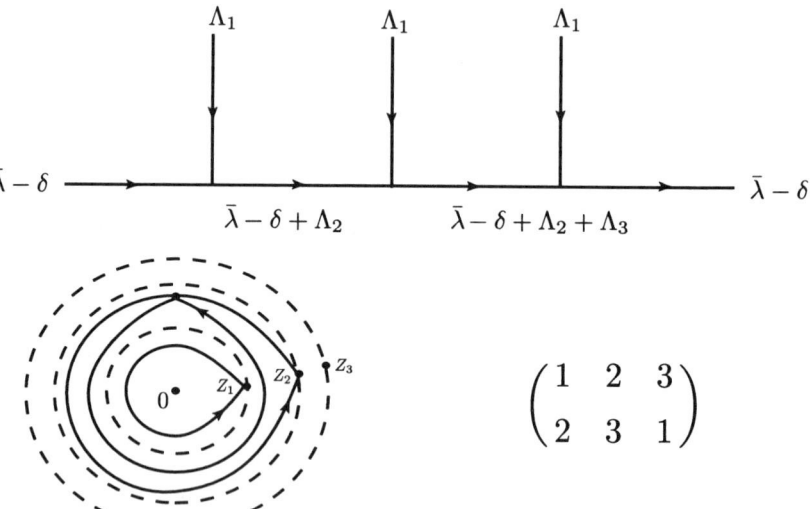

FIGURE 3D.

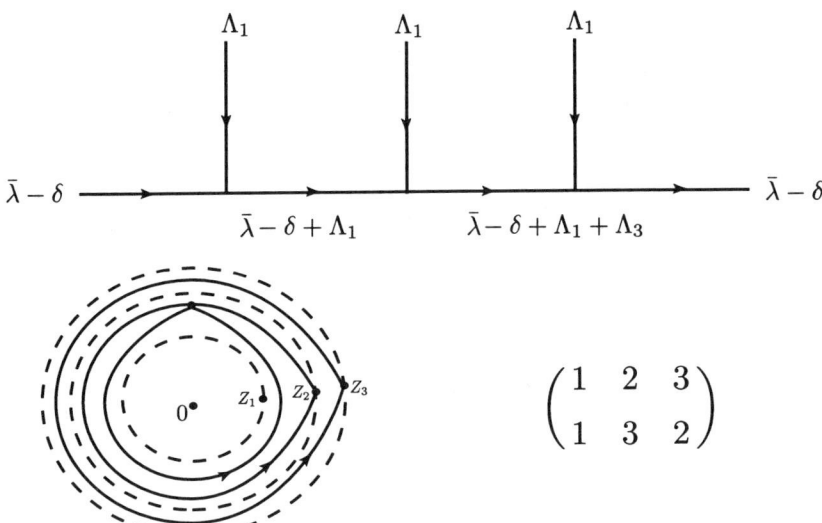

FIGURE 3E.

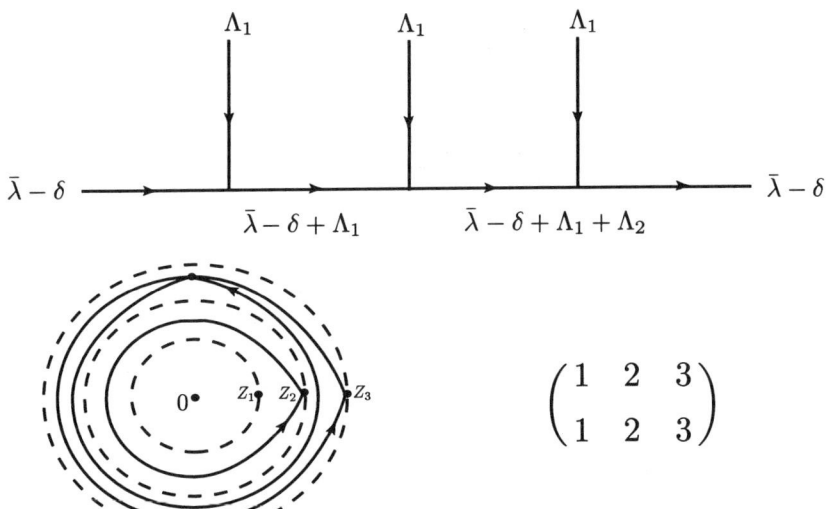

FIGURE 3F.

The theorem is an application of the two following lemmas, also section 2 of [11] is useful.

Lemma 2.3. *(Elementary decomposition)* Let $z_1, z_2, \ldots, z_n$ be real and $0 < z_1 < z_2 < \ldots < z_n$. Consider the following integral:

$$\int t^{a_0-1}(z_1 - t)^{a_1-1}(z_2 - t)^{a_2-1} \ldots (z_n - t)^{a_n-1} dt$$

and consider contours $\gamma_1(t), \gamma_2(t), \gamma(t)$, $t \in [0, 1]$ as follows.

$$\gamma(t) = tz_{i-1} + (1 - t)z_i$$

$\gamma_1(t)$ is a loop which starts and ends at z_{i-1} and goes counterclockwise s.t. the following inequalities are fullfilled for all $t \in [0, 1]$:

$$z_{i-2} < |\gamma_1(t)| \leq z_{i-1}$$

$\gamma_2(t)$ is a loop which starts and ends at z_i and goes counterclockwise ,s.t. the following inequalities are fullfilled for all $t \in [0, 1]$:

$$z_{i-1} < |\gamma_2(t)| \leq z_i$$

as indicated on fig. 4, phases of the factors should be appropriately chosen. The following is the specific choice of the phases : if all $a_0, a_1, \ldots, a_n$ are real, then the phase of the integrand along $\gamma_1(t), \gamma_2(t), \gamma(t)$ is chosen to be zero for small values of t.

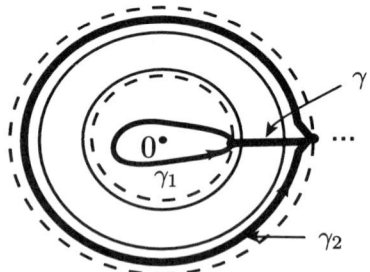

FIGURE 4. Elementary decomposition.

Then we have the following relation between γ_1, γ_2 and γ:

$$\gamma = -\gamma_1 \times \frac{e^{-\pi i(a_0 + a_1 + \ldots + a_{i-1})}}{(2i) \sin \pi(a_0 + a_1 + \ldots + a_i)} + \gamma_2 \times \frac{e^{-\pi i(a_0 + a_1 + \ldots + a_i)}}{(2i) \sin \pi(a_0 + a_1 + \ldots + a_i)}$$

Lemma 2.4. *(Elimination of 'wrong' diagrams).* Integrals of the form $\omega(z, t)$ $(\omega_w(z, t))$, such that contours for integration of t_{ij}, $t_{i,j+1}$, $t_{i+1,j+1}$ are shown on fig. 5, provided k is not an integer, are equal to zero. We suppose that t_{ij} goes from $t_{i,j+1}$ to $t_{i+1,j+1}$, $t_{i,j+1}$ goes from $t_{i+1,j+2}$ to $t_{i+1,j+2}$, and $t_{i+1,j+1}$ goes from $t_{i+1,j+2}$ to $t_{i+1,j+2}$ cf. fig. 5a.

The same holds true for $t_{i-1,n-1}, t_{i-1,n}, t_{i,n},$ and z_i correspondingly, cf. fig. 5b.

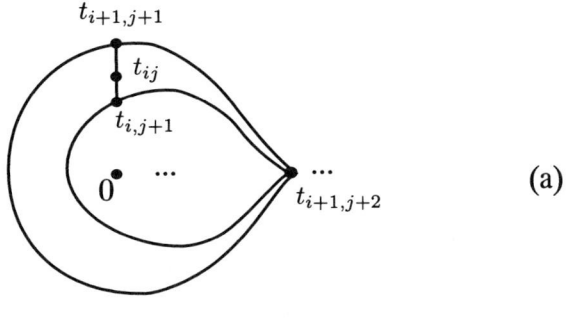

(a)

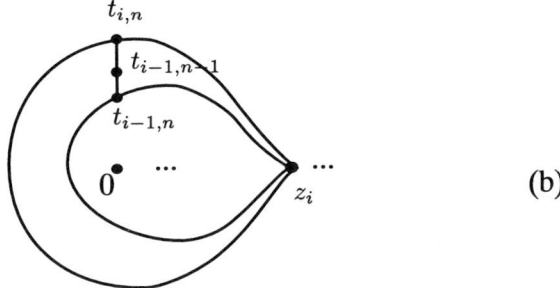

(b)

FIGURE 5. (a), (b), Cycles of this type are homological to zero.

By the 'wrong' diagrams we mean diagrams ,where the two arrows have the same target, see figs. 5 and 6 of [34] .

Remark 2.5. Lemma 2.4 is equivalent to quantum Serre's relations in the form given in [3], see also [2].

3. NORMALIZATION

Let

$$F_w(z) = (\prod_{\alpha \in R_+} \frac{\Gamma((-w\lambda, \alpha^\vee)) \sin(\pi(-w\lambda, \alpha^\vee))}{\Gamma((-w\lambda, \alpha^\vee) + k)}$$

$$\times e^{-2\pi i(\lambda, \delta)} e^{-\pi i(k-1)l(w)} \Gamma(k)^{\frac{n(n+1)}{2}} (2i)^{\frac{n(n+1)}{2}})^{-1} \int_{\Delta_w(z)} \omega_w(z, t)$$

Then

$$F_w(z) = z^{w\lambda + \rho}(1 + \dots)$$

cf. [11] theorem 6.1.

Also, let

$$F_\Delta(z) = \frac{\Gamma(k)\Gamma(2k)\ldots\Gamma((n+1)k)}{\Gamma(k)^{\frac{(n+1)(n+2)}{2}}} \int_{\Delta(z)} \omega_\Delta(z,t)$$

Then $F_\Delta(1,1,\ldots,1) = 1$, cf. [10] theorem 1.5.

After this normalization theorem 1 reads as usual Harish-Chandra decomposition cf. [12,15].

Theorem 3.1. *In the above normalization we have:*

$$F_\Delta(z) = \sum_{w\in S_{n+1}} c(w\lambda, k) F_w(z),$$

where $c(w\lambda, k)$ is a c-function of Harish-Chandra:

$$c(w\lambda, k) = \frac{\prod_{\alpha\in R_+} \frac{\Gamma((\rho,\alpha^\vee)+k)}{\Gamma((\rho,\alpha^\vee))}}{\prod_{\alpha\in R_+} \frac{\Gamma((-w\lambda,\alpha^\vee)+k)}{\Gamma((-w\lambda,\alpha^\vee))}}$$

I.e. $F_\Delta(z)$ is identified with zonal spherical function.

Corollary 3.2. *Suppose $z_1(t), z_2(t),\ldots,z_{n+1}(t)$, $t\in[0,1]$ are closed loops on a complex plane, i.e. $z_1(0) = z_1(1), z_2(0) = z_2(1),\ldots,z_{n+1}(0) = z_{n+1}(1)$, such that $z_i(t) \neq z_j(t)$ for $i\neq j$. Let also $Re(z_i(t)) > 0$ for each $i = 1,\ldots,n+1$. Then the homological class of the cycle Δ is preserved under the monodromy along paths $z_i(t)$.*

Remark 3.3. In this approach multiplicative structure of c-function of Harish-Chandra gets a very simple explanation. Namely:

$$c(\lambda, k) = \frac{\prod_{1\leq i<j\leq n} \frac{\Gamma((\rho,e_i-e_j)+k)}{\Gamma((\rho,e_i-e_j))}}{\prod_{1\leq i<j\leq n} \frac{\Gamma((-w\lambda,e_i-e_j)+k)}{\Gamma((-w\lambda,e_i-e_j))}} \times \frac{\prod_{1\leq i<n+1} \frac{\Gamma((\rho,e_i-e_{n+1})+k)}{\Gamma((\rho,e_i-e_{n+1}))}}{\prod_{1\leq i<n+1} \frac{\Gamma((-w\lambda,e_i-e_{n+1})+k)}{\Gamma((-w\lambda,e_i-e_{n+1}))}}$$

Here $\{e_i - e_j|\ 1 \leq i < j \leq n+1\}$ are positive roots of root system of type A_n. Multiplicative properties of c-function of Harish-Chandra were observed by Bhanu-Murti in the case of $SL(n,\mathbb{R})$ and in general case by Gindikin and Karpelevich [17]. c-function of Harish-Chandra is equal to the product of elements of 6j-symbols , see fig. 6. Multilpicative structure of c-function of Harish-Chandra amounts to simple combinatorics related to positive roots , in this case:

$$\{e_i-e_j|1 \leq i < j \leq n+1\} = \{e_i-e_j|1 \leq i < j \leq n\}\bigcup\{e_i-e_{n+1}|1 \leq i < n\}.$$

This combinatorics is both very instructive and restrictive.

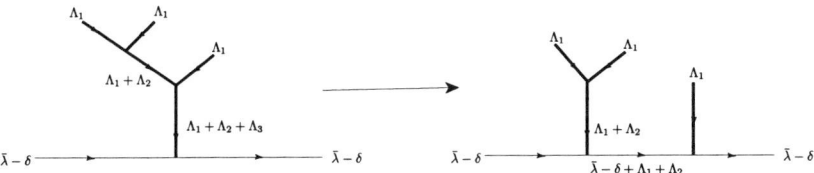

FIGURE 6. c-function of Harish-Chandra as a product of elements of $6j$-symbols .

Remark 3.4. We have also checked the monodromy properties of the cycle Δ using quantum group argument, see [34].

Remark 3.5. Harish-Chandra decomposition for zonal spherical function might be considered as an analogue of Bernstein-Gelfand-Gelfand resolution.

Concluding remark. We would like to point out once more that the distinguished cycle Δ appeared in the classical calculation of Gelfand and Naimark [16] of zonal spherical function for $SL(n, \mathbb{C})$, originates in the so-called elliptic coordinates and provides a materialization of the flag manifold.

Acknowledgments. I am grateful to I. Gelfand for stimulating discussions concerning the theory of spherical functions and the theory of hypergeometric functions, to S. Lukyanov for stimulating discussions concerning conformal field theory, to V. Brazhnikov for helpful discussions.

References

1. Gelfand I.M., Tsetlin M.L., *Finite-dimensional representations of the group of unimodular matrices*, Dokl. Akad. Nauk SSSR **71** (1950), 825-828.
2. Schechtman V., Varchenko A., *Quantum groups and homology of local systems.*, IAS preprint (1990).
3. Bouwknegt P.,McCarthy J., Pilch K., *Quantum group structure in the Fock space resolutions of $SL(n)$ representations*, Comm. Math. Phys. **131**, 125-156.
4. Felder G., *BRST approach to minimal models*, Nucl. Phys. **B 317** (1989), 215-236.
5. Varchenko A., *The function $(t_i - t_j)^{\frac{a_{ij}}{k}}$ and the representation theory of Lie algebras and quantum groups*, manuscript (1992).
6. Schechtman V., Varchenko A., *Arrangements of hyperplanes and Lie algebra homology*, Invent.Math **106** (1991), 139.
7. Fateev V., Lukyanov S., *Vertex operators and representations of Quantum Universal enveloping algebras*, preprint Kiev (1991).
8. Lukyanov S., Fateev V., *Additional Symmetries and exactly soluble models in two-dimensional conformal field theory*, Sov.Sci.Rev.A Phys. **Vol 15** (1990), 1-17.

9. Matsuo A., *Integrable connections related to zonal spherical functions*, Invent. math. **110** (1992), 95-121.

10. Kazarnovski-Krol A., *Value of generalized hypergeometric function at unity*, preprint hep-th 9405122(1994);this volume.

11. Kazarnovski-Krol A., *Cycles for asymptotic solutions and the Weyl group*, preprint 1994; to appear in Gelfand seminars.

12. Heckman G., Opdam E., *Root systems and hypergeometric functions I*, Comp. Math. **64** (1987), 329-352, ...

13. Harish-Chandra, *Spherical functions on a semisimple Lie group I*, Amer. J. of Math **80** (1958), 241-310.

14. Helgason S., *Groups and geometric analysis*, Academic Press, Inc. (1984).

15. Opdam E., *An analogue of the Gauss summation formula for hypergeometric functions related to root systems*, preprint (July 1991).

16. Gelfand I.M., Naimark M.A., *Unitary representations of classical groups*, Tr.Mat.Inst. Steklova **36** (1950), 1-288.

17. Gindikin S.G.,Karpelevich F.I., *Plancherel measure for Riemannian symmetric spaces of nonpositive curvature*, Dokl.Akad. Nauk SSSR **145** (1962), no. 2, 252-255.

18. Aomoto K., *Sur les transformation d'horisphere et les equations integrales qui s'y rattachent*, J.Fac.Sci.Univ.Tokyo **14** (1967), 1-23.

19. Rosso M., *An analogue of P.B.W. Theorem and the universal R-matrix for $U_h sl(N+1)$*, Comm. math. phys. **124** (1989), 307 - 318.

20. Drinfeld V.G., *Quantum groups*, Proc. ICM **vol. 1** (Berkeley, 1986), 798-820.

21. Jimbo M., *A q-analogue of $U(gl(N+1))$, Hecke algebra and Yang-Baxter equation*, Lett. in Math. Phys. **11** (1986).

22. Kohno T., *Quantized universal enveloping algebras and monodromy of braid groups*, preprint (1988).

23. Gomez C.,Sierra G., *Quantum group meaning of the Coulomb gas*, Phys. Lett. B **240** (1990), 149 - 157.

24. Ramirez C., Ruegg H., Ruiz-Altaba M., *The Contour picture of quantum groups: Conformal field theories*, Nucl. Phys. B **364** (1991), 195-233.

25. Alvarez-Gaume L., Gomez C., Sierra G., *Quantum group interpretation of some conformal field theories*, Phys. Lett. B (1989), 142- 151.

26. Ramirez C., Ruegg H., Ruiz-Altaba M., *Explicit quantum symmetries of WZNW theories*, Phys. Lett. B (1990), 499 - 508.

27. Kirillov A.N., Reshetikhin N., *q-Weyl group and a Multiplicative Formula for Universal R-Matrices*, Commun. Math. Phys. **134** (1990), 421-431.

28. Feigin B., Fuchs D., *Representations of the Virasoro Algebra*, in Representations of infinite-dimensional Lie groups and Lie algebras (1989), 465-554.

29. Heckman G., *Hecke algebras and hypergeometric functions*, Invent. Math. **100** (1990), 403-417.

30. Cherednik I., *Monodromy representations of generalized Knizhnik-Zamolodchikov equations and Hecke algebras*, Publ.RIMS Kyoto Univ. **27**

(1991), 711-726.

31. Schechtman V., Varchenko A., *Hypergeometric solutions of Knizhnik-Zamolodchikov equations*, Letters in Math.Phys. **20** (1990), 279-283.

32. Cherednik I., *Integral solutions of trigonometric Knizhnik-Zamolodchikov equations and Kac-Moody algebras*, Publ.RIMS Kyoto Univ. **27** (1991), 727-744.

33. Dotsenko Vl.,Fateev V., *Conformal algebra and multipoint correlation functions in 2D statistical models*, Nucl. Phys. **B240** (1984), 312-348.

34. Kazarnovski-Krol A, *Cycle for integration for zonal spherical function of type A_n*, this volume, pp. 000-000.

35. Belavin A.A, Polyakov A.M., Zamolodchikov A.B., *Infinite conformal symmetry in two-dimensional quantum field theory*, Nucl. Phys. **B241** (1984), 333-380.

36. Kirillov A.N., Reshetikhin N.Yu, *Representations of the algebra $U_q(sl(2))$, q-orthogonal polynomials and invariants of links*, in Infinite-dimensional Lie algebas and groups. Kac V.G. (ed.) (1989).

37. Finkelberg M., Schechtman V., *Localization of u-modules I. Intersection cohomology of real arrangements*, hep-th 9411050.

38. Kazarnovski-Krol A., *A generalization of Selberg integral*, preprint July 1995, q-alg 9507011.
March 1996

Positive paths in the linear symplectic group

François Lalonde[*]   *Dusa McDuff*[†]

1   Introduction

A positive path in the linear symplectic group $\mathrm{Sp}(2n)$ is a smooth path which is everywhere tangent to the positive cone. These paths are generated by negative definite (time-dependent) quadratic Hamiltonian functions on Euclidean space. A special case are autonomous positive paths, which are generated by time-independent Hamiltonians, and which all lie in the set $\mathcal{U}$ of diagonalizable matrices with eigenvalues on the unit circle. However, as was shown by Krein, the eigenvalues of a general positive path can move off the unit circle. In this paper, we extend Krein's theory: we investigate the general behavior of positive paths which do not encounter the eigenvalue 1, showing, for example, that any such path can be extended to have endpoint with all eigenvalues on the circle. We also show that in the case $2n = 4$ there is a close relation between the index of a positive path and the regions of the symplectic group that such a path can cross. Our motivation for studying these paths came from a geometric squeezing problem [16] in symplectic topology. However, they are also of interest in relation to the stability of periodic Hamiltonian systems [9] and in the theory of geodesics in Riemannian geometry [4].

Main results

We consider $\mathbf{R}^{2n}$ equipped with the standard (linear) symplectic form

$$\omega(X, Y) = Y^T J X,$$

where J is multiplication by i in $\mathbf{R}^{2n} \cong \mathbf{C}^n$. Thus if $e_1, \ldots, e_{2n}$ form the standard basis $J e_{2i-1} = e_{2i}$. The Lie algebra $\mathbf{sp}(2n)$ of $G = \mathrm{Sp}(2n, \mathbf{R})$ is the set of all matrices which satisfy the equation

$$A^T J + J A = 0.$$

---

[*]Partially supported by NSERC grant OGP 0092913 and FCAR grant ER-1199.

[†]Partially supported by NSF grant DMS 9401443. Both authors supported by NSERC grant CPG 0163730.

Hence $A \in \mathbf{sp}(2n)$ if and only if JA is symmetric. The symplectic gradient (or Hamiltonian vector field) X_H of a function $H : \mathbf{R}^{2n} \to \mathbf{R}$ is defined by the identity

$$\omega(X_H, Y) = Y^T J X_H = dH(Y) \quad \text{for all } Y \in \mathbf{R}^{2n}.$$

Given a symmetric matrix P, consider the associated function

$$\mathcal{Q}(x) = -\frac{1}{2} x^T P x$$

on $\mathbf{R}^{2n}$. Then the calculation

$$d\mathcal{Q}_x(Y) = -Y^T P x = Y^T J(JPx)$$

shows that the symplectic gradient of $\mathcal{Q}$ is the vector field

$$X_{\mathcal{Q}}(x) = JP(x), \quad \text{at} \quad x \in \mathbf{R}^{2n},$$

which integrates to the linear flow $x \mapsto e^{JPt}x$. We call these paths $\{e^{JPt}\}_{t \in [0,1]}$ **autonomous**, since they are generated by autonomous (i.e. time-independent) Hamiltonians.

We are interested in paths $A_t \in \mathrm{G}$ which are **positive** in the sense that their tangent vector at time t has the form $JP_t A_t$ for some symmetric, positive definite matrix P_t. Thus they are generated by a family of time-dependent quadratic Hamiltonians[1] of the form $\mathcal{Q}_t = -\frac{1}{2} x^T P_t x$. These paths are everywhere tangent to the **positive cone field** $\mathcal{P}$ on G given by

$$\cup_{A \in \mathrm{G}} \mathcal{P}_A = \cup_{A \in \mathrm{G}} \{ JPA : P = P^T, P > 0 \} \subset \cup_A T_A \mathrm{G} = T\mathrm{G}.$$

In particular, an autonomous path e^{JPt} is positive if P is.

Note that an autonomous path e^{JPt} always belongs to the set $\mathcal{U}$ of elements in G which are diagonalizable (over $\mathbf{C}$) and have spectrum on the unit circle. The following result shows that this is not true for all positive paths.

Proposition 1.1 *Any two elements in G may be joined by a positive path.*

This is a special case of a general result in control theory known as Lobry's theorem: see Lobry [17], Sussmann [19] and Grasse–Sussmann [11]. We give a proof here in §4.2 as a byproduct of our other results.

The structure of positive paths $A_{t \in [0,1]}$ in G was studied by Krein in a series of papers during the first half of the 50's. Later Gelfand–Lidskii [9]

---

[1]It is somewhat unfortunate that our sign conventions imply that $\mathcal{Q}_t$ is actually negative definite.

took up the study of the topological properties of general paths in connection with the stability theory of periodic linear flows. Almost simultaneously, and surely independently, Bott [3, 4, 5] studied positive flows in the complexified linear group in connection with the geometry of closed geodesics. As a result of these works, the structure of positive paths is well-understood provided that the spectrum of A_t remains on the unit circle. However, we need to investigate their behavior outside this circle. Surprisingly, a thorough study of the behavior of these paths outside the unit circle does not seem to exist. The aim of this article is to undertake such a study, at least in the case of generic positive paths, that is to say those paths which meet only the codimension 0 and codimension 1 strata of the matrix group. For the convenience of the reader we will develop the theory from scratch, even though some of our first results are well-known.

Our main result is motivated by the geometric application in [16]. Before stating this result, we introduce some notation. As above, $\mathcal{U}$ denotes the set of elements in G which are diagonalizable and have spectrum on the unit circle (unless indicated to the contrary, "diagonalizable" is considered over $\mathbf{C}$). Further,

$$\mathcal{S}_1 = \{A \in G : \det(A - \mathbb{1}) = 0\}$$

is the set of elements with at least one eigenvalue equal to 1. A path $\gamma = A_{t \in [0,1]}$ in G which starts at the identity $A_0 = \mathbb{1}$ and is such that $A_t \in G - \mathcal{S}_1$ for all $t > 0$ will be called **short**. Equivalently, it has Conley-Zehnder index equal to 0.[2] Moreover, an element $A \in G$ is called **generic** if all its eigenvalues (real or complex) have multiplicity 1. In the next theorem, all paths begin at $\mathbb{1}$.

Theorem 1.2 (i) *An element of* $G - \mathcal{S}_1$ *is the endpoint of a short positive path if and only if it has an even number of real eigenvalues* λ *with* $\lambda > 1$.

(ii) *Any short positive path may be extended to a short positive path with endpoint in* $\mathcal{U}$.

(iii) *The space of short positive paths with endpoint in* $\mathcal{U}$ *is path-connected.*

We can interpret these results in terms of a "conical" subRiemannian geometry. SubRiemannian geometry is usually a study of paths in a manifold M which are everywhere tangent to some distribution ξ in the tangent bundle TM. For example, there is recent interest in the case when ξ is a contact structure: see [12]. One could generalize this to the consideration

---

[2]The Conley–Zehnder index (or Maslov index) of a path measures how many times it goes through the eigenvalue 1 and characterizes its homotopy class: see Gelfand–Lidskii [9], Ekeland [7, 8], Robbin–Salamon /citeROSA1.

of paths which are everywhere tangent to some fixed convex conical neighborhood $\mathcal{N}_\varepsilon(\xi)$ of ξ. The geometry of positive paths more or less fits into this context: one just has to replace the distribution ξ by a distribution of rays. Thus we can consider the positive cone to be a neighborhood of the ray generated by the right invariant vector field $JA \in T_A G$ corresponding to the element J in the Lie algebra. Proposition 1.1 shows that G remains path-connected in this geometry. Next, one might try to understand what happens to the fundamental group. More precisely, let us define the **positive fundamental group** $\pi_{1,\text{pos}} G$ to be the semigroup generated by positive loops based at $\mathbb{1}$, where two such loops are considered equivalent if they may be joined by a smooth family of positive loops. It would be interesting to calculate this semigroup. In particular it is not at present clear whether or not the obvious map

$$\pi_{1,\text{pos}} G \;\to\; \pi_1(G)$$

is injective in general. (When $n = 1$ this is an easy consequence of Proposition 2.4 and Lemma 3.1.)

The proof of Theorem 1.2 is based on studying the relation between the positive cone and the fibers of the projection $\pi : G \to \mathcal{C}onj$, where $\mathcal{C}onj$ is the space of conjugacy classes of elements in G with the quotient (non-Hausdorff) topology. The space $\mathcal{C}onj$ has a natural stratification coming from the stratification of G. Recall that a symplectic matrix A is similar to its transpose inverse $(A^T)^{-1}$ (in fact, $(A^T)^{-1} = -JAJ$). Thus its eigenvalues occur either in pairs $\lambda, \bar{\lambda} \in S^1$ and $\lambda, 1/\lambda \in \mathbf{R} - 0$, or in complex quadruplets $\lambda, \bar{\lambda}, 1/\lambda, 1/\bar{\lambda}$. In particular, the eigenvalues ± 1 always occur with even multiplicity. Roughly speaking, the open strata of G consist of generic elements (ie diagonalizable elements with all eigenvalues of multiplicity 1), and the codimension 1 strata consist of non-diagonalizable elements which have one pair of eigenvalues of multiplicity 2 lying either on $\mathbf{R} - \{\pm 1, 0\}$ or on $S^1 - \{\pm 1\}$, or which have one pair of eigenvalues equal to ± 1: see [1] and §3 below. Note that a conjugacy class which lies in an open stratum is a submanifold of $\mathrm{GL}(2n, \mathbf{R})$ of codimension n since each pair of distinct eigenvalues (in $\mathbf{R} \cup S^1$) has one degree of freedom, and each quadruplet has two.

We write G_0 for the set of all generic elements in G, G_1 for the union of all codimension 1 strata, and similarly $\mathcal{C}onj_0, \mathcal{C}onj_1$ for their projections in $\mathcal{C}onj$. We shall consider in detail only generic paths in $\mathcal{C}onj$. By definition, these intersect all lower strata transversally. Hence the elements of these paths lie in $\mathcal{C}onj_0$ for all except the finite number of times at which they cross $\mathcal{C}onj_1$. Note that the codimension of a stratum in $\mathcal{C}onj$ always refer to the codimension of its lift in G.

The main ingredient of the proof of the theorem is to characterize the (generic) paths in $\mathcal{C}onj$ which lift to positive paths in G. We shall see in

Proposition 3.6 that the only significant restriction on the path comes from the way that eigenvalues enter, move around and leave S^1. We will not state the general result here since it needs a certain amount of notation. However some of the essential features are present in the case $n = 1$ which is much easier to describe. In this case $Conj$ is the union of the circle with the intervals $(-\infty, -1] \cup [1, \infty)$ with the usual topology, except that the points ± 1 are each tripled. To see this, first consider matrices with an eigenvalue pair $\lambda, \bar{\lambda} \in S^1$. It is possible to distinguish between these eigenvalues and hence to label the pair with an element in S^1. In higher dimensions this is accomplished by the *splitting number* described in §2. Here it suffices to note that the rotations through angles θ and $2\pi - \theta$ are not conjugate in $SL(2, \mathbf{R})$, the only conjugating matrices being reflections. We also claim that the point 1 occurs with three flavors: plain $\mathbb{1}$ (the conjugacy class of the identity) which has codimension 3, and two nilpotent classes $\mathbb{1}^{\pm} \in Conj_1$. A similar statement holds at -1. As explained in Lemma 3.1, a positive path has to project to a path in $Conj$ which goes round the circle anticlockwise, but it can move along the real axes in either direction. Moreover, the projection of a generic positive path has to leave the circle at $(-\mathbb{1})^-$ or $\mathbb{1}^-$ and then, after wandering around $(-\infty, -1) \cup (1, \infty)$ for a while, enter the circle again at $(-\mathbb{1})^+$ or $\mathbb{1}^+$ and continue on its way.

The next step is to carry out a similar analysis in the case $n = 2$. We shall see that the restrictions in the behavior of positive paths in this case all stem from Krein's lemma which says that simple eigenvalues on S^1 with positive splitting number must flow anti-clockwise round the circle. This gives rise to restrictions in the way in which four eigenvalues on the unit circle can move to a quadruplet outside the unit circle (the Krein–Bott bifurcation). As a consequence, we show in Proposition 5.4 that there are restrictions on the regions in $Conj$ which can be visited by a generic positive path of bounded Conley-Zehnder index.

In the last section of the paper, we will present a brief survey of the use of positive paths in stability theory, in Riemannian geometry and in Hofer geometry, and will give some new applications.

We wish to thank Nancy Hingston for describing to us her ideas about perambulations in the symplectic group, which were a direct inspiration for our work, and Hector Sussmann for explaining to us some basic results in control theory.

2 Basic facts

2.1 General results on positive paths

First, two elementary lemmas. A piecewise smooth path is said to be piecewise positive if the tangent vectors along each smooth segment (including those at the endpoints) are in the positive cone.

Lemma 2.1 (i) *The set of positive paths is open in the C^1-topology.*

(ii) *Any piecewise positive path may be C^0-approximated by a positive path.*

(iii) *Let $\{A_t\}, \{B_t\}$ be two positive paths with the same initial point $A_0 = B_0$. Then, given $\varepsilon > 0$, there is $\delta > 0$ and a C^0-small deformation of $\{A_t\}$ to a positive path $\{A'_t\}$ which coincides with $\{B_t\}$ for $t \in [0, \delta]$ and with $\{A_t\}$ for $t > \varepsilon$.*

Proof: The first statement follows from the openness of the positive cone, the second from its convexity and the third from a combination of these two facts.   □

Note that statement (ii) above is a special case of a much more general result in control theory: see, for example, Sussmann [19] and Grasse–Sussmann [11].

Lemma 2.2 (i) *Any positive path can be extended to a positive path which ends at a generic element.*

(ii) *Any positive path $\{A_t\}$ starting at $A_0 = \mathbb{1}$ can be perturbed fixing A_0 so that $\{A_t\}$ is generic. Moreover, if A_1 is generic we may fix that also during the perturbation.*

Proof: This follows immediately from openness.   □

We continue with some simple remarks about the relation between positive paths in G and their projections (also called positive paths) in *Conj*.

Lemma 2.3 (i) *The positive cone is invariant under conjugation.*

(ii) *The conjugate $\{X^{-1}A_t X\}$ of any positive path $\{A_t\}$ is positive.*

(iii) *There is a positive path from $\mathbb{1}$ to A in G if and only if there is a positive path in Conj from $\mathbb{1}$ to $\pi(A)$.*

Proof: (i) The positive vector JPA at A is taken by conjugation to the positive vector $X^{-1}JPAX = X^{-1}JPXX^{-1}AX = J(X^TPX)B$ at $B = X^{-1}AX$.
(ii) follows immediately from (i).
(iii) Since the "only if" statement is obvious, we consider the "if" statement. By definition, a positive path from $\pi(\mathbb{1})$ to $\pi(A)$ lifts to a positive path from $\mathbb{1}$ to $Y^{-1}AY$ for some $Y \in G$. Now conjugate this path by Y^{-1} and use (ii).   □

Warning Given a positive path $\{A_t\}$ it is *not* true that all paths of the form $\{X_t^{-1}A_tX_t\}$ are positive. One can easily construct counterexamples in $\mathrm{SL}(2,\mathbf{R})$ using the methods of §3.

Proposition 2.4 *Let $\{A_t\}$ in G be a generic positive path joining two generic points A_0, A_1. Then the set of positive paths in G which lift $\gamma = \{\pi(A_t)\}_{t\in[0,1]} \subset Conj$ is path-connected. Moreover, the set of these paths with one fixed endpoint is also path-connected. Finally, if $A_0 = \mathbb{1}$, the set of these paths with both endpoints fixed is path-connected.*

Proof: Let $\{B_t\}$ be another path which lifts $\gamma = \{\pi(A_t)\}$. By genericity we may suppose that $\{A_t\}$ and $\{B_t\}$ are disjoint embedded paths. Moreover, when $n \geq 2$, the projection in $Conj$ of the positive cone at any point $X \in G_0 \cup G_1$ of a generic path is always at least 2-dimensional (see Proposition 3.6). Hence we may assume that the paths $\{A_t\}$ and $\{B_t\}$ are never tangent to the fiber of the projection $G \to Conj$. Our first aim is to define positive piecewise smooth vector fields ξ_A, ξ_B that are tangent to $\pi^{-1}(\gamma)$ and which extend the tangent vector fields to the paths $\{A_t\}, \{B_t\}$. To make this easier we first normalise the path B_t near the finite number of times $t_1 < \ldots < t_k$ at which A_t crosses a codimension 1 stratum. If T is one of these times t_i, there is a matrix X_T such that $B_T = X_T^{-1}A_TX_T$, and by Lemma 2.1 (iii) we may suppose that $B_t = X_T^{-1}A_tX_T$ for t near T.

On the other hand, since the set of tangent vectors $Y \in TG$ which projects to the tangent vector $\pi_*(\dot{A}_T)$ is an affine subbundle of the restriction $TG\mid_{\pi^{-1}(\pi(A_T))}$, we can find a section of this subbundle whose value at A_T coincides with $\dot{A}_T$. This defines ξ_A over neighbourhoods of all times t_i, which we choose small enough so that ξ_A is positive. We then define ξ_B over each such neighbourhood by the adjoint map $X_T^{-1}\xi_AX_T$.

At all other times A_t is generic and the map $t \mapsto \pi(A_t)$ is an immersion. Therefore, the set

$$Y_i = \{\pi^{-1}(\pi(A_t)) : t_i \leq t \leq t_{i+1}\}$$

is a submanifold. We now extend ξ_A, ξ_B to the whole of Y_i by choosing sections of the intersection of the positive cone field with TY_i which project to the tangent vector field of γ. This is possible because this intersection is an open convex cone. Observe that π projects all integral curves of ξ_A and ξ_B to γ. Indeed all integral curves of the vector field $s\xi_A + (1-s)\xi_B$ (for $s \in [01]$) are positive paths which project to γ. Clearly there is a path of such curves joining $\{A_t\}$ to $\{B_t\}$.

It remains to check the statement about the endpoints. We will assume, again using Lemma 2.1, that if $\{A_t\}$ and $\{B_t\}$ have the same endpoint they agree for t near that endpoint. Then, if $B_0 = A_0$ the above construction gives a family of paths starting at A_0. Therefore the result for fixed initial

endpoint is obvious. To get fixed final endpoint one can use a similar argument applied to the reversed (negative) paths $\{A_{1-t}\}, \{B_{1-t}\}$. Finally consider the case of two fixed endpoints, with $A_0 = B_0 = \mathbb{1}$. The endpoints A_1^λ of the family of paths constructed above are all conjugate to A_1 and equal A_1 at $\lambda = 0, 1$. Because A_1 is generic, there is a smooth family of elements $X_\lambda \in G$ with $X_0, X_1 = \mathbb{1}$ such that $X_\lambda^{-1} A_1^\lambda X_\lambda = A_1$. The desired family of paths is then $\{X_\lambda^{-1} A_t^\lambda X_\lambda\}_{t \in [0,1]}$: these are positive by Lemma 2.3.
□

2.2 Splitting numbers

This section summarizes what is known as Krein theory, that is the theory of positive paths in $\mathcal{U}$. The theory is nicely described by Ekeland in [7, 8] where it is used to develop an index theory for closed orbits. Interestingly enough, this analysis is very closely related to work by Ustilovsky [20] on conjugate points on geodesics in Hofer geometry on the group of Hamiltonian symplectomorphisms.

It will be convenient to work over $\mathbf{C}$ rather than $\mathbf{R}$. We extend J and ω to $\mathbf{C}^{2n} = \mathbf{R}^{2n} \otimes \mathbf{C}$ by complex linearity, so that $\omega(v, w) = w^T J v$ is complex bilinear. Let $\langle v, w \rangle$ denote the standard Hermitian inner product on $\mathbf{C}^n$, namely

$$\langle v, w \rangle = \bar{w}^T v = \omega(Jv, \bar{w}).$$

We will also use the form β given in our notations by

$$\beta(v, w) = -i\omega(v, \bar{w}) = -i\langle Jv, w \rangle = -i\bar{w}^T Jv.$$

Lemma 2.5 (i) *β is a nondegenerate Hermitian symmetric form.*

(ii) $i\beta(v, Jv) = \langle v, v \rangle$.

(iii) *An element $A \in \mathrm{GL}(2n, \mathbf{C})$ belongs to G if and only if it preserves β and is real, i.e.*
$$\overline{Av} = A\bar{v}, \quad v \in \mathbf{C}^{2n}.$$

(iv) *The invariant subspaces E_λ, E_μ corresponding to eigenvalues λ, μ are β-orthogonal unless $\lambda\bar{\mu} = 1$.*

(v) *If v is an eigenvector with eigenvalue $\lambda \in S^1$ of multiplicity 1 then $\beta(v, v) \in \mathbf{R} - \{0\}$.*

Proof: β is Hermitian because

$$\overline{\beta(w, v)} = \overline{-i\omega(w, \bar{v})} = i\omega(\bar{w}, v) = -i\omega(v, \bar{w}) = \beta(v, w).$$

Statements (ii) and (iii) are obvious. To prove (iv), consider $A \in G = \mathrm{Sp}(2n, \mathbf{R})$ with eigenvalues λ, μ and corresponding eigenvectors v, w and observe that

$$\beta(v, w) = \beta(Av, Aw) = \beta(\lambda v, \mu w) = \lambda \bar{\mu} \beta(v, w).$$

The proof when v, w belong to the invariant subspace but are not eigenvectors is similar (see Ekeland [8]). To prove (v), suppose $\lambda \in S^1$ is an eigenvalue of multiplicity 1 with eigenvector v. Then $\beta(v, v) \neq 0$, since β is non-degenerate and $\beta(v, w) = 0$ whenever w belongs to any other eigenspace of A. Further, $\beta(v, v) \in \mathbf{R}$ by (i).   □

It follows from (v) that simple eigenvalues λ on S^1 may be labelled with a number $\sigma(\lambda) = \pm 1$ called the **splitting number** chosen so that

$$\beta(v, v) \in \sigma(\lambda) \mathbf{R}^+.$$

For example, when $n = 1$, the rotation matrix

$$\rho(t) = \begin{pmatrix} \cos t & -\sin t \\ \sin t & \cos t \end{pmatrix}$$

has eigenvectors

$$v^+ = \begin{pmatrix} 1 \\ -i \end{pmatrix}, \quad v^- = \begin{pmatrix} 1 \\ i \end{pmatrix}$$

corresponding to the eigenvectors e^{it}, e^{-it}, and it is easy to see that e^{it} has splitting number $+1$, while e^{-it} has splitting number -1.

More generally, if $\lambda \in S^1$ has multiplicity > 1 we define $\sigma(\lambda)$ to be the signature of the form β on the corresponding invariant subspace. Note that $\sigma(\lambda) = -\sigma(\bar{\lambda})$. Hence the splitting number of ± 1 is always 0. It is not hard to check that the conjugacy class of an element in $\mathcal{U}$ is completely described by its spectrum together with the corresponding set of splitting numbers. The following result is central to our argument.

Lemma 2.6 (Krein) *Under a positive flow simple eigenvalues on S^1 labelled with $+1$ must move anti-clockwise and those labelled with -1 must move clockwise.*

Proof: We repeat the proof from Ekeland's book [8] for the convenience of the reader. For all t in some neighborhood of t_0, let $e^{i\theta_t}$ be a simple eigenvalue of A_t with corresponding eigenvector $x_t \in \mathbf{C}$. For simplicity when $t = t_0$ we use the subscript 0 instead of t_0 (writing A_0, x_0 etc), and will denote the derivative of A_t at t_0 by JPA_0, where P is positive definite. It is easy to check that for any x

$$\langle A_t x, J x_t \rangle = \langle x, A_t^T J x_t \rangle = \langle x, J A_t^{-1} x_t \rangle = \langle e^{i\theta_t} x, J x_t \rangle.$$

Applying this with $x = x_0$ and differentiating at $t = t_0$, we find that

$$\langle JPA_0 x_0, J x_0 \rangle = \langle i\frac{d\theta_t}{dt}e^{i\theta_0}x_0, J x_0 \rangle \Big|_{t=t_0},$$

from which it readily follows that

$$\frac{d\theta_t}{dt}\Big|_{t=t_0} = \frac{\langle Px_0, x_0 \rangle}{\beta(x_0, x_0)}.$$

Since the right hand side has the same sign as $\beta(x_0, x_0)$, the result follows. $\square$

Observe, however, that not all flows whose eigenvalues move in this way are positive. Further, this result shows that there may not be a short positive path between an arbitrary pair of elements in a given conjugacy class, even if one allows the path to leave the conjugacy class.

The next lemma shows that if a (simple) eigenvalue leaves the circle it must do so at a point with splitting number 0. This observation highlights the importance of the splitting number. It is not relevant to the case $n = 1$ of course, but is a cornerstone of the argument in higher dimensions.

Lemma 2.7 *Let A_t be any path in* G *and $\lambda(t) \in \operatorname{Spec} A_t$ a continuous family of eigenvalues such that*

$$\lambda(t) \in S^1, \quad for \ t \leq T,$$
$$\lambda(t) \notin S^1, \quad for \ t > T.$$

Suppose also that $\lambda(t)$ has multiplicity 1 when $t > T$ and multiplicity 2 at $t = T$. Then

$$\sigma(\lambda_T) = 0.$$

Proof: For $t \geq T$ let $V_t \subset \mathbf{C}^{2n}$ be the space generated by the eigenspaces $E_{\lambda(t)}, E_{1/(\bar\lambda(t))}$. By hypothesis, this is 2-dimensional for each such t, and it clearly varies continuously with t. As above, β is non-degenerate on each V_t. Therefore the splitting number can be $2, 0$ or -2 and will be 0 if and only if there are non-zero v such that $\beta(v, v) = 0$. But when $t > T$, $\beta(v, v) = 0$ on the eigenvectors in V_t. Therefore, β has signature 0 when $t > T$ and so must also have signature 0 on V_T. $\square$

3 Conjugacy classes and the positive cone

3.1 The case $n = 1$

We will write $\mathcal{O}_{\mathcal{R}}^{\pm}$ for the set of all elements in $\operatorname{SL}(2, \mathbf{R})$ with eigenvalues in $\mathbf{R}^{\pm} - \{\pm 1\}$, and will also divide $\mathcal{U} - \{\pm 1, \pm 1^{\pm}\}$ into two sets $\mathcal{O}_{\mathcal{U}}^{\pm}$, distinguished by the splitting number of the eigenvalue λ in the upper half-plane

$$\mathbf{H} = \{z \in \mathbf{C} : \Im z > 0\}.$$

Thus, for elements of $\mathcal{O}_{\mathcal{U}}^+$, the eigenvalue in $\mathbf{H}$ has splitting number $+1$, and, by considering the rotations

$$\rho(t) = \begin{pmatrix} \cos t & -\sin t \\ \sin t & \cos t \end{pmatrix},$$

it is easy to check that

$$\mathcal{O}_{\mathcal{U}}^+ = \{A : \mathrm{tr}A < 2,\ A_{21} > 0\}, \quad \mathcal{O}_{\mathcal{U}}^- = \{A : \mathrm{tr}A < 2,\ A_{21} < 0\}.$$

These sets project to the open strata in $\mathcal{C}onj$.

We now work out in detail the structure of conjugacy classes near the element $-\mathbb{1}$. The structure at $\mathbb{1}$ is similar, but will not be so important to us since we are interested in paths which avoid the eigenvalue 1.

It is not hard to check that there are 3 conjugacy classes with spectrum $\{-1\}$, the diagonalizable class consisting only of $\{-\mathbb{1}\}$, and the classes $\mathcal{N}^\pm$ containing the nilpotent elements

$$\begin{pmatrix} -1 & 0 \\ \mp 1 & -1 \end{pmatrix}.$$

Again, the classes $\mathcal{N}^\pm$ may be distinguished by the sign of the 21-entry. Moreover, if we write the matrices near $-\mathbb{1}$ as

$$A = \begin{pmatrix} -1+x & y \\ z & \frac{1+yz}{-1+x} \end{pmatrix},$$

the elements in $\mathcal{N}^- \cup \{-\mathbb{1}\} \cup \mathcal{N}^+$ are those with trace $= -2$ and so form the boundary of the cone $x^2 + yz = 0$. The interior of the cone is the set where $x^2 < -yz$, and has two components

$$\mathcal{U}^+ = \{y < 0,\ z > 0\}, \quad \mathcal{U}^- = \{y > 0,\ z < 0\}.$$

The rest of the space consists of points in $\mathcal{O}_{\mathcal{R}}^-$ with $x^2 > -yz$. Observe that the labelling is chosen so that $\mathcal{N}^-$ is in the closure of $\mathcal{O}_{\mathcal{U}}^+$. (This class $\mathcal{N}^-$ was labelled by $-$ rather than $+$ because, as we shall see, it is the place where positive paths leave $\mathcal{U}$.)

The next lemma describes the structure of short positive paths in $\mathrm{SL}(2, \mathbf{R})$.

Lemma 3.1 *Let $A_{t\in[0,1]}$ be a short positive path in $\mathrm{SL}(2,\mathbf{R})$, and for each $t > 0$, let*

$$\lambda_t = r(t)e^{i\theta_t}, \quad \theta_t \in (0, 2\pi), \quad |r(t)| \geq 1,$$

be an eigenvalue of A_t, chosen so that it has splitting number $+1$, when $\theta \neq \pi$. Then:

(i) *there is $\varepsilon > 0$ such that $A_t \in \mathcal{O}_\mathcal{U}^+$ for $t \in [0, \varepsilon]$;*

(ii) *the function θ_t is increasing, and is strictly increasing except perhaps at the point $\theta_t = \pi$;*

(iii) *if $A_T = -\mathbb{1}$ for any T then A_t remains in $\mathcal{U}$ for $t > T$;*

(iv) *if A_t enters $\mathcal{O}_\mathcal{R}^-$ it does so through a point of $\mathcal{N}^-$ and if it reenters $\mathcal{U}$ it must go through $\mathcal{N}^+$ to $\mathcal{O}_\mathcal{U}^-$.*

(v) $A_t \notin \mathcal{O}_\mathcal{R}^+$ *for $t > 0$.*

Proof: We first claim that at every point N of $\mathcal{N}^-$ all vectors in the positive tangent cone at N point into $\mathcal{O}_\mathcal{R}^-$. For if not, by openness (Lemma 2.1(i)), there would be a positive vector pointing into $\mathcal{O}_\mathcal{U}^+$ and hence a positive path along which a positive eigenvalue would move clockwise, in contradiction to Krein's lemma. Since conjugation by an orthogonal reflection interchanges $\mathcal{N}^+$ and $\mathcal{N}^-$ and positive and negative, it follows that the positive tangent cone points in $\mathcal{O}_\mathcal{U}^-$ at all points of $\mathcal{N}^+$. Since every neighbourhood of $-\mathbb{1}$ contains points in $\mathcal{N}^\pm$, it follows from openness that positive paths starting at $-\mathbb{1}$ must remain in $\mathcal{U}$. Since a similar result holds for negative paths, there is no short positive path which starts with eigenvalues on $\mathbf{R}^-$ and ends at $-\mathbb{1}$. This proves (iii) and (iv). A similar argument proves (i) and (v). Statement (ii) is a direct consequence of Lemma 2.6). $\qquad\square$

Remark 3.2 (i) The above proof shows that the structure of these positive paths in $\mathrm{SL}(2, \mathbf{R})$ is determined by Krein's lemma and the topology of the conjugacy classes. However the result may also be proved by direct computation. For example, if

$$N = \begin{pmatrix} -1 & 0 \\ -1 & -1 \end{pmatrix} \in \mathcal{N}^+$$

and P is symmetric and positive definite, then

$$\frac{d}{dt}\Big|_{t=0} \operatorname{tr}\left(e^{tJP}N\right) = \operatorname{tr}\left(JPN\right) > 0.$$

Hence the trace lies in $(-2, 2)$ for $t > 0$, which means that every positive path starting at N moves into $\mathcal{U}^-$. By invariance under conjugacy, this has be to true for every element in $\mathcal{N}^+$.

(ii) It is implicit in the above proof that under a positive flow in $\mathrm{SL}(2, \mathbf{R})$ eigenvalues on $\mathbf{R}$ can flow in either direction. To see this explicitly, let A be the diagonal matrix $\mathrm{diag}(\lambda, 1/\lambda)$ (where $|\lambda| > 1$) and consider the positive matrices

$$P_1 = \begin{pmatrix} 2 & -1 \\ -1 & 1 \end{pmatrix}, \quad P_2 = \begin{pmatrix} 2 & 1 \\ 1 & 1 \end{pmatrix}.$$

Then the derivative of the trace (and hence the flow direction of λ) has different signs along the flows $t \mapsto e^{JP_i t} A, i = 1, 2$.

3.2 The case $n = 2$

This case contains all the complications of the general case. We will see that the behavior of eigenvalues along positive paths as they enter or leave the circle at a value in $S^1 - \{\pm 1\}$ is much the same as the behavior that we observed in the case $n = 1$ at the triple points ± 1 in *Conj*. When $n = 2$ it is also possible for two pairs of real eigenvalues to come together and then leave the real axis. However, we shall see that the positivity of the path imposes no essential restiction here. One indication of this is that there is no relevant notion of splitting number when λ is not on the unit circle.

We first consider the structure of elements $A \in \mathrm{Sp}(4, \mathbf{R})$. As remarked above, the eigenvalues of A consist either of pairs $\lambda, \bar{\lambda} \in S^1$ and $\lambda, 1/\lambda \in \mathbf{R} - \{0\}$, or of quadruplets $\lambda, \bar{\lambda}, 1/\lambda, 1/\bar{\lambda}$ where $\lambda \notin S^1 \cup \mathbf{R}$. We will label these pairs and quadruplets by the element λ with $|\lambda| \geq 1$ and $\Im \lambda \geq 0$.

A generic element of G lies in one of the following open regions:

(i) $\mathcal{O}_\mathcal{C}$, consisting of all matrices whose spectrum is a quadruplet in $\mathbf{C} - (\mathbf{R} \cup S^1)$;

(ii) $\mathcal{O}_\mathcal{U}$, consisting of all matrices whose eigenvalues lie in $S^1 - \{\pm 1\}$ and either all have multiplicity 1 or have multiplicity 2 and non-zero splitting numbers[3];

(iii) $\mathcal{O}_\mathcal{R}$, consisting of all matrices whose eigenvalues have multiplicity 1 and lie in $\mathbf{R} - \{0, \pm 1\}$ (divided into two components: $\mathcal{O}_\mathcal{R}^\pm$);

(iv) $\mathcal{O}_{\mathcal{U},\mathcal{R}}$, consisting of all matrices with 4 distinct eigenvalues, one pair on $S^1 - \{\pm 1\}$ and the other on $\mathbf{R} - \{0, \pm 1\}$.

Note that the region $\mathcal{O}_\mathcal{C}$ is connected, but the others are not.

Lemma 3.3 *The codimension 1 part of the boundaries of the above regions are:*

(v) $\mathcal{B}_\mathcal{U}$, *consisting of all non-diagonalizable matrices whose spectrum consists of a pair of conjugate points in $S^1 - \{\pm 1\}$ each of multiplicity 2 and splitting number 0.*

---

[3]By Lemma 2.7 the latter elements must be diagonalizable.

(vi) $\mathcal{B}_\mathcal{R}$, *consisting of all non-diagonalizable matrices whose spectrum is a pair of distinct points* $\lambda, 1/\lambda \in \mathbf{R} - \{\pm 1, 0\}$ *each of multiplicity* 2.

(vii) $\mathcal{B}_{\mathcal{U},\mathcal{R}}$, *consisting of all non-diagonalizable matrices with spectrum* $\{\pm 1, \pm 1, \lambda, \bar{\lambda}\}$, *with* $\lambda \in S^1 - \{\pm 1\}$.

(viii) $\mathcal{B}_{\mathcal{R},\mathcal{U}}$, *consisting of all non-diagonalizable matrices with spectrum* $\{\pm 1, \pm 1, \lambda, 1/\lambda\}$, *with* $\lambda \in \mathbf{R} - \{\pm 1, 0\}$.

Proof: Use Lemma 2.7. □

The next step is to describe the conjugacy classes of elements in the above parts of G. It follows easily from Lemma 2.7 that the conjugacy class of an element $A \in \mathcal{O}_\mathcal{U} \cup \mathcal{O}_\mathcal{R} \cup \mathcal{O}_{\mathcal{U},\mathcal{R}}$ is determined by its spectrum and the splitting numbers of those eigenvalues lying on S^1. The next lemma deals with the other cases.

Lemma 3.4 (i) *The conjugacy class of* $A \in \mathcal{O}_\mathcal{C}$ *is entirely determined by its eigenvalue* $\lambda \in \mathbf{H}$ *with* $|\lambda| > 1$.

(ii) *There are two types of conjugacy class in* $\mathcal{B}_\mathcal{U}$. *More precisely, for each eigenvalue* $\lambda \in S^1 \cap \mathbf{H}$ *of multiplicity* 2 *and splitting number* 0, *there are two conjugacy classes of non-diagonalizable matrices, namely the two nilpotent classes* $\mathcal{N}_\lambda^\pm$.

(iii) *For each* $\lambda \in (-\infty, -1) \cup (1, \infty)$, *there is one type of conjugacy class in* $\mathcal{B}_\mathcal{R}$.

(iv) *A matrix in* $\mathcal{B}_{\mathcal{U},\mathcal{R}}$ *or* $\mathcal{B}_{\mathcal{R},\mathcal{U}}$ *is conjugate to a matrix which respects the splitting* $\mathbf{R}^4 = \mathbf{R}^2 \oplus \mathbf{R}^2$. *Thus the conjugacy classes of* $\mathcal{B}_{\mathcal{U},\mathcal{R}}$ *are determined by* $\lambda \in (0, \pi)$, *its splitting number, and the conjugacy class* $\mathcal{N}^-$ *or* $\mathcal{N}^+$ *of the case* $n = 1$. *Those of* $\mathcal{B}_{\mathcal{R},\mathcal{U}}$ *are determined by* $\lambda \in (-\infty, -1) \cup (1, \infty)$ *and the conjugacy class* $\mathcal{N}^-$ *or* $\mathcal{N}^+$ *of the case* $n = 1$.

Proof: First consider an element $A \in \mathcal{O}_\mathcal{C}$. By Lemma 2.5(iv), the subspace V spanned by the eigenvectors v, w with eigenvalues $\lambda, 1/\bar{\lambda}$ is β-orthogonal to its complex conjugate $\bar{V}$, because this is the span of the eigenvectors $\bar{v}, \bar{w}$ with eigenvalues $\bar{\lambda}, 1/\lambda$. By the same Lemma, $\beta(v, v) = \beta(w, w) = 0$. Hence, because it is non-degenerate, the form β is non-zero on (v, w). Thus β has zero signature on the subspaces $V, \bar{V}$. Conversely, suppose given a 2-dimensional subspace V of $\mathbf{C}^4$ which is β-orthogonal to its complex conjugate $\bar{V}$. Let B be any complex linear β-preserving automorphism of V and extend B to $\bar{V}$ by complex conjugation. Then B is an element of $\mathrm{Sp}(4, \mathbf{R})$ because it is real and preserves β. In particular, if β has signature 0 on V, the subspace V is spanned by vectors v, w such that $\beta(v, v) =$

$\beta(w, w) = 0$. Then $\beta(v, w) \neq 0$ since β is non-degenerate, and if we define $B = B(V, \lambda)$ by setting:

$$Bv = \lambda v, \quad Bw = (1/\bar{\lambda})w,$$

B preserves β because

$$\beta(Bv, Bw) = \beta(\lambda v, 1/\bar{\lambda} w) = \beta(v, w).$$

The corresponding element of $\mathrm{Sp}(4, \mathbf{R})$ has spectrum $\lambda, \bar{\lambda}, 1/\lambda, 1/\bar{\lambda}$. Moreover, every element of $\mathrm{Sp}(4, \mathbf{R})$ with such spectrum has this form. In particular, every such element is determined by the choice of λ with $|\lambda| > 1, \Im \lambda > 0$ and of the (ordered) decomposition $V = \mathbf{C}v \oplus \mathbf{C}w$ of V. Hence there is exactly one conjugacy class for each such λ. This proves (i).

To prove (ii), first observe that under the given hypotheses the eigenspace $V = E_\lambda$ has the same properties as in (i). Namely, $\beta|_V$ is a non-degenerate form of signature 0, and V is β-orthogonal to $\bar{V}$. Moreover, the restriction of A to V can be any β-invariant complex linear map with eigenvalue λ of multiplicity 2 and splitting number 0. Suppose that $A|_V$ has an eigenvector v with $\beta(v, v) \neq 0$. Then there is a β-orthogonal vector w such that $\beta(w, w) = -\beta(v, v)$, and because

$$\beta(w, w) = \beta(Aw, Aw) = \beta(\lambda w + \mu v, \lambda w + \mu v) = \beta(w, w) + |\mu|^2 \beta(v, v)$$

we must have $\mu = 0$. Thus A is diagonalizable (over $\mathbf{C}$) in this case. If no such v exists, we may choose a basis v, w for V such that $\beta(v, v) = \beta(w, w) = 0$ and so that $Av = \lambda v, Aw = \lambda w + \mu v$, for some $\mu \neq 0$. If we also set $\beta(v, w) = i$, it is easy to see that $\beta(Aw, Aw) = \beta(w, w)$ only if $\lambda \bar{\mu}$ is real. Given v, our choices have determined w up to a transformation of the form $w \mapsto w' = w + \kappa v$, where $\kappa \in \mathbf{R}$ since $\beta(w', w') = 0$. It is easy to check that if we make this alteration in w then μ does not change. On the other hand, if we rescale v, w replacing them by $v/\kappa, \bar{\kappa} w$ then μ changes by the positive factor $|\kappa|^2$. Hence we may assume that $\mu = \pm \lambda$. This shows that $A|_V$ is conjugate to exactly one of the matrices

$$D_\lambda = \begin{pmatrix} \lambda & 0 \\ 0 & \lambda \end{pmatrix}, \quad N_\lambda^- = \begin{pmatrix} \lambda & 0 \\ -\lambda & \lambda \end{pmatrix}, \quad N_\lambda^+ = \begin{pmatrix} \lambda & 0 \\ \lambda & \lambda \end{pmatrix}.$$

To prove (iii), observe that the eigenspace W of λ (where $|\lambda| > 1$) is Lagrangian, and by suitably conjugating A by an element of G we may suppose that the other eigenspace is JW. It then follows that A has the form $\lambda C \oplus \lambda^{-1} J^T (C^{-1})^T J$ for some $C \in \mathrm{SL}(2, \mathbf{R})$ with trace $= 2$. In fact, any linear map of the form $B \oplus J^T (B^{-1})^T J$, where $B \in \mathrm{GL}(2, \mathbf{R})$, is in G. It follows easily that there are two conjugacy classes, one in which C is diagonalizable, and one in which C is not.

(iv) Finally, consider a matrix A in $\mathcal{B}_{\mathcal{U},\mathcal{R}}$ with a pair of eigenvalues $\lambda, \bar{\lambda}$, with $\lambda \in S^1 \cap \mathbf{H}$ where $\mathbf{H}$ is the upper half plane, and the eigenvalue say -1, with multiplicity 2. Assume say that the splitting number of λ is positive. Let V be the eigenspace generated by the eigenvectors v, w corresponding to the eigenvalues $\lambda, \bar{\lambda}$, and V' the invariant subspace associated to the double eigenvalue -1. By Lemma 2.5, v is β-orthogonal to V' and w, and $\beta(v, v) > 0$ by assumption. Similarly, w is β-orthogonal to V' and v, and $\beta(w, w) < 0$. Thus the restriction of β to V' is a also a non-degenerate symmetric Hermitian form with zero signature. Because A is non-diagonalizable, its restriction to V' is non-diagonalizable. Note that both eigenspaces V and V' are invariant under conjugation, and therefore that the restrictions of A to each subspace is real. This shows that we can identify $V \oplus V'$ with $\mathbf{C}(\mathbf{R}^2) \oplus \mathbf{C}(\mathbf{R}^2)$ by a linear map f which respects the factors, preserves β and is real, and such that fAf^{-1} is the direct sum of a rotation on the first factor and of an element in $\mathcal{N}^\pm$ on the other factor.

The proof is similar in the other cases. □

We next investigate the relationship between the positive cone field and the projection $\pi : G \to \mathcal{C}onj$. Krein's Lemma 2.6 shows that the movement of eigenvalues along positive paths in $\mathcal{O}_{\mathcal{U}}$ and $\mathcal{O}_{\mathcal{U},\mathcal{R}}$ is constrained. However, this is not so for the other open strata.

Lemma 3.5 *The projection* $\pi : G \to \mathcal{C}onj$ *maps the positive cone* $\mathcal{P}_A$ *at* $A \in \mathcal{O}_{\mathcal{C}}$ *onto the tangent space to* $\mathcal{C}onj$ *at* $\pi(A)$. *A similar statement is true for* $A \in \mathcal{O}_{\mathcal{R}}$.

Proof: Recall from Lemma 3.4 that an element A of $\mathcal{O}_{\mathcal{C}}$ defines a unique splitting $V \oplus \bar{V}$, where $\beta|_V$ has signature 0 and where $A|_V$ has eigenvalues $\lambda, 1/\bar{\lambda}$ for some $\lambda \in \{\Im z > 0\}$. Moreover, we may choose the eigenvectors v, w for $\lambda, 1/\bar{\lambda}$ respectively so that

$$\beta(v, v) = \beta(w, w) = 0, \quad \beta(v, w) = i.$$

Therefore, if we fix such v, w we get a unique representative of each conjugacy class by varying λ.

Most such splittings $V \oplus \bar{V}$ are not J-invariant, and so it is hard to describe the positive flows at A. However, because the positive cone field on G is invariant under conjugation (see Lemma 2.3), we only need to prove this statement for one representative of each conjugacy class, and so we choose one where this splitting is J-invariant. For example, if we set $v = 1/\sqrt{2}(1, 0, i, 0)$ and $w = -Jv$, all the required identities are satisfied. Moreover, the transformation X which takes the standard basis to $v, w, \bar{v}, \bar{w}$ is unitary. Therefore, any linear automorphism B of V which is represented by a positive definite matrix with respect to the basis v, w extends by complex conjugation to a linear automorphism $B \oplus \bar{B}$ of $\mathbf{C}^4$ which is also positive definite with respect to the usual basis. This means that there are positive tangent vectors to G at A which preserve V and restrict to JB there. We are now essentially reduced to the 2-dimensional case. Just as in Lemma 3.1, if B is a real positive definite matrix with negative 21-entry then the eigenvalue λ moves along a ray towards S^1 and if its 21-entry is positive λ moves along this ray away from S^1. Since π maps the positive cone onto an open convex cone, this means that the image has to be the whole tangent space to $\mathcal{C}onj$.

The argument for $A \in \mathcal{O}_{\mathcal{R}}$ is similar but easier. $\qquad\square$

The next step is to study generic positive paths through the various parts of the codimension 1 boundary components. We will see that Krein's lemma forces the behavior of positive paths at the boundary $\mathcal{B}_{\mathcal{U}}$ to mimic that that near $-\mathbb{1}$ in the case $n = 1$, while the boundary $\mathcal{B}_{\mathcal{R}}$ is essentially not seen by positive paths. Our results may be summarized in the following proposition.

Proposition 3.6 (i) *If $A \in \mathcal{N}_\lambda^- \subset \mathcal{B}_{\mathcal{U}}$, then π maps the positive cone $\mathcal{P}_A$ into the set of vectors at $\pi(A)$ which point into $\pi(\mathcal{O}_C)$. Similarly, if $A \in \mathcal{N}_\lambda^+$, then $\pi_*(\mathcal{P}_A)$ consists of vectors pointing into $\pi(\mathcal{U})$.*

(ii) *If $A \in \mathcal{B}_{\mathcal{R}}$ then $\pi_*(\mathcal{P}_A)$ contains vectors which point into $\pi(\mathcal{O}_C)$ as well as vectors pointing into $\pi(\mathcal{O}_{\mathcal{R}})$.*

(iii) *If $A \in \mathcal{B}_{\mathcal{U},\mathcal{R}} \cup \mathcal{B}_{\mathcal{R},\mathcal{U}}$ then A is conjugate to an element which preserves the splitting $\mathbf{R}^4 = \mathbf{R}^2 \oplus \mathbf{R}^2$, and π_* takes the subcone in $\mathcal{P}_A$ formed by vectors which preserve this splitting onto the whole image $\pi_*(\mathcal{P}_A)$. In other words, positive paths through $\mathcal{B}_{\mathcal{U},\mathcal{R}}$ and $\mathcal{B}_{\mathcal{R},\mathcal{U}}$ behave just like paths in the product $\mathrm{SL}(2, \mathbf{R}) \times \mathrm{SL}(2, \mathbf{R})$.*

Proof: (i) Suppose first that $A \in \mathcal{N}_\lambda^-$. We claim that there is a neighbourhood of $A \in$ G whose intersection with $\mathcal{U}$ consists of elements which have an eigenvalue λ' with splitting number $+1$ which is close to λ and to the right of it, i.e with $\arg \lambda' < \arg \lambda$. To see this, observe first that by conjugacy invariance it suffices to prove this for a neighbourhood of A in the subgroup

$$G_V = \{A' \in G : A'(V) = V\}.$$

If $A' \in G_V$, then $A'|_V$ may be written as $\lambda'Y$ where λ' is close to λ, and Y has determinant 1 and preserves β. If

$$Y = \begin{pmatrix} a & b \\ c & d \end{pmatrix}$$

with respect to the basis v, w, then we must have $a\bar{c} \in \mathbf{R}$, $b\bar{d} \in \mathbf{R}$ and $a\bar{d} - c\bar{b} = 1$ in order to preserve β and $\mathrm{tr} Y = 2$. This implies that Y

must be in $SL(2, \mathbf{R})$. It is also close to the element $\frac{1}{\lambda} A|_V \in \mathcal{N}^- = \mathcal{N}^-_{-1}$. Therefore, the result follows from the corresponding result in the case $n = 1$: see the discussion just before Lemma 3.1.

Since eigenvalues with positive splitting number flow anticlockwise under a positive path (by Krein), every positive path through $A \in \mathcal{N}^-_\lambda$ must have all eigenvalues off S^1 for $t > T$ and so must enter $\mathcal{O}_C$. Similarly, positive paths through any point of $\mathcal{N}^+_\lambda$ flow into $\mathcal{U}$. As in the case when $\lambda = -1$, this implies that positive paths through points of $\mathcal{D}_\lambda$ remain in $\mathcal{U}$.

(ii) Suppose first that A in the closure $\bar{\mathcal{B}}_\mathcal{R}$ of $\mathcal{B}_\mathcal{R}$ is diagonalizable, say $A = \mathrm{diag}(\lambda, 1/\lambda, \lambda, 1/\lambda)$. Then, as in Remark 3.2, there is a positive flow starting at A which keeps A diagonalizable and in $\mathcal{B}_\mathcal{D}$, while decreasing $|\lambda|$. Since any sufficiently C^1-small perturbation of this flow is still positive, we can clearly flow positively from A into both regions $\mathcal{O}_\mathcal{R}$ and $\mathcal{O}_C$.

Next observe that if we fix $\lambda \in \mathbf{R}$ and consider only those elements of $\mathcal{B}_\mathcal{R}$ with fixed λ-eigenspace W and $1/\lambda$-eigenspace W', then these form a cone $C(\lambda)$ whose vertex is the diagonalizable element D_λ. (The structure here is just like that near $-\mathbb{1}$ which was discussed in §3.1.) Therefore, if v is a positive vector at D_λ which points into $\mathcal{O}_C$, there is a nearby vector v' at a nearby (non-diagonalizable) point A which also points into $\mathcal{O}_C$. Moreover, we may assume that v' is positive by the openness of the positive cone. Therefore one can move from A into $\mathcal{O}_C$, and similarly into $\mathcal{O}_\mathcal{R}$. Since there is only one conjugacy class of such A for each λ, this completes the proof.

(iii) As shown in Lemma 3.1, the behavior of positive paths at the two conjugacy classes $-\mathbb{1}^\pm$ is dictated by Krein's lemma and the topology of $SL(2, \mathbf{R})$. A similar argument applies here. □

Remark 3.7 (i) One can construct explicit paths from $A \in \mathcal{B}_\mathcal{R}$ into $\mathcal{O}_C$ and $\mathcal{O}_\mathcal{R}$ as follows. Note that the matrices $A \in \mathcal{B}_\mathcal{R}$ all satisfy the relation

$$\sigma_2 = \frac{\sigma_1^2}{4} + 2,$$

where σ_j denotes the jth symmetric function of the eigenvalues. Similarly, one can easily check that the matrices in $\mathcal{O}_C$ satisfy the condition $\sigma_2 > \frac{\sigma_1^2}{4} + 2$ while those in $\mathcal{O}_\mathcal{R}$ satisfy $\sigma_2 < \frac{\sigma_1^2}{4} + 2$. Therefore, to see where the positive path $A_t = (1 + tJP + \ldots)A$ goes as t increases from 0 we just have to compare the derivatives

$$\sigma_2' = \frac{d}{dt}\big|_{t=0}\, \sigma_2, \quad \frac{\sigma_1 \sigma_1'}{2} = \frac{\sigma_1}{2} \frac{d}{dt}\big|_{t=0}\, \sigma_1.$$

For example, if we take

$$
A = \begin{pmatrix} \lambda & 0 & 0 & 0 \\ 0 & \frac{1}{\lambda} & 0 & \alpha \\ -\frac{\alpha}{\lambda^2} & 0 & \lambda & 0 \\ 0 & 0 & 0 & \frac{1}{\lambda} \end{pmatrix}, \quad P = \begin{pmatrix} 1 & 0 & 0 & 0 \\ 0 & 5 & -2 & 0 \\ 0 & -2 & 1 & 0 \\ 0 & 0 & 0 & 1 \end{pmatrix},
$$

then

$$
(1 + tJPA) - \mu \mathbb{1} = \begin{pmatrix} \lambda - \mu - \frac{2\alpha t}{\lambda^2} & -\frac{5t}{\lambda} & 2t\lambda & -5\alpha t \\ t\lambda & \frac{1}{\lambda} - \mu & 0 & \alpha \\ -\frac{\alpha}{\lambda^2} & 0 & \lambda - \mu & -\frac{t}{\lambda} \\ -\frac{\alpha t}{\lambda^2} & -\frac{2t}{\lambda} & t\lambda & \frac{1}{\lambda} - \mu - 2\alpha t \end{pmatrix}.
$$

Thus

$$
\frac{\sigma_1 \sigma_1'}{2} = -2\alpha \left(\frac{2}{\lambda} + \lambda + \frac{1}{\lambda^3} \right)
$$

while

$$
\sigma_2' = -2\alpha \left(2\lambda + \frac{2}{\lambda^3} \right).
$$

Therefore

$$
\sigma_2' - \frac{\sigma_1 \sigma_1'}{2} = -2\alpha \left(\lambda + \frac{1}{\lambda^3} - \frac{2}{\lambda} \right) = -2\alpha (\lambda^{1/2} - \lambda^{-3/2})^2,
$$

and the path goes into $\mathcal{O}_C$ if $\alpha < 0$ and into $\mathcal{O}_R$ if $\alpha > 0$.

(ii) Observe also that we did *not* show that π takes the positive cone at a point of $\mathcal{B}_R$ onto the full tangent space of $\pi(\mathcal{O}_C) = \mathbf{H}$. However, because of Lemma 2.1(iii), all that matters to us is that movement between these zones is possible.

4 Proof of the main results

4.1 The case $n \leq 2$

For the convenience of the reader we now restate Theorem 1.2.

Theorem 4.1 *Let $n \leq 2$.*

(i) *An element of $\mathbf{G} - \mathcal{S}_1$ is the endpoint of a short positive path (from $\mathbb{1}$) if and only if it has an even number of real eigenvalues λ with $\lambda > 1$.*

(ii) *There is a positive path between any two elements $A, B \in \mathbf{G}$. Moreover, any short positive path from $\mathbb{1}$ may be extended to a short positive path with endpoint in $\mathcal{U}$.*

(iii) *The space of short positive paths (from $\mathbb{1}$) with endpoint in $\mathcal{U}$ is path-connected.*

Proof: (i) There are two ways eigenvalues can reach the positive real axis $\mathbf{R}^+$. Either a pair of eigenvalues moves from S^1 through $+1$ to a pair on $\mathbf{R}^+$, or a quadruplet moves from $\mathbf{C} - \mathbf{R}$ to $\mathbf{R}$. The first scenario cannot happen along a short path: for the only time a short path has an eigenvalue $+1$ is at time $t = 0$ and then, by Lemma 3.1, the eigenvalues have to move into S^1. In the second case, one gets an even number of eigenvalues on $(1, \infty)$.

(ii) Using Proposition 3.6 it is not hard to see that there is a positive path in $\mathcal{C}onj$ from $\pi(A)$ to $\mathbb{1}$ and from $\mathbb{1}$ to $\pi(B)$. The result now follows from Lemma 2.3. The second statement is proved similarly.

(iii) We have to show that any two short positive paths $\{A_t\}, \{B_t\}$ beginning at $\mathbb{1}$ and ending in $\mathcal{U}$ can be joined by a homotopy of short positive paths beginning at $\mathbb{1}$ and ending in $\mathcal{U}$. We may clearly assume that both paths are generic. (Note however that the homotopy of paths may have to go through codimension 2 strata.) Moreover, by Lemmas 2.1 and 2.6 we may assume that there is $\delta > 0$ such that $A_t = B_t \in \mathcal{U}$ for $t \in [0, \delta]$. Therefore, it will suffice to consider the case when the second path $\{B_t\}$ is (a reparametrization of) $A_t, 0 \leq t \leq \delta$. Thus we have to show how to "shrink" the path $\{A_t\}$ down to its initial segment in $\mathcal{U}$, keeping its endpoint in $\mathcal{U}$.

We do this by constructing a homotopy γ_t^μ in $\mathcal{C}onj$ between the projections of the two paths which at each time obeys the restrictions stated in Proposition 3.6. Indeed, such a homotopy means the existence of a homotopy $\tilde{\gamma}_t^\mu$ of short positive paths in G from $\mathbb{1}$ and ending in $\mathcal{U}$ which satisfies: $\tilde{\gamma}_t^0 = A_t$ for all t, $\tilde{\gamma}_0^\mu = \mathbb{1}$ and $\tilde{\gamma}_1^\mu \in \mathcal{U}$ for all μ, and $\pi(\tilde{\gamma}_t^1) = \pi(B_t)$. By Proposition 2.4, there is a homotopy of paths (with starting points equal to $\mathbb{1}$ and endpoints free in the conjugacy class of B_1) between $\{\tilde{\gamma}_t^1\}$ and $\{B_t\}$. The composition of these two homotopies is the desired path from $\{A_t\}$ to $\{B_t\}$.

We construct the homotopy γ^μ in $\mathcal{C}onj$ by means of a smooth map $r : \mu \mapsto r^\mu$ from the interval $[0, 1]$ to the space of short positive paths in $\mathcal{C}onj$ as follows.[4] Suppose that we have found r^μ which satisfy:
(i) $r^\mu(\mu) = \pi(A_\mu)$ and $r^\mu(1) \in \pi(\mathcal{U})$;
(ii) the paths r^μ are constant (with respect to t) for μ near 0.
Then it is not hard to check that we may take γ^μ to be:

$$\gamma^\mu(t) = \begin{array}{ll} \pi(A_t) & \text{if } t \leq \mu \\ r^\mu(t) & \text{if } t \geq \mu. \end{array}$$

We will construct the r^μ backwards, starting at $\mu = 1$. Intuitively, r^μ should be the simplest path from $\pi(A_\mu)$ to $\pi(\mathcal{U})$ (i.e. it should cross

---

[4]As will become clear, we are only interested in $r^\mu(t)$ on the time interval $t \in [\mu, 1]$.

the stratum $Conj_1$ as few times as possible), and we should think of the homotopy γ^μ as shrinking away the kinks in $\{A_t\}$ as μ decreases. To be more precise, let us define the **complexity** $c(\gamma)$ of a generic path $\gamma = \{C_t\}$ to be the number of times that $\pi(C_t)$ crosses $Conj_1$. Then we will construct the r^μ so that $c(\gamma^\mu)$ does not increase as μ decreases. Moreover it decreases by 2 every time that A_μ moves (as μ decreases) from $\mathcal{O}_\mathcal{R}$ to $\mathcal{O}_\mathcal{C}$, or from $\mathcal{O}_{\mathcal{U},\mathcal{R}}$ to $\mathcal{O}_\mathcal{U}$ or from $\mathcal{O}_\mathcal{C}$ to $\mathcal{O}_\mathcal{U}$.

To this end we assume that, except near places where $\pi(A_\mu) \in Conj_1$, the paths r^μ have the following form:

- if $A_\mu \in \mathcal{U}$ then r^μ is constant;
- if $A_\mu \in \mathcal{O}_\mathcal{C}$ is labelled with $\lambda = se^{i\theta}$ then r_μ goes down the ray $s'e^{i\theta}$ until it meets the circle in $\pi(\mathcal{N}^+_{e^{i\theta}})$ and enters $\pi(\mathcal{U})$;
- if $A_\mu \in \mathcal{O}_\mathcal{R}$ then r^μ moves the two eigenvalue pairs together, pushes them into $\pi(\mathcal{O}_\mathcal{C})$ and then follows the previous route to $\pi(\mathcal{U})$;
- if $A_\mu \in \mathcal{O}_{\mathcal{U},\mathcal{R}}$ then r^μ fixes the eigenvalues in S^1 and moves the real ones through $\pi(\mathcal{N}^+_{-1})$ to S^1.

We now describe how to extend r over each type of crossing. While doing this, there will be times μ at which we will want to splice different positive paths r^μ together. More precisely, at these μ_j we will have two different choices for the path r^{μ_j} from A_{μ_j} to $\mathcal{U}$. But these choices will be homotopic (by a homotopy which respects the endpoint conditions) and so, if we reparametrize the path A_t so that it stops at $t = \mu_j$ for a little while, we may homotop from one choice to the other. To be more precise, we choose a nondecreasing function $\rho : [0,1] \to [0,1]$ which is bijective over all points except the μ_j and is such that $\rho^{-1}(\mu_j)$ is an interval, and then change the relation between r^μ, γ^μ and A_t by requiring that

- $r^\mu(t)$ is defined for $t \in [\rho(\mu), 1]$ and $r^\mu(\rho(\mu)) = A_{\rho(\mu)}$,
- $\gamma^\mu(t) = A_t, t \leq \rho(\mu)$.

First let us consider how to handle crossings of the stratum $\mathcal{B}_\mathcal{U}$. Such crossings take place either at $\mathcal{N}^+_\lambda$ or $\mathcal{N}^-_\lambda$. However, because positive paths starting at $\mathcal{N}^+_\lambda$ point into $\mathcal{U}$ there is no problem extending r smoothly over this type of crossing. Observe that at this crossing the complexity remains unchanged. The problem comes when $A_\mu \in \mathcal{N}^-_\lambda$. To deal with this, first let C_t be a positive path which starts in $\mathcal{U}$, crosses $\mathcal{N}^-_\lambda$ at time μ, goes a little into $\mathcal{O}_\mathcal{C}$ and then crosses back into $\mathcal{U}$ through $\mathcal{N}^+_\lambda$. We may choose this path so that it is homotopic through positive paths with fixed endpoints to a positive path lying entirely in $\mathcal{U}$. (In fact, we could start with a suitable path in $\mathcal{U}$ and then perturb it.) By Lemma 2.1, we may suppose that $\pi(A_t) = \pi(C_t)$ for t in some interval $(\mu - \varepsilon, \mu + \varepsilon)$ and so (using the splicing technique described above) we may choose r so that, for some $\mu' \in (\mu, \mu + \varepsilon)$, we have $r^{\mu'}(t) = \pi(C_t)$ for $t \geq \mu'$. We now use this formula to extend r over the interval $(\mu - \varepsilon, \mu')$. Then $r^{\mu-\varepsilon} = \{C_t\}$ starts and ends in $\mathcal{U}$ and goes only a little way outside $\mathcal{U}$. We now reparametrize $\{A_t\}$ stopping it at time $t = \mu - \varepsilon$ so that there is time first to homotop $r^{\mu-\varepsilon} = \{C_t\}$ with fixed endpoints to a path in $\mathcal{U}$ and then to shrink it, fixing its first endpoint $A_{\mu-\varepsilon}$, to a constant path. This completes the crossing. Note that the complexity does decrease by 2.

A little thought shows that the same technique may be used to deal with crossings of the other boundaries $\mathcal{B}_{\mathcal{U},\mathcal{R}}, \mathcal{B}_{\mathcal{R},\mathcal{U}}$ and $\mathcal{B}_{\mathcal{R}}$. For example, if as μ decreases A_μ passes from $\mathcal{O}_{\mathcal{C}}$ to $\mathcal{O}_{\mathcal{R}}$, one easily extends r but does not change the complexity of γ^μ. On the other hand, if A_μ passes from $\mathcal{O}_{\mathcal{R}}$ to $\mathcal{O}_{\mathcal{C}}$ one needs to take more trouble in extending r but in exchange one decreases the complexity. $\square$

4.2   The case $n > 2$

As previously explained, the eigenvalues of an element of $\mathrm{Sp}(2n, \mathbf{R})$ form complex quadruplets or pairs on S^1 or $\mathbf{R}$. It is easy to check that the only singularities (or bifurcations) encountered by a generic 1-dimensional path $\gamma = A_{t \in [0,1]}$ are those in which

(a) a pair of eigenvalues on S^1 or $\mathbf{R}$ become equal to ± 1;

(b) two pairs of eigenvalues coincide on S^1 or $\mathbf{R}$ and then move into $\mathbf{C}$ (or conversely).

In particular, two quadruplets or three pairs do not coincide generically. Therefore, for each t, the space $\mathbf{R}^{2n}$ decomposes into a sum of 2- and 4-dimensional eigenspaces $E_1(t) \oplus \ldots \oplus E_k(t)$. Moreover, the interval $[0, 1]$ may be divided into a finite number of subintervals over which this decomposition varies smoothly with t. (The type of the decomposition may be different in the different pieces.) We claim that within each such piece the eigenvalue flow is just the same as it would be if the decomposition were fixed rather than varying. The reason for this is that all the restrictions on the eigenvalue flow are forced by Krein's lemma, which is valid for arbitrary variation of decomposition, together with topological data concerning Conj (i.e. topological information on the way the types of conjugacy classes fit together.) With these remarks it is not hard to adapt all the above arguments to the general case. In particular, the proof of (iii) is not essentially more difficult when $n > 2$ since in this case we reduce the complexity of the path as we procede with the homotopy. It is important to note that there still are essentially unique ways of choosing the paths r from the different components of the top stratum $\mathcal{C}_0$ into $\pi(\mathcal{U})$. ("Essentially unique" means that the set of choices is connected.) For example, if A^μ were on a stratum with 3 eigenvalue pairs on $\mathbf{R}^-$, then one might combine two to make a quadruplet in $\mathbf{C}$ which then moves down a ray to S^1 leaving one pair to move through $\mathcal{N}_{-1}^+$ to S^1, or one might move all 3 pairs directly down $\mathbf{R}$ to S^1. But these paths are homotopic (through positive paths with endpoints on $\pi(\mathcal{U})$) and so it is immaterial which choice we make. □

5 Positive paths in Hamiltonian systems and in Hofer's geometry

We present here a brief outline of the various ways in which positive paths intervene in the stability theory of Hamiltonian systems, and in Hofer's geometry. Actually, they are also a crucial ingredient of the theory of closed geodesics as developed by Bott in a series of papers ([3, 4, 5]). But since the application of positive paths to closed geodesics (and in particular to the computation of the index of iterates of a given closed geodesic) can also be presented within the framework of positive Hamiltonian systems via the geodesic flow (where Ekeland's formula appears as a generalization of Bott's iteration formula), this application is, at least theoretically, reducible to the following one.

5.1 Periodic Hamiltonian systems

Many Hamiltonian systems are given as periodic perturbations of autonomous flows. When both the autonomous Hamiltonian and the perturbation are quadratic maps $\mathcal{Q}_t$, the fundamental solution of the periodic system

$$\begin{aligned} \dot{x}(t) &= -J\nabla_X \mathcal{Q}_t(x(t)), \quad \text{where} \quad \mathcal{Q}_t = \mathcal{Q}_{t+1}, \\ x(0) &= \xi \end{aligned}$$

is a path of matrices $A_t \in \mathrm{G}$ as in §1. (In other words, the trajectory $x(t)$ which starts at ξ is $A_t(\xi)$.) By the periodicity of the generating Hamiltonian, it is clear that $A_{k+t} = A_t A_1^k$ for all integers k and real numbers $t \in [0, 1)$.

Definition 5.1 The above periodic system is **stable** if all solutions $x(t)$ remain bounded for all times (in other words, it is stable if there is a constant C such that $\|A_t\| \leq C$ for all $t > 0$). A matrix A is **stable** if there is C such that $\|A^k\| \leq C$ for all positive integers k.

Clearly, the above system is stable if and only if $A = A_1$ is stable. But it is easily seen that a matrix A is stable if and only if all its eigenvalues lie on the unit circle. Thus the stability of the periodic Hamiltonian system is determined by the spectrum of its time-1 flow A.

Definition 5.2 The Hamiltonian system is **strongly stable** if any C^2-small periodic (and quadratic) perturbation of it remains stable. A matrix A is **strongly stable** if it has a neighbourhood which consists only of stable matrices.

Since a C^2-small perturbation $\mathcal{Q}'_t$ of the Hamiltonian $\mathcal{Q}_t$ leads to a flow $A'_{t\in[0,1]}$ which is C^1-close (as a path) to the unperturbed flow, the time-1 map $A' = A'(1)$ is C^0-close to the time-1 map $A = A(1)$. Conversely any matrix A' close enough to A in the symplectic group is the time-1 map of a C^2-small quadratic perturbation of $\mathcal{Q}_{t\in[0,1]}$. This means that a periodic Hamiltonian system is strongly stable if and only if the time-1 map A is strongly stable. Hence such a system is strongly stable exactly when its time-1 flow belongs to the *interior* of the set of symplectic matrices with spectrum in S^1. But we have seen in the previous sections that this interior consists of all matrices with spectrum in $S^1 - \{\pm 1\}$ and maximal splitting numbers. (More precisely, at each multiple eigenvalue the absolute value of the splitting number must equal the multiplicity of the eigenvalue.) This is the Stability Theorem due to Krein and Gelfand-Lidskii.

Now consider the case $n = 2$ and assume that the stable periodic Hamiltonian $\mathcal{Q}_t$ is negative and that its flow $\{A_t\}$ is short (that is the index of $\{A_t\}$ is zero). Further, let $\mathcal{NU}$ be the union of all *open* strata for which none of the eigenvalues lie on S^1, and let $e_A(s)$ be the number of times that the path $\{A_t\}$ enters $\mathcal{NU}$ and comes back to $\mathcal{O}_{\mathcal{U}}$ during the time interval $(0, s]$.

Lemma 5.3 *For every short positive path* $A_{t\in[0,1]}$, *beginning at* $\mathbb{1}$, $e_A(1) \leq 1$.

Proof: Because $\mathcal{NU}$ and $\mathcal{U}$ are open, it suffices to prove this when the path $\{A_t\}$ is generic. If $e_A(1) = 0$ there is nothing to prove. So suppose that $e_A(s) = 1$ for some $s \leq 1$. Then Proposition 3.6 implies that on the interval $[0, s]$ either A_t moves from $\mathcal{O}_{\mathcal{C}}$ to $\mathcal{O}_{\mathcal{U}}$ passing through some conjugacy class $\mathcal{N}^+_\lambda$ at some time $t_0 < s$ or A_t moves from $\mathcal{O}_{\mathcal{R}}^-$ into $\mathcal{O}_{\mathcal{U}}$ via matrices with all eigenvalues on $(-\infty, -1) \cup S^1$. In either case, it follows from Krein's lemma 2.6 that the first eigenvalue of A_s that one encounters when traversing S^1 anticlockwise from $\mathbb{1}$ has negative splitting number. Therefore, by Lemma 2.7, the eigenvalues of $A_t, t \geq s$, can only leave S^1 after a pair has crossed $+1$.   □

More generally, this argument shows that for a stable positive path $A_{t\in[0,1]}$, which is not necessarily short or generic, $e_A(1) \leq i + 1$, where $i = i_A(1)$ is the Conley-Zehnder index of $A_{t\in[0,1]}$. (In this context, $i_A(1)$ is defined to be the number of times $t > 0$ that a generic positive perturbation $\{A'_t\}$ of $\{A_t\}$ (with $A'_0 = A_0 = \mathbb{1}$) crosses the set $\mathcal{S}_1 = \{A : \det(A - \mathbb{1}) = $

0}. Observe that because $\{A'_t\}$ is positive each such crossing occurs with positive orientation.)

Combining this with Ekeland's version of Bott's iterated index formula (see [8]), we get:

Proposition 5.4 *Let $A_{t\geq 0}$ be a stable positive path generated by a 1-periodic Hamiltonian in $\mathbf{R}^4$. Then there is a positive real number a such that for any time $t > 0$:*

$$e_A(t) \leq a\,[t]\,i_A(1) + 1$$

where $[t]$ is the greatest integer less than or equal to t.

Thus some aspects of the qualitative behavior of such paths are linearly controlled by the Conley-Zehnder index. Similarly, the number of times that such a path A_t leaves $\mathcal{O}_\mathcal{U}$, enters any other open stratum and comes back to $\mathcal{O}_\mathcal{U}$ during the time interval $(0, t]$, is bounded above by

$$3(i_A(t) + 1) \leq 3(a\,[t]\,i_A(1) + 1).$$

As a last application, noted by Krein in [10], Chap VI, consider the system $\dot{x}(t) = -\mu J \nabla_X \mathcal{Q}_t(x(t))$, $\mathcal{Q}_{t+1} = \mathcal{Q}_t$, with a parameter $\mu \in \mathbf{R}$. Then:

Proposition 5.5 (Krein) *If each $\mathcal{Q}_t$ is negative definite, there is $\mu_0 > 0$ such that the periodic Hamiltonian system*

$$\dot{x}(t) = -\mu J \nabla_X \mathcal{Q}_t(x(t))$$

is strongly stable for all $0 < \mu < \mu_0$.

Actually, the value μ_0 is the smallest μ such that the time-1 flow A_μ of the system has at least one eigenvalue equal to -1. Hence each time-1 flow A_μ for $\mu < \mu_0$ must be strongly stable.

5.2 Hofer's geometry

Hofer's geometry is the geometry of the group $\mathrm{Ham}(M)$ of all smooth Hamiltonian diffeomorphisms, generated by Hamiltonians $H_{t\in[0,1]}$ with compact support, of a given symplectic manifold M, endowed with the bi-invariant norm

$$\|\phi\| = \inf_{H_t} \int_0^1 (\max_M H_t - \min_M H_t)\,dt$$

where the infimum is taken over all smooth compactly supported Hamiltonians $H_{t\in[0,1]}$ on M whose time-1 flows are equal to ϕ. This norm defines a Finslerian metric (which is L^∞ with respect to space and L^1 with respect to time) on the infinite dimensional Fréchet manifold Ham(M). In [15] and [2], it is shown that a path ψ_t generated by a Hamiltonian H_t is a geodesic in this geometry exactly when there are two points $p, P \in M$ such that p is a minimum of H_t and P is a maximum of H_t for all t. We then showed in [16] that the path $\psi_{t\in[0,1]}$ is a stable geodesic (that is a local minimum of the length functional) if the linearized flows at p and P have Conley-Zehnder index equal to 0. (See also [20].) Note that the linear flow at p is positive while the one at P is negative. The condition on the Conley-Zehnder index means that both flows are short. The proof of the above sufficient condition for the stability of geodesics in [16] relies on the application of holomorphic methods which reduce the problem to the purely topological Main Theorem 1.2. Hence the result of the present paper can be considered as the topological part of the proof of the stability criterion in Hofer geometry. We refer the reader to [16] for the equivalence between this stability criterion and the local squeezability of compact sets in cylinders.

References

[1] V.I. Arnold and A. Givental, Symplectic Topology, in *Dynamical Systems*, vol IV, Springer-Verlag, Berlin, 1990.

[2] M. Bialy and L. Polterovich, Geodesics of Hofer's metric on the group of Hamiltonian diffeomorphisms, *Duke J. Math.* **76** (1994), 273–292.

[3] R. Bott, Non-degenerate critical manifolds, *Ann. Math.* **60** (1954), 248–261.

[4] R. Bott, On the iteration of closed geodesics and Sturm intersection theory, *Comm. Pure Appl. Math* **9** (1956), 173–206.

[5] R. Bott, The periodicity theorem for classical groups, *Ann. Math.* **70** (1959), 179–203.

[6] C. Conley and E. Zehnder, Morse type index theory for flows and periodic solutions for Hamiltonian equations, *Comm. Pure Appl. Math* **27** (1984), 211–253.

[7] I. Ekeland, An index theory for periodic solutions of convex Hamiltonian systems, *Proc. Symp. Pure Math* **45** (1986), 395–423.

[8] I. Ekeland, *Convexity Methods in Hamiltonian Mechanics*, Ergebnisse Math **19**, Springer-Verlag Berlin, 1989.

[9] I. Gelfand and V. Lidskii, On the structure of the regions of stability of linear canonical systems of differential equations with periodic coefficients, *Uspekhi Math. Nauk. USSR* **10** (1955), 3 – 40 (AMS translation **8** (1958), 143–181).

[10] I.C. Gohberg and M.G. Krein, *Theory and Applications of Volterra Operators in Hilbert Space*, Transl. of Math. Monogr. vol. 24, American Mathematical Society, 1970.

[11] K. Grasse and H.J. Sussmann, Global controllability by nice controls, *Nonlinear Controllability and Oprimal Control*, ed. H. J. Sussmann, Dekker, New York, 1990, pp. 38–81.

[12] M. Gromov, Carnot-Caratheodory geometry, preprint IHES, 1993.

[13] N. Hingston, Perambulations in the symplectic group, in preparation.

[14] W. Klingenberg, Lectures on Closed Geodesics, Springer-Verlag, Berlin, 1981.

[15] F. Lalonde and D. McDuff, Hofer's L^∞-geometry: energy and stability of Hamiltonian flows parts I and II, to appear in *Invent. Math.*.

[16] F. Lalonde and D. McDuff, Local non-squeezing theorems and stability, *Geom. Funct. Anal.* **5** (1995), 364–386.

[17] C. Lobry, Controllability of nonlinear systems on compact manifolds, *SIAM Journ on Control* **12** (1974), 1–4.

[18] J.W. Robbin and D.A. Salamon, The Maslov index for paths, *Topology*, **32** (1993), 827–44.

[19] H.J. Sussmann, Reachability by means of nice controls, *Proc 26th IEEE conference on Decision and Control*, Los Angeles, (1987), 1368–1373.

[20] I. Ustilovsky, Conjugate points on geodesics of Hofer's metric, preprint, Tel Aviv, 1994.

François Lalonde
Univ. de Québec à Montréal
Dept. Math.
Montreal PQ H3C 3P8, Canada
flalonde@math.uqam.ca

Dusa McDuff
SUNY
Dept. Math.
Stony Brook, NY 11794-3651
dusa@math.sunysb.edu

August 25, 1995

Invariants of submanifolds in Euclidean space

V.D. Sedykh

ABSTRACT. The topology of the set of singular support hyperplanes and hyperspheres to a smooth submanifold in Euclidean space is studied. As a corollary, some relations between differential-geometric characteristics of a manifold are obtained. In particular, if a simple closed embedded generic curve in a plane has C global vertices (where the curvature circles are support circles to the curve) and T support circles touching the curve at three points, then $C - T = 4$. Similar invariants are also obtained for submanifolds in higher-dimensional spaces.

INTRODUCTION

A tangent hyperplane or hypersphere to a smooth submanifold in Euclidean space is called a support one if the manifold lies on one side of it. A nonsingular support hyperplane (hypersphere) is tangent to the manifold at just one point and has at this point first-order tangency with the germ of any curve on the manifold. All other support hyperplanes and hyperspheres are singular.

In the paper, we compute the homology groups of the set of singular support hyperplanes and hyperspheres to a closed connected manifold which is not contained in any hyperplane and in any hypersphere in the ambient space. The Euler number of the set of their tangent elements to the manifold is computed as well. As a corollary, we obtain numerical relations between different classes of singular support hyperplanes and hyperspheres to a submanifold in Euclidean space.

Take, for example, a simple closed embedded generic curve in a plane. A vertex of a curve (the local extremum point of its curvature) is said to be global if the curvature circle at this vertex is a support circle to the curve. Let C be the number of global vertices of a curve and T be the number of support circles tangent to the curve at three points. Then

$$C - T = 4.$$

Support circles lying inside (outside) of a curve are called internal (external). Vertices of a curve where the curvature circles are internal (external) support circles are said to be internal (external). Suppose that a curve has C_1 internal and C_2 external vertices, T_1 internal and T_2 external support circles tangent to the curve at three points. Then

The work was partially supported by Russian Foundation for Basic Researches (project 94-01-01203) and INTAS grant no.4373.

$$C_1 - T_1 = 2, \qquad C_2 - T_2 = 2.$$

The first from the above relations implies the classical 4-vertex theorem (in the generic case): any smooth closed embedded plane curve has at least four vertices. Indeed, the number of all vertices is greater than or equal to the number C of global vertices of the curve, and $C = T + 4 \geq 4$. We generalize this relation for the case of generic submanifolds in higher-dimensional spaces.

1. SINGULAR SUPPORT HYPERPLANES AND HYPERSPHERES

Let M be a C^∞-smooth k-dimensional submanifold ($k > 0$) in Euclidean space $\mathbb{R}^n$. Consider affine hyperplanes and hyperspheres (i.e. standard $(n-1)$-dimensional spheres) in $\mathbb{R}^n$ tangent to M.

Definition. A tangent hyperplane (hypersphere) to a manifold M is said to be a *support* one if M lies on one side of this hyperplane (hypersphere).

Definition. A support hyperplane (hypersphere) π to a manifold M is called *nonsingular* if
   1) π is tangent to M at just one point P, and
   2) any smooth curve lying on M and passing through P has at this point first-order tangency with π.
   All other support hyperplanes and hyperspheres to M are said to be *singular*.

Definition. A manifold M is called *nonflat* if it is not contained in any hyperplane and in any hypersphere in $\mathbb{R}^n$.

Denote by Σ the set of all singular support hyperplanes and hyperspheres to a manifold M. If M is closed then Σ is a compact subset in the smooth $(n+1)$-dimensional manifold of all hyperplanes and hyperspheres in $\mathbb{R}^n$.

Theorem 1. *Let M be a smooth closed connected nonflat k-dimensional submanifold in $\mathbb{R}^n$. Then the homology groups and the Euler number of the set Σ are given by the following formulas:*

$$H_0(\Sigma, \mathbb{Z}) = \begin{cases} \mathbb{Z} & \text{if } k < n-1, \\ \mathbb{Z} \oplus \mathbb{Z} & \text{if } k = n-1, \end{cases}$$

$$H_m(\Sigma, \mathbb{Z}) \cong \begin{cases} H^{n-m-1}(M, \mathbb{Z}) & \text{if } 0 < m < n-1, \\ 0 & \text{if } m \geq n-1, \end{cases}$$

(1) $$\chi(\Sigma) = \chi(S^n) - (-1)^{n-k}\chi(M).$$

In particular, for $k = n - 1$ the set Σ has two connected components and $H_m(\Sigma, \mathbb{Z}) \cong H_m(M, \mathbb{Z})$, $0 < m < n - 1$. All hyperspheres from a given component lie on one side of M.

The detailed proof of Theorem 1 will be published in [S3]. We consider the stereographic image of the manifold M on the unit sphere in $\mathbb{R}^{n+1}$ and apply to it the results of [S1] ([S2]) on topology of the set of singular support hyperplanes to a *convex* submanifold (lying on the boundary of its convex hull) of codimension greater than 1 in an affine space.

2. TANGENT ELEMENTS OF SINGULAR SUPPORT HYPERPLANES AND HYPERSPHERES

Let us consider a hyperplane (hypersphere) π in $\mathbb{R}^n$ tangent to a submanifold M at a point P.

Definition. The pair (π, P) is called a *tangent element* of π to M at P.

The set L of tangent elements of all tangent hyperplanes and hyperspheres to a k-dimensional manifold M is a smooth n-dimensional manifold. The mapping that assigns to a tangent element the point of tangency defines a smooth fibration of L over M with fiber $\mathbb{R}P^{n-k} \setminus \{\cdot\}$.

Denote by Ω the set of tangent elements of singular support hyperplanes and hyperspheres to M. If M is closed then Ω is a compact subset in L.

Theorem 2. *Let M be a smooth closed connected nonflat k-dimensional submanifold in $\mathbb{R}^n$. Then the mapping*

$$\Omega \longrightarrow M, \ (\pi, P) \longmapsto P$$

defines a C^0-fibration of Ω over M with fiber S^{n-k-1}. In particular, the Euler number of the set Ω is given by the following formula:

$$(2) \qquad \chi(\Omega) = \chi(S^{n-k-1})\chi(M).$$

For $k = n-1$, the set Ω consists of two connected components each of which is homeomorphic to M.

The proof of this statement will be given in [S3]. It is also based on the results of [S1].

3. SUPPORT $A_{\mu_1}...A_{\mu_m}$-PLANES AND SPHERES

We consider the space of all smooth embeddings of a closed k-dimensional $(k > 0)$ manifold M into $\mathbb{R}^n$ equipped with the C^∞-topology. The submanifolds corresponding to the embeddings from a certain open everywhere dense subset in this space are said to be *generic*. Generic submanifolds in $\mathbb{R}^n$ are nonflat.

Let us take a sequence of odd natural numbers $\mu_1 \leq ... \leq \mu_m$.

Definition. A support hyperplane (hypersphere) π to a k-dimensional sub-manifold M in $\mathbb{R}^n$ is called $A_{\mu_1}...A_{\mu_m}$-*plane* (*sphere*) if

1) π is tangent to M at m points $A_{\mu_1}, ..., A_{\mu_m}$ that are vertices of a $(m-1)$-dimensional simplex, and

2) the germ at $A_{\mu_i}, i = 1, ..., m$, of the restriction to M of any smooth function in $\mathbb{R}^n$ equal to 0 on π, nonnegative on M and having noncritical value at A_{μ_i} is given by the formula

$$t_1^{\mu_i+1} + t_2^2 + ... + t_k^2$$

in suitable local coordinates $t_1, ..., t_m$ on M.

The number $d = \mu_1 + ... + \mu_m$ is called the *degree* of a support $A_{\mu_1}...A_{\mu_m}$-plane (sphere).

Let M be a smooth closed k-dimensional generic submanifold in $\mathbb{R}^n$ where either $k = 1$ or $n \le 6$. Then the set A of all support hyperplanes and hyperspheres to M consists of support $A_{\mu_1}...A_{\mu_m}$-planes of degree $d \le n$ and support $A_{\mu_1}...A_{\mu_m}$-spheres of degree $d \le n + 1$ (see [A], [Z]).

The set $A(\mu_1, ..., \mu_m)$ of all support $A_{\mu_1}...A_{\mu_m}$-planes and spheres to a manifold M is a smooth submanifold of codimension $d = \mu_1 + ... + \mu_m$ in the manifold of all hyperplanes and hyperspheres in $\mathbb{R}^n$. The partition of the set A into the connected components (*strata*) of the manifolds $A(\mu_1, ..., \mu_m)$ for all possible sequences $\mu_1 \le ... \le \mu_m$ is a minimal finite Whitney C^∞-stratification.

This stratification determines a minimal finite Whitney C^∞-stratification of the set Σ of singular support hyperplanes and hyperspheres to the manifold M. In turn, this induces a finite Whitney C^∞-stratification of the set Ω of tangent elements of singular support hyperplanes and hyperspheres to M. Namely, a stratum σ in Σ determines strata in Ω consisting of the tangent elements of those hyperplanes and hyperspheres that belong to σ.

Let $\chi(\mu_1, ..., \mu_m)$ be the Euler number of the manifold $A(\mu_1, ..., \mu_m)$. Denote by $N(d)$ the set of all sequences of odd natural numbers $\mu_1 \le ... \le \mu_m$ such that $\mu_1 + ... + \mu_m = d$. Then the equalities (1) and (2) imply

Corollary 1. *For any smooth closed connected k-dimensional generic sub-manifold in $\mathbb{R}^n$, where either $k = 1$ or $n \le 6$, the following relations hold:*

$$\sum_{d=2}^{n+1}(-1)^{n+1-d} \sum_{(\mu_1,...,\mu_m)\in N(d)} \chi(\mu_1, ..., \mu_m) = \chi(S^n) - (-1)^{n-k}\chi(M),$$

$$\sum_{d=2}^{n+1}(-1)^{n+1-d} \sum_{(\mu_1,...,\mu_m)\in N(d)} m\chi(\mu_1, ..., \mu_m) = \chi(S^{n-k-1})\chi(M).$$

In particular, if n is even and k is odd then

(3) $$\sum_{d=2}^{n+1}(-1)^d \sum_{(\mu_1,...,\mu_m)\in N(d)} \left(m - \frac{n}{2} - 1\right)\chi(\mu_1, ..., \mu_m) = n + 2.$$

From Theorems 1 and 2, we have also

Corollary 2. *Let* M *be a smooth closed connected generic hypersurface in* $\mathbb{R}^n$, *where* $2 \le n \le 6$. *Then the set* Σ (*of singular support hyperplanes and hyperspheres to* M) *has two connected components* Σ_1 *and* Σ_2. *The Euler number of each component is given by*

(4)

$$
\chi(\Sigma_i) = \frac{1}{2}\left(\chi(M) + \sum_{d=3}^{n+1}(-1)^{n-d} \sum_{(\mu_1,...,\mu_m)\in N(d)} (m-2)\chi_i(\mu_1,...,\mu_m) \right),
$$

where $\chi_i(\mu_1,...,\mu_m)$ *is the Euler number of the manifold* $\Sigma_i \cap A(\mu_1,...,\mu_m)$, $i = 1, 2$.

4. APPLICATION TO PLANE CURVES

Let M be a simple closed embedded generic curve in a plane. Then the relation (3) implies

$$
C - T = 4,
$$

where $C = \chi(3)$ is the number of *global* vertices of M (where the curvature circles are support circles to the curve) and $T = \chi(1,1,1)$ is the number of support circles tangent to M at three points.

According to Theorem 1, the set of singular support lines and circles to the curve M has two connected components Σ_1 and Σ_2. They are simply-connected, and all circles from a given component lie on one side of M.

Definition. A support circle to a curve M is called *internal* (*external*) if it lies inside (outside) M.

A vertex of a curve is said to be *internal* (*external*) if the curvature circle at this vertex is an internal (external) support circle to the curve.

Corollary 3. *Suppose that a simple closed embedded generic curve* M *in a plane has* C_1 *internal and* C_2 *external vertices,* T_1 *internal and* T_2 *external support circles tangent to* M *at three points. Then*

(5) $$ C_1 - T_1 = 2, \qquad C_2 - T_2 = 2. $$

This statement follows from Corollary 2. Indeed, if $\Sigma_1(\Sigma_2)$ contains internal (external) support circles to the curve M then $C_i = \chi_i(3)$, $T_i = \chi_i(1,1,1)$, $i = 1, 2$. Now, the relations (5) follow from the equalities (4) and $\chi(\Sigma_i) = 1$.

5. ON SUPPORT A_5-SPHERES

A smooth submanifold in a 4-space can have support A_5-spheres. The points of tangency of these support hyperspheres with the manifold are called its *global vertices*. The above results allow to obtain some equalities connecting the number of global vertices of a manifold with other its differential-geometric characteristics.

Let M be a smooth closed connected generic submanifold in $\mathbb{R}^4$. Then the relation (3) implies
(6)
$$2\chi(5) - 2\chi(1,1,1,1,1) - \chi(1,3) + \chi(1,1,1,1) + 2\chi(3) - \chi(1,1) = 6 - 3\chi(M),$$

where $\chi(\mu_1, ..., \mu_m)$ is the Euler number of the manifold of support $A_{\mu_1}...A_{\mu_m}$-planes and spheres to M. On the other hand, we have two equalities

$$\chi(1,3) = \chi(5) + \chi(1,1,3),$$
$$2\chi(1,1,1,1) = 5\chi(1,1,1,1,1) + \chi(1,1,3),$$

which follow from the incidence relations of the graph consisting of 0-dimensional and 1-dimensional strata of the Maxwell set of a generic 4-parameter family of functions (see [A]). Therefore, (6) can be rewritten in the form

$$\chi(5) - \frac{1}{2}\chi(1,1,3) + \frac{1}{2}\chi(1,1,1,1,1) = 6 - 2\chi(3) + \chi(1,1) - 3\chi(M).$$

This relation implies the interesting invariant of *strongly convex* curves in a 4-space (the strong convexity means that every hyperplane intersects the curve at most at 4 points counted with the multiplicities).

Corollary 4. *Suppose that a simple closed embedded strongly convex generic curve M in a $\mathbb{R}^4$ has V global vertices, U support hyperspheres touching M at three points and having a contact of third order at one of them, and P support hyperspheres tangent to M at five points. Then*

$$2V - U + P = 12.$$

In particular, $2V + P$ is greater than or equal to 12.

Indeed, $V = \chi(5)$, $U = \chi(1,1,3)$, $P = \chi(1,1,1,1,1)$, and $\chi(3) = \chi(1,1) = \chi(M) = 0$ since M is a closed strongly convex generic curve (the manifold $A(3)$ and the double covering of the manifold $A(1,1)$ are the total spaces of locally-trivial fibrations over M).

Remark. Using the results of [S1], one can obtain a similar invariant for some class of curves in a 5-space.

Let us consider a simple closed embedded generic curve γ in $\mathbb{R}^5$ having a strongly convex projection to a hyperplane. Suppose that γ has V support A_5-planes, U support $A_1 A_1 A_3$-planes, and P support $A_1 A_1 A_1 A_1 A_1$-planes. Then $2V - U + P = 12$ (in particular, $2V + P \geq 12$). The proof is analogous to that of Corollary 4.

The points of tangency of support A_5-planes with the curve γ are its flattening points (where the osculating Frenet frame is degenerate). These points can be called *global vertices* of the curve. The above formula implies that the number of global vertices of the curve γ is equal to $V = 6 + (U - P)/2$. We conjecture that $V \geq 6$. It should be noted that the number of all flattening points of a smooth closed curve in $\mathbb{R}^5$ having a strongly convex projection to a hyperplane is greater than or equal to 6 (see [Ar]).

References

[A] V. I. Arnold, V. A. Vassiliev, V. V. Goryunov, and O. V. Lyashko, *Singularities. II. Classification and applications.* Itogi Nauki i Tekhniki, Sovremennye Problemy Matematiki, Fundamental'nye Napravleniya, Vol. 39, VINITI, Moscow, 1989; English transl. in Encyclopaedia of Math. Sci., Vol.39 (Dynamical Systems, VIII), Springer-Verlag, Berlin, 1993.

[Ar] V. I. Arnold, *On the number of flattening points on space curves*, Preprint of the Mittag-Leffler Institute (1994).

[S1] V. D. Sedykh, *Invariants of convex manifolds*, Dokl. Akad. Nauk **326** (1992), no. 6, 948–952; English transl. in Russian Acad. Sci. Dokl. Math. **46** (1993).

[S2] V. D. Sedykh, *Invariants of strictly convex manifolds*, Functsional. Anal. i Prilozhen. **27** (1993), no. 3, 67–75; English transl. in Functional Anal. Appl. **27** (1993).

[S3] V. D. Sedykh, *Invariants of nonflat manifolds*, Functsional. Anal. i Prilozhen. **29** (1995) (to appear).

[Z] V. M. Zakalyukin, *Singularities of convex hulls of smooth manifolds*, Functsional. Anal. i Prilozhen. **11** (1977), no. 3, 76–77; English transl. in Functional Anal. Appl. **11** (1977).

Received July 1995

DEPARTMENT OF MATHEMATICS, MOSCOW STATE UNIVERSITY OF TECHNOLOGY "STANKIN", VADKOVSKI PER. 3A, MOSCOW 101472, RUSSIA
E-mail address: sedykh@stanmat.mian.su

On Combinatorics and Topology
of Pairwise Intersections of
Schubert Cells in SL_n/B

Boris Shapiro, Michael Shapiro,[1] Alek Vainshtein

0. Introduction

0.1. Topological properties of intersections of pairs and, more generally, of k-tuples of Schubert cells belonging to distinct Schubert cell decompositions of a flag space are of particular importance in representation theory and have been intensively studied during the last 15 years, see e.g. [BB, KL1, KL2, De1, GS]. Intersections of certain special arrangements of Schubert cells are related directly to the representability problem for matroids, see [GS]. Most likely, for a somewhat general class of arrangements of Schubert cells their intersections are too complicated to analyze. Even the nonemptyness problem for such intersections in complex flag varieties is very hard. However, in the case of pairs of Schubert cells in the space of complete flags one can obtain a special decomposition of such intersections, and of the whole space of complete flags, into products of algebraic tori and linear subspaces. This decomposition generalizes the standard Schubert cell decomposition. The above strata can be also obtained as intersections of more than two Schubert cells originating from the initial pair. The decomposition considered is used to calculate (algorithmically) natural additive topological characteristics of the intersections in question, namely, their Euler $E^{p,q}$-characteristics (see [DK]). Generally speaking, this decomposition of the space of complete flags does not stratify all pairwise intersections of Schubert cells, i.e. the closure of a stratum is not necessary a union of strata of lower dimensions. Still there exists a natural analog of adjacency, and its combinatorial description is available, see Theorem D. We discuss combinatorics of this special decomposition and some rather simple consequences for the cohomology and the mixed Hodge structure of intersections of Schubert cells in $SL_n/\mathcal{B}$.

0.2. In order to formulate the main results, let us recall some standard

---

[1]The paper was partially written during the visit of this author to the Department of Mathematics, University of Stockholm, supported by the NFR-grant R-RA 01599-306.

notions. Let F_n denote the space of complete flags in $\mathbb{C}^n$. Each compete flag f can be interpreted both as a Borel subgroup and as a sequence of enclosed linear subspaces of all dimensions from 0 to n. The *Schubert cell decomposition* D_f of F_n relative to f consists of cells formed by all flags having a given set of dimensions of intersections with subspaces of f. Thus, we have a family of Schubert cell decompositions parameterized by F_n. Cells of any such decomposition are in 1-1-correspondence with permutations on n elements (see, e.g., [FF]).

0.3. For any k-tuple of flags $f, g, h, \ldots$ in F_n we introduce the k-*tuple Schubert decomposition* $D_{f,g,h,\ldots}$ consisting of all nonempty intersections of k-tuples of cells one taken from each decomposition $D_f, D_g, D_h \ldots$. Generally speaking, a k-tuple Schubert decomposition is not a stratification of the space of complete flags, and its strata can have very complicated topology, cp. [GS].

Given a family of linear subspaces, their +-*completion* is defined as the set of sums of all possible subfamilies of these subspaces. The +-completion of a pair of flags is just the +-completion of the family of all subspaces constituting these flags.

The *refined double decomposition* $RD_{f,g}$ of the space of complete flags relative to a given pair of flags (f, g) is the decomposition into pieces formed by all flags with given dimensions of intersections with all subspaces of the +-completion of (f, g). The pieces of this decomposition are called *refined double strata*.

0.4. Remark. The refined double decomposition coincides with some special k-tuple decomposition, see §1.

0.5. Remark. The refined double decomposition $RD_{f,g}$ subdivides the standard decompositions D_f, D_g and the double decomposition $D_{f,g}$, i.e. each Schubert cell with respect to f or g, as well as their pairwise intersections, consists of some number of refined double strata.

0.6. Remark. If g' belongs to the same Schubert cell of D_f as g, then $RD_{f,g}$ is isomorphic to $RD_{f,g'}$, and the isomorphism is induced by a linear operator preserving f.

Theorem A. *Each refined double stratum is biholomorphically equivalent to the product of a complex torus by a complex linear space.*

0.7. Given a system of coordinates in $\mathbb{C}^n$, a flag is called *coordinate* if all its subspaces are spanned by coordinate vectors. Each coordinate flag is identified naturally with the corresponding permutation of coordinates. The *standard* coordinate flag is the one which is identified with the unit permutation, i.e. for each i its i-dimensional subspace is spanned by the first i coordinate vectors.

For any two flags f and g in F_n one can always choose a *standartizing* system of coordinates such that f becomes the standard coordinate flag and g becomes a coordinate flag identified with some permutation σ. We then say that the pair (f, g) is in *relative position* σ. Observe that the permutation corresponding to the cell containing g in the Schubert cell decomposition D_f is exactly σ^{-1}.

0.8. In what follows we need some additional notation related to permutations. We write an arbitrary permutation π in the form $\pi = i_1 \ldots i_n$, which means that $\pi(1) = i_1, \ldots, \pi(n) = i_n$.

A *decreasing subsequence* in π is a subsequence $s = (i_{j_1}, i_{j_2}, \ldots, i_{j_k})$ such that $1 \leqslant j_1 < j_2 < \cdots < j_k \leqslant k$ and $i_{j_1} > i_{j_2} > \cdots > i_{j_k}$.

The *reduced length* of a decreasing subsequence is equal to the number of its elements minus one. The *domination* of a decreasing subsequence is equal to the number of elements $i_j \in \pi$ for which there exists an element $i_l \in s$ such that $j < l$ and $i_j < i_l$.

The *cyclic shift* of π wrt a decreasing subsequence $s = (i_{j_1}, i_{j_2}, \ldots, i_{j_k})$ is the transformation sending i_{j_1} onto i_{j_k}, i_{j_2} onto $i_{j_1}, \ldots$, i_{j_k} onto $i_{j_{k-1}}$ and preserving the rest of the elements. (If s consists of just one element, then the transformation is identical.)

Example. Consider $\pi = 6723451$ and its decreasing subsequence $(7, 3, 1)$. The reduced length is 2 and the domination is also 2, namely, element 6 is dominated by 7 belonging to the decreasing subsequence, and element 2 is dominated by 3. The cyclic shift of π wrt $(7, 3, 1)$ takes π to 6321457.

0.9. Till the end of this section we assume that the pair of flags (f, g) is in relative position σ. Let us enumerate (algorithmically) all strata of $RD_{f,g}$ using the permutation σ. To do this, we apply to σ the following n-step procedure.

Main algorithm.

Step 1. Find all decreasing subsequences in σ. Apply to σ cyclic shifts wrt each of these decreasing subsequences and obtain the set of resulting permutations. In each of these permutations block the largest element of the corresponding decreasing subsequence. (To *block* just means that this element is ignored on all subsequent steps of the algorithm.)

Step i. To each permutation obtained on Step $i - 1$ apply the same procedure that was applied to σ on Step 1. Namely, find all its decreasing subsequences (disregarding all blocked elements). Make cyclic shifts wrt each of these decreasing subsequences, and finally in each of the obtained permutations block the largest element of the corresponding decreasing subsequence.

0.10. Remark. The algorithm stops exactly after n steps, since in each permutation obtained after i steps we get exactly i elements blocked (i.e., on Step $i+1$ we work actually with permutations on $n-i$ elements). Moreover, each permutation with at least one nonblocked element contains at least one decreasing subsequence (possibly consisting of just one element).

The whole procedure for $\sigma = 321$ is illustrated in Fig. 1. The underlined numbers are blocked and the sequences of numbers on edges present the decreasing subsequences chosen.

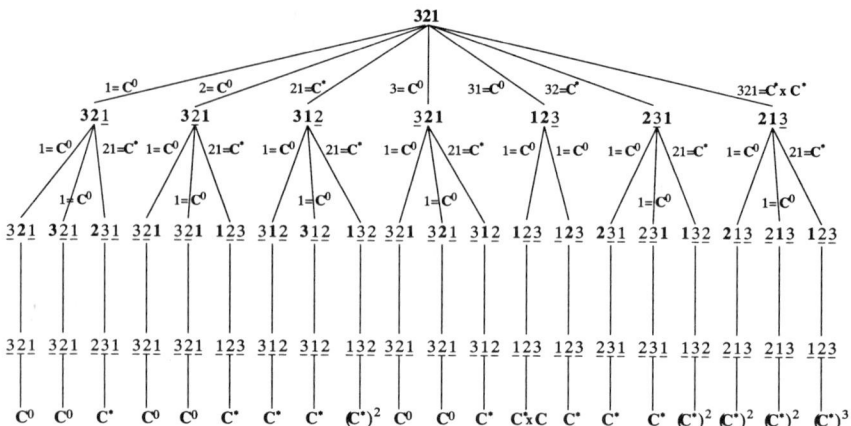

Figure 1. Illustration of the main algorithm for the case $\sigma = 321$

0.11. A *chain* of permutations is a sequence of $n+1$ permutations starting with σ and such that each consequent permutation is obtained from the preceding one as the result of a cyclic shift wrt a decreasing subsequence on the corresponding step of the above procedure.

So, Fig.1 contains 20 chains, which are just paths in the presented tree starting from the top element σ and going down to the bottom.

Let us assign to each chain two numbers, namely, its *total length* equal to the sum of the reduced lengths for all the permutations involved (actually, for the corresponding decreasing subsequences of these permutations), and its *total domination* equal to the sum of all dominations. (When we calculate the domination for a permutation with blocked elements we just disregard them completely.)

Theorem B. *Let (f,g) be a pair of flags in relative position σ. Then the strata of $RD_{f,g}$ are in 1-1-correspondence with the chains in the above procedure starting with σ. The stratum corresponding to a given chain is isomorphic to $(\mathbb{C}^*)^l \times \mathbb{C}^d$, where l is the total length of this chain and d is its total domination.*

Example. The structure of each refined double stratum is given in the bottom line of Fig.1.

0.12. For any two permutations α and β denote by $C_{1,\alpha}$ the Schubert cell consisting of all flags which are in relative position α with respect to f and by $C_{\sigma,\beta}$ the Schubert cell of all flags in relative position β with respect to g. (Warning: in the standard notation $C_{1,\alpha} = \mathcal{B}\alpha^{-1} \cdot \mathcal{B}$ and $C_{\sigma,\beta} = \sigma^{-1}\mathcal{B}\sigma\beta^{-1} \cdot \mathcal{B}$, see 0.7, where $\mathcal{B}$ is the subgroup of upper triangular matrices.)

By definition, the cell $C_{1,\alpha}$ belongs to the decomposition D_f, the cell $C_{\sigma,\beta}$ belongs to the decomposition D_g, and their intersection belongs to the double decomposition $D_{f,g}$.

Since the refined double decomposition $RD_{f,g}$ subdivides the decompositions D_f, D_g and $D_{f,g}$, it is useful to describe all refined double strata included in the Schubert cells $C_{1,\alpha}$, $C_{\sigma,\beta}$ and their intersection $C_{1,\alpha} \cap C_{\sigma,\beta}$.

Let us assign to a chain of permutations the following two new permutations. The *first blocking* permutation of a chain is the sequence of the successively blocked elements, i.e., its ith entry is the element blocked on the ith step of the procedure. The *second blocking* permutation is the sequence of positions on which the successively blocked elements stand, i.e., its ith element is the number of the position where the ith blocked element stands.

Example, see Fig.1. For the chain $\mathbf{321} \to \mathbf{123} \to \mathbf{12\underline{3}} \to \underline{\mathbf{123}}$, the first blocking permutation is 321 and the second blocking permutation is 321. For the chain $\mathbf{321} \to \mathbf{312} \to \mathbf{1\underline{32}} \to \underline{\mathbf{132}}$, the first blocking permutation is 231 and the second blocking permutation is 321. For the chain $\mathbf{321} \to \underline{\mathbf{321}} \to \underline{\mathbf{312}} \to \underline{\mathbf{312}}$, the first blocking permutation is 321 and the second blocking permutation is 132.

Theorem C. *A stratum of $RD_{f,g}$ belongs to the Schubert cell $C_{1,\alpha}$ if and only if the first blocking permutation of its chain coincides with α; a stratum of $RD_{f,g}$ belongs to the Schubert cell $C_{\sigma,\beta}$ if and only if the second blocking permutation of its chain coincides with β. Therefore, a stratum belongs to the intersection $C_{1,\alpha} \cap C_{\sigma,\beta}$ if and only if its first blocking permutation is α and its second blocking permutation is β.*

0.13. This theorem leads us to the following modifications of the described algorithm producing the refined double decompositions of the Schubert cells $C_{1,\alpha}$, $C_{\sigma,\beta}$ and of their intersection.

Modification 1. In order to obtain the decomposition of $C_{1,\alpha}$, one must consider on Step i, $i = 1, \dots, n$, only decreasing subsequences starting at the ith element of the permutation α.

Modification 2. In order to obtain the decomposition of $C_{\sigma,\beta}$, one must consider on Step i, $i = 1, \dots, n$, only decreasing subsequences ending at the position whose number is equal to the ith element in the permutation β.

Modification 3. Finally, in order to obtain the decomposition of $C_{1,\alpha} \cap C_{\sigma,\beta}$, one must consider on Step i, $i = 1, \dots, n$, only decreasing subsequences starting at the ith element of permutation α and ending at the position whose number is equal to the ith element in the permutation β.

The corresponding three examples are presented on Figs. 2,3.

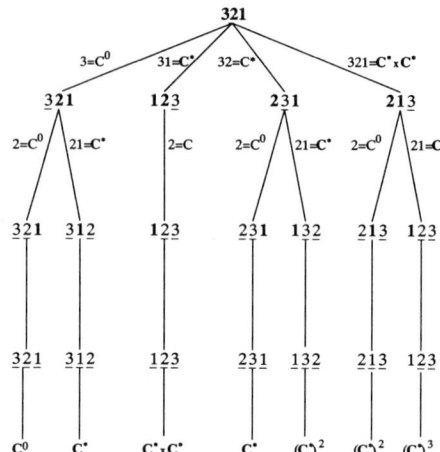

Figure 2. Modified algorithm for the case $\sigma = \alpha = 321$

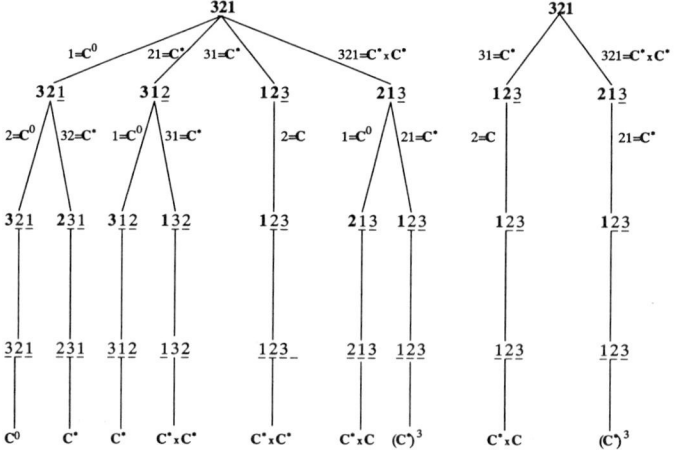

Figure 3. Modified algorithm for the cases $\sigma = \beta = 321$
and $\sigma = \alpha = \beta = 321$

A similar process was proposed by Francesco Brenti [Br1] in the case $\sigma = \omega_0$.

0.14. The following remark is valid for all versions of the main algorithm, i.e. for the refined double decomposition of the whole space of complete flags, of some particular Schubert cell, or of a pairwise intersection of Schubert cells.

Remark. Each step of the above algorithm is interpreted geometrically as the projection of flags in the considered set onto a linear subspace of the corresponding codimension, see §1. This means, in particular, that restricted chains, i.e. those starting at a permutation with blocked elements obtained after Step i, represent the decomposition of F_{n-i} (a Schubert cell in F_{n-i}, or a pairwise intersection of Schubert cells in F_{n-i}, depending on the modification of the algorithm).

0.15. To be more precise, we introduce the following operation on permutations with blocked elements. Let us consider a permutation on n elements with i blocked entries. The *reduction* of the blocked part from a given permutation is the operation which forms a new permutation on $n-i$ elements in the following way: we exclude all blocked elements and subtract from each nonblocked element the number of all blocked elements which are less than it.

Example. The reduction of the blocked part from the permutation 7563241 gives **4312**.

0.16. Proposition. i) *The set of all restricted chains, i.e. chains starting at some permutation $\tilde{\sigma}$ obtained after i steps of the algorithm (and thus containing i blocked elements), geometrically presents:*

    (1) *for the main algorithm, the refined double decomposition of F_{n-i} relative to the pair in relative position $\tilde{\sigma}'$, where $\tilde{\sigma}'$ is obtained from $\tilde{\sigma}$ by the reduction of all blocked elements;*

    (2) *for the 1st modification, the refined double decomposition of the Schubert cell $C_{1,\alpha'}$ in F_{n-i}, where α' is obtained by the reduction of the first i elements from α;*

    (3) *for the 2nd modification, the refined double decomposition of the Schubert cell $C_{\tilde{\sigma}',\beta'}$ in F_{n-i}, where $\tilde{\sigma}'$ is the same as above and β' is obtained by the reduction of the first i elements from β;*

    (4) *for the 3rd modification, the refined double decomposition of $C_{1,\alpha'} \cap C_{\tilde{\sigma}',\beta'}$.*

    ii) *Moreover, the geometrical meaning of Step i for any modification of the algorithm is the decomposition of the initial object (F_n, $C_{1,\alpha}$, $C_{\sigma,\beta}$ or $C_{1,\alpha} \cap C_{\sigma,\beta}$) into a disjoint union of products of similar objects in F_{n-i}, enumerated by the set of all permutations $\tilde{\sigma}$ obtained after Step i, by $(\mathbb{C}^*)^{l(\tilde{\sigma})} \times \mathbb{C}^{d(\tilde{\sigma})}$. Here $l(\tilde{\sigma})$ is equal to the sum of reduced lengths for all permutations in the chain starting at σ and ending at $\tilde{\sigma}$, and $d(\tilde{\sigma})$ is the sum of all dominations in this chain.*

Example. The set of all reduced chains passing through any permutation obtained after the first step except for 12**3** in Fig.1 presents the refined decomposition of the space of complete flags $F_2 = \mathbb{C}P^1$ relative to $\sigma = 21$ into three strata, namely, two points and $\mathbb{C}^*$. The chains passing through 12**3** (or $\sigma = 12$) present the standard Schubert cell decomposition of F_2 into a point and $\mathbb{C}$.

0.17. Remark. The above results hold in complete generality for the spaces of complete flags over any algebraically closed field, or $\mathbb{R}$. In this case $\mathbb{C}^*$ must be substituted by the multiplicative and $\mathbb{C}$ by the additive group of the field, cp. [Cu3].

0.18. Our next result gives sufficient conditions for the topological "adjacency" of strata in $C_{1,\alpha} \cap C_{\sigma,\beta}$ in terms of combinatorial adjacency, i.e. enumerates strata that can have nonempty intersection with the closure of some given stratum in $C_{1,\alpha} \cap C_{\sigma,\beta}$. Since the refined double decomposition in general is not a stratification, one has to use this modified notion of adjacency.

Consider two chains of permutations. One of them is said to be *less or equal* than the other one if each permutation of the former is less or equal in the Bruhat order than the corresponding permutation of the latter (for the notion of the Bruhat order see e.g. [Hu]). The above partial order on the set of all refined double strata (or their chains of permutations) in $C_{1,\alpha} \cap C_{\sigma,\beta}$ is called the *adjacency partial order*, or the *generalized Bruhat order*.

By Theorem C, each nonempty refined double stratum of $RD_{f,g}$ is contained in a single pairwise intersection $C_{1,\alpha} \cap C_{\sigma,\beta}$, namely, in the one for which α equals to the first and β to the second blocking permutation of the corresponding chain of permutations.

0.19. Theorem D. *The closure of a given refined double stratum in the corresponding pairwise intersection of Schubert cells belongs to (but in general does not coincide with) the union of all refined double strata included in the same pairwise intersection such that their chains of permutations are less or equal in the adjacency partial order than the chain of the stratum considered.*

This theorem enables us to construct rather simple examples of pairwise intersections whose refined double decompositions fail to be stratifications. In particular, the refined double decomposition of $C_{1234,4231} \cap C_{4231,4231}$ consists of three strata, namely, $\mathbb{C}^* \times (\mathbb{C})^2$ and two copies of $(\mathbb{C}^*)^3 \times \mathbb{C}$, see Fig. 9 below. In notations of Fig.9 the closure of the stratum $C = (\mathbb{C}^*)^3 \times \mathbb{C}$ is nonempty in the stratum $B = (\mathbb{C}^*)^3 \times \mathbb{C}$; moreover, the closure of $C = (\mathbb{C}^*)^3 \times \mathbb{C}$ does not contain the whole stratum $A = \mathbb{C}^* \times (\mathbb{C})^2$; see §7 for more examples of this kind.

0.20. Let us consider the *closure pattern* of a given refined double stratum, i.e. the set of all strata that have nonempty intersection with the closure of the given one. (The question about the combinatorial description of the closure pattern was raised for a similar situation in [Cu3].) Theorem D gives us a necessary combinatorial condition for a stratum to belong to the closure pattern of the other one. In §7 we present an example showing that this necessary condition is not sufficient. The relation between necessary and sufficient conditions motivates the following definition.

We call a pairwise intersection of Schubert cells $C_{1,\alpha} \cap C_{\sigma,\beta}$ *nice* if the refined double decomposition $RD_{f,g}$ gives its stratification, and *almost nice* if for any pair $\mathrm{St}_1 \prec \mathrm{St}_2$ of refined double strata one has $\dim \mathrm{St}_1 \leqslant \dim \mathrm{St}_2$. The rest of $C_{1,\alpha} \cap C_{\sigma,\beta}$ are called *hard*. An example of a hard $C_{1,\alpha} \cap C_{\sigma,\beta}$

is given in §7.

0.21. Let V denote an arbitrary complex quasiprojective variety. Let us denote by h_k^{pq} the Hodge numbers for the usual mixed Hodge structure in $H_c^*(V;\mathbb{C})$, see e.g. [Dl1, Dl2, Dl3].

Let us define generalized Euler characteristics depending on p and q:

$$\chi^{pq} = \sum_k (-1)^k h_k^{pq},$$

and form their generating function called the $E^{p,q}$-*polynomial*, or just the *E-polynomial* of V:

$$E_V(u,v) = \sum_{p,q} \chi^{pq} u^p v^q.$$

Let us denote by $E_\sigma^{\alpha,\beta}$ the E-polynomial of $C_{1,\alpha} \cap C_{\sigma,\beta}$, and by $\mathrm{CH}(\alpha,\beta,\sigma)$ the set of chains of permutations corresponding to the strata contained in $C_{1,\alpha} \cap C_{\sigma,\beta}$ (see Theorem B).

0.22. Corollary (of Theorems B and D). i) *The Hodge numbers $h_k^{p,q}$ of any intersection $C_{1,\alpha} \cap C_{\sigma,\beta}$ can be positive only if $p = q$.*

ii)

$$E_\sigma^{\alpha,\beta} = \sum_{\mathrm{ch}\in\mathrm{CH}(\alpha,\beta,\sigma)} z^{d(\mathrm{ch})}(z-1)^{l(\mathrm{ch})},$$

where $z = uv$, $d(\mathrm{ch})$ and $l(\mathrm{ch})$ are the total domination and the total length of a chain ch, respectively.

0.23. The same expression ii) can be rewritten as an inductive formula using the above remark on the geometrical meaning of our algorithm (see 0.14). More precisely, let $\mathcal{S}$ denote the set of all decreasing subsequences in σ starting at element $\alpha(1)$ and ending at the position with number $\beta(1)$, (i.e., the set of subsequences used on the first step of the construction of the refined double decomposition for $C_{1,\alpha} \cap C_{\sigma,\beta}$). For any decreasing subsequence $s \in \mathcal{S}$, let $l(s)$ and $d(s)$ denote its reduced length and domination, respectively. Finally, let α' and β' denote the results of the reduction of the first elements $\alpha(1)$ and $\beta(1)$ from α and β, respectively, σ_s denote the result of the cyclic shift of σ wrt s and the reduction of the first element of s (see the description of the algorithm and its modifications above).

Corollary (of Theorems B and D).

$$E_\sigma^{\alpha,\beta} = \sum_{s\in\mathcal{S}} z^{d(s)}(z-1)^{l(s)} E_{\sigma_s}^{\alpha',\beta'}.$$

Example.

$$E_{4321}^{4321,4321} = (z-1)E_{132}^{321,321} + (z-1)^2 E_{231}^{321,321}$$
$$+ (z-1)^2 E_{312}^{321,321} + (z-1)^3 E_{321}^{321,321}.$$

0.24. Corollaries 0.22 and 0.23 follow directly from the refined double decomposition of $C_{1,\alpha} \cap C_{\sigma,\beta}$. By results of [Cu3], their natural analogs are also valid for any $G/\mathcal{B}$, where G is a semisimple group and $\mathcal{B}$ is its Borel subgroup. Part i) of 0.22 holds as well for all quasiprojective varieties that can be decomposed into quasiprojective pieces satisfying $h_k^{p,q} = 0$ if $p \neq q$.

0.25. The adjacency partial order (see 0.18) allows us to consider different filtrations of $C_{1,\alpha} \cap C_{\sigma,\beta}$ by closed subsets consisting of refined double strata. Moreover, one gets the standard filtration of such a kind as follows.

Let P be a finite poset. We define the *height* of an element $a \in P$ as the maximum length of a chain having a as the maximal element. The *standard filtration* of P is its filtration by the subsets $P_i = \{a \in P \mid h(a) \leqslant i\}$, called *standard subposets*.

Consider now the refined double decomposition of $C_{1,\alpha} \cap C_{\sigma,\beta}$. Refined double strata are enumerated by chains of permutations, which form a finite poset with respect to the adjacency partial order. Theorem D shows that the standard filtration of this poset is a filtration by closed subsets.

Any filtration of $C_{1,\alpha} \cap C_{\sigma,\beta}$ by closed subsets leads to its Leray spectral sequence converging to the cohomology of $C_{1,\alpha} \cap C_{\sigma,\beta}$ with compact supports. In our case we are primarily interested in filtrations by closed quasiprojective subvarieties. Along with the above standard filtration, it is often convenient to consider "filtration" by dimension: its ith term is the union of all strata of dimension at most i. (Warning: for an arbitrary $C_{1,\alpha} \cap C_{\sigma,\beta}$ this filtration is apparently not a filtration by closed subsets.) In the case of an (almost) nice $C_{1,\alpha} \cap C_{\sigma,\beta}$ the filtration by dimension is a filtration by closed quasiprojective subvarieties. Moreover, the following statement is valid.

Theorem E. *The Leray spectral sequence associated with the filtration by dimension of the refined double decomposition of any (almost) nice pairwise intersection degenerates at the second page.*

0.26. Remark. A combinatorial description of d_1 is unavailable at the present moment and apparently is rather complicated.

0.27. As an application of the E-polynomials we prove the following combinatorial result.

Proposition. *If at least one of the permutations α, β or σ is the longest element w_0, then there are no gaps in (complex) dimensions of strata in $C_{1,\alpha} \cap C_{\sigma,\beta}$, i.e. there exist refined double strata of all intermediate (complex) dimensions between the minimal and the maximal ones.*

0.28. A similar family of decompositions was introduced by V.V.Deodhar in the case of intersections $C_{1,\alpha} \cap C_{w_0,\beta}$, where w_0 denotes the longest element in an arbitrary finite Coxeter system, see [De1, De2], and was extended to all intersections $C_{1,\alpha} \cap C_{\sigma,\beta}$ by Ch.Curtis in [Cu3]. These decompositions depend on a reduced expression of the element α as a product of simple reflections, and different choices of such expressions lead to different decompositions. The combinatorial data that codes strata in the approach of Deodhar–Curtis is similar to chains of permutations corresponding to refined double strata, but more lengthy. We have found the correspondence between the strata of these two decompositions and proved that the refined double decomposition coincides with one of decompositions suggested by Deodhar for some particular choice of reduced expression.

Let us define for any $\alpha \in \mathfrak{S}_n$ a reduced expression of α^{-1} of a special form, namely, $\alpha^{-1} = t_{\alpha(1)-1}t_{\alpha(1)-2}\cdots t_1\tilde{\alpha}^{-1}$, where $\tilde{\alpha}$ belongs to $\mathfrak{S}_{n-1}$ and t_i is the simple transposition interchanging i and $i+1$. This enables us to define inductively a reduced expression, which we call the *standard expression*.

Proposition . *The refined double decomposition of the Schubert cell $C_{1,\alpha} = \mathcal{B}\alpha^{-1} \cdot \mathcal{B}$ coincides with the decomposition suggested by Deodhar if one chooses the standard reduced expression of α^{-1}.*

0.29. Remark. There exist other natural refinements of double decompositions. These other decompositions of geometrical origin apparently coincide with Deodhar's decompositions corresponding to other choices of reduced expressions.

0.30. The starting point of this study was an attempt to calculate the cohomology of pairwise intersections of Schubert cells of the maximal dimension (see [SV]).

The authors are very grateful to Francesco Brenti for discussions of the material of this article. Sincere thanks are due to Prof. T.Springer who was first to point out to the authors the necessity to check the property of being a stratification and mentioned the paper [Boe].

1. Refined double decompositions as k-tuple decompositions and the very weak Bruhat order

1.1. In order to study the properties of refined double decompositions, let us fix a pair of flags (f, g) in relative position σ and consider its $+$-completion. By the definition (see 0.2), the $+$-completion consists of all sums of pairs of all possible subspaces from the flags f and g, see the example in Fig.4. Obviously, all the subspaces in the $+$-completion of (f, g) are coordinate (in the corresponding coordinate system, see 0.7) and partially ordered by inclusion.

By the *flag completion* of the pair (f, g) we mean the set of all complete flags such that each subspace of each of these flags belongs to the $+$-completion of (f, g).

The number of flags in the flag completion of a pair of flags in relative position σ is denoted by $k(\sigma)$; it is just the number of all different paths in the partially ordered +-completion going from the top to the bottom. For example, there are 10 paths (and therefore flags in the flag completion) in the diagram shown in Fig.4.

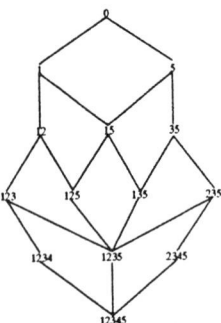

Figure 4. The +-completion of the pair of flags in relative position 53241.
Each set of numbers presents the coordinate subspace spanned
by the corresponding coordinate vectors.
Subspaces in each row are lexicographically ordered

1.2. Proposition (an alternative definition of the refined double decomposition). *The refined double decomposition $RD_{f,g}$ is the k-tuple decomposition $D_{f_1,f_2,\ldots}$, where f_i runs over the set of all complete flags in the flag completion of (f,g), and hence $k = k(\sigma)$ is the number of flags in this flag completion.*

1.3. Remark. A special $n!$-tuple Schubert decomposition into intersections of $n!$-tuples of cells taken from Schubert cell decomposition relative to each of coordinate flags was introduced in [GS] and called the decomposition into small cells. This decomposition and its projections on Grassmanians play an important role in matroid theory, since each small cell on some Grassmanian is the space of linear presentations of the corresponding matroid. The decomposition into small cells subdivides each of our refined double decompositions.

1.4. Let us define the following partial order on $\mathfrak{S}_n$. By an *elementary descent* in a permutation $i_1 \ldots i_n$ we call any inversion of two neighboring numbers i_l, i_{l+1} such that
    i) it decreases the inversion length of the permutation, and
    ii) i_{l+1} is less than all the elements $i_{l+2}, \ldots, i_n$.
The transitive closure of elementary descents defines a partial order on $\mathfrak{S}_n$, which we call the *very weak Bruhat order*. It is, obviously, a suborder

of the weak Bruhat order.

The only minimal element in the very weak Bruhat order is the identity, while the set of maximal elements consists of all elements having 1 in the last position. Figure 5 presents the very weak Bruhat order for the cases of $\mathfrak{S}_3$ and $\mathfrak{S}_4$.

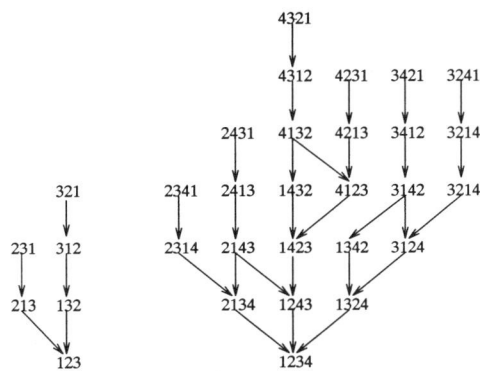

Figure 5. The very weak Bruhat order on $\mathfrak{S}_3$ and $\mathfrak{S}_4$

1.5. Proposition. *The number $k(\sigma)$ equals the number of permutations in the lower interval of σ in the very weak Bruhat order.*

Proof. Follows immediately from the definition (see 1.1 and 1.4).   ■

1.6. Below we present a simple scheme for finding the number $k(\sigma)$, which we call a *generalized Pascal triangle*. Recall that the standard Pascal triangle (of size n) can be represented as a box diagram having n rows and n columns. Each box is a square. The first row contains n boxes, the second row contains $n-1$ boxes, and so on; the first column contains 1 box, the second column contains 2 boxes, and so on. The diagram is filled by integers in the following way. We start at the unique box of the first column and fill it with 1. Next, the contents of any other box is defined to be the sum of the contents of its western and north-western neighbors (if any). Proceeding in accordance with this rule, we first fill all the boxes of the second column by ones; this gives a possibility to fill the boxes of the third column, and so on (see Fig. 6). The elements in the ith column of the triangle are just the binomial coefficients $\binom{i-1}{j}$, $0 \leqslant j \leqslant i-1$, and their sum is 2^{i-1}.

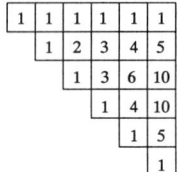

Figure 6. Pascal triangle for $n = 6$

To construct the generalized Pascal triangle of type σ, we define $L_\sigma(l)$, $1 \leqslant l \leqslant n$, as the number of elements i_j of σ such that $j > l$ and $i_j > i_l$. We now start from the diagram of the standard Pascal triangle and remove the horizontal partitions between the last (from left to right) $L_\sigma(1)$ boxes of the first and the second row, then the horizontal partitions between the last $L_\sigma(2)$ boxes of the second and the third row, and so on. As a result, we get a new box diagram; its boxes are no longer squares, but rectangles of width 1 and arbitrary height. It is easy to see that the number of boxes in the diagram equals n plus the inversion length of σ.

Example. Let $n = 6$, $\sigma = 426135$. Then the vector of L_σ's is $(2, 3, 0, 2, 1, 0)$. The diagram of the corresponding generalized Pascal triangle is presented in Fig.7a. The inversion length of σ is 7, and the number of boxes in the diagram is $6 + 7 = 13$.

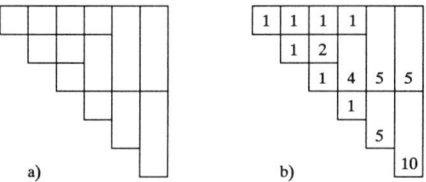

Figure 7. Generalized Pascal triangle of type 426135

The diagram thus constructed is now filled by integers according to the same rule as above. The only difference is that a box can have more than one western neighbor; in this case we have to sum the contents of all western neighbors, and that of the north-western one. As before, the table is filled column by column, starting from 1 placed in the unique box of the first column. The result of this procedure for the above example is presented in Fig.7b.

1.7. Proposition. *The number $k(\sigma)$ equals the sum of the numbers in the last column of the generalized Pascal triangle of type σ.*

Proof. Let us define two projections $\pi_1, \pi^1 \colon \mathfrak{S}_n \to \mathfrak{S}_{n-1}$. The first of them takes $i_1 \ldots i_k 1 i_{k+2} \ldots i_n$ to $i_1 - 1 \ldots i_k - 1\, i_{k+2} - 1 \ldots i_n - 1$; the second one takes $a i_2 \ldots i_n$ to $i_2^a \ldots i_n^a$, where $i_j^a = i_j$ if $i_j < a$ and $i_j^a = i_j - 1$ if $i_j > a$. It is easy to see that for $\sigma = i_1 \ldots i_n$ one has

$$(*) \qquad k(\sigma) = \begin{cases} k(\pi_1(\sigma)) + k(\pi^1(\sigma)) & \text{if } i_1 \neq 1, \\ k(\pi_1(\sigma)) = k(\pi^1(\sigma)) & \text{otherwise} \end{cases}.$$

Indeed, let $i_1 \neq 1$ (the case $i_1 = 1$ is trivial). By Proposition 1.5, we have to count the number of permutations in the lower interval of σ in the very weak Bruhat order. There are two types of such permutations: those in which i_1 and 1 form an inversion, and those in which they do not. By the definition, any path from σ to a permutation of the first type in the very weak Bruhat order avoids elementary descents transposing i_1 (otherwise condition ii) of 1.4 would be violated). Thus, i_1 remains all the time at the first position of any permutation in such a path and does not influence any of the descents used. Therefore, the number of the permutations of the first type equals $k(\pi^1(\sigma))$.

On the other hand, any path from σ to a permutation of the second type contains elementary descents involving i_1, and the first such descent on any of these paths is the transposition of i_1 and 1. Thus, 1 occurs at the first position, and remains there all the time. Therefore, the number of the permutations of the second type equals $k(\pi_1(\sigma))$.

Let now $b(\sigma)$ denote the sum of the elements of the last column in the generalized Pascal triangle of type σ; we shall prove that $b(\sigma)$ satisfies the same equation as $k(\sigma)$. Again we first assume that 1 is not in the first position of σ. Hence, the second column of the generalized Pascal triangle consists of two square boxes, and each of the boxes contains 1. Let us consider two other generalized Pascal triangles with the same diagram that the initial one; the first of them contains 0 in the first box of the second column and 1 in the second box, while the second triangle contains 1 in the first of these boxes and 0 in the other one. All the other columns of the both triangles ate filled according to the rule 1.6. Since the rule is additive, we get that the contents of any box of the initial triangle (except for the box in the first column) equals the sum of the contents of the corresponding boxes in the two new triangles constructed. However, the first of these triangles corresponds bijectively to the generalized Pascal triangle of type $\pi^1(\sigma)$ (it is sufficient to shorten by 1 the heights of all uppermost boxes; if such a box is of height 1, then its contents is 0, and we just remove it). In a similar way, the second of the triangles corresponds bijectively to the generalized Pascal triangle of type $\pi_1(\sigma)$ (we again remove certain square boxes containing 0 and shorten by 1 certain boxes of height > 1). Therefore, $b(\sigma)$ satisfies the first line of $(*)$. The case $\sigma = 1\,i_2 \ldots i_n$ is again trivial, since in this case the second column of the corresponding generalized Pascal triangle contains only one box.

Finally, for the trivial permutation $1 \in \mathfrak{S}_1$ one has $k(1) = 1$, while the generalized Pascal triangle of type 1 consists of one box containing 1, and so $b(1) = 1$. Therefore $b(\sigma)$ and $k(\sigma)$ satisfy the same equation with the same initial values. $\blacksquare$

2. Quotients of cells and refined double strata

2.1. Lemma. *Let K, L, and M be a triple of linear subspaces in some linear space. Then*

$$(*) \quad \dim((K/L) \cap (M/L))$$
$$= \dim((K + M) \cap L) - \dim(K \cap L) - \dim(M \cap L) + \dim(K \cap M).$$

Moreover, if M contains L, the above formula is simplified to

$$(**) \qquad \dim((K/L) \cap (M/L)) = \dim(K \cap M) - \dim(K \cap L).$$

The proof is obtained by a straightforward usage of inclusion and exclusion of vectors of an appropriate basis.

2.2. Given a complete flag f and a subspace L in $\mathbb{C}^n$, we denote by f_L the *complete quotient flag*, that is, the flag in the quotient space $\mathbb{C}^n/L$ obtained by taking consequent quotients of all subspaces constituting f (and ignoring occasional coincidences).

2.3. Assume that C is a cell of the Schubert cell decomposition D_f and $L \subseteq \mathbb{C}^n$ is a linear subspace. We denote by C_L the set of all flags in C containing L as a flag subspace.

Lemma . *If $C_L \neq 0$, then the set C_L/L of quotients of all flags in C_L is isomorphic to some cell in the Schubert cell decomposition D_{f_L} of the space of complete flags in $\mathbb{C}^n/L$. Moreover, if the flags in C are in relative position α to f and the dimension of L is i, then flags in C_L/L are in relative position α' to f_L, where α' is the permutation on $n - i$ elements obtained by the reduction of the first i elements of α.*

Proof. The set C_L/L belongs to some cell C' of the decomposition D_{f_L}. Indeed, let g be some flag in C_L, let M be its subspace containing L and let K be a subspace of f. Then $\dim((K/L) \cap (M/L))$ is equal to $\dim(K \cap M) - \dim(K \cap L)$ by $(**)$, and thus does not depend on the choice of g in C_L. Conversely, any flag $f' \in C'$ is the quotient of some flag $\tilde{f} \in C_L$. The corresponding flag $\tilde{f} \in C_L$ can be constructed as follows. Its subspaces included in L can be taken from an arbitrary flag in C_L, while those containing L are the inverse images of the subspaces of the considered quotient flag f'. An easy check shows that $\tilde{f}$ belongs to C_L, since it contains L and has the necessary dimensions of intersections with the subspaces of f. In order to prove that α' is obtained from α by the reduction of its first i elements, one must consider a standardtizing basis for the (f, g) (see 0.7) and take the quotient by the i-dimensional subspace of g. In this case everything is obvious. ∎

2.4. Corollary. *The assertion of Lemma 2.3 holds for any k-tuple Schubert decomposition, i.e. if* St *is a nonempty stratum of $D_{f,g,h\ldots}$ and* St$_L$ *is its nonempty subset of all flags containing a linear subspace L, then the set* St$_L$ /L *of quotients of flags in* St$_L$ *is isomorphic to a stratum in the k-tuple decomposition $D_{f_L,g_L,h_L,\ldots}$.*

2.5. Remark. The same holds for refined double decompositions, since they are particular cases of k-tuple decompositions, see Proposition 1.2.

2.6. Remark. If L is one-dimensional, then $C_L/L \cong C_L$.

2.7. Proposition. *Let h^1 and h^2 be two flags lying in the same refined double stratum of $RD_{f,g}$, and let L^1 and L^2 denote the subspaces in h^1 and h^2, respectively, of the same dimension. Then there exists an isomorphism $\Phi \colon \mathbb{C}^n/L^1 \to \mathbb{C}^n/L^2$ of the quotient spaces that sends the quotient flag f_{L^1} onto f_{L^2} and the quotient flag g_{L^1} onto g_{L^2}. Moreover, the quotient flags $\Phi(h^1_{L^1})$ and $h^2_{L^2}$ belong to the same refined double stratum of the refined double decomposition $RD_{f_{L^2},g_{L^2}}$.*

Proof. Such an isomorphism Φ exists if the pair of quotient flags (f_{L^1}, g_{L^1}) is in the same relative position as the pair (f_{L^2}, g_{L^2}). Thus, we must check that for any two subspaces K and M of f and g, respectively, $\dim((K/L^1) \cap (M/L^1)) = \dim((K/L^2) \cap (M/L^2))$. Indeed, if we substitute L^1 for L in (∗), then we get the same four terms in the right hand side as if we substitute L^2 (this follows immediately from the definition of refined double strata).

To prove that such an isomorphism Φ takes the quotient flag $h^1_{L^1}$ to the same refined double stratum of $RD_{f_{L^2},g_{L^2}}$ where $h^2_{L^2}$ lies, we must show the following. Given any subspace K in the +-completion of the pair (f,g) and subspaces $\tilde{L}^1 \supset L^1$ and $\tilde{L}^2 \supset L^2$ of the same dimension that belong to h^1 and h^2, respectively, prove that $\dim((K/L^1) \cap (\tilde{L}^1/L^1)) = \dim((K/L^2) \cap (\tilde{L}^2/L^2))$. Using (∗∗), we rewrite the above relation as $\dim(K \cap \tilde{L}^1) - \dim(K \cap L^1) = \dim(K \cap \tilde{L}^2) - \dim(K \cap L^2)$, which follows immediately from the definition of the refined double decomposition. ∎

3. Stabilizer of a pair of flags
and its action on the projective space

3.1. Let us consider the subgroup $G_\sigma = \mathcal{B} \cap \sigma^{-1}\mathcal{B}\sigma$, which is the stabilizer of both flags f and g (recall that $\mathcal{B}$ denotes the subgroup of upper triangular matrices). The subgroup G_σ is given by the following relations: the matrix entry $a_{i,j}$, $i < j$, vanishes if and only if the numbers i and j form an inversion in σ, i.e. i occurs in σ further than j.

3.2. The subgroup G_σ acts on strata of all four decompositions D_f, D_g, $D_{f,g}$ and $RD_{f,g}$, since it preserves both f and g. We first consider the natural action of G_σ on the projective space P^{n-1} of all lines in $\mathbb{C}^n$.

Lemma. *The set of all G_σ-orbits on P^{n-1} corresponds bijectively to the set S of all decreasing subsequences in σ. The orbit $O_{s,\sigma}$ corresponding to a decreasing subsequence s is biholomorphically equivalent to $(\mathbb{C}^*)^{l(s)} \times \mathbb{C}^{d(s)}$, where $l(s)$ is the reduced length of s and $d(s)$ is its domination. Moreover, for each orbit $O_{s,\sigma}$ there exists a subgroup $G_{s,\sigma} \subseteq G_\sigma$ biholomorphically equivalent to $(\mathbb{C}^*)^{l(s)+1} \times \mathbb{C}^{d(s)}$ such that its quotient by the scalar matrices (isomorphic to $O_{s,\sigma}$) acts on $O_{s,\sigma}$ freely and transitively.*

Proof. Let $s = (i_1, \ldots, i_{l(s)+1})$ be a decreasing subsequence in σ and $D = (j_1, j_2, \ldots, j_{d(s)})$ be the set of elements in σ dominated by elements of s. Then the orbit $O_{s,\sigma}$ consists of all lines whose spanning vectors in an appropriate standarizing basis for (f, g) have arbitrary nonvanishing coordinates with the numbers $i_1, \ldots, i_{l(s)+1}$ and arbitrary (possibly vanishing) coordinates with the numbers $j_1, \ldots, j_{d(s)}$; the rest of the coordinates vanish identically. Obviously, $O_{s,\sigma}$ is an orbit of G_σ and has the form $(\mathbb{C}^*)^{l(s)} \times \mathbb{C}^{d(s)}$, see 3.1. The orbit corresponding to the decreasing subsequence s contains the distinguished line p_s whose spanning vector has 1's for all $i_j \in s$ and 0's for the rest of coordinates. In order to describe $G_{s,\sigma} \subseteq G_\sigma$ we consider for each element i_k from s the (possibly empty) subset J_k of elements in D dominated precisely by i_k, i.e. all the elements that are dominated by i_k but not dominated by $i_1, \ldots, i_{k-1}$. Consider now a subset in G_g consisting of the matrices with arbitrary entries in positions (i_k, J_k), arbitrary nonvanishing entries in positions (i_k, i_k), 1's on the rest of the main diagonal and zeros elsewhere. This set of matrices forms a subgroup $G_{s,\sigma}$ in G_g, which topologically is $(\mathbb{C}^*)^{l(s)+1} \times \mathbb{C}^{d(s)}$. The subgroup $G_{s,\sigma}$ acts freely and transitively on the set of all spanning vectors of lines in $O_{s,\sigma}$; one can check it by multiplying elements of $G_{s,\sigma}$ by a spanning vector of the distinguished line in $O_{s,\sigma}$. Thus, the quotient of $G_{s,\sigma}$ by the scalar matrices acts freely and transitively on $O_{s,\sigma}$.

A straightforward check shows that for any line (which corresponds to a point on P^{n-1}) there exists an appropriate decreasing subsequence s in σ such that the orbit $O_{s,\sigma}$ contains this line. ■

3.3. Remark. All lines belonging to an orbit $O_{s,\sigma}$ have the same dimensions of intersections with any subspace of the +-completion of (f, g). And conversely, lines with a given set of dimensions of intersections with the subspaces of the +-completion belong to the same orbit.

3.4. Lemma. i) *For any line L in an orbit $O_{s,\sigma}$ the quotient flags f_L and g_L are in the same relative position σ_s defined by the cyclic shift of σ wrt s and the reduction of the first element of s.*

ii) *The set F^s of all flags in F_n containing lines of $O_{s,\sigma}$ is diffeomorphic to $O_{s,\sigma} \times F_{n-1}$.*

Proof. i) The independence of the relative position of f_L and g_L of the choice of L in $O_{s,\sigma}$ follows from the transitivity of the action of $G_{s,\sigma}$ on $O_{s,\sigma}$. The relative position of f_L and g_L can be calculated easily if one takes L equal to the distinguished line $p_s \in O_{s,\sigma}$ (see the proof of Lemma 3.2).

ii) The set of all flags containing a fixed line L is isomorphic obviously to the set of all flags in $\mathbb{C}^n/L$, i.e. to F_{n-1}. Let us denote by $\nu \colon F_n \to P^{n-1}$ the natural map sending each flag onto its 1-dimensional subspace. The restriction of the projection ν onto the set F^s defines the structure of a fiber bundle with the fiber F_{n-1} and the base $O_{s,\sigma}$. This bundle is trivial for the following reason. The transitive and free action of $G_{s,\sigma}$ modulo scalar matrices on F^s defines another structure of a fiber bundle on F_s, which is transversal to the first one. Namely, its fibers are diffeomorphic to $O_{s,\sigma}$ and its base can be chosen as the set of all flags containing the distinguished line in $O_{s,\sigma}$. ∎

4. Proofs of Theorems A and B

4.1. Proof of Theorem A. We proceed by induction on n. The space F_2 coincides with $\mathbb{C}P^1$; any pair of flags is just a pair of two (possibly coinciding) points. Each refined double stratum is a point, $\mathbb{C}^*$, or $\mathbb{C}$ (if the points coincide).

Let us consider now some refined double stratum St consisting of all flags with a fixed set of dimensions of intersections with all the subspaces of the $+$-completion of (f, g). We focus first on conditions satisfied by the 1-dimensional subspace of any flag in St. This 1-dimensional subspace must belong to some coordinate subspace and avoid in it at least one (but possibly more) coordinate hyperplanes. The set of all such lines is diffeomorphic to the product of a complex torus by a linear space. Actually, each of these lines belong to the image $\nu(\mathrm{St})$, i.e. is the 1-dimensional subspace of some flag in St. Moreover, the restricted projection $\nu \colon \mathrm{St} \to \nu(\mathrm{St})$ is a trivial fiber bundle, see Lemma 3.4. By Remarks 2.5–2.6, the set all flags in St with a fixed line is isomorphic to a refined double stratum in the space F_{n-1} of all complete flags in $\mathbb{C}^{n-1}$. Thus, St is diffeomorphic to the product of a complex torus by a linear space by some refined double stratum in F_{n-1}. Since any refined double stratum in F_{n-1} has the necessary form by the inductive hypothesis, Theorem A is proved. ∎

4.2. Proof of Theorem B. Let us consider the space of complete flags

F_n and a fixed standard pair (f, g) in relative position σ. We present an inductive construction of the refined double decomposition $RD_{f,g}$ of the space of complete flags, which exactly fits the main algorithm. The space F_n is decomposed into the disjoint union of F^s, where F^s is the union of all flags whose 1-dimensional subspaces belong to the orbit $O_{s,\sigma}$. Thus, the set of all F^s's is in 1-1-correspondence with the set of all decreasing subsequences in σ, and hence with the branches of Step 1 of the algorithm. Each F^s is diffeomorphic to $O_{s,\sigma} \times F_{n-1}$, by Lemma 3.4. Let us now subdivide each F^s. Namely, take a fiber (which is isomorphic to F_{n-1}) in F^s over some line $L \in O_{s,\sigma}$ and take the quotient flags f_L and g_L in this fiber. We consider their stabilizer G_{σ_s} and its orbits on the space P^{n-2}. Choosing an G_{σ_s}-orbit on P^{n-2} corresponding to a decreasing subsequence s' of σ_s, we fix the dimensions of intersections of the 1-dimensional subspace in a variable quotient flag with the +-completion of (f_L, g_L). Subsets of fibers that are mapped to this orbit form a fibration over $O_{s,\sigma}$. Moreover, the set $F_{s,s'}$ of flags in all fibers that are projected on this orbit forms a fibration over $O_{s,\sigma} \times O_{s',\sigma_s}$. Since G_σ modulo scalar matrices acts freely on $O_{s,\sigma}$, see Lemma 3.2, this fibration is trivial. Fixing the orbits of G_σ and G_{σ_s} is equivalent to fixing the dimensions of intersections of 1- and 2-dimensional subspaces of a variable flag with the +-completion of (f, g). Arguing inductively along the same lines, we get a sequence of fibrations $F^{s,s',\cdots}$ over the products of tori and linear subspaces. Each of these sets is characterized by the following conditions: the dimensions of intersections of the consecutive subspaces (from 0 up to some dimension) of a variable flag belonging to such a subset with the +-completion of (f, g) are fixed. Therefore, after n steps one sees that the sets $F^{s,s',\cdots}$ are precisely the refined double strata diffeomorphic to products of tori and linear subspaces, and they are in 1-1-correspondence with chains of permutations in the main algorithm. ∎

4.3. Remark. The main algorithm describes this sequence of operations using the description of G_σ-orbits in Lemmas 3.2, 3.4. To avoid the reduction of elements and to keep track of elements in the original permutation we prefer to use blocking instead of the reduction.

5. The image of a Schubert cell in the projective space
and Theorem C

The proof of Theorem C and the modifications of the main algorithm follows easily from Theorem B and several additional statements.

5.1. Lemma. *Let $C_{1,\alpha}$ be a Schubert cell in D_f and let $\alpha(1) = k$. Then the image $\nu(C_{1,\alpha})$ is the $(k-1)$-dimensional affine subspace in P^{n-1}*

consisting of all lines that belong to the k-dimensional subspace of f but do not belong to its $(k-1)$*-dimensional subspace.*

Proof. Follows immediately from the definition of $C_{1,\alpha}$. ∎

5.2. Remark. Quite similarly, if $C_{\sigma,\beta} \in D_g$ and $\beta(1) = m$, then the image $\nu(C_{\sigma,\beta}) \subset P^{n-1}$ is the $(m-1)$-dimensional affine subspace of all lines lying in the m-dimensional subspace of g, but not in its $(m-1)$-dimensional subspace.

5.3. Lemma. *Under the assumptions of Lemma* 5.1 *the set* $\mathcal{S}_\alpha$ *of all* G_σ*-orbits constituting* $\nu(C_{1,\alpha})$ *corresponds to the set of all decreasing subsequences in* σ *having the following two additional properties:*

    i) *any subsequence* $s \in \mathcal{S}_\alpha$ *contains the number* k;

    ii) *any* $s \in \mathcal{S}_\alpha$ *does not contain any number greater than* k.

Proof. By Lemma 5.1, we must choose only those orbits that contain lines lying in the k-dimensional subspace of f, but not in its $(k-1)$-dimensional subspace. This is guaranteed exactly by conditions a) and b) above. ∎

5.4. Remark. Similarly, let $\sigma = i_1 \ldots i_n$ and $\beta(1) = m$, then the set $\mathcal{S}^\beta$ of all G_σ-orbits constituting $\nu(C_{\sigma,\beta})$ corresponds to the set of all decreasing subsequences in σ having the following two additional properties:

    i) any $s \in \mathcal{S}^\beta$ contains the number i_m;

    ii) any $s \in \mathcal{S}^\beta$ does not contain the numbers $i_{m+1}, \ldots, i_n$.

5.5. Corollary. *The set* $\mathcal{S}_\alpha^\beta$ *of all* G_σ*-orbits constituting* $\nu(C_{1,\alpha}) \cap \nu(C_{\sigma,\beta})$ *corresponds to the set of all decreasing subsequences in* σ *located on the interval between the numbers* k *and* i_m *and containing both of them.*

So, if $k = i_m$, then there is only one decreasing subsequence consisting of the element $k = i_m$. If $k > i_m$, then the set $\mathcal{S}_\alpha^\beta$ is empty. The topology and the dimensions of these orbits are described in Lemma 3.2.

5.6. Proof of Theorem C. By definition, the refined double decomposition $RD_{f,g}$ of the space F_n subdivides each Schubert cell $C_{1,\alpha}$, $C_{\sigma,\beta}$, and their intersection $C_{1,\alpha} \cap C_{\sigma,\beta}$. Statements 5.1-5.3 describe explicitly the decreasing subsequences that must be used on a step of the algorithm in order to select those refined double strata which are contained in one of the above three types of spaces. We also know what happens to the permutations α and β after a single step of the algorithm. The modifications given in Introduction just present this sequence of operations formally, using blocking instead of the reduction. ∎

5.7. Proof of Proposition 0.16. Part i) of the proposition is just an obvious corollary of Theorem B and the description of the modifications of the algorithm. Indeed, all restricted chains starting at some permutation

$\tilde{\sigma}$ are obtained by the same procedure as the one for σ itself ignoring all the blocked elements in $\tilde{\sigma}$ and (in modifications of the main algorithm) the first i elements in α and β. Part ii) is obtained by a slight generalization of Lemma 3.4 and the proof of Theorem B via an inductive argument. ∎

6. Adjacency of orbits in P^{n-1} and Theorem D

6.1. Let us describe combinatorially the adjacency of G_σ-orbits on P^{n-1}. We consider the following partial order on the set of decreasing subsequences in σ. A decreasing subsequence s_1 is said to be *less or equal* than a decreasing subsequence s_2 (notation $s_1 \vdash s_2$) if for each element of s_1 there exists an element in s_2 that either dominates it or coincides with it.

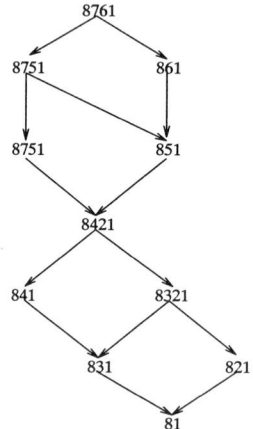

Figure 8. The natural partial order on the set
of decreasing subsequences in 83427561

The restrictions of this partial order to the sets $\mathcal{S}_\alpha$, $\mathcal{S}^\beta$, and $\mathcal{S}^\beta_\alpha$ define the corresponding partial orders on these subsets.

Example. The partial order on the set of the decreasing subsequences starting at 8 and ending at the last position in the permutation $\sigma = 83427561$ is shown in Fig. 8.

The *maximal* decreasing subsequence of σ, denoted by $s_{\max}(\sigma)$, is obtained in the following way. Its first element is the maximal element of σ. The ith element of $s_{\max}(\sigma)$ is the maximal element of σ standing to the right of the $(i-1)$th element of $s_{\max}(\sigma)$. Similarly one defines the maximal decreasing subsequence of any subset of elements in σ.

6.2. Lemma. *The partial order $\vdash$ describes the adjacency of the G_σ-orbits on P^{n-1}, $\nu(C_{1,\alpha})$, $\nu(C_{\sigma,\beta})$, and $\nu(C_{1,\alpha}) \cap \nu(C_{\sigma,\beta})$, respectively. In* *each of these cases there is no gaps in dimensions, i.e. there exist strata of all complex dimensions between the minimal and the maximal.*

Proof. The statement concerning the adjacency follows immediately from the description of orbits, see Lemma 3.2. The absence of gaps in dimensions is settled by the following observation. It suffices to consider the first case; the other cases are similar. Take any decreasing subsequence s corresponding to an orbit of a dimension bigger than one. In this case s contains either more than one element, or a single element having a nonempty set of dominated elements. We now construct a decreasing subsequence s' corresponding to an orbit whose dimension is less by one, and which lies in the closure of the first orbit. If s consists of just one element, then s' is the maximal decreasing subsequence of the set of dominated elements, see the definition in 6.1. If the number of elements in s is bigger than one, we replace the last element by the maximal decreasing subsequence of the set of elements dominated precisely by the last element, but not by the previous ones. If this set is empty then we just drop the last element. All checks are straightforward. ∎

6.3. Consider a 1-parameter family of subspaces $L(t)$, $t \in [0, 1]$, such that for all $t \in [0, 1)$ the subspaces $L(t)$ have a given set of dimensions of intersections with all subspaces in the +-completion of (f, g), and $L(1)$ has the same set of dimensions of intersections with the individual subspaces of f and g, but possibly bigger dimensions of intersections with the rest of the subspaces in the +-completion. Denote by $\pi(t)$ the permutation presenting the position of $g_{L(t)}$ relative to $f_{L(t)}$.

Lemma. *For any $t \in [0, 1)$ one has $\pi(t) \succeq \pi(1)$.*

Proof. The Bruhat order $\succeq$ describes the adjacency of Schubert cells. Thus, the assertion of the lemma is equivalent to the following inequality: for any two subspaces K and M of f and g, respectively, $\dim((K/L(1)) \cap (M/L(1))) \geqslant \dim((K/L(t)) \cap (M/L(t)))$. Applying formula $(*)$ of §2 to both sides of the inequality and taking into account the conditions $\dim(K \cap L(1)) = \dim(K \cap L(t))$ and $\dim(M \cap L(1)) = \dim(M \cap L(t))$, we get the necessary result. ∎

6.4. Proof of Theorem D. Let us show that the closure of any given stratum St $\in C_{1,\alpha} \cap C_{\sigma,\beta}$ is contained in the union of strata whose chains are less than the chain of St in the adjacency partial order, see 0.18. Indeed, consider a path $\gamma \colon [0, 1] \to C_{1,\alpha} \cap C_{\sigma,\beta}$ such that $\gamma[0, 1)$ belongs to St and $\gamma(1)$ belongs to the boundary of St. This means that there exists at least one subspace $L(1)$ in $\gamma(1)$ such that the dimension of its intersection with some subspace in the +-completion of (f, g) is bigger than that for the subspaces $L(t)$ of the same dimension in $\gamma(t)$, $t \in [0, 1)$. Now, applying

Lemma 6.3 we get precisely the necessary statement.   ∎

7. Combinatorial versus topological adjacency

7.1. Theorem D gives an obvious sufficient combinatorial condition for the refined double decomposition of $C_{1,\alpha} \cap C_{\sigma,\beta}$ to define its stratification (i.e., for the closure of each stratum to belong to the union of strata of lower dimensions).

Criterion. *If for any two refined double strata* St_1 *and* St_2 *in* $C_{1,\alpha} \cap C_{\sigma,\beta}$ *such that* $St_1 \prec St_2$ *the dimension of* St_1 *is less than that of* St_2, *then the refined double decomposition provides the stratification of* $C_{1,\alpha} \cap C_{\sigma,\beta}$.

7.2. Examples. All intersections of top-dimensional Schubert cells in F_4 are stratifications. On the contrary, the simplest example of nonstratification is provided by the refined double decomposition of the intersection $C_{1234,4231} \cap C_{4231,4231}$; it consists of the three strata, namely, of $\mathbb{C}^* \times \mathbb{C}^2$ and of the two copies of $(\mathbb{C}^*)^3 \times \mathbb{C}$. The intersection of the two opposite top-dimensional Schubert cells in F_5 contains the fragment shown in Fig.9a, which apparently shows that it is not a stratification. Still another example is shown in Fig.9b. This example shows also that not all branches of the algorithm reach the bottom level.

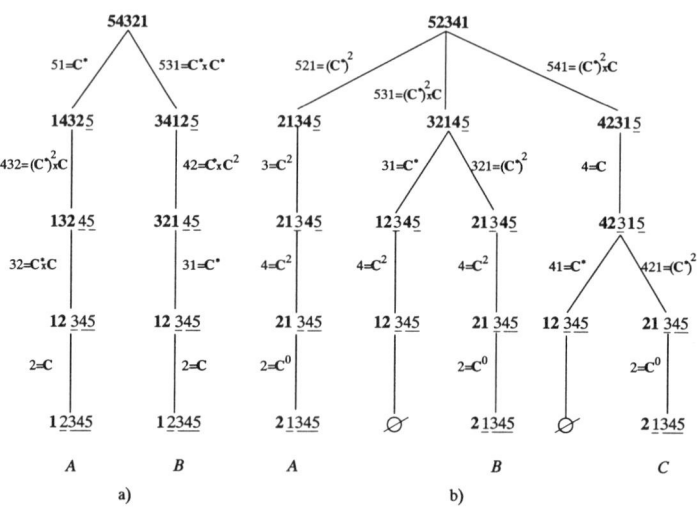

Figure 9. Examples of nonnice refined double decompositions.
a) a nonnice fragment for $\alpha = \beta = \sigma = 54321$;
b) nonnice decomposition for $\alpha = 53421$, $\beta = 53412$, $\sigma = 52341$

Let us describe also in more detail the above simplest example. We use projective complete flags on $\mathbb{C}P^3$ instead of complete flags in the linear space $\mathbb{C}^4$. The strata are defined by the following conditions:

$A = (\mathbb{C}^*) \times \mathbb{C}^2$ corresponding to $\mathbf{4231} \to \mathbf{1234} \to 1\underline{234} \to 1\underline{234} \to$ $1\underline{234}$: the point of the variable flag h belongs to the line containing the points of f and g, and the line of h belongs to the 2-plane containing the lines of f and g.

$B = (\mathbb{C}^*)^3 \times \mathbb{C}$ corresponding to $\mathbf{4231} \to \mathbf{2134} \to 1\underline{234} \to 1\underline{234} \to$ $1\underline{234}$: the line o h belongs to the 2-plane containing the lines of f and g, and the point of h does not belong to the line containing the points of f and g.

$C = (\mathbb{C}^*)^3 \times \mathbb{C}$ corresponding to $\mathbf{4231} \to \mathbf{3214} \to 3\underline{214} \to 1\underline{234} \to$ $1\underline{234}$: the line of h passes through the origin and the point of h does not belong to the 2-plane containing the lines of f and g.

If we want to get a stratification, we must split B into B_1 and B_2, where B_1 is given by an additional condition that the line of h passes through the origin and B_2 is given by the condition that the line of h does not pass through the origin. We also split A into A_1 and A_2 given precisely by the same additional conditions as B_1 and B_2, respectively. Then A_1, which is diffeomorphic to $\mathbb{C}^* \times (\mathbb{C} \setminus \{2 \text{ points}\})$, coincides with the closure of C in A and B_1 coincides with the closure of C in B. Pay attention to the fact that A_1 is no longer the product of a complex torus by a complex linear space.

7.3. Using the previous example one can construct the following example, which reveals an essential difference between combinatorial and topological adjacencies. Namely, we take the refined double decomposition of the intersection $C_{12345,54231} \cap C_{54321,54231}$ and consider the two strata A and B defined by the following conditions:

$A = (\mathbb{C}^*)^6 \times \mathbb{C}$ corresponding to $\mathbf{54321} \to \mathbf{42315} \to 32\underline{145} \to 3\underline{2145} \to$ $1\underline{2345} \to 1\underline{2345}$;

$B = (\mathbb{C}^*)^4 \times \mathbb{C}^2$ corresponding to $\mathbf{54321} \to \mathbf{41325} \to 21\underline{345} \to$ $1\underline{2345} \to 1\underline{2345} \to 1\underline{2345}$.

Then $B \prec A$, but B does not intersect the closure of A, which means that A is not adjacent to B.

7.4. To complete the section, we provide the only known to the authors example of a hard $C_{1,\alpha} \cap C_{\sigma,\beta}$. The intersection $C_{123456789,987654321} \cap C_{987654321,987654321}$ contains the following two strata:

$$\mathbf{987654321} \xrightarrow{9751} \mathbf{785614329} \xrightarrow{86432} 765413\underline{289} \xrightarrow{76532}$$

$$653412\underline{789} \xrightarrow{6542} 543216\underline{789} \xrightarrow{5431} 4312\underline{56789} \xrightarrow{42}$$

$$2314\underline{56789} \xrightarrow{31} 213\underline{456789} \xrightarrow{21} 123\underline{456789}$$

and

$$987654321 \xrightarrow{971} 781654329 \xrightarrow{865432} 761543289 \xrightarrow{765432}$$

$$651432789 \xrightarrow{654342} 541326789 \xrightarrow{5432} 431256789 \xrightarrow{42}$$

$$231456789 \xrightarrow{31} 213456789 \xrightarrow{21} 123456789.$$

(The numbers above the arrows present decreasing subsequences used on the corresponding steps.) The first chain is bigger in the adjacency partial order than the second chain, i.e. its permutations are bigger or equal in the usual Bruhat order than the corresponding permutations in the second chain. The first stratum has the form $(\mathbb{C}^*)^{20} \times \mathbb{C}^8$, while the second has the form $(\mathbb{C}^*)^{22} \times \mathbb{C}^7$. Thus, we get an example of a hard $C_{1,\alpha} \cap C_{\sigma,\beta}$.

7.5. Remark. Several natural conjectures about the combinatorics of refined double decompositions turned out to be wrong. For example, let us consider the set $\widetilde{S} \subset S$ used on the first step of the algorithm in the decomposition of $C_{1,\alpha} \cap C_{\sigma,\beta}$ such that for any decreasing subsequence $s \in \widetilde{S}$ the intersection $C_{1,\alpha'} \cap C_{\sigma_s,\beta'}$ is nonempty. A natural conjecture would be that $\widetilde{S}$ is an interval in the partial order $\vdash$ on S, see 6.1. This turns to be wrong in the examples $C_{12345,546213} \cap C_{654321,654213}$ and $C_{12345,53412} \cap C_{52341,53412}$. Moreover, in the latter case the set $\widetilde{S}$ consists of $51 = \mathbb{C}^*$, $531 = (\mathbb{C})^{*2} \times \mathbb{C}$ and $541 = (\mathbb{C})^{*2} \times \mathbb{C}^2$, and thus even has a gap in dimensions of its elements.

Several combinatorial conjectures based on our consideration and questions connected with refined double decompositions are presented in §11.

8. On the mixed Hodge structure in the cohomology
of intersections of Schubert cells

8.1. Remark. All the elements in the pth cohomology with compact supports of $(\mathbb{C}^*)^k \times \mathbb{C}^l$ have Hodge indices $p - (k + l), p - (k + l)$, and its E-polynomial (see 0.21) equals $(z - 1)^k z^l$, where $z = uv$.

8.2. The proof of Corollary 0.22 is based on the following lemma.

Lemma (see e.g. [Du]. *Let $Y \subset X$ be a closed quasiprojective submanifold of a quasiprojective manifold X and let $U = X \backslash Y$. Then the exact sequence of the pair (X, Y) for the cohomology with compact supports*

$$(\dagger) \qquad \cdots \to H_c^k(U) \to H_c^k(X) \to H_c^k(Y) \to H_c^{k+1}(U) \to \cdots$$

is an exact sequence of Hodge structures, that is, all differentials respect Hodge indices.

8.3. Proof of Corollary 0.22i. Applying Lemma 8.2, one gets a simple inductive proof of the first part of Corollary 0.22. Indeed, let us consider the standard filtration $F^1 \subset F^2 \subset \cdots \subset F^m = C_{1,\alpha} \cap C_{\sigma,\beta}$, see 0.24, where F^1 is the union of all minimal refined double strata in the adjacency partial order, F^2 is the union of all minimal strata and all strata that are adjacent only to strata in F^1, i.e. they form two bottom levels in the adjacency order, and so on. The difference between any F^i and F^{i-1} is the disjoint union of strata, by Theorem D. Now we prove (by induction on i) that all F^i have $h_k^{p,q} = 0$ if $p \neq q$. Indeed, by Remark 8.1, F^1, which is the disjoint union of refined double strata, has this property. Next, we consider the long exact sequence (†) for the pair (F^2, F^1) and use Lemma 8.2 and the fact that $F^2 \setminus F^1$ is again the disjoint union of strata to get the necessary property for F^2, and so on. ∎

Remark. Exactly the same argument works for any quasiprojective $V = \bigcup V_i$ where V_i satisfy $h_k^{p,q}(V_i) = 0$ for $p \neq q$.

8.4 Lemma (see [Du]. *If a quasiprojective V is decomposed in the disjoint union $V = \bigcup_i V_i$, where V_i are also quasiprojective, then $\chi^{p,q}(V) = \sum_i \chi^{p,q}(V_i)$ for all p, q, or equivalently, $E_V(u,v) = \sum_i E_{V_i}(u,v)$.*

8.5. Proof of Corollary 0.22ii. To prove 0.22ii it suffices to apply Lemma 8.4 and Remark 8.1 to the refined double decomposition of any $C_{1,\alpha} \cap C_{\sigma,\beta}$. ∎

8.6. To prove Corollary 0.23 we need the following additional statement. Suppose X is an arbitrary complex manifold; denote

$$\Pi_X(z) = \sum_i \chi^{ii}(X) z^i.$$

Lemma. *Let $X = U \times Y$ and the following assumptions be true:*
  1) $\chi^{ij}(Y) = 0$ for $i \neq j$;
  2) $h_k^{ij}(U) = 0$ for $i \neq j$.
Then i) $\Pi_X(z) = \Pi_U(z) \Pi_Y(z)$,
  ii) $\chi^{ij}(X) = 0$ for $i \neq j$.

Proof. Evidently, assumption 2) yields

$$h_k^{ij}(X) = \sum_{p,q,l} h_l^{pq}(U) h_{k-l}^{i-p,j-q}(Y) = \sum_{p,l} h_l^{pp}(U) h_{k-l}^{i-p,j-p}(Y).$$

Hence,

$$\chi^{ij}(X) = \sum_k (-1)^k h_k^{ij}(X) = \sum_{p,l} (-1)^l h_l^{pp}(U) \sum_k (-1)^{k-l} h_{k-l}^{i-p,j-p}(Y)$$

$$= \sum_{p,l} (-1)^l h_l^{pp}(U) \chi^{i-p,j-p}(Y) = \sum_p \chi^{pp}(U) \chi^{i-p,j-p}(Y).$$

Taking into account this relation, we see that assertion ii) follows immediately from assumption 1). Next,

$$\Pi_X(z) = \sum_i \sum_p \chi^{pp}(U)\chi^{i-p,i-p}(Y)z^i$$

$$= \left(\sum_p \chi^{pp}(U)z^p\right)\left(\sum_j \chi^{jj}(Y)z^j\right) = \Pi_U(z)\Pi_Y(z),$$

and assertion i) follows. ∎

8.7. Proof of Corollary 0.23. To prove the inductive formula of 0.23 we use part ii) of Proposition 0.16 for one step of the algorithm, and get

$$C_{1,\alpha} \cap C_{\sigma,\beta} = \coprod_{s\in S} O_{s,\sigma} \times (C_{1,\alpha'} \cap C_{\sigma_s,\beta'}),$$

where $O_{s,\sigma}$ is the orbit in P^{n-1} corresponding to the decreasing subsequence s and $C_{1,\alpha'} \cap C_{\sigma_s,\beta'}$ is the intersection of Schubert cells in F_{n-1}, see Lemma 3.2. Then, from Corollary 0.22i, Lemma 8.6, and Remark 8.1 we see that the E-polynomial of $O_{s,\sigma} \times (C_{1,\alpha'} \cap C_{\sigma_s,\beta'})$ is equal to the product of $z^{d(s)}(z-1)^{l(s)}$ and $E_{C_{1,\alpha'} \cap C_{\sigma_s,\beta'}}$. Thus, we obtain the following inductive formula for the E-polynomials:

$$E_\sigma^{\alpha,\beta} \equiv E_{C_{1,\alpha} \cap C_{\sigma,\beta}}(u,v) = \sum_{s\in S} z^{d(s)}(z-1)^{l(s)} E_{\sigma_s}^{\alpha',\beta'}.$$

∎

8.8. The Leray spectral sequence can be used successfully for a filtration of a topological space X satisfying the following condition. Let $F \subset X$ be a closed subspace of the filtration considered. One has the following natural short exact sequence of sheaves: $0 \to {}_{j!}\mathbb{C} \to \mathbb{C} \to {}_{i_*}\mathbb{C} \to 0$, where i and j denote the obvious inclusions $F \xrightarrow{i} X \xleftarrow{j} X \setminus F$ and $\mathbb{C}$ is the usual constant sheaf on X. The inclusion $F \hookrightarrow X$ is called *good* if the sheaves ${}_{i_*}\mathbb{C}$ and ${}_{j!}\mathbb{C}$ coincide with constant sheaves on F and $X \setminus F$, respectively. An example of a bad inclusion is the inclusion of the line $x = 0$ in the space $\mathbb{R}^2 \setminus \{y \neq 0\}$. Inclusions of quasiprojective varieties are always good. Consider now a topological space that is an arbitrary union of refined double strata in some $C_{1,\alpha} \cap C_{\sigma,\beta}$. Warning: arbitrary unions of refined double strata are not always quasiprojective varieties. At the same time, all lower intervals in the adjacency partial order, unions of lower intervals and differences between such unions present quasiprojective varieties. The adjacency partial order defines the stricture of a poset on any subset of the strata involved. Any union of lower intervals in this poset defines a closed subset in the above topological space and any sequence of embedded systems of lower intervals defines a filtration by closed subspaces.

8.9. Proof of Theorem E. Let us first cover the nice case. We consider the Leray spectral sequence for the cohomology with compact supports associated with the standard filtration of a nice $C_{1,\alpha} \cap C_{\sigma,\beta}$. By definition, we consider the filtration of $C_{1,\alpha} \cap C_{\sigma,\beta}$ by closed subsets F^i such that F^i is formed by the union of all strata of (complex) dimension less or equal to i. The (p,q)th entry of the first page contains the $(p+q)$th cohomology with compact supports of F^{p+1} mod F^p, which coincides with the direct sum of the $(p+q)$th cohomology with compact supports of all refined double strata of dimension p. Let $(\mathbb{C}^*)^k \times (\mathbb{C})^l$ be one of the strata of minimal dimension in the considered nice $C_{1,\alpha} \cap C_{\sigma,\beta}$. It follows from Proposition 9.1 below that the $(p+q)$th term is the direct sum of several copies of the $(p+q)$th cohomology of some $(\mathbb{C}^*)^{k+2(p-1)} \times \mathbb{C}^{l-(p-1)}$. The differential d_1 coincides with the connecting homomorphism in the exact sequence of triples $H_c^{p+q}(F^{p+1}, F^p) \to H_c^{p+q+1}(F^{p+2}, F^{p+1})$. The crucial property of differentials in this spectral sequence for the filtration by quasiprojective subvarieties is that they respect the Hodge weights (filtrations) in the cohomology, see [Dl2,Dl3]. By Remark 8.1, the Hodge weights of all elements of the (p,q)th term are equal to $p-(k+l), p-(k+l)$. Thus, the only nontrivial differential in this spectral sequence can be d_1, since only d_1 respects the weights.

Now, let us modify this proof for the almost nice case. As it was mentioned before, the filtration of an almost nice $C_{1,\alpha} \cap C_{\sigma,\beta}$ by dimensions is a filtration by closed subsets, and rows of the corresponding spectral sequence contain the cohomology with compact supports of unions of strata of a given dimension. The union of all strata of a given dimension in the almost nice case is a quasiprojective variety with the standard filtration. The corresponding Leray spectral sequence, obviously, degenerates at E_1, since by Proposition 9.1 below all refined double strata of the same dimension have the same form and none of the differentials in this additional spectral sequence respects the Hodge weights. Thus, one gets the same cohomology with compact supports for the union of these strata as for their disjoint union. Therefore, exactly the same degeneracy arguments as above apply. ∎

8.10. Examples. The authors have calculated Betti numbers for all intersections of top-dimensional Schubert cells in F_3 and F_4, which are all nice, using a different method. (In dimension 5 the intersection of the opposite top-dimensional Schubert cells is only almost nice, see Fig.9a.) The only interesting cases in these dimensions are $\sigma = 321$, $\sigma = 4231, 3421, 4312$ or 4321. These Betti numbers show that in all these cases the differential d_1 acts in the maximal way, i.e. that all rows of the first page form acyclic

complexes except for the top dimension, see an example in Fig.10. The
second page in all these examples contains only one nontrivial column, and
the cohomology in this column has pure weights; thus, the mixed Hodge
structure is pure.

Let us consider the filtration $F^3 \supset F^2 \supset F^1$, where $F^1 = A$, $F^2 = A \cup B$, $F^3 = C_{1234,4231} \cap C_{4231,4231}$, and use the Leray sequence, see Fig.11
and Example 7.2. This filtration is precisely the one obtained from the
adjacency partial order. From the geometrical description of the strata A,
B, C one can see that the nontrivial differential d_1 acting from the first
column to the second one (pay attention to the weights) kills the whole
first column, thus leaving us with the second page, which again contains
the pure Hodge structure.

8.11. Counterexample to the purity of the mixed Hodge structure of $C_{1,\alpha} \cap C_{\sigma,\beta}$.

In all previous examples the so-called *purity* of the mixed
Hodge structure holds, which in the case of $C_{1,\alpha} \cap C_{\sigma,\beta}$ means that the only
nonvanishing Hodge numbers are $h_i^{i,i}$ and they are equal to $h_i^{i,i} = (-1)^i \chi_c^{i,i}$.
However, this turned out to be false in general (for arbitrary $C_{1,\alpha} \cap C_{\sigma,\beta}$)
for the following reason. If purity holds for some $C_{1,\alpha} \cap C_{\sigma,\beta}$, then the
coefficients of $E_\sigma^{\alpha,\beta}(z)$ must have strictly alternating signs. Recently, using
computer, B. Boe has found 4 intersections in $\mathfrak{S}_6$ whose E-polynomials do
not have strictly alternating signs. Two simplest of these examples have

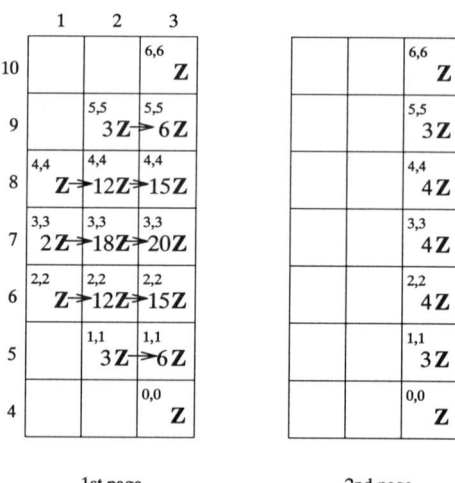

Figure 10. Leray spectral sequence for the nice intersection
$\alpha = \beta = \sigma = 4321$. Arrows show nontrivial differentials,
pairs of numbers in the left upper corners are Hodge weights

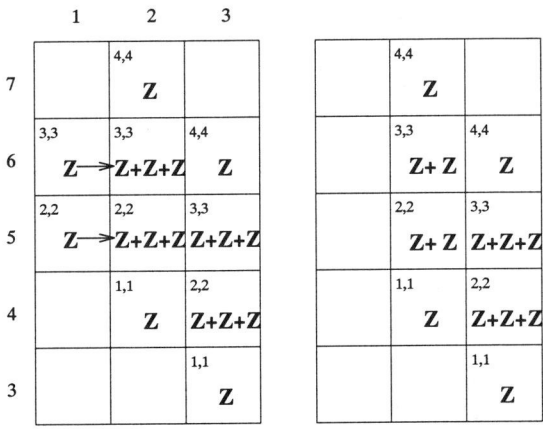

1st page   2nd page

Figure 11. Leray spectral sequence for the simplest nonnice
intersection $\alpha = \beta = \sigma = 4231$; arrows show nontrivial differentials

been checked by the authors using the above method, and here is their
description.

1) $C_{123456,563412} \cap C_{654312,654321}$ consists of the following 24 refined double strata: 1 copy of $(\mathbb{C}^*)^3 \times \mathbb{C}^5$, 6 copies of $(\mathbb{C}^*)^5 \times \mathbb{C}^4$, 10 copies of $(\mathbb{C}^*)^7 \times \mathbb{C}^3$, 6 copies of $(\mathbb{C}^*)^9 \times \mathbb{C}^2$, and 1 copy of $(\mathbb{C}^*)^{11} \times \mathbb{C}$. The corresponding E-polynomial is equal to $q(-1 + 5q - 11q^2 + 13q^3 - 7q^4 - q^5 + q^6 + 7q^7 - 13q^8 + 11q^9 - 5q^{10} + q^{11})$, and its coefficients do not have strictly alternating signs.

2) $C_{123456,563412} \cap C_{654321,654321}$ consists of the following 41 refined double strata: 1 copy of $(\mathbb{C}^*)^2 \times \mathbb{C}^5$, 5 copies of $(\mathbb{C}^*)^4 \times \mathbb{C}^4$, 12 copies of $(\mathbb{C}^*)^6 \times \mathbb{C}^3$, 15 copies of $(\mathbb{C}^*)^8 \times \mathbb{C}^2$, 7 copies of $(\mathbb{C}^*)^{10} \times \mathbb{C}$, and 1 copy of $(\mathbb{C}^*)^{12}$. This E-polynomial is equal to $1 - 5q + 11q^2 - 13q^3 + 8q^4 - q^5 - q^6 - q^7 + 8q^8 - 13q^9 + 11q^{10} - 5q^{11} + q^{12}$. Considerations of the adjacency partial order reveal that both of these examples are almost nice, but not nice. Since by Theorem E the corresponding Leray spectral sequence degenerates at E_2, it will be interesting to calculate d_1, and therefore the cohomology of these intersections.

9. Combinatorial properties and applications of E-polynomials

9.1. Proposition. *For any nonempty stratum, the sum of the (complex) dimension of the torus and the doubled dimension of the linear space equals* $\lng(\alpha) + \lng(\beta) - \lng(\sigma)$, *where* $\lng$ *is the inversion length of a permutation. In particular, all strata of the same dimension have the same form.*

Example. For the intersection $C_{1234,4231} \cap C_{4231,4231}$ we get $\lng(\alpha) =$

$\ln g(\beta) = \ln g(\sigma) = 5$, and $l(\mathrm{St}) + 2d(\mathrm{St}) = 5$ for all strata.

9.2. In order to prove Proposition 9.1, we start with the following lemma relating the length of a permutation σ to the length of σ_s (see Lemma 3.4).

Lemma. *Let $s = (i_1, \ldots, i_k)$, then*

$$\ln g(\sigma_s) - \ln g(\sigma) = l(s) + 2d(s) - i_1 - \sigma^{-1}(i_k) + 2,$$

where $l(s)$ and $d(s)$ are the reduced length and the domination of s, respectively.

Proof. Indeed, let us denote by I_j the set of consequtive elements of σ lying between i_j and i_{j-1} (we set $i_0 = 0$). When passing from σ to σ_s, we loose all the inversions containing i_1, and, for each i_j, $2 \leqslant j \leqslant k$, all the inversions (i, i_j) such that $i \in I_j$ and $i > i_j$. On the other hand, for each i_j, $2 \leqslant j \leqslant k$, we acquire new inversions of the form (i_j, i) such that $i \in I_j$ and $i < i_j$. So,

$$\ln g(\sigma_s) - \ln g(\sigma) = \sum_{j=2}^{k} \#\{i \in I_j : i < i_j\} -$$

$$\sum_{j=2}^{k} \#\{i \in I_j : i > i_j\} - \#\{i \in I_1 : i > i_1\} - \#\{i \notin I_1 : i < i_1\} =$$

$$\sum_{j=1}^{k} \#\{i \in I_j : i < i_j\} - \sum_{j=1}^{k} \#\{i \in I_j : i > i_j\} - \#\{i : i < i_1\}.$$

Evidently, the first term in the right hand side equals $d(s)$, while the last term equals $i_1 - 1$. Next, the sum of the first and the second terms is $\sum_{j=1}^{k} \#\{i : i \in I_j\}$, and hence it equals $n - k - (n - \sigma^{-1}(i_k)) = \sigma^{-1}(i_k) - k$. Taking into account that $l(s) = k - 1$ (see 0.8), we get the statement of the lemma. ∎

Example. Let $\sigma = 46237518$ and the decreasing subsequence in it be $(6, 3, 1)$, then $\sigma_s = 4321657$. We have $\ln g(\sigma) = 12$, $\ln g(\sigma_s) = 7$. On the other hand, $d(s) = 2$, $l(s) = 2$, $i_1 = 6$, $\sigma^{-1}(i_k) = 7$. So, $7 - 12 = 4 + 2 - 6 - 7 + 2$.

9.3. Proof of Proposition 9.1. The proof is done by induction. In the cases $n = 1$ and $n = 2$ the formula is trivially true.

Step of induction. For any nonempty stratum $\mathrm{St} \in C_{1,\alpha} \cap C_{\sigma,\beta}$ we want to prove that $l(\mathrm{St}) + 2d(\mathrm{St}) = \ln g(\alpha) + \ln g(\beta) - \ln g(\sigma)$, where $l(\mathrm{St})$ is the total length of St and $d(\mathrm{St})$ is its total domination (see 0.11). Assume that the chain corresponding to St starts with the permutation σ, and that the next permutation of the chain is σ_s. By the inductive hypothesis, we have

$l(\text{St}') + 2d(\text{St}') = \ln g(\alpha') + \ln g(\beta') - \ln g(\sigma_s)$, where St$'$ is the refined double stratum in $C_{1,\alpha'} \cap C_{\sigma_s,\beta'}$ corresponding to the subchain of the initial chain starting with σ_s, α' and β' are obtained from α and β, respectively, by the reduction of the first element, see Proposition 0.16. Subtracting the above two formulas and applying Theorem B, we arrive at

$$l(s) + 2d(s) = \ln g(\alpha) - \ln g(\alpha') + \ln g(\beta) - \ln g(\beta') - \ln g(\sigma) + \ln g(\sigma_s).$$

Obviously, $\ln g(\alpha) - \ln g(\alpha') = \alpha(1) - 1$ and $\ln g(\beta) - \ln g(\beta') = \beta(1) - 1$. Hence, by Lemma 9.2, we have to prove that $\alpha(1) + \beta(1) = i_1 + \sigma^{-1}(i_k)$. However, by Theorem C, $i_1 = \alpha(1)$ and $\sigma^{-1}(i_k) = \beta(1)$, and the result follows. ∎

9.4. Remark. The assertion of Proposition 9.1 remains valid for intersections of Schubert cells in all flag spaces G/B.

9.5. Lemma. *The intersection $C_{1,\alpha} \cap C_{\sigma,w_0}$ is isomorphic to $C_{1,\alpha} \cap C_{w_0,w_0\sigma w_0} \times C_{1,\sigma w_0}$. Thus, $C_{1,\alpha} \cap C_{\sigma,w_0}$ is nonempty if and only if $C_{1,\alpha} \cap C_{w_0,w_0\sigma w_0}$ is nonempty, that is, in the case $\alpha \succeq w_0\sigma$, where $\succeq$ denotes the usual Bruhat order. If $C_{1,\alpha} \cap C_{\sigma,w_0}$ is nonempty, then its dimension equals $\ln g(\alpha)$.*

Proof. See [KL2] 1.4, p.187. ∎

9.6. Corollary. *Let $s_1 \vdash s_2$ be two decreasing subsequences with coinciding first and last elements, then $\sigma_{s_1} \preceq \sigma_{s_2}$ and $w_0\sigma_{s_1} \succeq w_0\sigma_{s_2}$. Thus, if $C_{1,\alpha'} \cap C_{\sigma_{s_1},w_0}$ is nonempty, then $C_{1,\alpha'} \cap C_{\sigma_{s_2},w_0}$ is also nonempty.*

Proof. The set of all decreasing subsequences in σ with fixed first and last elements describes the set of all G_σ-orbits that have a fixed set of dimensions of intersections with the subspaces of f and g, see Corollary 5.5. Moreover, by Lemma 6.2, the partial order $\vdash$ describes their adjacency. Thus, any 1-parameter family of lines $l(t)$, $t \in [0,1]$, such that $l(t) \in O_{\sigma,s_2}$ for $t \in [0,1)$ and $l(1) \in O_{\sigma,s_1}$ satisfies the assumptions of Lemma 6.3 and gives the necessary assertion. ∎

Example. For $\sigma = 83427561$ and $s_1 = (8,3,2,1)$, $s_2 = (8,7,5,1)$ we get $\sigma_{s_1} = 3241756 \vdash \sigma_{s_2} = 7342516$.

9.7. We now start proving Proposition 0.27 in the case $\beta = w_0$ (or $\alpha = w_0$, which is the same), i.e. for intersections of pairs of Schubert cells one of which has the top dimension. This class of pairwise intersections is closed under the projections involved in our algorithm, since the subset of flags in a top-dimensional cell containing some fixed line is isomorphic to the top-dimensional cell in the quotient space.

We proceed by induction on n. Cases $n = 1$ or 2 are obvious.

Step of induction. Assume that the assertion holds for the refined decomposition of F_{n-1} and consider the refined double decomposition of $C_{1,\alpha} \cap C_{\sigma,w_0}$ in F_n. By Theorem C, refined double strata in $C_{1,\alpha} \cap C_{\sigma,w_0}$ are diffeomorphic to Cartesian products of orbits O_s (where $s \in S_\alpha^{w_0}$) by refined double strata in $C_{1,\alpha'} \cap C_{\sigma_s,w_0'}$, where w_0' is the longest permutation on $n - 1$ elements and α' has the same sense as before. It suffices to show that nonempty refined top–dimensional strata corresponding to all $s \in S_\alpha^{w_0}$ have no gaps in dimensions. The dimension of any nonempty intersection $C_{1,\alpha'} \cap C_{\sigma_{sub},w_0}$, and thus the dimension of the top-dimensional strata, equals $\lng(\alpha')$ independently of s, see Lemma 9.5. Thus, we have to show that the numbers $\lng(\alpha') + \dim(O_s)$ have no gaps when s runs over the set of all decreasing subsequences in $S_\alpha^{w_0}$ such that $C_{1,\alpha'} \cap C_{\sigma_s,w_0}$ are nonempty. Since $\lng(\alpha')$ does not depend on s, we must show that the dimensions of all orbits corresponding to the decreasing subsequences s such that $\alpha' \succeq w_0 \sigma_s$ have no gaps, cp. Lemma 9.5. The latter statement follows from Lemmas 6.2, 6.3 and Corollary 9.6. ∎

9.8. Now we cover the second case of Proposition 0.27, $\sigma = w_0$.

By Lemma 9.5, the intersection $C_{1,\alpha} \cap C_{\sigma,w_0}$ is isomorphic to $C_{1,\alpha} \cap C_{w_0,w_0\sigma w_0} \times C_{1,\sigma w_0}$. Thus, the absence of gaps in dimensions of refined double strata in the case $\sigma = w_0$ will follow from the preceding considerations in the case $\beta = w_0$ and the following

Lemma. *There exists a bijection between the refined double strata in $C_{1,\alpha} \cap C_{\sigma,w_0}$ and those in $C_{1,\alpha} \cap C_{w_0,w_0\sigma w_0} \times C_{1,\sigma w_0}$. Strata in $C_{1,\alpha} \cap C_{\sigma,w_0}$ are diffeomorphic to Cartesian products of the corresponding strata in $C_{1,\alpha} \cap C_{w_0,w_0\sigma w_0}$ by $C_{1,\sigma w_0}$.*

The proof is based on the following observation.

9.9. Proposition. *Given the polynomial $E_\alpha^{\sigma,\beta}(z)$ of some intersection $C_{1,\alpha} \cap C_{\sigma,\beta}$ and two numbers k and l such that the top-dimensional strata in the refined double decomposition of $C_{1,\alpha} \cap C_{\sigma,\beta}$ have the form $(\mathbb{C}^*)^k \times (\mathbb{C})^l$, one restores the number of strata in $C_{1,\alpha} \cap C_{\sigma,\beta}$ in all dimensions.*

Proof. Since the dimension of $C_{1,\alpha} \cap C_{\sigma,\beta}$ is $k + l$, the degree of $E_\alpha^{\sigma,\beta}(z)$ is also $k + l$, and the leading coefficient a_{k+l} is equal to the number of top-dimensional strata. Consider $E_\alpha^{\sigma,\beta}(z) - a_{k+l}(z-1)^k z^l$ and assume that its degree is $p < k + l$. Then its leading coefficient a_p gives the number of strata of dimension p in $C_{1,\alpha} \cap C_{\sigma,\beta}$. By Proposition 9.1, the strata of dimension p have the form $(\mathbb{C}^*)^{2d-k-l} \times (\mathbb{C})^{2l+k-d}$. Subtracting from $E_\alpha^{\sigma,\beta}(z) - a_{k+l}(z-1)^k z^l$ the polynomial $a_p(z-1)^{2d-k-l} z^{2l+k-d}$, one gets the number of strata in the next dimension, and so on. ∎

9.10. Proof of Lemma 9.8. By Lemma 9.5 and the definition of $E_\alpha^{\sigma,\beta}(z)$

one has $E_\alpha^{w_0,w_0\sigma w_0}(z) = z^{\lg(\sigma w_0)}E_\alpha^{\sigma w_0}(z)$. Moreover, it is known that both $C_{1,\alpha} \cap C_{w_0,w_0\sigma w_0}$ and $C_{1,\alpha} \cap C_{\sigma w_0}$ are irreducible varieties, and one can show that their only top-dimensional strata have the form $(\mathbb{C}^*)^{\lg(\alpha)-\lg(\sigma w_0)}$ and $(\mathbb{C}^*)^{\lg(\alpha)-\lg(\sigma w_0)} \times \mathbb{C}^{\lg(\sigma w_0)}$, respectively. (These strata are obtained by using maximal decreasing subsequences on all steps of the algorithm.) Now, applying the procedure of defining the number of strata of each dimension described in Proposition 9.9 and using the result for the first special case, one gets the necessary statement. ∎

9.11. Here we recall the relation between the values of structure constants $c_{w_1,w_2}^{w_3}(z)$ in the Hecke algebra and the E-polynomials, cp. [KL1, Ka, Cu1]. This relation holds for any flag space G/B where G is a semisimple group.

The Hecke algebra $\mathcal{H}$ is the deformation of the group algebra of $\mathfrak{S}_n$. As a linear space $\mathcal{H}$ has the standard basis $\{T_w\}$, $w \in \mathfrak{S}_n$. As an algebra $\mathcal{H}$ is spanned by $\{T_{t_i}\}$, where t_i is the ith simple transposition, satisfying the Hecke algebra relations. Namely,

$$\begin{cases} T_{t_i} * T_{t_{i+1}} * T_{t_i} = T_{t_{i+1}} * T_{t_i} * T_{t_{i+1}}, \\ T_{t_i} * T_{t_j} = T_{t_j} * T_{t_i}, \\ T_{t_i} * T_{t_i} = (z-1)T_{t_i} + zT_1. \end{cases}$$

The structure constants $c_{w_1,w_2}^{w_3}$ are the coefficients in the decomposition $T_{w_1} * T_{w_2} = \sum_{w_3} c_{w_1,w_2}^{w_3} T_{w_3}$.

Proposition. $E_{C_{1,\alpha} \cap C_{\sigma,\beta}}(z) = c_{\alpha,\beta^{-1}}^\sigma(z)$.

Proof. If z is a power of prime, then the structure constant $c_{w_1,w_2}^{w_3}(z)$ calculates the number of points in the intersection $C_{1,w_1} \cap C_{w_2,w_3^{-1}}$, see e.g. [Cu1]. In an appropriate basis of $\mathbb{C}^n$, all refined double strata in F_n are defined over $\mathbb{Z}$, and thus the refined double decomposition is defined simultaneously over any finite field GF. Any refined double stratum over GF is isomorphic to an appropriate $(\text{GF}^*)^k \times (\text{GF})^l$ where GF^* is the multiplicative group of GF. Both $E_{C_{1,\alpha} \cap C_{\sigma,\beta}}(z)$ and $c_{\alpha,\beta^{-1}}^\sigma(z)$ are polynomials in z with integer coefficients coinciding when z is a power of a prime, since they both count the number of points in $C_{1,\alpha} \cap C_{\sigma,\beta}$ over the corresponding finite field. Thus, $E_{C_{1,\alpha} \cap C_{\sigma,\beta}}(z) = c_{\alpha,\beta^{-1}}^\sigma(z)$. ∎

10. Relation to the decomposition of Curtis–Deodhar

10.1. In order to present to the readers certain similar results due to V.V.Deodhar and Ch. Curtis, we quote several statements of [De1] constructing a special decomposition of $C_{1,\alpha}$ and $C_{1,\alpha} \cap C_{w_0,\beta}$ in terms of reduced expressions for α^{-1}. We preserve mostly the original notation. Let $W(y)$ denote the lower interval of y in W in the usual Bruhat order.

10.2. Proposition ([Del], Theorem 1.1). *The Bruhat cell $\mathcal{B}y \cdot \mathcal{B}$ can be decomposed into disjoint nonempty subsets $\{D_{\underline{\sigma}}\}_{\underline{\sigma} \in \mathcal{D}}$ ($\mathcal{D}$ is the indexing set, which can be described explicitly) such that*

(i) $\mathcal{B}y \cdot \mathcal{B} = \cup_{\underline{\sigma} \in \mathcal{D}} D_{\underline{\sigma}}$.

(ii) *For each $\underline{\sigma} \in \mathcal{D}$ there exist unique non-negative integers $m(\underline{\sigma})$ and $n(\underline{\sigma})$ such that $D_{\underline{\sigma}} \simeq \mathbb{C}^{m(\underline{\sigma})} \times (\mathbb{C}^*)^{n(\underline{\sigma})}$.*

(iii) *For each $\underline{\sigma} \in \mathcal{D}$ there exists $x \in W$ (recall that in our case $W = \mathfrak{S}_n$, and so $D_{\underline{\sigma}} \subseteq \mathcal{B}^- x \cdot \mathcal{B}$ with $\mathcal{B}^- = w_0 \mathcal{B} w_0$). This element x is unique, belongs to $W(y)$ and is denoted by $\pi(\underline{\sigma})$. Thus one gets a map $\pi \colon \mathcal{D} \to W(y)$.*

10.3. Corollary ([Del], Corollary 1.2). $\mathcal{B}y \cdot \mathcal{B} \cap \mathcal{B}^- x \cdot \mathcal{B} = \cup_{\underline{\sigma} \in \mathcal{D}, \pi(\underline{\sigma}) = x} D_{\underline{\sigma}}$. *(In particular, $\mathcal{B}y \cdot \mathcal{B} \cap \mathcal{B}^- x \cdot \mathcal{B} \neq \varnothing$ if and only if $x \in W(y)$.)*

10.4. Let U^+ and U^- denote the maximal unipotent subgroups in SL_n corresponding to the set of all positive and negative roots, respectively. (In our case they are the usual upper- and lower triangular unipotent subgroups.) Fixing $y \in W$ and its reduced expression $y = s_1 \ldots s_k$, where $s_j = t_{i_j}$, let us define for all $1 \leqslant j \leqslant k$ the subsets $U_j = U^+ \cap {}^{s_j \cdots s_k} U^-$ and $U^j = U^+ \cap {}^{s_k \cdots s_j} U^+$. (For a subset $A \in SL_n$ and $g \in SL_n$, ${}^g A = gAg^{-1}$.)

Proposition ([Del], Lemma 2.2).

(i) $U_1 \supset {}^{s_1} U_2 \supset {}^{s_1 s_2} U_3 \supset \cdots \supset {}^{s_1 \cdots s_k} U_{k+1} = \{\mathrm{id}\}$.

(ii) $U^1 \subset U^2 \subset \cdots \subset U^{k+1} = U^+$.

(iii) *For any j, $1 \leqslant j \leqslant k+1$, one has $U^+ = U_j \cdot {}^{s_j \cdots s_k} U^j$ and such an expression is unique.*

(iv) *Any element $\xi \in \mathcal{B}y \cdot \mathcal{B}$ can be uniquely written as $us_1 \ldots s_k$ with $u \in U_1$.*

10.5. The indexing set $\mathcal{D}$ in Proposition 10.2 is described in terms of a reduced expression $y = s_1 \ldots s_k$, where k is the length of y. It consists of subexpressions $s_1 \ldots \hat{s}_{p_1} \ldots \hat{s}_{p_m} \ldots s_k$ with some additional properties, which are called distinguished. More precisely, one can regard a subexpression as a sequence $\underline{\sigma} = (\sigma_0, \ldots, \sigma_k)$ of elements of W such that (i) $\sigma_0 = \mathrm{id}$ and (ii) $\sigma_{j-1}^{-1} \sigma_j \in \{\mathrm{id}, s_j\}$ for all $1 \leqslant j \leqslant k$. Note that for any j one gets that σ_{j-1} and σ_j are always comparable in the Bruhat order, i.e. one has a trichotomy $\sigma_{j-1} < \sigma_j$, or $\sigma_{j-1} = \sigma_j$, or $\sigma_{j-1} > \sigma_j$. Let us denote by $n(\underline{\sigma})$ the number of all positions j such that $\sigma_{j-1} = \sigma_j$ and by $m(\underline{\sigma})$ the number of positions j such that $\sigma_{j-1} > \sigma_j$. Finally, let $\pi(\underline{\sigma}) = \sigma_k$.

Now we can define the index set $\mathcal{D}$ in Proposition 10.2. A subexpression $\underline{\sigma}$ is called *distinguished* if it satisfies the additional condition (iii) $\sigma_j \leqslant \sigma_{j-1} s_j$ for all $1 \leqslant j \leqslant k$.

Proposition ([Del], Proposition 3.1) *Let $u_1 \in U_1$ be fixed. For $0 \leqslant j \leqslant k$, let $\sigma_j \in W$ be the unique element such that $u_1 s_1 \ldots s_j \in \mathcal{B}^- \sigma_j \mathcal{B}$. Then*

$\underline{\sigma} = (\sigma_0, \sigma_1, \ldots, \sigma_k)$ *is a distinguished subexpression.*

In this way one gets a map $\nu : U_1 \to \mathcal{D}$ defined in an obvious way.

10.6. Proof of Proposition 0.28. Let us define the standard reduced expression of $y \in \mathfrak{S}_n$ by the following rule: $y = t_{y^{-1}(1)-1} t_{y^{-1}(1)-2} \cdots t_1 \tilde{y}$, where $\tilde{y}$ belongs to $\mathfrak{S}_{n-1}$, i.e. $\tilde{y}(1) = 1$, and t_i is defined in 3.1. We consider $\tilde{y}$ as an element of $\mathfrak{S}_{n-1}$ acting on the set $\{2, 3, \ldots, n\}$ and define the standard reduced expression inductively.

Example. $3241 = (4,3)(3,2)(2,1) \cdot (1324) = (4,3)(3,2)(2,1)(32)$.

Let $y = t_1 \ldots t_k$ be the standard reduced expression and $q = y^{-1}(1)$. Let us consider the result of Deodhar's decomposition of $\mathcal{B}y \cdot \mathcal{B}$ after the first $q - 1$ steps, which corresponds to the sequence $U_1 \supset {}^{s_1} U_2 \supset \cdots \supset {}^{s_1 \cdots s_{q-2}} U_{q-1}$ (see Proposition 10.4 above). We denote by v_1 the spanning vector of the one-dimensional space of the flag $f \in \mathcal{B}y \cdot \mathcal{B}$. It has the following properties: $v_1^q = 1$, $v_1^j = 0$ if $j > q$. The first $q - 1$ steps of Deodhar's procedure decompose the cell into the disjoint union of 2^{q-1} sets enumerated by all possible subsets of nonvanishing coordinates of v_1. At the same time, our main algorithm for $C_{1,\alpha}$ (with $\alpha = y^{-1}$) produces the same result after the first step. Moreover, since the elementary transpositiopns $s_q, \ldots s_k$ do not contain the element 1, one may assume that the vector v_1 is fixed on all the subsequent steps. An easy reformulation of Deodhar's procedure presents the subsequent steps of his algorithm as an application of the previous steps to the quotient space and quotient flags mod v_1. Thus, it gives exactly the same result as our main algorithm. ∎

11. Final remarks

Below we formulate several combinatorial and topological problems related to the material of the preceding sections with some comments.

11.1. Combinatorial questions. Problem 1. *Calculate the number of strata in the flag completion of (f, g).*

One can use the very weak Bruhat order on $\mathfrak{S}_n$ defined in §1. By Proposition 1.5, the number of flags in the flag completion of (f, g) in relative position σ equals the number of elements in the lower interval of σ in this order. Proposition 1.7 provides an inductive way to find this number.

Problem 2. *Calculate the number of strata in the refined double decomposition $RD_{f,g}$ of the whole flag space F_n* (see Fig. 1 and the description of the main algorithm).

As a preliminary question, one needs to find the number of decreasing subsequences in a permutation. Fig. 12 contains the number of decreasing subsequences and refined double strata for all permutations in $\mathfrak{S}_4$.

The next group of questions deals with the properties of the adjacency partial order. Let us introduce the following natural linear lexicographic order on permutations and chains of permutations. Namely, a permutation α is said to be *bigger than* β if the leftmost element of α that is distinct from the corresponding element of β is bigger than the latter one. This lexicographic order can be extended naturally to chains of permutations. (The chains on all figures are arranged from left to right according to this lexicographic linear order.)

Conjecture 1. *For any nonempty $C_{1,\alpha} \cap C_{\sigma,\beta}$, the refined double stratum that is the maximal element in the lexicographic linear order is also the maximal element in the adjacency partial order. Moreover, it is one of the top-dimensional strata.*

Conjecture 2. *For any nonempty $C_{1,\alpha} \cap C_{\sigma,\beta}$, the refined double stratum that is the minimal element in the lexicographic linear order is one of the strata of the minimal dimension.*

permutation	number of flags in flag completion	number of decreasing subsequences	number of refined double strata
1234	1	4	24
2134	2	5	36
1324	2	5	36
1243	2	5	36
2314	3	6	52
2143	4	6	54
3124	3	6	52
1342	3	6	52
1423	3	6	52
3214	4	8	80
2341	4	7	73
2413	3	7	75
3142	4	7	75
4123	4	7	73
1432	4	8	80
3241	4	9	99
2431	6	9	112
4213	5	9	112
3412	6	8	107
4132	5	9	112
3421	7	11	165
4231	6	11	167
4312	6	11	163
4321	8	15	261

Figure 12. Some combinatorial characteristics for permutations in $\mathfrak{S}_4$

Conjecture 3. *Let S and T be two refined double strata such that $S \succeq T$ (in the adjacency partial order), and assume that* $\dim S - \dim T > 1$. *Then there exists a sequence of intermediate strata* $S = S_1 \succeq S_2 \succeq S_3 \succeq \cdots \succeq S_i = T$ *such that* $\dim S_i - \dim S_{i-1} = 1$ *for all i.*

Consider the sequence of numbers of strata (contained in some $C_{1,\alpha} \cap C_{\sigma,\beta}$) ordered by dimensions. For example, these sequences for 2 counterexamples of B. Boe are $1, 6, 10, 6, 1$ and $1, 5, 12, 15, 7, 1$.

Conjecture 4. *The above sequence is unimodal, that is, nonstrictly increasing up to some place and nonstrictly decreasing after that.*

B.Boe has verified this conjecture for $\mathfrak{S}_4$ and $\mathfrak{S}_5$, as well as for the Coxeter groups B_4, C_4, and D_5.

11.2. Topological questions. **Problem 3.** *Describe combinatorially the closure pattern of refined double strata, that is, enumerate all strata* St_i *such that* $\overline{\mathrm{St}} \cap \mathrm{St}_i \neq \varnothing$.

Problem 4. *Give a combinatorial description of the differential d_1 in the Leray spectral sequence of §8 for (almost) nice* $C_{1,\alpha} \cap C_{\sigma,\beta}$.

Consider the standard filtration of any (e.g., hard) $C_{1,\alpha} \cap C_{\sigma,\beta}$ by closed subsets, see 0.25.

Conjecture 5. *The associated Leray spectral sequence degenerates at E_2.*

Another interesting topic concerning intersections of Schubert cells is the relation between their topological properties over $\mathbb{R}$ and $\mathbb{C}$. We say that a real algebraic variety V enjoys the *M-property* if the sum of its Betti numbers with coefficients in $\mathbb{Z}/2\mathbb{Z}$ coincides with that of its complexification. It is shown in [Sh] that the intersection $\mathbb{R}C_{1,w_0} \cap \mathbb{R}C_{w_0,w_0}$ enjoys such a property.

Conjecture 6. *Any intersection $\mathbb{R}C_{1,\alpha} \cap \mathbb{R}C_{\sigma,\beta}$ enjoys the M-property.*

References

[BB] Bialynicki–Birula, A., *Some theorems on actions of algebraic groups*, Ann. Math. **98** (1973), 480–497.

[Boe] Boe, B., *A counterexample to the Gabber–Joseph conjecture*, Kazhdan-Lusztig theory and Related topics, Contemp. Maths, vol. 139, 1993, pp. 1–3.

[Bo] Bourbaki, N., *Groupes et algebres de Lie, II*, Hermann, Paris, 1968.

[Br1] Brenti, Fr., *Combinatorial properties of the Kazhdan-Lusztig R-polynomials for S_n*, Preprint # 37, Institut Mittag-Leffler (1992), 1–27.

[Br2] Brenti, Fr., *Combinatorial properties of the Kazhdan-Lusztig and R-polynomials for S_n*, Proceedings of the 5th conference on formal power series and algebraic combinatorics (1993), 109–116.

[Cu1]   Curtis, Ch. W., *Representation theory of Hecke algebras*, Vorlesungen aus dem Fachbereich Mathematik der Universität Essen **15** (1987), 1– 86.

[Cu2]   Curtis, Ch. W., *Algebres de Hecke et representationes des groupes finis de type Lie*, Publ. Math. Univ. Paris VII **8** (1987), 1–59.

[Cu3]   Curtis, Ch. W., *A further refinement of the Bruhat decomposition*, Proc. AMS **102** (1988), no. 1, 37–42.

[DK]   Danilov, V. I., Khovansky, A. G., *Newton polyhedra and an algorithm for calculating Hodge-Deligne numbers*, Izv. Acad. Nauk SSSR, ser. mat. **50** (1986), no. 5, 925–945.

[Dl1]   Deligne, P., *Theorie de Hodge, I*, Proc. Inter. Congress Math, vol. 1, 1970, pp. 425–430.

[Dl2]   Deligne, P., *Theorie de Hodge, II*, Publ. Math. IHES **40** (1971), 5–58.

[Dl3]   Deligne, P., *Theorie de Hodge, III*, Publ. Math. IHES **44** (1974), 5–77.

[De1]   Deodhar, V. V., *On some geometric aspects of Bruhat orderings, I*, Inv. Math. **79** (1985), 499–511.

[De2]   Deodhar, V. V., *A combinatorial setting for questions in Kazhdan-Lusztig theory*, Geom. Dedicata, **36** (1990), 95–120.

[Du]   Durfee, A., *Algebraic varieties which are a disjoint union of subvarieties*, Geometry & Topology. Manifolds, Varieties & Knots (C.McCrory and T.Schifrin, eds.), Marcel Dekker, New York & Boston, 1987, pp. 99–102.

[El]   ElZein, F., *Mixed Hodge structures*, Trans. of the AMS **275** (1983), no. 1, 71–106.

[FF]   Fomenko, A. T., Fuchs, D. B., *Homotopic topology*, Nauka, Moscow, 1989, English transl., Akad. Kiado, Budapest, 1986..

[GS]   Gelfand, I. M., Serganova, V. V., *Combinatorial geometries and strata of torus on a homogeneous compact manifold*, Uspekhi Math. Nauk **42** (1987), no. 2, 107–134.

[GH]   Griffiths, P., Harris, J., *Principles of algebraic geometry*, vol. 1, John Wiley & Sons, 1978.

[GSc]   Griffiths, P., Schmid, W., *Recent development in Hodge theory, a discussions of techniques and results*, Proc. Internat. Colloq. on Discrete Subgroups of Lie Groups (1973), 31–127.

[Hu]   Humphreys, J., *Reflection groups & Coxeter groups*, Cambridge Univ. Press, 1990.

[IM]   Iwahori, N., Matsumoto, H., *On some Bruhat decompositions and the structure of the Hecke ring of a p-adic group*, Publ. Math. IHES **25** (1965), 5–48.

[Ka]   Kawanaka, N., *Unipotent elements and characters of finite Chevalley groups*, Osaka J. Math **12** (1975), 523–554.

[KL1]   Kazhdan, D., Lusztig, G., *Representations of Coxeter groups and Hecke algebras*, Inv. Math. **53** (1979), 165–184.

[KL2]   Kazhdan, D., Lusztig, G., *Schubert varieties and Poincare duality*, Proc. Symp. Pure Math., vol. 36, 1980, pp. 185–203.

[Sh]   Shapiro, M., *Nonoscillating differential equations*, Ph.D. Thesis, Moscow State University, 1992.

[SV]   Shapiro, B., Vainshtein, A., *Euler characteristics for links of Schubert cells in the space of complete flags*, Adv. Sov. Math., vol. 1, AMS, Providence, 1990, pp. 273–286.

[SSV]   Shapiro, B., Shapiro, M., Vainshtein, A., *Topology of intersections of Schubert cells and Hecke algebra*, to appear in Discr. Math. (1993).

[So]   Soergel, A., *Kategorie $\mathcal{O}$, perverse Garben und Moduln über den Koinvarianten zur Weylgruppe*, J. Amer. Math. Soc. **3** (1990), no. 2, 421–445.

B. Shapiro: Department of Mathematics, University of Stockholm, S-10691, Sweden, `shapiro@matematik.su.se`

M. Shapiro: Department of Theoretical Mathematics, The Weizmann Institute of Science, Rehovot, 76100, Israel, `shapiro@wisdom.weizmann.ac.il`

A. Vainshtein: School of Mathematical Sciences, Tel Aviv University, Ramat Aviv, Israel 69978, `alek@math.tau.ac.il`

August 1995